Introduction to Systems Biology

(?) (1) scale of phenotype & complexity of bioinformatics / investment / resources?

(2) which problem / aspect of CAD prediction?

– unique

– leverage disease

Introduction to Systems Biology

Edited by

Sangdun Choi, PhD

Department of Biological Sciences
Ajou University
Suwon, Korea

HUMANA PRESS ✻ TOTOWA, NEW JERSEY

999 Riverview Drive, Suite 208
Totowa, New Jersey 07512

www.humanapress.com

For additional copies, pricing for bulk purchases, and/or information about other Humana titles, contact Humana at the above address or at any of the following numbers: Tel.: 973-256-1699; Fax: 973-256-8341, E-mail: order@humanapr.com; or visit our Website: http://www.humanapress.com

This publication is printed on acid-free paper. ♾
ANSI Z39.48-1984 (American National Standards Institute) Permanence of Paper for Printed Library Materials

10 9 8 7 6 5 4 3 2 1

Library of Congress Control Number: 2006940362

ISBN: 978-1-58829-706-8 e-ISBN: 978-1-59745-531-2

Preface

Introduction to Systems Biology is intended to be an introductory text for undergraduate and graduate students who are interested in comprehensive biological systems. Because genomics, transcriptomics, proteomics, interactomics, metabolomics, phenomics, localizomics, and other omics analyses provide enormous amounts of biological data, systematic instruction on how to use computational methods to explain underlying biological meanings is required to understand the complex biological mechanisms and to build strategies for their application to biological problems.

The book begins with an introductory section on systems biology. The experimental omics tools are briefly described in Part II. Parts III and IV introduce the reader to challenging computational approaches that aid in understanding biological dynamic systems. These last two parts provide ideas for theoretical and modeling optimization in systemic biological researches by presenting most algorithms as implementations, including the up-to-date, full range of bioinformatic programs, as well as illustrating available successful applications.

The authors also intend to provide a broad overview of the field using key examples and typical approaches to experimental design (both wet-lab and computational). The format of this book makes it a great resource book and provides a glimpse of the state-of-the-art technologies in systems biology. I hope that this book presents a clear and intuitive illustration of the topics on biological systemic approaches and further introduces ideal computational methods for the reader's own research.

Sangdun Choi
Department of Biological Sciences,
Ajou University, Suwon, Korea

P. 575

Principle component Analysis

PC1 - direction

PC2 vector orthogonal (how determine?)

eigenvalues?

loading vector for individual genes.

Contents

Appendices

Contributors

Ping Ao
Department of Mechanical Engineering, University of Washington, Seattle, WA, USA

Matthew J. Beal
Department of Computer Science and Engineering, State University of New York at Buffalo, Buffalo, NY, USA

Dieter Beule
MicroDiscovery GmbH, Berlin, Germany

Nils Blüthgen
Institute of Theoretical Biology, Humboldt University, Berlin, Germany

Benjamin Bornstein
Machine Learning Systems Group, Jet Propulsion Laboratory, California Institute of Technology, Pasadena, CA, USA

Fabien Campagne
Institute for Computational Biomedicine and Department of Physiology and Biophysics, Weill Medical College of Cornell University, New York, NY, USA

Eliza Chan
Institute for Computational Biomedicine and Department of Physiology and Biophysics, Weill Medical College of Cornell University, New York, NY, USA

Shann-Ching Chen
Department of Biomedical Engineering, Carnegie Mellon University, Pittsburgh, PA, USA

Sangdun Choi
Department of Biological Sciences, Ajou University, Suwon, Korea

Marcio Rosa da Silva
Research Group Systems Biology, GBF—German Research Centre for Biotechnology, Braunschweig, Germany

Yves Deville
Computing Science and Engineering Department, Université Catholique de Louvain, Louvain-la-Neuve, Belgium

Pawan K. Dhar
RIKEN Genomic Sciences Centre, Yokohama, Kanagawa, Japan

Andrew Finney
Physiomics PLC Oxford, Oxford, UK

Stephen S. Fong
Department of Chemical and Life Science Engineering, Virginia Commonwealth University, Richmond, VA, USA

Akira Funahashi
ERATO-SORST Kitano Symbiotic Systems Project, Japan Science and Technology Agency, Shibuya-ku, Tokyo, Japan

David Galas
Institute for Systems Biology, Seattle, WA, USA

Chris Garwood
School of Computer Science, University of Manchester, Manchester, UK

Kevin Garwood
School of Computer Science, University of Manchester, Manchester, UK

Zoubin Ghahramani
Department of Engineering, University of Cambridge, Cambridge, UK

Yuki Hasegawa
Laboratory for Genome Exploration Research Group, RIKEN Genomic Sciences Center (GSC), Yokohama Institute, Tsurumi-ku, Yokohama, Kanagawa, Japan

Yoshihide Hayashizaki
Laboratory for Genome Exploration Research Group, RIKEN Genomic Sciences Center (GSC), Yokohama Institute, Tsurumi-ku, Yokohama, Kanagawa, Japan

Hanspeter Herzel
Institute of Theoretical Biology, Humboldt University, Berlin, Germany

Noriko Hiroi
ERATO Kitano Symbiotic Systems Project, Japan Science and Technology Agency, Shibuya-ku, Tokyo, Japan

Leroy Hood
Institute for Systems Biology, Seattle, WA, USA

Michael Hucka
Division of Control and Dynamical Systems and Biological Network Modeling Center, California Institute of Technology, Pasadena, CA, USA

Takashi Ito
Department of Computational Biology, Graduate School of Frontier Sciences, University of Tokyo, Kashiwa, Japan

Michael C. Jewett
Center for Microbial Biotechnology, BioCentrum-DTU, Technical University of Denmark, Lyngby, Denmark

Ji-Bin Sun
Research Group Systems Biology, GBF—German Research Centre for Biotechnology, Braunschweig, Germany

Andrew R. Jones
School of Computer Science, University of Manchester, Manchester, UK

Akiya Jouraku
ERATO-SORST Kitano Symbiotic Systems Project, Japan Science and Technology Agency, Shibuya-ku, Tokyo, Japan

Sarah M. Keating
Science and Technology Research Institute, University of Hertfordshire, Hatfield, UK

Boris N. Kholodenko
Department of Pathology and Cell Biology, Daniel Baugh Institute for Functional Genomics/Computational Biology, Thomas Jefferson University, Philadelphia, PA, USA

Szymon M. Kielbasa
Max Planck Institute for Molecular Genetics, Computational Molecular Biology, Berlin, Germany

Hiroaki Kitano
Sony Computer Science Laboratories, Inc., Shinagawa, Tokyo, Japan

Hiroaki Kitano
ERATO-SORST Kitano Symbiotic Systems Project, Japan Science and Technology Agency, Shibuya-ku, Tokyo, Japan

Keiji Kito
Department of Computational Biology, Graduate School of Frontier Sciences, University of Tokyo, Kashiwa, Japan

Bharani Kumar
Research Group Systems Biology, GBF—German Research Centre for Biotechnology, Braunschweig, Germany

Ursula Kummer
Bioinformatics and Computational Biochemistry, EML Research, Heidelberg, Germany

Koji Kyoda
RIKEN Genomic Sciences Center (GSC), Yokohama Institute, Tsurumi-ku, Yokohoma, Kanagawa, Japan

Stéphane Le Crom
INSERM U368, Ecole Normale Supérieure, Paris, France

Nicolas Le Novère
Computational Neurobiology, EMBL-EBI, Wellcome-Trust Genome Campus, Hinxton, UK

Stefan Legewie
Institute of Theoretical Biology, Humboldt University, Berlin, Germany

Gaëlle Lelandais
CNRS UMR 8541, Ecole Normale Supérieure, Paris, France

Christian Lemer
Unité de Conformation des Macromolécules Biologiques, Université Libre de Bruxelles, Bruxelles, Belgium

Juan Li
Department of Computer Science and Engineering, State University of New York at Buffalo, Buffalo, NY, USA

Hong-wu Ma
Research Group Systems Biology, GBF—German Research Centre for Biotechnology, Braunschweig, Germany

Yukiko Matsuoka
ERATO-SORST Kitano Symbiotic Systems Project, Japan Science and Technology Agency, Shibuya-ku, Tokyo, Japan

Joanne Matthews
Science and Technology Research Institute, University of Hertfordshire, Hatfield, UK

Mineo Morohashi
Human Metabolome Technologies, Inc., Tsuruoka, Yamagata, Japan

Robert F. Murphy
Department of Biomedical Engineering, Carnegie Mellon University, Pittsburgh, PA, USA

Vijayalakshmi H. Nagaraj
BioMaPS Institute, Rutgers University, The State University of New Jersey, Piscataway, NJ, USA

Jens Nielsen
Center for Microbial Biotechnology, BioCentrum-DTU, Technical University of Denmark, Lyngby, Denmark

Ana Paula Oliveira
Licenciada, Center for Microbial Biotechnology, BioCentrum-DTU, Technical University of Denmark, Lyngby, Denmark

Stephen Oliver
School of Life Sciences, University of Manchester, Manchester, UK

Lars F. Olsen
Department of Biochemistry and Molecular Biology, Syddansk Universitet, Syddansk, Denmark

Shuichi Onami
RIKEN Genomic Sciences Center (GSC), Yokohama Institute, Tsurumi-ku, Yokohama, Kanagawa, Japan

Bernhard Ø. Palsson
Department of Bioengineering, University of California, San Diego, La Jolla, CA, USA

Norman W. Paton
School of Computer Science, University of Manchester, Manchester, UK

Howard R. Petty
Department of Ophthalmology and Visual Sciences, University of Michigan Medical School, Ann Arbor, MI, USA

Roberto Romero
Perinatology Research Branch, National Institute of Child Health and Human Development, Bethesda, MD, *and* Hutzel Hospital, Detroit, MI, USA

Maria J. Schilstra
Science and Technology Research Institute, University of Hertfordshire, Hatfield, UK

Kumar Selvarajoo
Institute of Advanced Biosciences, Keio University, Tsurouka, Yamagata, Japan

Anirvan M. Sengupta
BioMaPS Institute, Rutgers University, The State University of New Jersey, Piscataway, NJ, USA

Bruce E. Shapiro
Division of Biology and Biological Network Modeling Center, California Institute of Technology, Pasadena, CA, USA

Ines Thiele
Bioinformatics Program, University of California, San Diego, La Jolla, CA, USA

Masa Tsuchiya
Institute of Advanced Biosciences, Keio University, Tsurouka, Yamagata, Japan

David L. Wild
Keck Graduate Institute, Claremont, CA, USA

Shoshana Wodak
Department of Biochemistry and Structural Biology, Department of Medical Genetics, University of Toronto, Toronto, Ontario, Canada

Lan Yin
School of Physics, Peking University, Beijing, People's Republic of China

An-Ping Zeng
Research Group Systems Biology, GBF—German Research Centre for Biotechnology, Braunschweig, Germany

Ting Zhao
Department of Biomedical Engineering, Carnegie Mellon University, Pittsburgh, PA, USA

Xiaomei Zhu
GenMath Corp., Seattle, WA, USA

Hao Zhu
Division of Applied Mathematics, School of Mathematical Sciences, University of Nottingham, Nottingham, UK

Part I

Introduction

1

Scientific Challenges in Systems Biology

Hiroaki Kitano

Summary

Systems biology is the study of biological systems at the system level. Such studies are made possible by progress in molecular biology, genomics, computer science, and other fields that deal with the complexity of systems. For systems biology to grow into a mature scientific discipline, there must be basic principles or conceptual frameworks that drive scientific inquiry. The author argues that understanding the robustness of biological systems and the principles behind such phenomena is critically important for establishing the theoretical foundation of systems biology. It may be a guiding principle not only for basic scientific research but also for clinical studies and drug discovery. A series of technologies and methods need to be developed to support investigation of such theory-driven and experimentally verifiable research.

Key Words: Systems biology; robustness; trade-offs; technology platforms.

1. Introduction

Systems biology aims at a system-level understanding of biological systems (1,2). The investigation of biological systems at the system level is not a new concept. It can be traced back to homeostasis by Canon (3), cybernetics by Norbert Weiner (4), and general systems theory by von Bertalanffy (5). Also, several approaches in physiology have taken a systemic view of the biological subjects. The reason why "systems biology" is gaining renewed interest today is, in my view, due to emerging opportunities to solidly connect system-level understanding to molecular-level understanding, as well as the possibility of establishing well-founded theory at the system level. This is only possible today because of the progress of molecular biology, genomics, computer science, modern control theory, nonlinear dynamics theory, and other relevant fields, which had not sufficiently matured at the time of early attempts.

However, "system-level understanding" is a rather vague notion and is often hard to define. This is because a system is not a tangible object. Genes and proteins are more tangible because they are identifiable matter. Although systems are composed of this matter, the system itself cannot be made tangible. Often, a diagram of the gene regulatory networks and protein interaction networks are shown as a representation of systems. It is certainly true that such diagrams capture one aspect of the structure of the system, but they are still only a static slice of the system. The heart of the system lies in the dynamics it creates and the logic behind it. It is science on the dynamic state of affairs.

There are four distinct phases that lead us to system-level understanding at various levels. First, system structure identification enables us to understand the structure of the systems. Although this may be a static view of the system, it is an essential first step. Structure is ultimately identified in both physical and interaction structures. Interaction structures are represented as gene regulatory networks and biochemical networks that identify how components interact within and between cells. Physical details of a specific region of the cell, overall structure of cells, and organisms are also important because such physical structures impose constraints on possible interactions, and the outcome of interactions impacts the formation of physical structures. The nature of an interaction could be different if the proteins involved move by simple diffusion or under specific guidance from the cytoskeleton.

Second, system dynamics need to be understood. Understanding the dynamics of the system is an essential aspect of study in systems biology. This requires integrative efforts of experiments, measurement of technology development, computational model development, and theoretical analysis. Several methods, such as bifurcation analysis, have been used, but further investigations are necessary to handle the dynamics of systems with very high dimensional space.

Third, methods to control the system have to be investigated. One of the implications is to find a therapeutic approach based on system-level understanding. Many drugs have been developed through extensive effect-oriented screening. It is only recently that a specific molecular target has been identified, and leading compounds are designed accordingly. Success in control methods of cellular dynamics may enable us to exploit intrinsic dynamics of the cell, so that its effects can be precisely predicted and controlled.

Finally, designing the system—i.e., modifying and constructing biological systems with designed features. Bacteria and yeast may be redesigned to yield the desired properties for drug production and alcohol production. Artificially created gene regulatory logic could be introduced and linked to innate genetic circuits to attain the desired functions (6).

Several different approaches can be taken within the systems biology field. One may decide to carry out large-scale, high-throughput experiments and try to find the overall picture of the system at coarse-grain resolution (7–10). Alternatively, working on precise details of specific signal transduction (11,12), the cell cycle (13,14), and other biological issues to find out the logic behind them are also viable research approaches. Both approaches are essentially complementary, and together reshape our understanding of biological systems.

2. Robustness as a Fundamental Organizational Principle

Although systems biology is often characterized by the use of massive data and computational resources, there are significant theoretical elements that need to be addressed. After all, efforts to digest large data sets are designed to deepen understanding of biological systems, as well as to be applied for medical practices and other issues. In either case, there must be hypotheses to test using these data and computational practices.

Stunning diversity and robustness of biological systems are the most intriguing features of living systems, and can be observed across an astonishingly broad range of species. Robustness is the fundamental feature that enables diverse species to generate and evolve. It is ubiquitous, as it can be observed in virtually all species across different aspects of biological systems. Therefore, one of the central themes of systems biology is to understand robustness and its trade-offs in biological systems and the principle behind them (15).

Why is robustness so important? First, it is a feature that is observed to be ubiquitous in biological systems, from such a fundamental process as phage fate-decision switch (16) and bacterial chemotaxis (17–19) to developmental plasticity (20) and tumor resistance against therapies (21,22), which implies that it may be a basis of principles that are universal in biological systems. These principles may lead to opportunities for finding cures for cancer and other complicated diseases. Second, robustness and evolvability are tightly coupled. Robustness against environmental and genetic perturbation is essential for evolvability (23–25), underlying a basis of evolution. Evolution tends to select individuals with more robust traits against environmental and genetic perturbations than less robust individuals. Third, robustness is a distinctively system-level property that cannot be observed by just looking at its components. Fourth, diseases may be manifestations of trade-offs between robustness and fragility that are inevitable in evolvable robust systems. Therefore, an in-depth understanding of robustness trade-offs is expected to provide us with insights for better preventions and countermeasures for diseases such as cancer, diabetes, and immunological disorders.

Robustness is a property of the system that maintains a specific function against certain perturbations. A specific aspect of the system, function to be maintained, and type of perturbation that the system is robust against must be well defined to make solid arguments. For example, modern airplanes (system) have a function to maintain its flight path (function) against atmospheric perturbations (perturbations). Across engineering and biological systems, there are common mechanisms that make systems robust against various perturbations.

First, extensive system control is used (most obviously negative feedback loops) to make the system dynamically stable around the specific site of the system. An integral feedback used in bacterial chemotaxis is a typical example (17–19). Because of integral feedback, bacteria can sense changes in chemoattractant and chemorepellent activity independent of absolute concentration, so that proper chemotaxis behavior is maintained over a wide range of ligand concentration. In addition, the same mechanism makes it insensitive to changes in rate constants involved in the circuit. Positive feedbacks are often used to create bistability in signal transduction and cell cycles, so that the system is tolerant against minor perturbation in stimuli and rate constants (11,13,14).

Second, alternative (or fail-safe) mechanisms increase tolerance against component failure and environmental changes by providing alternative components or methods to ultimately maintain a function of the system. Occasionally, there are multiple components that are similar to each other that are redundant. In other cases, different means are used to cope with perturbations that cannot be handled by the other means. This is often called phenotypic plasticity (26,27) or diversity. Redundancy and phenotypic plasticity are often considered as opposite events, but it is more consistent to view them as different ways to meet an alternative fail-safe mechanism.

Third, modularity provides isolation of perturbation from the rest of the system. The cell is the most significant example. Less obvious examples are modules of biochemical and gene regulatory networks. Modules also play important roles during developmental processes by buffering perturbations so that proper pattern formation can be accomplished (20,28,29). The definition of a module, and how to detect such modules, are still controversial, but the general consensus is that modules do exist and play an important role (30).

Fourth, decoupling isolates low-level noise and fluctuations from functional level structures and dynamics. One example is genetic buffering by Hsp90, in which misfolding of proteins caused by environmental stresses is fixed; thus, effects of such perturbations are isolated from functions of circuits. This mechanism also applies to the genetic variations, where genetic changes in a coding region that may affect protein structures are masked because protein folding is fixed by Hsp90, unless such masking is removed by extreme stress (24,31,32). Emergent behaviors of complex networks also exhibit such buffering properties (33). These effects may constitute the canalization proposed by Waddington (34).

Apart from these basic mechanisms, there is a global architecture of networks that is characteristic of evolvable robust systems. The bow-tie network is an architecture that has diverse and overlapping inputs and output cascades connected by a "core" network (15,35). Such a structure is observed in metabolic pathways (36) and signal transductions (37,38), and can be considered to play an important role.

In addition, there is an interesting tendency in living organisms to enhance robustness through acquisition of "nonself" biologic entities into "self," namely, self-extending symbiosis, such as horizontal gene transfer, serial endosymbiosis, oocyte-mediated vertical transfer of symbionts, and bacterial flora (39).

3. Intrinsic Nature of Robust Systems

Robustness is a basis of evolvability. For the system to be evolvable, it must be able to produce a variety of nonlethal phenotypes (40). At the same time, genetic variations need to be accumulated as neutral networks, so that pools of genetic variants are exposed when the environment changes suddenly. Systems that are robust against environmental perturbations entail mechanisms such as system control, alternative modularity, and decoupling, which also support, by congruence, the generation of a nonlethal phenotype and genetic buffering. In addition, the capability to generate flexible phenotype and robustness requires emergence of bow-tie structures as an architectural motif (35). One of the reasons why robustness in biological systems is so ubiquitous is because it facilitates evolution, and evolution tends to select traits that are robust against environmental perturbations. This leads to successive addition of system controls.

Given the importance of robustness in biological systems, it is important to understand the intrinsic properties of such systems. One such property is the intrinsic trade-offs among robustness, fragility, performance, and resource demands. Carlson and Doyle argued, using simple examples from physics and forest fires, that systems that are optimized for specific perturbations are extremely fragile against unexpected perturbations (41,42). This means when robustness is enhanced against a range of perturbations, then it must be countered by fragility elsewhere, compromised performance, and increased resource demands. Highly optimized tolerance model systems are successively optimized/designed (although not necessarily globally optimized) against perturbations, in contrast to self-organized criticality (43) or scale-free networks (44), which are unconstrained stochastic additions of components without design or optimizations involved. Such differences actually affect failure patterns of the systems, and thus have direct implications for understanding of the nature of disease and therapy design.

Disease often reflects an exposed fragility of the system. Some diseases are maintained to be robust against therapies because such states are maintained or even promoted through mechanisms that support robustness of normal physiology of our body.

Diabetes mellitus is an excellent example of how systems that are optimized for near-starving, intermittent food supply, high-energy utilization lifestyle, and highly infectious conditions are exposed to fragility against unusual perturbations, in evolutionary time scale (i.e., high energy content foods, and low energy utilization lifestyle) (45). Because of optimization to near-starving condition, extensive control to maintain minimum blood glucose level has been acquired so that activities of central neural systems and innate immunity are maintained. However, no effective regulatory loop has been developed against excessive energy intake, so that blood glucose level is chronically maintained higher than the desired level, leading to cardiovascular complications.

Cancer is a typical example of robustness hijacking (21,22). Tumor is robust against a range of therapies because of genetic diversity, feedback loop for multidrug resistance, and tumor–host interactions. Tumor–host interactions, for example, are involved in HIF-1 up-regulation that then up-regulates VEGF, uPAR, and other genes that trigger angiogenesis and cell motility (46). HIF-1 up-regulation takes place because of hypoxia in tumor clusters and dysfunctional blood vessels caused by tumor growth. This feedback regulation enables tumor to grow further or cause metastasis. However, HIF-1 up-regulation is important for normal physiology under oxygen-deprived conditions, such as breathing at high altitudes and lung dysfunctions (47). This indicates that mechanisms that provide protection for our body are effectively hijacked.

Mechanisms behind infectious diseases, autoimmune disorders, and immune deficiencies, and why certain countermeasures work and others do not, can be properly explained from the robustness perspective (48).

I would consider three theoretically motivated countermeasures for such diseases. First, robustness of epidemic state should be controlled by systematically perturbing biochemical and gene regulatory circuits using low-dose drugs. Second, robust epidemic state implies that there is a point of fragility somewhere. Identification or active induction of such a point may lead to novel therapeutic approaches with dramatic effects. Third, one may wish to retake control of feedback loops that give rise to robustness in the epidemic state. One possible approach is to introduce a decoy that effectively disrupts feedback control or invasive mechanisms of the epidemic.

How we can systematically identify such strategic therapy is yet unknown, and will be a subject of major research in the future (49). However, it is important to emphasize that a conceptual foundation to view robustness as a fundamental principle of biological systems is the critical aspect of this research program. Without such perspective, the search for cures is, at best, a random process.

4. Technology Platforms in Systems Biology

For theoretical analysis to be effective, it is essential that a range of tools and resources are made available. One of the issues is to create a standard for representing models. Systems Biology Mark-up Language (SBML; http://www.sbml.org/) was designed to enable standardized representation and exchange of models among software tools that comply with SBML standards (50). The project was started in 1999, and has now grown into a major community effort. SBML Level-1 and Level-2 have been released and used by over 110 software packages (as of March 2007). Systems Biology Workbench is an attempt to provide a framework where different software modules can be seamlessly integrated, so that researchers can create their own software environment (51). A recent addition to such standardization efforts is Systems Biology Graphical Notation (http://www.sbgn.org/), which aims at the formation of standard and solidly defined visual representations of molecular interaction networks.

In addition to standard formation efforts, technologies to properly measure and compute cellular dynamics are essential. One of the major interests in computational aspects of systems biology is how numerical simulations can be used for deeper understanding of organisms and medical applications. There is no doubt that simulation, if properly used, can be a powerful tool for scientific and engineering research. Modern aircraft cannot be developed without the help of computational fluid dynamics (CFD). There are at least two issues that shall be carefully examined in computational simulation. First, the purpose of simulation has to be well defined, and the model has to be constructed to maximize the purpose of the simulation. This affects the choice of modeling technique, levels of abstractions, scope of modeling, and parameters to be varied. Second, simulation needs to be well placed in the context of the entire analysis procedure. In most cases, simulation is not the only method of analysis, so that the part of analysis that uses numerical simulation and the other parts that use nonsimulation methods will be well coordinated to maximize overall analysis activity.

An example from racing car design illustrates these issues. CFD is extensively used in Formula 1 car design to obtain optimal aerodynamics, i.e., higher down-force and lower drag. Particular interests are placed on the effects of various aerodynamic components, such as front wings, rear wings, and ground effects, but complicated interference between front wings, suspension members, wheels, and brake air-intake ducts is also investigated. Combustion in the engine is the other issue where simulation studies are often used, but it is simulated separately from the CFD model. The success of CFD relies upon the fact that basic principles of fluid dynamics are relatively well understood, although there are still issues that remain to be resolved, so that simulation can be done with relative confidence. This exemplifies practice of proper focus and abstraction. When receptor dynamics is being investigated, transcription machinery will not be modeled, as it is only remotely related.

CFD is not the only tool for aerodynamic design. Formula 1 racing cars are initially designed using CFD (*in silico*), then further investigated using a wind tunnel (*in physico*), followed by an actual run at the test course (*in vitro*) before being deployed in actual races (*in vivo*). CFD, in this case, is used for initial search of candidate designs that are subject to further investigation using a wind tunnel.

There are three major reasons why CFD is now widely accepted. First, the Navier–Stokes equation has been well established to provide computational basis for fluid dynamics with reasonable accuracy. Although there are unresolved issues on how to accurately compute tabular flows, the Navier–Stokes equation provides an acceptable, practical solution for most needs. Second, many CFD results are compared and calibrated against wind tunnel experiments that are highly controlled and extensively monitored. Because of the existence of the wind tunnel, CFD models can be improved for their accuracy and reliability of predictions. Third, decades of effort have been spent on improving CFD and related fluid dynamics research. The current status of CFD is a result of decades of effort.

For computer simulation and analysis in biology to parallel the success of CFD, it must establish a fundamental computing paradigm comparable to the Navier–Stokes equation, create the equivalent to a wind tunnel in biological experiments, and keep working on the problems for decades. Of course, biological systems are much more heterogeneous and complex than fluids, but a set of basic equations must be established so that the fundamental principles behind the computing are pointing in the right direction. It is essential that not only interaction networks but also physical structures be modeled together so that they provide improved reality, particularly for high-resolution modeling of complex mammalian cells. Such an approach may be called computational cellular dynamics (52). Second, highly controlled and high-precision experimental systems are essential; these will be "wind-tunnels" in biology. Microfluidics and other emerging technologies may provide us with experimental setups that have remarkably high precision (53).

One caution that has to be made on the use of computational modeling in biology is to make clear scientific questions that have to be answered by using the computational approach. Mere attempts to create computational models that behave like actual cells do not constitute good scientific practice. Simulation and modeling is the abstraction of actual phenomena. Without proper scientific questions, the correct level of abstraction and scope of the model to be created cannot be determined. This is also the case in CFD. CFD in racing car design has a clear and explicit optimization goal, which is high down-force and low drag. The problem for simulation in biology is that what needs to be discovered by the simulation is not as straightforward as racing car design. Here, the importance of a guiding principle, such as robustness, shall be remembered. The guiding principle provides a view of what needs to be investigated and identified, which can be the starting point of a broad range of applications. One goal of computational simulation is to understand the nature and degree of robustness, and to find out through a set of perturbations how such robustness can be compromised in a controlled manner.

In summary, emphasis shall be placed on the importance of research to identify fundamental system-level principles of biological systems, where numerous insights in both basic science and applications can come out. There are emerging opportunities now because of massive data that are being generated in large-scale experimental projects, but such data are best utilized when processed with certain hypotheses behind them that capture essential aspects of system-level properties. Robustness is one principle that is ubiquitous and fundamental. Investigation on robustness of biological systems will provides us with guiding principles for understanding biological systems and diseases, as well as the effective use of computational tools.

Acknowledgments: The author wishes to thank members of Sony Computer Science Laboratories, Inc., and the Exploratory Research for Advanced Technology (ERATO) Kitano Symbiotic Systems Project for valuable discussions.

This research is supported, in part, by the ERATO and the Solution-Oriented Research for Science and Technology (SORST) programs (Japan Science and Technology Organization), the NEDO Grant (New

Energy and Industrial Technology Development Organization)/Japanese Ministry of Economy, Trade and Industry (METI), the Special Coordination Funds for Promoting Science and Technology, and the Center of Excellence Program for Keio University (Ministry of Education, Culture, Sports, Science, and Technology), the Rice Genome and Simulation Project (Ministry of Agriculture), and the Air Force Office of Scientific Research (AFOSR).

References

1. Kitano H. Systems biology: a brief overview. *Science* 2002;295(5560):1662–1664.
2. Kitano H. Computational systems biology. *Nature* 2002;420(6912):206–210.
3. Cannon WB. The Wisdom of the Body, 2nd edition. New York: W.W. Norton; 1939.
4. Wiener N. Cybernetics: Or Control and Communication in the Animal and the Machine. Cambridge: The MIT Press; 1948.
5. Bertalanffy LV. General System Theory. New York: George Braziller; 1968.
6. Hasty J, McMillen D, Collins JJ. Engineered gene circuits. *Nature* 2002; 420(6912):224–230.
7. Guelzim N, Bottani S, Bourgine P, et al. Topological and causal structure of the yeast transcriptional regulatory network. *Nat Genet* 2002;31(1):60–63.
8. Ideker T, Ozier O, Schwikowski B, et al. Discovering regulatory and signalling circuits in molecular interaction networks. *Bioinformatics* 2002;18 Suppl 1: S233–S240.
9. Ideker T, Thorsson V, Ranish JA, et al. Integrated genomic and proteomic analyses of a systematically perturbed metabolic network. *Science* 2001; 292(5518):929–934.
10. Ihmels J, Friedlander G, Bergmann S, et al. Revealing modular organization in the yeast transcriptional network. *Nat Genet* 2002;31(4):370–377.
11. Ferrell JE, Jr. Self-perpetuating states in signal transduction: positive feedback, double-negative feedback and bistability. *Curr Opin Cell Biol* 2002; 14(2):140–148.
12. Bhalla US, Iyengar R. Emergent properties of networks of biological signaling pathways. *Science* 1999;283(5400):381–387.
13. Tyson JJ, Chen K, Novak B. Network dynamics and cell physiology. *Nat Rev Mol Cell Biol* 2001;2(12):908–916.
14. Chen KC, Calzone L, Csikasz-Nagy A, et al. Integrative analysis of cell cycle control in budding yeast. *Mol Biol Cell* 2004;15(8):3841–3462.
15. Kitano H. Biological robustness. *Nat Rev Genet* 2004;5(11):826–837.
16. Little JW, Shepley DP, Wert DW. Robustness of a gene regulatory circuit. *EMBO J* 1999;18(15):4299–4307.
17. Alon U, Surette MG, Barkai N, et al. Robustness in bacterial chemotaxis. *Nature* 1999;397(6715):168–171.
18. Barkai N, Leibler S. Robustness in simple biochemical networks. *Nature* 1997;387(6636):913–917.
19. Yi TM, Huang Y, Simon MI, et al. Robust perfect adaptation in bacterial chemotaxis through integral feedback control. *Proc Natl Acad Sci USA* 2000;97(9):4649–4653.
20. von Dassow G, Meir E, Munro EM, Odell GM. The segment polarity network is a robust developmental module. *Nature* 2000;406(6792):188–192.
21. Kitano H. Cancer as a robust system: implications for anticancer therapy. *Nat Rev Cancer* 2004;4(3):227–235.
22. Kitano H. Cancer robustness: tumour tactics. *Nature* 2003;426(6963):125.

23. Wagner GP, Altenberg L. Complex adaptations and the evolution of evolvability. *Evolution* 1996;50(3):967–976.
24. Rutherford SL. Between genotype and phenotype: protein chaperones and evolvability. *Nat Rev Genet* 2003;4(4):263–274.
25. de Visser J, Hermission J, Wagner GP, et al. Evolution and Detection of Genetics Robustness. *Evolution* 2003;57(9):1959–1972.
26. Agrawal AA. Phenotypic plasticity in the interactions and evolution of species. *Science* 2001;294(5541):321–326.
27. Schlichting C, Pigliucci M. Phenotypic Evolution: A Reaction Norm Perspective. Sunderland: Sinauer Associates, Inc.; 1998.
28. Eldar A, Dorfman R, Weiss D, et al. Robustness of the BMP morphogen gradient in Drosophila embryonic patterning. *Nature* 2002;419(6904):304–308.
29. Meir E, von Dassow G, Munro E, et al. Robustness, flexibility, and the role of lateral inhibition in the neurogenic network. *Curr Biol* 2002;12(10):778–786.
30. Schlosser G, Wagner G. Modularity in Development and Evolution. Chicago: The University of Chicago Press; 2004.
31. Rutherford SL, Lindquist S. Hsp90 as a capacitor for morphological evolution. *Nature* 1998;396(6709):336–342.
32. Queitsch C, Sangster TA, Lindquist S. Hsp90 as a capacitor of phenotypic variation. *Nature* 2002;417(6889):618–624.
33. Siegal ML, Bergman A. Waddington's canalization revisited: developmental stability and evolution. *Proc Natl Acad Sci USA* 2002;99(16):10528–10532.
34. Waddington CH. The Strategy of the Genes: A Discussion of Some Aspects of Theoretical Biology. New York: Macmillan; 1957.
35. Csete ME, Doyle J. Bow ties, metabolism and disease. *Trends Biotechnol* 2004;22(9):446–450.
36. Ma HW, Zeng AP. The connectivity structure, giant strong component and centrality of metabolic networks. *Bioinformatics* 2003;19(11):1423–1430.
37. Oda K, Kitano H. A comprehensive pathway map of toll-like receptor signaling network. *Mol Syst Biol* 2:2006.0015. Epub 2006 Apr 18.
38. Oda K, Matsuoka Y, Funahashi, et al. A comprehensive pathway map of epidermal growth factor receptor signaling. *Mol Syst Biol* 2005;1:E1–17.
39. Kitano H, Oda K. Self-extending symbiosis: a mechanism for increasing robustness through evolution. *Biol Theory* 2006;1(1):61–66.
40. Kirschner M, Gerhart J. Evolvability. *Proc Natl Acad Sci USA* 1998;95(15):8420–8427.
41. Carlson JM, Doyle J. Highly optimized tolerance: a mechanism for power laws in designed systems. *Phys Rev E Stat Phys Plasmas Fluids Relat Interdiscip Topics* 1999;60(2 Pt A):1412–4127.
42. Carlson JM, Doyle J. Complexity and robustness. *Proc Natl Acad Sci USA* 2002;99 Suppl 1:2538–2545.
43. Bak P, Tang C, Wiesenfeld K. Self-organized criticality. *Phys Rev A* 1988;38(1):364–374.
44. Barabasi AL, Oltvai ZN. Network biology: understanding the cell's functional organization. *Nat Rev Genet* 2004;5(2):101–113.
45. Kitano H, Kimura T, Oda K, et al. Metabolic syndrome and robustness tradeoffs. diabetes. 2004;53(Supplment 3):S1–S10.
46. Harris AL. Hypoxia–a key regulatory factor in tumour growth. *Nat Rev Cancer* 2002;2(1):38–47.
47. Sharp FR, Bernaudin M. HIF1 and oxygen sensing in the brain. *Nat Rev Neurosci* 2004;5(6):437–448.

48. Kitano H, Oda K. Robustness trade-offs and host-microbial symbiosis in the immune system. *Mol Syst Biol* 2006;doi:10.1038/msb4100039.
49. Kitano H. A robustness-based approach to systems-oriented drug design. *Nat Rev Drug Discov* 2007;6(3):202–210.
50. Hucka M, Finney A, Sauro HM, et al. The systems biology markup language (SBML): a medium for representation and exchange of biochemical network models. *Bioinformatics* 2003;19(4):524–531.
51. Hucka M, Finney A, Sauro HM, Bolouri H, Doyle J, Kitano H. The ERATO Systems Biology Workbench: enabling interaction and exchange between software tools for computational biology. *Pac Symp Biocomput* 2002: 450–461.
52. Kitano H. Computational Cellular Dynamics: A Network-Physics Integral. *Nat Rev Mol Cell Biol* 2006;7:163.
53. Balagadde FK, You L, Hansen CL, Arnold FH, Quake SR. Long-term monitoring of bacteria undergoing programmed population control in a microchemostat. *Science* 2005;309(5731):137–140.

2

Bringing Genomes to Life: The Use of Genome-Scale *In Silico* Models

Ines Thiele and Bernhard Ø. Palsson

Summary

Metabolic network reconstruction has become an established procedure that allows the integration of different data types and provides a framework to analyze and map high-throughput data, such as gene expression, metabolomics, and fluxomics data. In this chapter, we discuss how to reconstruct a metabolic network starting from a genome annotation. Further experimental data, such as biochemical and physiological data, are incorporated into the reconstruction, leading to a comprehensive, accurate representation of the reconstructed organism, cell, or organelle. Furthermore, we introduce the philosophy of constraint-based modeling, which can be used to investigate network properties and metabolic capabilities of the reconstructed system. Finally, we present two recent studies that combine *in silico* analysis of an *Eschirichia coli* metabolic reconstruction with experimental data. While the first study leads to novel insight into *E. coli*'s metabolic and regulatory networks, the second presents a computational approach to metabolic engineering.

Key Words: Metabolism; reconstruction; constraint-based modeling; *in silico* model; systems biology.

1. Introduction

Over the past two decades, advances in molecular biology, DNA sequencing, and other high-throughput methods have dramatically increased the amount of information available for various model organisms. Subsequently, there is a need for tools that enable the integration of this steadily increasing amount of data into comprehensive frameworks to generate new knowledge and formulate hypotheses about organisms and cells. Network reconstructions of biological systems provide such frameworks by defining links between the network components in a bottom-to-top approach. Various types of "omics" data can be used to identify the list of network components and their interactions. These network reconstructions represent biochemically, genetically, and genomically

Table 1. Organisms and network properties for which genome-scale metabolic reconstructions have been generated.

	ORFs	SKI	N_G	N_M	N_R	Ref
BACTERIA						
Bacillus subtilis	4,225	4.8	614	637	754	19
Escherichia coli	4,405	55.1	904	625	931	20
			720	438	627	21
Geobacter sulfurreducens	3,530		588	541	523	22
Haemophilus influenzae	1,775	8.9	296	343	488	23
			400	451	461	24
Helicobacter pylori	1,632	13	341	485	476	25
			291	340	388	26
Lactococcus lactis	2,310		358	422	621	27
Mannheimia succiniproducens	2,463		335	352	373	28
Staphylococcus aureus	2,702	16	619	571	641	29
Streptomyces coelicolor	8,042	0.13	700	500	700	30
ARCHAEA						
Methanosacrcina barkerii	5,072		692	558	619	31
EUKARYA						
Mus musculus	28,287	15.6	1,156[b]	872	1,220	32
Saccharomyces cerevisiae	6,183	10.6	750	646	1,149	33
			708	584	1,175	34

Listed is the number of open reading frames (ORF) of each organism, the number of genes included in the reconstruction (N_G), as well as the number of metabolites (N_M) and reactions (N_R) in the metabolic network. The Species Knowledge Index (SKI) (1) is a measure of the amount of scientific literature available for an organism. Adapted from Reed (18).

(BIGG) structured databases that simultaneously integrate all component data, and can be used to visualize and analyze further high-throughput data, such as gene expression, metabolomics, and fluxomics data.

There are at least three ways to represent BIGG databases: (i) textual representation, which allows querying of its content; (ii) graphical representation, which allows the visualization of the network interactions and their components; and (iii) mathematical representation, which enables the usage of a growing number of analytical tools to characterize and study the network properties. Several metabolic reconstructions have been published recently, spanning all domains of life (Table 1), and most of them are publicly available.

In this chapter, we will first define the general properties of a biological system, and then learn to how to reconstruct metabolic networks. The second part of the chapter will introduce the philosophy of constraint-based modeling and highlight two recent research efforts that combined experimental and computational methods. Although this chapter concentrates on metabolic reconstructions, networks of protein–protein interactions, protein–DNA interactions, gene regulation, and cell signaling can be reconstructed using similar rules and techniques. The general scope of this chapter is illustrated in Figure 1, which represents the main process of "bringing genomes to life."

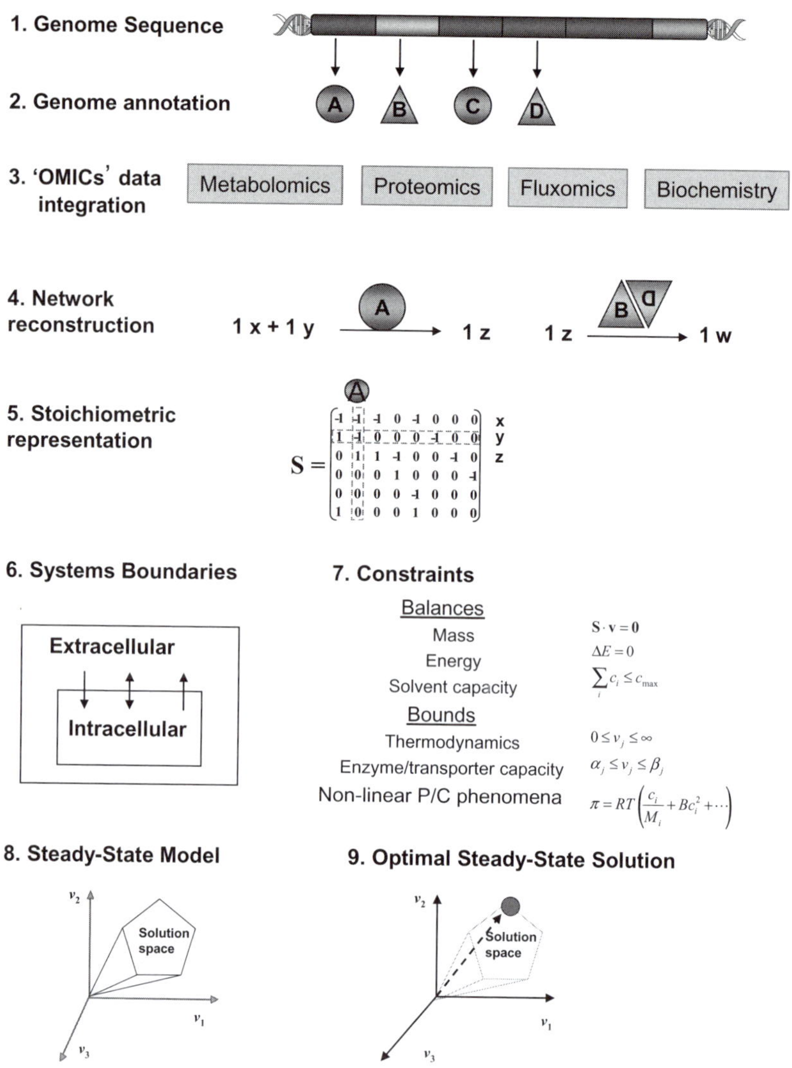

Figure 1. Bringing genomes to life. This figure illustrates the main outline of the chapter and the general approach to network reconstruction and analysis. Starting from the genome sequence, an initial component list of the network is obtained. Using additional data such as biochemical and other omics data the initial component list is refined as well as information about the links between the network components. Once the network links, or reactions, are formulated, the stoichiometric matrix can be constructed using the stoichiometric coefficients that link the network components. The definition of the system boundaries transforms a network reconstruction into a model of a biological system. Every network reaction is elementary balanced and may obey further constraints (e.g., enzyme capacity). These constraints allow the identification of candidate network solutions, which lie within the set of constraints. Different mathematical tools can be used to study these allowable steady-state network states under various aspects such as optimal growth, byproduct secretion and others.

2. Properties of Biological Networks

In this section, we will discuss general properties of biological systems and how these can be used to define a general scheme that describes biological systems in the terms of the components and links of the network.

2.1. General Properties of Biological Systems

The philosophy of network reconstruction and constraint-based modeling is based on the fact that there are general principles any biological system has to obey. Because the interactions, or links, between network components are chemical transformations, they are based on principles derived from basic chemistry. First, in living systems, the prototypical transformation is bilinear at the molecular level. This association involves two compounds coming together to either be chemically transformed through i) the breakage or formation of covalent bonds, as is typical for metabolic reactions and reactions of the macromolecular synthesis,

$$X + Y \leftrightarrow X - Y \quad \text{covalent bonds}$$

or ii) two molecules associate together to form a complex that may be held together by hydrogen bonds and/or other physical association forces to form a complex, which has a different functionality from the individual components:

$$X + Y \leftrightarrow X{:}Y \quad \text{association of molecules.}$$

An example of the latter association is the binding of a transcription factor to DNA to form an activated transcription site that enables the binding of the RNA polymerase.

Second, the reaction stoichiometry is fixed and described by integer numbers counting the molecules that react and that are formed as a consequence of the chemical reaction. Chemical transformations are constrained by elemental and charge balancing, as well as other features. The stoichiometry is invariant between organisms for the same reactions, and it does not change with pressure, temperature, or other conditions. Therefore, stoichiometry gives the primary topological properties of a biochemical reaction network.

Third, all reactions inside a cell are governed by thermodynamics. The relative rate of reactions, forward and backward, is therefore fixed by basic thermodynamic properties. Unlike stoichiometry, thermodynamic properties do change with physicochemical conditions, such as pressure and temperature. In addition, the thermodynamic properties of association between macromolecules can be changed, for example, by altering the sequence of a protein or the base-pair sequence of a DNA-binding site.

Fourth, in contrast to stoichiometry and thermodynamics, the absolute rates of chemical reactions inside cells are evolutionarily malleable. Cells can thus extensively manipulate the rates of reactions through changes in their DNA sequence. Highly evolved enzymes are very specific in catalyzing particular chemical transformations.

These rules dictate that cells cannot form new links at will, and candidate links are constrained by the nature of covalent bonds and by the thermodynamic nature of interacting macromolecular surfaces. All of these are subject to the basic rules of chemistry and thermodynamics. Furthermore, intracellular conditions restrict the activity of systems, such as physicochemical conditions, spatiotemporal organization of cellular components, and the quasicrystalline state of the cell.

2.2. Steady-State Networks

Biological systems exist in a steady state, rather than in equilibrium. In a steady-state system, flow into a node is equal to flow out of a node. Consequently, depletion or accumulation in a steady-state network is not allowed, which means that a produced compound has to be consumed by another reaction. If this is not the case, the corresponding compound represents a network gap (or dead end), and its producing reaction is called a blocked reaction because no flux through this reaction is possible.

3. Reconstruction of Metabolic Networks

The genome annotation, or 1D annotation, provides the most comprehensive list of components in a biological network. In metabolic network reconstructions, the genome annotation is used to identify all potential gene products involved in the metabolism of an organism. By using more types of information, such as biochemical, physiological, and phenotype data, the interaction of these components will be defined. Subsequently, we will refer to network reconstructions as 2D genome annotation because the network links defined in the network reconstruction represent a second dimension to the 1D genome annotation.

3.1. Sources of Information

1D genome annotations are one of the most important information sources for reconstructions because they provide the most comprehensive list of network components. However, one has to keep in mind that without biochemical or physiological verification, the 1D annotation is merely a hypothesis.

The links in metabolic networks are the reactions carried out by metabolic gene products. To assign cellular components with the metabolic reactions, different information is required and provided by various sources. Organism-specific and non–organism-specific databases contain a vast amount of data regarding gene function and associated metabolic activities. Especially valuable are organism-specific literature providing information on the physiological and pathogenic properties of the organism, along with biochemical characterization of enzymes, gene essentiality, minimal medium requirements, and favorable growth environments. Although biochemical data are used during the initial reconstruction effort to define metabolic reactions, organism-specific information such as medium requirements and growth environment can be used to derive transport reactions when not provided by the 1D genome annotations. In addition, gene essentiality data can be used during the network evaluation process to compare and validate the reconstruction. Physiological data, such as medium composition, secretion products, and growth performance, are also needed for the evaluation of the reconstruction and can be found in primary literature or can be generated experimentally. Phylogenetic data can substitute organism-specific information when a particular organism is not well studied, but has a close relative that is. In addition, cellular localization of enzymes can be found in studies that use immunofluorescence or GFP-tagging for individual proteins to identify their place of action. Alternatively, there are several algorithms predicting a protein's compartmentalization based on localization signal sequences.

Because some of these information sources are more reliable than others, a confidence scoring system may be used to distinguish them.

3.2. How to Choose an Organism to Reconstruct

The amount of information available differs significantly from organism to organism; therefore, the choice of organism to reconstruct is critical for the quality of the final reconstruction. Because the genome annotation serves as a first parts list in most reconstruction efforts, its availability and high quality are primary criteria. Furthermore, the quantity of primary and review publications available for metabolism should be considered. A good estimate of legacy data available for an organism can be obtained with the Species Knowledge Index (SKI) (1). This SKI value is a measure of the amount of scientific literature available for an organism, calculated as the number of abstracts per species in PubMed (National Center for Biotechnology Information) divided by the number of genes in the genome (see Table 1 for some SKI values of reconstructed organisms). Finally, organism-specific databases maintained by experts can be very valuable sources of information during the reconstruction process.

3.3. Formulation of Model

The translation of a 1D genome annotation into a metabolic network reconstruction can be done in a step-wise fashion by incorporating different types of data. First, relevant metabolic genes have to be identified from the 1D annotation. The gene functions have to be translated in elementary and charged balanced reactions. Next, the network is assembled by considering each metabolic pathway separately and by filling in missing reactions as necessary. When this first version of the network reconstruction is finished, the reconstruction will be tested *in silico* and compared with physiological data to ensure that it has the same metabolic capabilities as the cell *in vivo*. This latter step might identify further reactions that need to be included, whereas other ones will be replaced or their directionality might be changed. It is important to remember that the sequence-derived list of metabolic enzymes cannot be assumed to be complete because of the large numbers of open reading

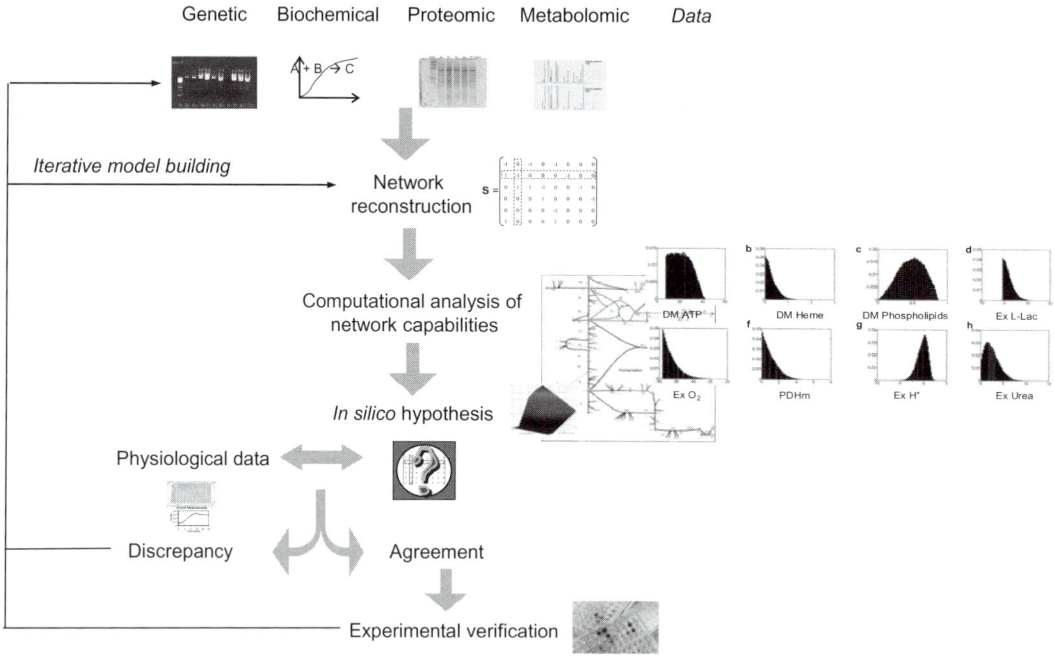

Figure 2. The iterative process of network reconstruction. Normally, several iterations of reconstruction are necessary to ensure quality and accuracy of the reconstructed network. After an initial reconstruction, accounting for the main components identified by the different sources of information, is obtained, the reconstruction will be tested for its ability to produce certain metabolites such as biomass precursors. Comparison with experimental data, like phenotypical and physiological data, will help to identify any discrepancy between *in silico* and *in vivo* properties. The iterative re-evaluation of legacy data and network properties will eventually lead to a refined reconstruction.

frames (ORFs) still having unassigned functions. The iterative process of network reconstruction and evaluation will lead to further refinement of reconstruction (Figure 2).

3.3.1. Defining Biochemical Reactions

The biochemical reaction carried out by a gene product can be determined in five steps (Figure 3). First, the substrate specificity has to be determined because it can differ significantly between organisms. In general, one can distinguish between two groups of enzymes based on their substrate specificity. The first group of enzymes can only act on a few highly similar substrates, whereas the second group recognizes a class of compounds with similar functional groups; thus, the enzymes have a broader substrate specificity. The substrate specificity of either type of these enzymes may differ across organisms for primary metabolites, as well as for coenzymes (such as NADH vs. NADPH and ATP vs. GTP). Often, it is very difficult to derive this information solely from the gene sequence because substrate- and coenzyme-binding sites might be similar for related compounds.

Once the metabolites and coenzymes of an enzyme are identified, the charged molecular formula at a physiologically relevant pH has to be calculated, as a second step. In general, a pH of 7.2 is used in the reconstruction. However, the pH in some organelles can differ from the rest of the cell, as is the case for peroxisomes, where the pH has been reported to be between 6 and 8 (2,3). The pK_a value for a given compound can be used to determine its degree of protonation.

Third, the stoichiometry of the reaction needs to be specified. As in basic chemistry, reactions need to be charge and mass balanced, which may lead to the addition of protons and water.

The fourth step adds basic thermodynamic considerations to the reaction, defining its reversibility. Biochemical characterization studies will sometimes test the reversibility of enzyme reactions, but the directionality can differ between *in vitro* and *in vivo* environments because of differences in temperature, pH, ionic strength, and metabolite concentrations.

The fifth step requires reactions and proteins to be assigned to specific cellular compartments. This task is relatively straightforward for

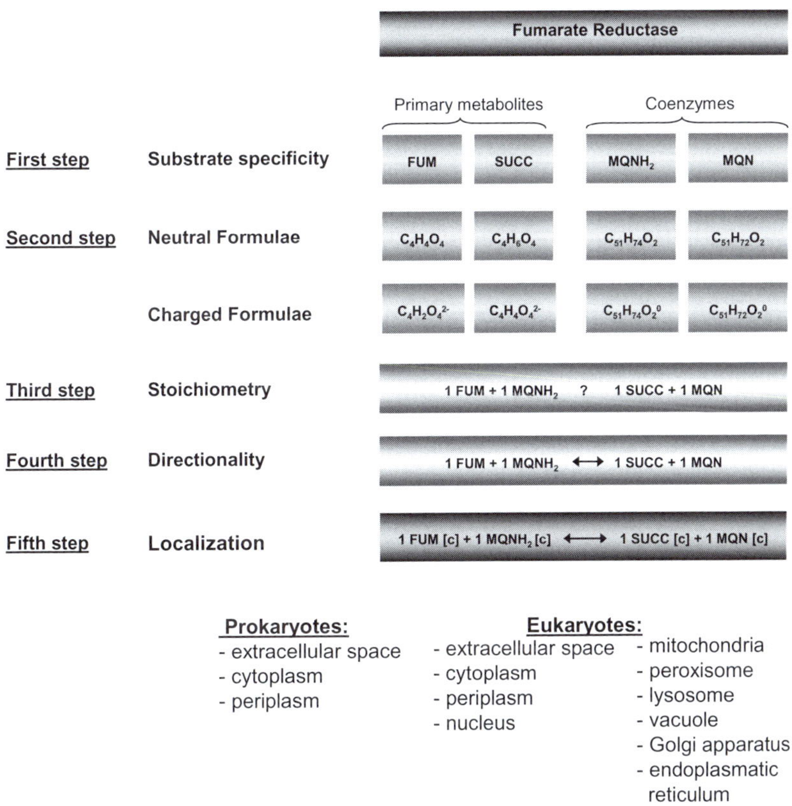

Figure 3. The five steps to formulate a biochemical reaction. The reaction carried out by a metabolic gene product can be determined by the five depicted steps. Here, we show the example of the fumarate reductase of *E. coli*, which converts fumarate (FUM) into succinate (SUCC) using menaquinone (MQN) as electron donor.

prokaryotes, which do not exhibit compartmentalization, but becomes challenging for eukaryotes, which may have up to 11 subcellular compartments (Figure 3). Incorrect assignment of the location of a reaction can lead to additional gaps in the metabolic network and misrepresentation of the network properties. In the absence of experimental data, proteins should be assumed to reside in the cytosol to reduce the number of intracellular transport reactions, which are also often hypothetical and therefore have a low confidence score.

3.3.2. Assembly of Metabolic Network Reconstruction

Once the network reactions are defined, the metabolic network can be assembled in a step-wise fashion by starting with central metabolism, which contains the fueling reactions for the cell, and moving on to the biosynthesis of individual macromolecular building blocks (e.g., amino acids, nucleotides, and lipids). The step-wise assembly of the network facilitates the identification of missing steps within the pathway that were not defined by the 1D annotation. Once well-defined metabolic pathways are assembled, reactions can be added that do not fit into these pathways, but are supported by the 1D annotation or biochemical studies. Such enzymes might be involved in the utilization of other carbon sources or connect different pathways.

3.3.3. Gap Analysis

Even genomes of well-studied organisms harbor genes of unknown functions (e.g., 20% for *E. coli*). Subsequently, metabolic networks constructed solely on genomic evidence often contain many network gaps, so-called blocked reactions. Physiological data may help to determine whether a pathway is functional in the organism, and thus may provide evidence of the missing reactions. This procedure is called gap filling, and it is a crucial step in network reconstruction. For example, if proline is a nonessential amino acid for an organism, then the metabolic network should contain a complete proline biosynthesis pathway, even if some of the enzymes are not in the current 1D annotation. In contrast, if another amino acid, let's say methionine, is known to be required in the medium, then the network gap should not be closed, even if only one gene is missing. In this case, filling the gap would significantly change the phenotypical *in silico* behavior of the reconstruction.

These examples show that physiological data of an organism provide important evidence for improving, refining, and expanding the quality and content of reconstructed networks. Reactions added to the network at this stage should be assigned low confidence scores if there are no genetic or biochemical data available to confirm them. Subsequently, for each added reaction, putative genes can be identified using homology-based and context-based computational techniques. Such added reactions and putative assignments form a set of testable hypotheses that are subject to further experimental investigation. Because the reconstructed network integrates many different types of data available for an organism, its completeness also reflects the knowledge about the organism's metabolism. Remaining unsolved network gaps involving blocked reactions or dead-end metabolites reflect these knowledge gaps.

3.3.4. Evaluation of a Network Reconstruction

Network evaluation is a sequential process (Figure 3). First, the network is examined to see if it can generate the precursor metabolites, such as biomass components, and metabolites the organism is known to produce or degrade. Second, network gaps have to be identified and metabolic pathways may need to be completed based on physiological information. Finally, the comparison of the network behavior with various experimental observations, such as secretion products and gene essentiality, will ensure similar properties and capabilities of the *in silico* metabolic network and the biological system. This sequential, iterative process of network evaluation is labor intensive, but it will ensure high accuracy and quality by network adjustments, refinements, and expansions.

3.4. Automating Network Reconstruction

The manual reconstruction process is laborious and can take up to a year for a typical bacterial genome, depending on the amount of literature available. Hence, efforts have been undertaken to automate the reconstruction process. Like most manually assembled reconstructions, most automatic reconstruction efforts start from the annotation. For example, Pathway Tools (4) is a program that can automate a network reconstruction using metabolic reactions associated with Enzyme Commision numbers (5) and/or enzyme names from a 1D genome annotation. To overcome missing annotations, Pathway Tools has the option to include missing gene products and their reactions in a pathway if a significant fraction of the other enzymes are functionally assigned to this pathway in the genome annotation. As for the manually curated reconstruction, the automated gap filling procedure has to be done with caution, as the inclusion of reactions without confidence may alter the phenotypical outcome of the reconstruction.

Although the automation of reconstruction is necessary on a larger scale, the results of these informatics approaches are limited by the quality of the information on which they operate. Therefore, automated reconstructions need detailed evaluation to assure their accuracy and quality. Frequent problems with these automated reconstructions involve incorrect substrate specificity, reaction reversibility, cofactor usage, treatment of enzyme subunits as separate enzymes, and missing reactions with no assigned ORF. Although an initial list of genes and reactions can be easily obtained by using the automated methods, a good reconstruction of biological networks demands the understanding of properties and characteristics of the organism or the cell. Because the number of experimentally verified gene products and reactions is limited for most organisms, knowledge about the metabolic capabilities of the organism is crucial.

4. Mathematical Characterization of Network Capabilities

In this section, we briefly illustrate the general philosophy of the constraint-based modeling approach that resulted in a growing number of mathematical tools to interrogate a reconstructed network. The method relies primarily on network stoichiometry, and thus it is not necessary to define kinetic rate constants and other parameters, which are difficult or impossible to determine accurately in the laboratory. A more comprehensive description of the different tools can be found in Palsson's work (6) and in a recently published review (7).

4.1. Stoichiometric Representation of Network

The stoichiometric matrix, denoted as S, is formed by the stoichiometric coefficients of the reactions that comprise a reaction network (Figure 1 and Figure 4). This matrix is organized such that every column corresponds to a reaction, and every row corresponds to a compound. The matrix entries are integers that correspond to the stoichiometric coefficients of the network reactions. Each column describes a reaction, which is constrained by the rules of chemistry, such as elementary balancing. Every row describes the reactions in which a compound participates, and therefore how the reactions are interconnected.

Mathematically, the stoichiometric matrix, S, transforms the flux vector v, which contains the reaction rates, into a vector that contains the time derivatives of the concentrations. The stoichiometric matrix, thus contains chemical and network information. Mathematically spoken, the stoichiometric matrix S is a linear transformation of the flux vector,

$$v = (v_1, v_2, \ldots, v_n),$$

to a vector of time derivatives of the concentration vector,

$$x = (x_1, x_2, \ldots x_n),$$

as

$$dx/dt = S.v.$$

At steady state, there is no accumulation or depletion of metabolites in a metabolic network, so the rate of production of each metabolite in the network must equal its rate of consumption. This balance of fluxes can be represented mathematically as

$$S.v = 0.$$

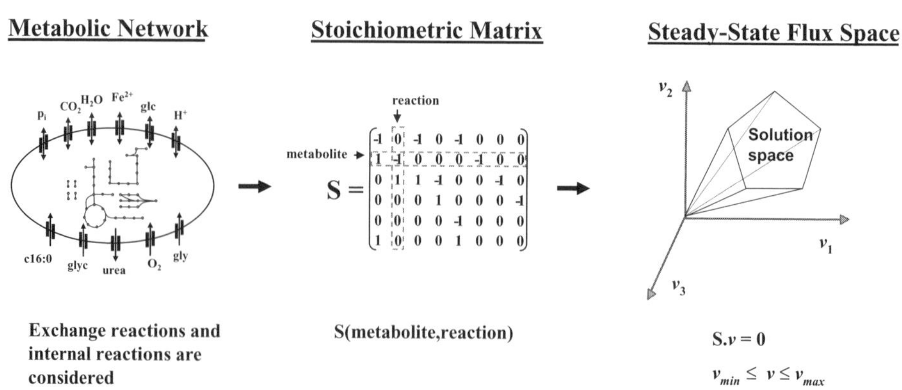

Figure 4. Matrix representation of metabolic network.

Bounds that further constrain the values of individual variables can be identified, such as fluxes, concentrations, and kinetic constants. Upper and lower limits can be applied to individual fluxes, such that

$$v_{i,min} \leq v_i \leq v_{i,max}.$$

For elementary (and irreversible) reactions, the lower bound is defined as $v_{min} = 0$. Specific upper limits (v_{max}) that are based on enzyme capacity measurements are generally imposed on reactions.

4.2. Reconstruction Versus Model

The network reconstruction represents the framework for a biological model. The definition of systems boundaries provides the transition from a network reconstruction to a model. These systems boundaries can be drawn in various ways (Figure 5). Typically, the systems boundaries are drawn around the cell, which is consistent with a physical entity, and the resulting model can be used to investigate properties and capabilities of the biological system. However, it might be useful to draw "virtual" boundaries to segment the network into subsystems (e.g., nucleic acid synthesis or fatty acid synthesis).

The "physical" systems boundaries are drawn to distinguish between the inside metabolites of the cell to the outside metabolites and thus, correspond to the cell membrane. Reactions that connect the cell and its environment are called exchange reactions. These exchange reactions allow the exchange of metabolites in and out of the cell boundaries.

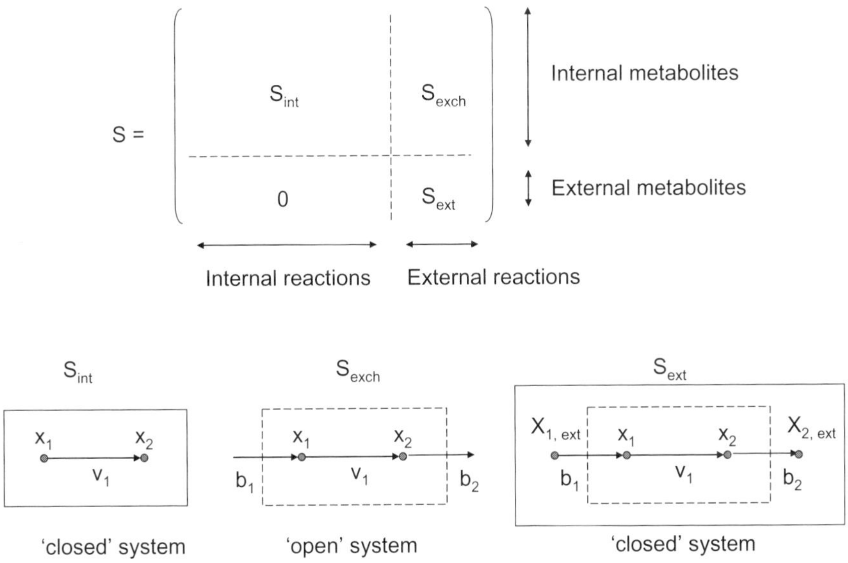

Figure 5. Systems Boundaries. The network reactions are partitioned in internal (int) and external (ext) reactions. The exchange fluxes are denoted by b_i and internal fluxes by v_i.

The stoichiometric matrix S (or S_{tot}) can be partitioned such that there are three fundamental subforms of S_{tot}: i) the exchange stoichiometric matrix (S_{exch}), which does not consider external metabolites and only contains the internal fluxes and the exchange fluxes with the environment; ii) the internal stoichiometric matrix (S_{int}), which considers the cell a closed system; and iii) the external stoichiometric matrix (S_{ext}), which only contains external metabolites and exchange fluxes (Figure 5). These different forms of S can be used to study topological properties of the network. For example, S_{exch} is frequently used in pathway analysis (extreme pathway analysis), whereas S_{int} is useful to define pools of compounds that are conserved within the network (e.g., currency or secondary metabolites such as ATP, NADH, and others).

4.3. Identification of Constraints

Cellular functions are limited by different types of constraints, which can be grouped in four general categories: fundamental physicochemical, spatial or topological, condition-dependent environmental, and regulatory or self-imposed constraints. Although the first two categories of constraints are assumed to be independent from the environment, the latter two may vary in the simulation.

4.3.1. Physicochemical Constraints

Many physicochemical constraints are found in a cell. These constraints are inviolable and provide "hard" constraints on cell functions because mass, energy, and momentum must be conserved. For example, the diffusion rates of macromolecules inside a cell are generally slow because the contents of a cell are densely packed and form a highly viscous environment. Reaction rates are determined by local concentrations inside the cell and are limited by mass transport beside their catalytic rates. Furthermore, biochemical reactions can only proceed in the direction of a negative free-energy change. Reactions with large negative free-energy changes are generally irreversible. These physicochemical constraints are normally considered when formulating the network reactions and their directions.

4.3.2. Spatial Constraints

The cell content is highly crowded, which leads to topological, or spatial, constraints that affect both the form and the function of biological systems. For example, bacterial DNA is about 1,000 times longer than the length of a cell. Thus, on one hand, the DNA must be tightly packed in a cell without becoming entangled; however, on the other hand, the DNA must also be accessible for transcription, which results in spatial-temporary pattern. Therefore, two competing needs, which are the packaging and the accessibility of the DNA, constrain the physical arrangement of DNA in the cell. Incorporating these constraints is a significant challenge for systems biology.

4.3.3. Environmental Constraints

Environmental constraints on cells are time and condition dependent. Nutrient availability, pH, temperature, osmolarity, and the availability of electron acceptors are examples of such environmental constraints. This group of constraints is of fundamental importance for the quantitative analysis of the capabilities and properties of organisms because it allows determining their fitness, or phenotypical properties, under various environmental settings. Because the performance of an organism varies under different environmental conditions, data from various laboratories can only be compared and integrated when the experimental conditions, such as medium composition, are well documented. In contrast, laboratory experiments with undefined media composition are often of limited use for quantitative *in silico* modeling.

4.3.4. Regulatory Constraints

Regulatory constraints differ from the three categories discussed above, as they are self-imposed and subject to evolutionary change. For this reason, these constraints may be referred to as regulatory constraints, in contrast to hard physicochemical constraints and time-dependent environmental constraints. On the basis of environmental conditions, regulatory constraints allow the cell to eliminate suboptimal phenotypic states. Regulatory constraints are implemented by the cell in various ways, including the amount of gene products made (transcriptional and translational regulation) and their activity (enzyme regulation).

4.4. Tools For Analyzing Network States

The analysis of an organism's phenotypic functions on a genome scale using constraint-based modeling has developed rapidly in recent years. A plethora of steady-state flux analysis methods can be broadly classified into the following categories: i) finding best or optimal states in the allowable range; ii) investigating flux dependencies; iii) studying all allowable states; iv) altering possible phenotypes as a consequence of genetic variations; and v) defining and imposing further constraints. In this section, we will discuss some of the numerous methods that have been developed (Table 2). A more comprehensive list of methods can be found in Price's work (7).

4.4.1. Optimal or Best States

Mathematical tools, such as linear optimization, can be used to identify metabolic network states that maximize a particular network function, such as biomass, ATP production, or the production of a desired secretion product. The objective function can be either a linear or nonlinear function. For linear functions, linear optimization or linear programming (LP) can be used to calculate one optimal reaction network state under the given set of constraints. Growth performance of an organism can be assessed by calculating the optimal (growth) solution under different medium conditions. Using visual tools, such as metabolic maps, the optimal network state can be easily accessed and compared. This mathematical tool has been widely used for the identification of optimal network states for the objective function of interest. Interestingly, for genome-scale networks in particular, there can be multiple network states or flux distributions with the same optimal value of the objective function; therefore the need for enumerating alternate optima arises.

Table 2. List of constraint-based modeling methods.

Analysis Method	Illustration	Applied metabolic networks	References
Optimal solutions	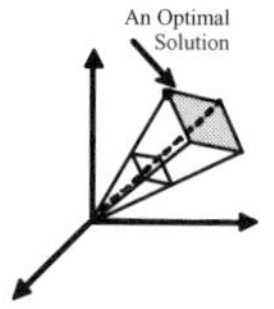	***Escherichia coli***	**35**
Alternate Optima	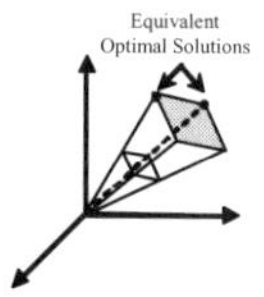	***Escherichia coli*, human cardiac myocyte mitochondrion**	**8, 36, 37**
OptKnock	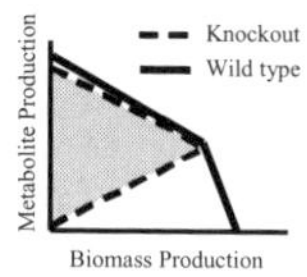	***Escherichia coli***	**12, 38**
Sampling	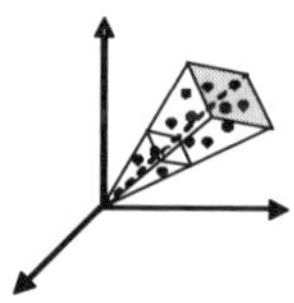	**Red blood cell, *Helicobacter pylori*, human cardiac myocyte mitochondrion**	**39, 42**

A myriad of analytical methods have arisen over recent years. The methods discussed in this chapter are depicted in this table along with some metabolic networks that have been applied to study network properties. Redrawn from Price (7).

4.4.2. Alternate Optima

Alternate optima are a set of flux distributions that represent equally optimal network states given any particular objective function. The number of such alternate optima varies depending on the size of the metabolic network, the chosen objective function, and the environmental conditions. In general, the larger and more interconnected the network, the higher the number of alternate optimal phenotypes. A recursive mixed-integer LP algorithm has been developed to exhaustively enumerate all alternate optima (8). Genome-scale metabolic networks contain several redundant pathways, which makes the enumeration of all optima computationally challenging.

4.4.3. OptKnock

OptKnock is a bilevel optimization algorithm to computationally predict gene deletion strategies for byproduct overproduction, such as succinate, lactate, and amino acids. The OptKnock algorithm calculates solutions that simultaneously optimize two objective functions, which are biomass formation and secretion of a target metabolite. Multiple gene deletions can be introduced in the metabolic network, such that the fluxes through reactions of the target metabolite are optimally used, while reactions leading to other byproducts from common precursors are deleted from the network. The premise underlying this bilevel optimization algorithm is that overproduction of target metabolites can be achieved by altering the structure of the metabolic network through gene deletions. With this direct stoichiometric coupling of target metabolite production to biomass, it is hypothesized that an increase in growth rate should concurrently result in an increase in the target metabolite production rate.

4.4.4. Unbiased Modeling

In addition to the above listed examples of optimization-based methods, non–optimization-based techniques have also been developed to study the full range of achievable metabolic network states that are provided by the solution spaces. These methods enable the user to determine not only the solutions selected by the statement of an objective, but all the solutions in the space. The results are therefore not biased by a statement of an objective, but indicate properties of the genome-scale network as a whole. Uniform random sampling is one example of an unbiased method. Here, the solution space is sampled by calculating uniform, random points within the space. The content of a solution space can be studied by the set of uniform random sampling of points within the space. The sampling points describe candidate metabolic states that are in agreement with the imposed constraints. The projection of the sampling point into a 2D diagram results in a flux distribution for every reaction in the network that can be understood as a probability distribution of flux values for every reaction.

The methods described in this section have been successfully used to characterize and investigate the network capabilities of numerous genome-scale metabolic networks. Until recently, it has focused on the steady-state flux distributions through a reconstructed network, but is now being used to study all allowable concentration and kinetic states (9).

5. Two Sample Studies

In this section, we will highlight two studies that combined *in silico* analysis and experimental data to gain new insight into the metabolism of *E. coli*.

5.1. "Integrating High-throughput and Computational Data Elucidates Bacterial Networks" (10) (Figure 6)

Regulatory constraints are used by cells to control the expression state of genes, leading to distinct sets of expressed genes under different environmental conditions. Assuming the expression state of a gene can be only on or off (expressed or depressed), the regulation of genes can be represented in the form of Boolean rules (on or off, 1 or 0).

For the purpose of this study, the regulatory rules for the metabolic genes included in *i*JR904 (11) were created and incorporated based on literature and databases. The resulting reconstruction, MC1010v1, was the first integrated genome-scale *in silico* reconstruction of a transcriptional regulatory and metabolic network. MC1010v1 accounted for 1,010 genes in *E. coli*, including 104 regulatory genes whose products, together with other stimuli, regulate the expression of 479 of the 904 genes in the reconstructed metabolic network.

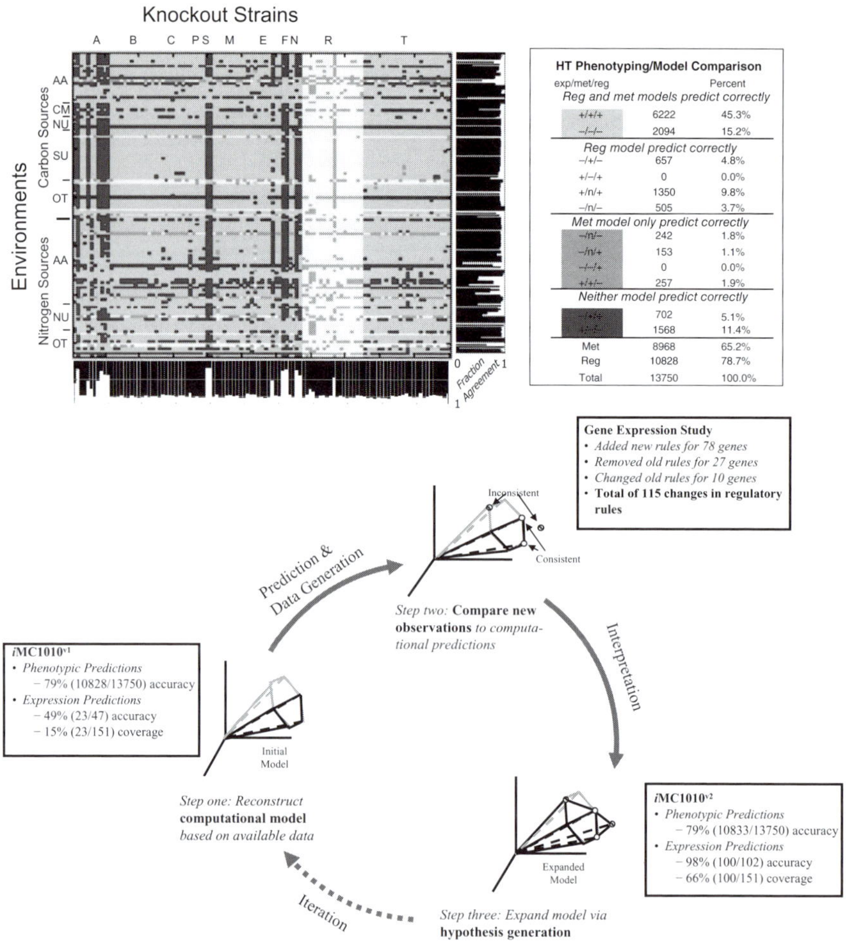

Figure 6. "Integrating high-throughput and computational data elucidates bacterial networks." Top Panel: Comparison of high-throughput phenotyping array data with the *in silico* predictions for the *E. coli* network, with (Reg) and without (Met) regulatory constraints. Each case lists the results of the experimental data (exp), metabolic model (met) and regulatory metabolic model (reg). "+": predicted or observed growth, "–": no growth, and 'n': for cases involving a regulatory gene knockout not predictable by the metabolic model.
Bottom Panel: Metabolic and regulatory networks may be expanded by using high-throughput phenotyping and gene expression data coupled with the predictions of a computational model. The accuracy refers to the percentage of model predictions that agreed with experimental data; the coverage indicates the percentage of experimental changes predicted correctly by the model. Redrawn from (10).

To determine the importance of regulatory rules on the predictive potential of the metabolic reconstruction, both reconstructions, *i*JR904 (unregulated metabolic network) and MC1010v1 (regulated metabolic network), were used to calculate *in silico* growth performance under different medium conditions and to assess the outcome of gene deletion to the growth performance. The *in silico* results were compared with the outcomes of high-throughput growth phenotyping and gene expression experiments (Figure 6). Based on these results, several substrates and knockout strains were found whose growth behavior did not match predictions. Further investigation of these conditions and strains led to the identification of five environmental conditions in which dominant, yet uncharacterized, regulatory interactions actively contributed to the observed growth phenotype. In addition, five environmental conditions and eight knockout strains were identified that highlight uncharacterized enzymes or noncanonical pathways and that are predicted to be used by this study. Furthermore, the results indicated that some transcription factors that were involved in the regulation differed from previously reported data. These new rules were incorporated in the reconstruction leading to a second version, MC1010v2, which could successfully predict the outcome of high-throughput growth phenotyping and gene expression experiments.

The results of this study, and the iterative modification of the regulatory rules, led to two main observations. First, some of the results of the knockout perturbation analysis are complex enough to make Boolean rule formulation difficult. Second, many of these gene expression changes involve complex interactions and indirect effects. Transcription factors may be affected, for example, by the presence of fermentation byproducts or the buildup of internal metabolites. Such effects would be extremely difficult to identify or account for without a computational model.

This study showed that the reconciliation of high-throughput data sets with genome-scale computational model predictions enables systematic and effective identification of new components and interactions in microbial biological networks.

5.2. "In Silico Design and Adaptive Evolution of Escherichia coli for Production of Lactic Acid" (12) (Figure 7)

In this study, OptKnock was used to design candidate knockout mutations *in silico*, which were subsequently analyzed and verified experimentally. The overall goal was to create an *E. coli* mutant that could overproduce lactic acid in minimal medium supplemented with glucose. In contrast, *E. coli* wild type produces only traces of lactate under this medium condition. Other studies already engineered lactate-overproducing *E. coli* mutants; however, in this study it was shown how to use metabolic reconstructions to successfully engineer stable mutants.

The most recent reconstruction of *E. coli*'s metabolism, *i*JR904 (11), was used by the OptKnock algorithm to identify the possible solutions that induce *E. coli* to secrete lactic acid as a byproduct during optimal cellular growth. For this purpose single, double, triple, and quadruple gene deletions were designed *in silico* and tested for bioptimal production of lactic acid and growth yield. Based on these calculations, three different designs for production of lactate were selected for experimental verification: (i) pta-adhE double-deletion strain, (ii) pta-pfk double-deletion strain, and (iii) pta-adhE-pfk-glk quadruple deletion strains (pta, phosphate acetyltransferase; adhE, acetaldehyde dehydrogenase; pfk, 6-phosphofructokinase; glk, glucokinase) (Figure 7).

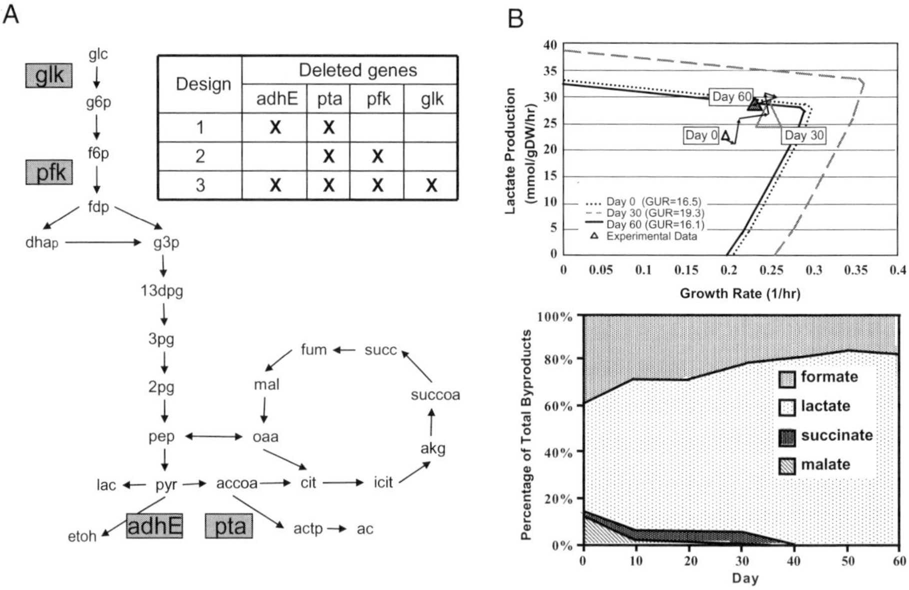

Figure 7. "In Silico Design and Adaptive Evolution of *Escherichia coli* for Production of Lactic Acid." A: schematic picture of the pathways in which the gene deletions are involved. B: Strain design 1 (adhE$^-$,pta$^-$). **Top panel:** growth performance and lactate secretion at the beginning of adaptive evolution (day 0), in the middle (day 30) and at the end (day 60). The computational predictions (lines) were done based on the glucose uptake rate (GUR) measured in the deletion strain at the different time points. The **bottom panel** shows the byproduct secretion rates for the mutant strain during the course of adaptive evolution. It is easily visible that lactate becomes the main fraction of byproduct. Redrawn from (12).

Predicted strain designs were constructed *in vivo* and evolved over 60 d. Over this time period, the growth rates of constructed strains and the byproduct secretion rates were monitored. By measuring these growth rates and lactic acid secretion rates, as well as the glucose uptake rates, the experimental phenotypes could be directly compared to the computationally predicted possible solutions for each design. Both the pta-adhE strains and the pta-pfk strains showed good agreement with the computationally determined solution spaces. In all cases, the byproduct secretion profiles stabilized after approximately 20 d of adaptive evolution, with all strains showing sustained elevated lactic acid titers throughout the course of adaptive evolution over the wild-type strain.

The goal of this study was to experimentally test computationally predicted strain designs calculated from a genome-scale metabolic model using the OptKnock algorithm. For the generated designs, it was shown that this combination of computational approaches can prospectively and effectively calculate strain designs for lactic acid overproduction. The long-term adaptive evolution experiments showed that: i) the computationally predicted phenotypes are experimentally reproducible and consistent; ii) the process of adaptive evolution leads to increased secretion rates of a target metabolite and can lead to improved product titers;

and iii) the generation of stable production strains can be achieved through this method. Overall, all evolved strains exhibited secretion profiles that supported the OptKnock hypotheses, in which the metabolite overproduction was stoichiometrically coupled to biomass generation.

6. Further Levels of Annotation

The majority of this chapter focused on the second dimension of genome annotation that defines the network links between the components given by the 1D annotation. In this section, we will briefly look at the remaining two dimensions, i.e., space and time. Although no reconstruction exists to date that considers these two additional dimensions, further research will provide the basis, and thus enable such reconstructions.

6.1. 3D Annotation: Spatial Position and Orientation

In the previous sections, we saw that the 1D annotation delivers a list of genes and their functions, which can be translated into a table of gene products and their known interactions (2D annotation). These interaction networks must operate within the three dimensional structure of a cell. A growing number of studies indicate that both the genomic location (i.e., the linear allelic address), as well as the spatial location (i.e., the position of a gene within the cell) of a gene is important in genome function (13). In addition, the growth phase of a cell influences the geometrical, and therefore topological, organization of a genome. An explicit link between the geometrical organization of the genome and the expression level of individual genes has yet to be established. However, log phase growth clearly requires many genes to be expressed contemporaneously, which cannot be achieved with a condensed chromosome.

6.1.1. 4D Annotation: Evolutionary Changes

Genomes can undergo short-term adaptive changes; thus, one can think of a fourth dimension to the genome—time. Such changes can be caused epigenetically or genetically, leading to modification in genome function over time. Mechanisms and how they function during adaptation have been studied for individual loci (such as arcB [14], mglD [15], mglO [15], and glpR [16] in *E. coli*), but have not yet been elucidated on a genome scale, with the exception of genome rearrangements. It is becoming appreciated that the genome sequence we have are "snap-shots" of a genome that is continually evolving. Thus, a more detailed understanding of the plasticity and adaptation of genomes on a genome scale is needed. The genetic basis for adaptation of genomes may emerge from full genome resequencing, enabling us to fully determine all the sequence changes that occur in genomes. Furthermore, resequencing may have the potential to provide insights into the mechanisms and functions of these adaptive evolutionary changes of an entire genome.

7. Future Directions

The four dimensions of genome annotation are important for describing and capturing the functional capabilities of a cell. A detailed, quality-controlled, and quality-assessed process for genome-scale reconstruction of metabolic networks (as an example of a 2D annotation) has developed over the past 5–10 years (17,18). It is a laborious and detailed process that involves the manual curation of a wide range of data types. Somewhat similar to sequence assembly and 1D genome annotation, this process of 2D annotation is iterative, involving the successive addition of more and more detailed data as they become available for a particular organism. These high-quality reconstructions can be used as the basis for computation of phenotypic traits, and they represent a key step in the development of the burgeoning field of systems biology (6). The number of organisms with publicly available genome-scale reconstructions continues to grow (Table 1).

Although the focus of this chapter was on metabolic networks, other networks, such as protein interaction, signaling, and regulatory networks, can be reconstructed in a similar manner. The nature of these networks is often qualitative in nature; the description of its components and their interactions may lack the biochemical details of metabolic reconstructions. However, these networks abide by the same chemical laws governing metabolic networks, such as conservation of mass and energy. Thus, many of the reconstruction details presented in this chapter are transferable to these networks if the details, such as stoichiometry, are known.

References

1. Janssen P, Goldovsky L, Kunin V, et al. Genome coverage, literally speaking. The challenge of annotating 200 genomes with 4 million publications. *EMBO Rep* 2005;6(5):397–399.
2. Dansen TB, Wirtz KW, Wanders RJ, Pap EH. Peroxisomes in human fibroblasts have a basic pH. *Nat Cell Biol* 2000;2(1):51–53.
3. Nicolay K, Veenhuis M, Douma AC, et al. A 31P NMR study of the internal pH of yeast peroxisomes. *Arch Microbiol* 1987;147(1):37–41.
4. Karp PD, Riley M, Saier M, et al. The EcoCyc and MetaCyc databases. *Nucleic Acids Res* 2000;28(1):56–59.
5. (NC-IUBMB) NCotIUoBaMB. Enzyme Nomenclature. 6th ed. San Diego: Academic Press; 1992.
6. Palsson BO. Systems Biology: Properties of Reconstructed Networks. New York: Cambridge University Press; 2006.
7. Price ND, Reed JL, Palsson BO. Genome-scale models of microbial cells: evaluating the consequences of constraints. *Nat Rev Microbiol* 2004;2(11):886–897.
8. Reed JL, Palsson BO. Genome-scale in silico models of *E. coli* have multiple equivalent phenotypic states: assessment of correlated reaction subsets that comprise network states. *Genome Res* 2004;14(9):1797–1805.
9. Famili I, Mahadevan R, Palsson BO. k-Cone analysis: determining all candidate values for kinetic parameters on a network scale. *Biophys J* 2005; 88(3):1616–1625.

10. Covert MW, Knight EM, Reed JL, Herrgard MJ, Palsson BO. Integrating high-throughput and computational data elucidates bacterial networks. *Nature* 2004;429(6987):92–96.
11. Reed JL, Vo TD, Schilling CH, et al. An expanded genome-scale model of *Escherichia coli* K-12 (*i*JR904 GSM/GPR). *Genome Biol* 2003;4(9):R54.1–R.12.
12. Fong SS, Burgard AP, Herring CD, et al. In silico design and adaptive evolution of *Escherichia coli* for production of lactic acid. *Biotechnol Bioeng* 2005;91(5):643–648.
13. Thanbichler M, Viollier PH, Shapiro L. The structure and function of the bacterial chromosome. *Curr Opin Genet Dev* 2005;15(2):153–162.
14. Flores N, Flores S, Escalante A, et al. Adaptation for fast growth on glucose by differential expression of central carbon metabolism and gal regulon genes in an *Escherichia coli* strain lacking the phosphoenolpyruvate:carbohydrate phosphotransferase system. *Metab Eng* 2005;7(2):70–87.
15. Notley-McRobb L, Ferenci T. Adaptive mgl-regulatory mutations and genetic diversity evolving in glucose-limited *Escherichia coli* populations. *Environ Microbiol* 1999;1(1):33–43.
16. Raghunathan A, Palsson B. Scalable method to determine mutations that occur during adaptive evolution of *Escherichia coli*. *Biotechnol Lett* 2003; 25:435–441.
17. Reed JL, Palsson BO. Thirteen years of building constraint-based in silico models of *Escherichia coli*. *J Bacteriol* 2003;185(9):2692–2699.
18. Reed JL, Famili I, Thiele I, Palsson BO. Towards multidimensional genome annotation. *Nat Rev Genet* 2006;7(2):130–141.
19. Park SM, Schilling CH, Palsson BO. Compositions and methods for modeling *Bacillus subtilis* metabolism. USA. 2003. Available at http://www.freepatentsonline.com/20030224363.html.
20. Reed JL, Vo TD, Schilling CH, et al. An expanded genome-scale model of *Escherichia coli* K-12 (iJR904 GSM/GPR). *Genome Biol* 2003;4(9): R54.
21. Edwards JS, Palsson BO. The *Escherichia coli* MG1655 in silico metabolic genotype: its definition, characteristics, and capabilities. *Proc Natl Acad Sci USA* 2000;97(10):5528–5533.
22. Mahadevan R, Bond DR, Butler JE, et al. Characterization of metabolism in the Fe(III)-reducing organism *Geobacter sulfurreducens* by constraint-based modeling. *Appl Environ Microbiol* 2006;72(2):1558–1568.
23. Schilling CH, Palsson BO. Assessment of the metabolic capabilities of *Haemophilus influenzae* Rd through a genome-scale pathway analysis. *J Theor Biol* 2000;203(3):249–283.
24. Edwards JS, Palsson BO. Systems properties of the *Haemophilus influenzae* Rd metabolic genotype. J Biol Chem 1999;274(25):17410–17416.
25. Thiele I, Vo TD, Price ND, et al. An expanded metabolic reconstruction of *Helicobacter pylori* (*i*IT341 GSM/GPR): An in silico genome-scale characterization of single and double deletion mutants. *J Bacteriol* 2005; 187:5818–5830.
26. Schilling CH, Covert MW, Famili I, Church GM, Edwards JS, Palsson BO. Genome-scale metabolic model of *Helicobacter pylori* 26695. *J Bacteriol* 2002;184(16):4582–4593.
27. Oliveira AP, Nielsen J, Forster J. Modeling *Lactococcus lactis* using a genome-scale flux model. *BMC Microbiol* 2005;5(1):39.
28. Hong SH, Kim JS, Lee SY, et al. The genome sequence of the capnophilic rumen bacterium *Mannheimia succiniciproducens*. *Nat Biotechnol* 2004; 22(10):1275–1281.

29. Becker SA, Palsson BO. Genome-scale reconstruction of the metabolic network in *Staphylococcus aureus* N315: an initial draft to the two-dimensional annotation. *BMC Microbiol* 2005;5(1):8.
30. Borodina I, Krabben P, Nielsen J. Genome-scale analysis of *Streptomyces coelicolor* A3(2) metabolism. Genome Res 2005;15(6):820–829.
31. Feist AM, Scholten JCM, Palsson BO, et al. Modeling methanogenesis with a genome-scale metabolic reconstruction of *Methanosarcina barkeri*. 2006; 2(1):msb4100046-E1-msb-E14.
32. Sheikh K, Forster J, Nielsen LK. Modeling hybridoma cell metabolism using a generic genome-scale metabolic model of Mus musculus. *Biotechnol Prog* 2005;21(1):112–121.
33. Duarte NC, Herrgard MJ, Palsson BO. Reconstruction and validation of *Saccharomyces cerevisiae* iND750, a fully compartmentalized genome-scale metabolic model. *Genome Res* 2004;14(7):1298–1309.
34. Forster J, Famili I, Fu P, et al. Genome-scale reconstruction of the *Saccharomyces cerevisiae* metabolic network. *Genome Res* 2003;13(2):244–253.
35. Majewski RA, Domach MM. Simple constrained optimization view of acetate overflow in *E. coli*. *Biotechnol Bioeng* 1990;35:732–738.
36. Lee S, Phalakornkule C, Domach MM, et al. Recursive MILP model for finding all the alternate optima in LP models for metabolic networks. *Comp Chem Eng* 2000;24:711–716.
37. Vo TD, Greenberg HJ, Palsson BO. Reconstruction and functional characterization of the human mitochondrial metabolic network based on proteomic and biochemical data. *J Biol Chem* 2004;279(38):39532–39540.
38. Burgard AP, Pharkya P, Maranas CD. OptKnock: a bilevel programming framework for identifying gene knockout strategies for microbial strain optimization. *Biotechnol Bioeng* 2003;84(6):647–657.
39. Thiele I, Price ND, Vo TD, Palsson BO. Candidate metabolic network states in human mitochondria: Impact of diabetes, ischemia, and diet. *J Biol Chem* 2005;280(12):11683–11695.
40. Price ND, Thiele I, Palsson BO. Candidate states of *Helicobacter pylori's* genome-scale metabolic network upon application of "loop law" thermodynamic constraints. *Biophys J* 2006;90(11):3919–3928.
41. Price ND, Schellenberger J, Palsson BO. Uniform sampling of steady state flux spaces: means to design experiments and to interpret enzymopathies. *Biophysl J* 2004;87(4):2172–2186.
42. Wiback SJ, Famili I, Greenberg HJ, et al. Monte Carlo sampling can be used to determine the size and shape of the steady state flux space. *J Theor Biol* 2004;228(4):437–447.

3

From Gene Expression to Metabolic Fluxes

Ana Paula Oliveira, Michael C. Jewett, and Jens Nielsen

Summary

The development of genome-wide high-throughput technologies to identify and map cellular components and to quantify different types of cellular molecules has offered new possibilities for the study of biological systems. In the field of metabolic engineering, which deals with the rational redirection of metabolic fluxes toward a product of interest through the introduction of targeted genetic modifications, it is of interest to have tools and models that relate genotype and phenotype. Here, we illustrate how systems biology approaches are being used in metabolic engineering to explore properties and capabilities of microbial cells, to uncover hidden regulatory mechanisms, and to design enhanced microbial cell factories. Several "omics" technologies that are particularly useful for metabolic engineering are described, including methods for quantification of mRNA levels, metabolite concentrations, and fluxes through reactions. Furthermore, we review classical and integrative methods for analysis of omics data and describe several mathematical models used to predict phenotypic behavior based on the metabolic network structure. Because metabolic networks and metabolic fluxes are at the core of metabolic engineering, a brief introduction to the characteristics of genome-scale metabolic networks and to key aspects of regulation and control of metabolic fluxes are also referred.

Key Words: Systems biology, metabolic engineering, metabolic network, regulatory networks, transcriptome analysis, data integration, phenotypic behavior, predictive models.

1. Introduction

Cells are complex systems encoding and executing the functions of life, toward their survival, growth, and reproduction. The many different functionalities in living cells are organized in hierarchical levels of information, which are controlled by intricate regulatory structures. The genome of a cell specifies the total potential inventory of cellular resources, and the genes within the genome are expressed and translated into proteins that are responsible for the various functions operating within the cell. Proteins can have a catalytic role (enzymes), catalyzing reactions, a structural role, such as actin, or a regulatory role, e.g., inhibiting/activating the function of another protein or the expression of a gene. Often, proteins interact in regulatory networks, contributing to the robustness of a given cellular response. These regulatory structures are responsible for the efficient utilization of the available cellular resources and relocation of these resources under stress conditions.

To grow, cells require nutrients (or substrates), which in turn will be used to supply the cell with the free energy, redox power, and precursor metabolites needed to fuel cellular processes such as growth and maintenance. Substrates are converted to precursor metabolites, and then to macromolecules (preceding the assembly of biomass), through reactions catalyzed by enzymes. This process is called metabolism, and the manifestation of the operation of metabolism is herein referred to as phenotype. Box 1 introduces essential concepts on metabolism.

With the advent of genome sequencing, the post-genomic era has offered the possibility of identifying the elements of the cellular inventory: which genes are in the genome, which proteins they encode for, and

Box 1. Metabolism, Fluxes, and Networks

It is through **metabolism** that cells generate the energy and precursor metabolites needed to fuel all cellular processes. Typically, a microbial cell needs several substrates to grow: a carbon source (e.g., glucose), a nitrogen-source (e.g., NH_3), and trace amounts of other compounds, such as phosphate, sulfur, and calcium. Once the carbon source enters the cell, it is degraded and oxidized in a sequence of enzymatic steps toward 12 precursor metabolites, a process referred to as **catabolism**. During this process, free Gibbs energy is generated in the form of ATP, and redox power is produced in the form of NADH and NADPH. Catabolism is often accompanied by the formation and secretion of byproducts (e.g., CO_2, ethanol, acetate, glycerol), which are products associated with the production of more ATP and/or involved in recycling NAD(P)H surplus not used in other processes. Once the precursor metabolites, energy, and redox power are available, they are used in biosynthetic reactions to produce the building blocks of the cell, i.e., the preceding elements for the assembly of macromolecules, such as proteins, DNA, and RNA. This process leading to the assembly of biomass components is also known as **anabolism**, and is an ATP-demanding process.

Enzymatic reactions catalyze the conversion of chemical compounds (or metabolites) into others, and the rate of conversion is referred to as **metabolic flux**. Often, the term metabolic pathway is used to describe any sequence of observable enzymatic reactions connecting two metabolites in a related process. More generally, the term **metabolic network** refers to a part or the whole set of connected metabolic reactions.

what are their functions. It also triggered the development of several analytical techniques, allowing the simultaneous quantification of molecules such as mRNA and proteins, at a genome-scale level. Additionally, other large-scale analytical techniques are available that allow quantification of the metabolic state, i.e., that describe the metabolic fluxes through reactions and metabolite concentrations in the cell at a given time. These advances in analytical and computational methods have shifted the focus of modern biology from a traditional "local" reductionist approach to a "global" holistic perspective of the cellular processes (1,2).

Viewing the cell as a network of interacting genes, proteins, and reactions offers the opportunity to dissect cellular complexity and represent the cell as a simplified system that can be used to relate genotype and phenotype. The integration of these structural networks with quantified molecular and metabolic elements has been termed systems biology (1,3,4). Despite the widespread use of this term, methodologies and goals vary depending on the application. Figure 1 summarizes the different perspectives and goals underlying the concept of systems biology in three different areas of science: health sciences, basic sciences, and engineering.

In this chapter, we focus on the use of systems biology as a framework to develop models and tools to explore the emergent properties and capabilities of microbial systems, to elucidate hidden regulatory mechanisms and to design newly enhanced strategies for producing microbial cell factories with desired phenotypes. Several functional genomics tools are described, and their use in identifying metabolic traits of

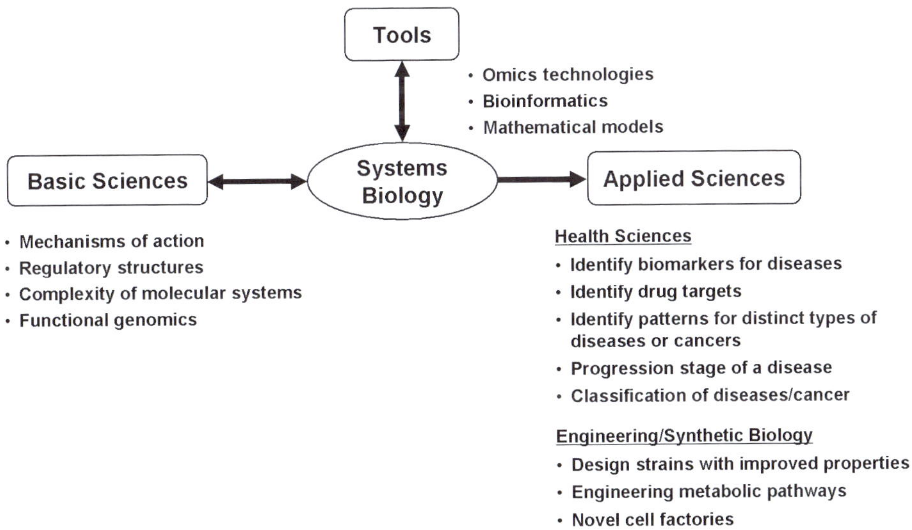

Figure 1. Impact of systems biology in different areas of science. Systems biology arose as a field that aims to answer questions posed by basic sciences using modern system-level biology tools derived from omics technologies. Today, systems biology approaches are becoming widely applied in engineering and health sciences.

interest is exemplified. Finally, we highlight the importance of selecting appropriate mathematical models for data analysis and experimental design.

2. Engineering a Cell: A Systems Approach

2.1. Strain Improvement and Metabolic Engineering

Microbial cells are widely exploited by the pharmaceutical and chemical industries for the production of a variety of compounds, ranging from metabolites to proteins and even to entire cells. Products derived from microbial fermentation include chemicals like ethanol and lactic acid, food ingredients like citric acid and glutamate, antibiotics, industrial enzymes (such as the ones used in the detergent and starch industries), and high-value therapeutic proteins such as insulin, growth hormone, and interferons. The number of products deriving from microbial fermentation is likely to grow in the future, not only because of the increasing tendency to replace petrochemical processes with more environmentally friendly processes, but also because of the growing interest in developing novel compounds with potential valuable commercial applications. Moreover, there is a continuous industrial interest in improving producing strains.

Traditionally, the selection of the "best" microorganism with the desired phenotype is made through the screening of a large collection of microbial cultures, followed by the exposure of the selected strain to random mutagenesis. Improved "randomly mutated" strains for the desired properties are selected this way for industrial purposes. There are numerous cases where yields of compounds of interest have been increased several-fold. For example, penicillin production by *Penicillium chrysogenum* has been increased more than 2,000-fold since the 1950s, through classic strain improvement programs and process optimization (5).

The advent of genetic engineering in the 1970s, and advances in molecular biology during the last few decades, has offered the opportunity of performing targeted genetic modifications in many organisms. This has paved the way for a more rational approach for strain improvement by allowing the introduction of direct genetic modifications that can be used to redesign the metabolic capabilities of a cell—an approach referred to as metabolic engineering (6–8). Through metabolic engineering, one can aim at redirecting fluxes and improving cellular activities by manipulating the enzymatic, regulatory, and transport functions of the cell using recombinant DNA technologies (6). These technologies offer not only the possibility of inserting, deleting or overexpressing homologous genes but also allow the insertion of heterologous genes, i.e., genes originating from other organisms. The number of potential metabolic modifications to be explored is, therefore, enormous.

To be successful, metabolic engineering requires an efficient interplay between the analysis of cellular function and genetic engineering. The effect of a genetic modification can be assessed at the metabolic level through physiological characterization and analysis of the metabolic state. Analytical techniques for flux analysis and metabolite profiling are typically used in metabolic engineering to evaluate metabolic changes between modified and reference strains. Other analytical measurements, such as enzymatic activities, mRNA levels, and protein levels give direct information on how the cellular inventory responds, but may not have a trivial relationship with the physiological response (9). Analysis of the metabolic state with the appropriate mathematical and metabolic models plays an important role in designing the next round of improvements. A successful metabolic engineering application typically results from the continuous improvement of the desired cellular property through several rounds of experiment, analysis, and design (8).

2.2. Metabolic Networks and Regulation of Metabolic Fluxes

Metabolic networks and metabolic fluxes are at the core of metabolic engineering. Metabolic networks represent the topology of metabolism, which is the set of possible conversion routes, whereas metabolic fluxes reflect the result of cellular component interactions, i.e., how the different metabolic pathways within the metabolic network are being used. As the objective of metabolic engineering is to optimize the fluxes through dedicated pathways leading to the product of interest, quantification of intracellular fluxes *in vivo* is a central issue in metabolic engineering. Because intracellular fluxes are difficult to obtain from direct measurements, indirect methods are called for. Quantification of intracellular fluxes typically requires the development of a metabolic model combining the structure of the network with the experimental measurements of extracellular fluxes (7,10). To further constrain the model-estimated fluxes, one may use labeled substrates, combined with analysis of the specific carbon labeling of different pathway metabolites (11). Stoichiometric metabolic models, derived from fundamental principles of conservation of mass and energy, are particularly useful for quantification of metabolic fluxes. Besides quantification of metabolic fluxes, it is desirable to understand how the fluxes in metabolic pathways are controlled and how the individual enzymes are regulated, as this may lead to rational modification of the network operation, and thereby lead to improved fluxes toward the product of interest, which is often the goal in metabolic engineering.

In the following sections, we introduce genome-scale metabolic networks and elaborate on how hierarchical regulation and control is exerted by the cell, contributing to the robustness of metabolic response under genetic and environmental perturbations.

2.2.1. Genome-Scale Metabolic Networks

In the last decade (1995–2005), more than 180 genomes from different organisms in all three kingdoms have been sequenced (www.genomenewsnetwork.org). These advances were accompanied by the development of powerful bioinformatics tools used for genome annotation and protein function prediction. As a result, it has been possible to identify nearly all genes (or, more correctly, Open Reading Frames: ORFs) in a genome and assign a function to many of the identified gene products. Annotation of the genome therefore offers the opportunity to explore the whole enzymatic and transport potential of a cell, allowing the reconstruction of the corresponding genome-scale metabolic network. These networks are simply the set of all potential reactions and transport steps occurring in a cell, written in a stoichiometric representation. For more details on reconstruction of genome-scale metabolic networks, see Chapter 2 in this book.

Genome-scale metabolic networks have played a key role in the systemic analysis of cellular function (12–14) and in the development of truly genome-wide integrative models relating genotype and phenotype (2,15–17). For instance, it has been shown that the network structure by itself can be used to deduce network functionality and regulatory information (12,18). A well-known application of genome-scale metabolic networks is flux balance analysis (FBA). This mathematical framework based on linear programming has been successfully used to model phenotypic behavior under (pseudo) steady-state conditions. More recently, a method combining the genome-scale network, represented as a graph, and gene expression data have been reported to identify co-regulated metabolic modules and key points in metabolism around which the most changes in expression occur, the so called *reporter metabolites* (15).

In a work by Jeong and coworkers, the representation of genome-scale metabolic networks as a graph has contributed to insight into the topological properties of these networks, while revealing common metabolic organization among species (14). In their graph representation, nodes correspond to metabolites, and edges connecting the nodes correspond to reactions. Using the networks of 43 organisms, they showed that metabolic networks are scale free and exhibit the typical characteristics of a small-world network. In other words, this means that most of the nodes have few neighbor edges, whereas few nodes have many neighbors, and that the average distance between any two metabolites is very small (approximately 3 nodes away from each other). This high degree of connectivity of metabolic networks is characteristic of robust and error-tolerant networks, suggesting that organisms have evolved toward robust systems to respond efficiently to external changes or internal errors (14,19).

Because of constraints posed by mass balances around all metabolites within the cell, the stoichiometry of the whole metabolic network defines the boundaries of metabolic capabilities. *In vivo*, however, the number of possible flux distributions is confined to a smaller set of fluxes, as metabolic fluxes are tightly controlled by complex and hierarchical regulatory mechanisms, as described in the following section.

2.2.2. Control and Regulation of Metabolic Fluxes

Metabolic fluxes reflect the final outcome of the cellular orchestration under defined genetic and environmental conditions. According to the central dogma of biology, the digital code of life is encoded in the DNA, and the encoding entities are called genes. When a gene is expressed, information flows from DNA, to mRNA, to protein, to functional protein (Figure 2). This dogmatic view suggests that, ultimately, the mRNA level of a gene correlates with the amount of functional protein and, consequently, with the flux through reaction (in case of an "enzymatic gene").

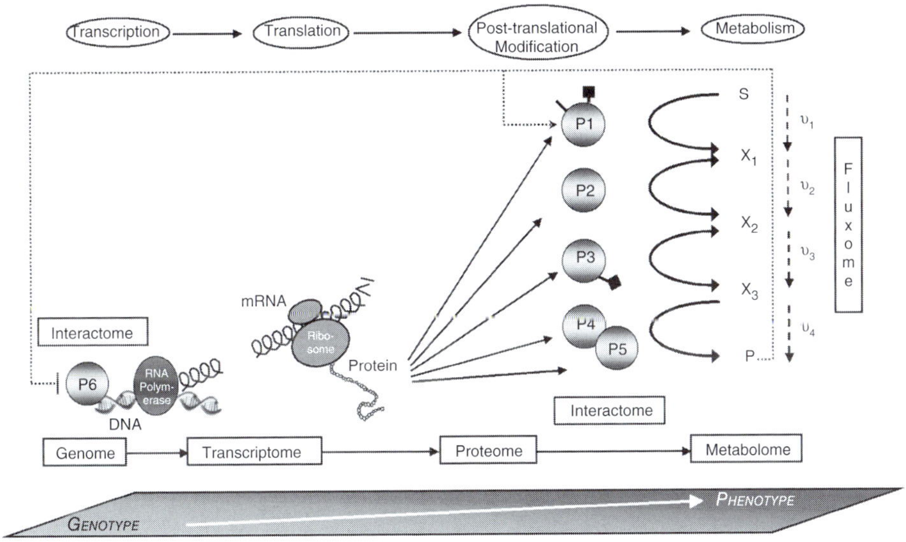

Figure 2. The central dogma. According to the central dogma of biology, information flows from DNA, to mRNA, to proteins, to functional proteins. Information is encoded in the DNA within entities called genes, which are transcripted into mRNA, and then translated into proteins. Proteins carry out the functions to operate the cell, such as catalyzing reactions (P1 to P5) and inhibiting/activating the expression of genes (P6). It is through metabolism that cells convert nutrients into free energy, redox power, and precursor metabolites needed to fuel cellular processes such as growth and maintenance. The fluxes through reactions are tightly controlled to keep the metabolite pools balanced. Metabolites can also exert cellular control (e.g., high levels of P activate P1 and repress P6).

However, because of different levels of cellular control and kinetic properties of enzymes, such correlations may not be present. These different levels of control give the cell the flexibility and robustness needed to closely balance the rates of synthesis and conversion of metabolites over a wide range of external conditions, avoiding the complete depletion of intracellular metabolite pools (7). The control strength exerted by the different enzymes of a pathway can be assessed through metabolic control analysis, a sensitivity analysis framework developed to quantify the response of fluxes and metabolite pools to changes in kinetic and biochemical parameters of enzymes (7,20,21).

Flux control can be exerted by the cell at the following different levels: transcriptional level, mRNA degradation level, translational level, by protein activation/inactivation, and by allosteric regulation of enzymes (22). This complex and (somehow) hierarchical control structure requires the existence of intricate mechanisms for sensing and signaling. Sensing of a "key" metabolite/protein concentration or metabolic flux can trigger a signaling cascade or a feedback loop, leading to a certain level of control. However, identification of these regulatory structures represents a major challenge in modern biology.

Mapping all elements involved in the control of metabolism offers the opportunity to establish metabolic engineering strategies based on relief of regulatory control. Ostergaard and co-workers (23) illustrated this by manipulating the GAL-regulon of *Saccharomyces cerevisiae*. The GAL system is a tightly regulated system involved in galactose utilization. In their work, they showed that the disruption of three genes encoding the negative regulators Gal6, Gal81, and Mig1 resulted in a significant increase in flux through the galactose utilization pathway.

The flow of information from genes to mRNA to proteins also suggests that, hierarchically, regulation may be dominant at the transcriptional level. In other words, it suggests that one of the main ways the cell has to respond to a perturbation may be through the expression or repression of some of its genes. But how can we identify other types of control, and how does mRNA level relate to protein abundance, enzyme activity, and metabolic fluxes? A simplified answer for these questions is that the lack of correlation between two types of measurements, e.g., between mRNA levels and protein abundance, indicate the intermediate occurrence of some type of control, in this case mRNA degradation or translational control. Ter Kuile and Westerhoff (24) proposed a method to identify if control is being dominated at the gene expression level (i.e., hierarchical regulation) or at the metabolic level. They proposed that regulation of fluxes through enzymes can be divided into a hierarchical regulation term, ρ_h, and a metabolic regulation term, ρ_m, whose sum should equal one. From their analysis on glycolytic enzymes, they conclude that regulation is rarely completely hierarchical. Nevertheless, their method does not enable differentiation of hierarchical regulation into different levels, e.g., transcriptional regulation and regulation at the posttranscriptional level.

Gygi et al. (25) and Futcher et al. (26) determined the correlation between mRNA and protein abundance in the yeast *S. cerevisiae*, and found that a linear correlation exists, although it is not very strong (in the later study, the Pearson correlation coefficient was 0.76). These analyses were based on less than 150 protein spots and corresponding mRNA abundances. Daran-Lapujade and co-workers (27) compared mRNA levels with flux distributions in the central carbon metabolism of *S. cerevisiae* growing in different carbon sources, under steady-state conditions. They observed three distinct types of correlation: (i) a very strong correlation for enzymes in gluconeogenesis and the glyoxylate cycle, whose encoding genes are known to be strongly regulated at the transcriptional level; (ii) a medium correlation for enzymes in the tricarboxylic acid cycle and pentose-phosphate pathway, suggesting a shared regulation between transcription control and other levels of control; and (iii) a lack of correlation for enzymes in the glycolysis, pointing to regulation at the posttranslational and/or metabolic level in this pathway. This work highlights the fact that transcriptional data by itself may have a limited capability in predicting phenotype, only describing what is happening at the transcriptional control level. Understanding other levels of flux control therefore requires the quantification of other molecular components and the identification of how these components interact. Nevertheless, measuring mRNA abundance is currently the only truly genome-wide analytical method available, and, although not exact, in the absence of more information it is often assumed that messenger RNA (mRNA) levels are proportional to the corresponding enzymatic fluxes.

2.3. Improving Metabolic Engineering Using Omics Data

Genome-wide or large-scale quantification of molecular components and experimental assessment on how these components interact have offered to metabolic engineering the possibility of getting a broader insight into cellular function and into the effects of genetic and environmental perturbations. Omics data include quantification of mRNA transcripts levels (transcriptome), protein abundance (proteome), metabolic fluxes (fluxome), intracellular and extracellular metabolites concentration (metabolome), and information on protein–protein and protein–DNA interactions (interactome). To efficiently extract relevant biological insight from these vast amounts of data, appropriate and goal-dependent tools and models are required. The use of omics data in metabolic engineering have been applied with different purposes, namely: (i) in the identification of a metabolic trait of interest, using reverse engineering; (ii) in the reconstruction of regulatory and signaling networks that can later be used as targets for metabolic engineering; (iii) in understanding emergent cellular properties; and iv) in the development of models to predict metabolic behavior (Figure 3).

2.3.1. Identifying Metabolic Traits of Interest

In metabolic engineering, it is generally of interest to identify the genes that confer a desired phenotype (for instance, to seek the rationale behind a strain improved by classic mutagenesis). This type of problem can be solved using a reverse engineering approach. Reverse metabolic engineering can be defined as the process of dissecting a cell and analyzing its components, with the intention of later reconstructing the

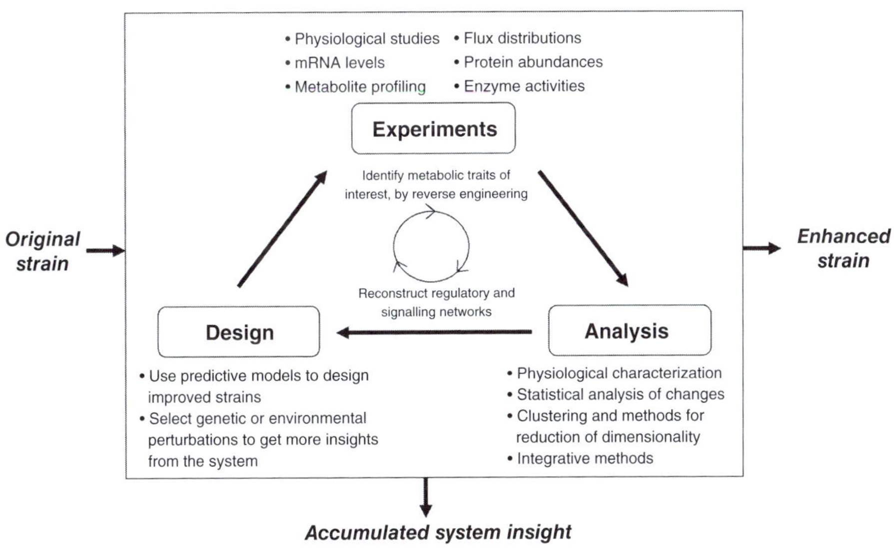

Figure 3. The metabolic engineering cycle in the systems biology era.

components that confer a desirable phenotype. Omics data, particularly gene expression data, are particularly useful for this purpose (28,29). Bro and co-workers followed a reverse engineering approach to improve the flux through the galactose utilization pathway in *S. cerevisiae* (30). They used DNA microarrays to analyze the gene expression levels of three selected strains: a wild type and two strains previously improved by metabolic engineering for galactose utilization (23). Looking at the significant changes in gene expression among the three strains, they identified the gene *PGM2* as being up-regulated in the two strains with improved galactose utilization capabilities. Further analysis of enzymatic activity supported the hypothesis that Pgm2 may be involved in the utilization of galactose. Subsequently, they overexpressed *PGM2* and observed an increase of 70% in the galactose uptake rate, as compared to the wild type.

Askenazi and co-workers planned a more sophisticated experiment to improve lovastatin production in *Aspergillus terreus*, integrating metabolite profiling and gene expression analysis (31). They first designed and engineered a collection of strains producing different amounts of lovastatin, ranging from very low to very high producers. Metabolite profiling was then conducted to quantify lovastatin and several other secondary metabolites in all strains. Next, genome-wide gene expression levels were measured for several selected strains (more than 10). The two types of data were analyzed together using clustering, principal component analysis (PCA), and other advanced statistical techniques. Several genes were identified by these methods as conferring improved lovastatin yields. Overexpression of one of those genes, *LOVF*, contributed for the increased production of this secondary metabolite.

2.3.2. Reconstruction of Regulatory and Signaling Networks

The example on how galactose consumption was increased upon deletion of negative regulators of the *GAL* system (23) elucidates how release of regulatory control can be important for increasing fluxes or redirecting them into other pathways. Reconstruction of regulatory and signaling networks therefore offers the opportunity to identify metabolic engineering targets. However, unlike metabolic networks, regulatory and signaling networks cannot be easily reconstructed from the annotated genome. Advances in high-throughput methodologies to identify protein–protein and protein–DNA interactions has impelled the development of models and automated systems to reconstruct signaling cascades and regulatory structures than to activate or inhibit transcription. Integration of other omics data, together with these structural networks, has contributed to identify co-regulated modules at different levels of cellular organization.

The reconstruction process usually starts by collecting all possible information from the literature. Because there are many regulatory mechanisms in the cell, reconstruction typically focuses on the mechanism(s) of interest. For example, Ideker and co-workers (32) put together a model on galactose utilization in *S. cerevisiae*, based on literature-derived information. The defined network included one permease that transports galactose into the cell, four enzymatic genes converting galactose intracellularly, and four regulatory proteins inhibiting or activating the transcriptional response of other proteins in the system. Once the regulatory model was established, they performed genetic and environmental perturbations corresponding to the single deletion of all defined genes in the model, both in the presence and absence of galactose, and further measured the global changes at the transcriptome and the proteome level for each perturbed strain (changes relative to a reference strain). Next, mRNA and protein responses were evaluated in the context of the model, together with global protein–protein and protein–DNA information. Based on the observations, new hypotheses were generated, tested, and used to improve the regulatory model.

The galactose utilization pathway in *S. cerevisiae* is a relatively well-characterized metabolic/regulatory system. However, many signaling and regulatory mechanisms cannot be easily reconstructed from literature because experimental "low-throughput" information is often incomplete. Several approaches have been proposed to model signal transduction networks and gene-regulatory pathways using large interactome databases that are mainly derived from high-throughput studies.

In a later work by Ideker et al. (33), they describe a graph theory–based method to discover regulatory and signaling circuits in molecular interactions networks. Here, they represent the global protein–protein and protein–DNA interactions network of the yeast *S. cerevisiae* as a graph and score each node (a protein or gene) with the corresponding significance of change in gene expression. Next, they apply their algorithm to score subnetworks and search for the high-scoring subnetwork. The final high-scoring subnetwork corresponds to a module of co-regulated genes that significantly change their transcriptional levels in response to a perturbation. The propagation effects of the perturbation are therefore hypothesized to correspond to physically connected proteins (signaling cascades) and to transcription factors controlling the expression of an affected gene. Other methods (34–36) have also been proposed to model transcriptional regulation and signal transduction networks.

2.3.3. Understanding Emergent Cellular Properties

In metabolic engineering, each round of improvements/experiments should be followed by a thorough analysis, bringing insights into the metabolic responses and their molecular bases. Omics quantification has contributed to a better understanding of whole cellular properties, and the acquired knowledge can be further used to design the next round of improvements. Typical analyses include the comparative profiling of transcriptome, proteome, fluxome, and/or metabolome data between a reference and a modified strain(s). Assessment of variation can be achieved either by analysis of ratios or by statistical significance tests. Integrated analyses, together with genome-scale metabolic networks, have yielded additional insights into the metabolic effects of genetic and environmental perturbations.

Bro and co-workers compared the genome-wide transcriptional responses of a reference strain of *S. cerevisiae* with the *GDH1*-deleted mutant, grown under anaerobic glucose-limited continuous cultures (37).

GDH1 encodes a NADPH-dependent glutamate dehydrogenase, which is an enzyme that plays an important role in ammonia assimilation and utilizes approximately 50% of all the NADPH required for biomass synthesis. Previous physiological characterization of the *GDH1*-deleted strain showed an altered redox metabolism, resulting in a considerable decrease of glycerol production and a slight increase in ethanol production (38). Transcriptional analysis was carried out using a statistical test to evaluate significance of differential gene expression. To reduce the focus of the analysis, only transcriptome levels of genes encoding NAD(P)H-dependent enzymes were analyzed, and 13 of these were identified as significantly altered. Interestingly, it was found that *GND1, ZWF1,* and *ALD6*, encoding the most important enzymes for regeneration of NADPH, were down-regulated in the mutant, suggesting a possible common redox-dependent regulation. These findings offer new possible targets for future metabolic engineering strategies.

In the aforementioned work, Bro et al. (37) report their challenge in defining "what is significant." Integrated approaches, such as the one introduced by Patil and Nielsen (15), offer the opportunity to look at the metabolism as a whole and to identify the parts more significantly affected in response to a perturbation. In their work, Patil and Nielsen (15) integrate the genome-scale metabolic network with transcriptome data in a graph representation, and identify high-scoring metabolic subnetworks and reporter metabolites (*see* section 2.2.1 for a definition). Using this approach, analysis of GDH1 gene expression data immediately reveals a strong impact in redox-metabolism, with 10 genes encoding enzymes catalyzing oxidoreductive reactions popping up in the high-scoring metabolic subnetwork. Moreover, reporter metabolites include ammonia, glucose-6-phosphate, fructose-6-phosphate, and sedoheptulose-7-phosphate (all intermediates of the pentose phosphate pathways that serve as major sources for NADPH production), which is in good agreement with the known effects of Gdh1 in ammonia assimilation and NADPH usage. This integrative method allows a look into the metabolic effects of a perturbation without considerable *a priori* knowledge of the system. Such a method can also be used in functional genomics, contributing to the identification of the metabolic role of an unknown enzyme.

2.3.4. Prediction of Metabolic Behavior

Effective design tools of microbial cell factories are the ultimate goal of metabolic engineering. Although our understanding of cellular function is still far from complete, simplified models able to predict metabolic behavior under changing environmental and genetic conditions are called for. Genome-scale stoichiometric metabolic models have proven to be particularly useful for this purpose. Namely, FBA and related approaches have been successfully applied in predicting the lethal effect of gene deletions in several microorganisms, including *E. coli* (39) and *S. cerevisiae* (40), and in evaluating the maximum theoretical yield of a desired product at a certain growth rate, being a valuable tool in identifying metabolic engineering targets (16,17,41–43).

Omics-derived information can be applied to improve FBA predictions. For example, Åkesson et al. (44) used transcriptome data to identify the genes not being expressed under certain conditions, and included that information to impose additional on/off constraints into the FBA problem. This simple approach improved the FBA prediction significantly. Covert et al. have also proposed the introduction of regulatory constraints in the form of a Boolean on/off switch (45–47), this method defining which enzymes are active/inactive under certain conditions. In their studies, they showed that the addition of regulatory constraints significantly improves the predictive capacity of models in which regulatory effects play a dominant role in the metabolic response.

3. Omics Quantification

Identification and quantification of the cellular inventory are critical steps toward deciphering mechanisms that underlie cellular function. Numerous techniques that simultaneously detect multiple signals at the molecular level have been developed and continue to emerge. Ultimately, the objective of these strategies is to capture differences in profiles between different systems, cells, organisms, communities, etc., to crack the code underpinning regulatory schemes that control biological function. There are numerous analytical methods for systems-level cellular analysis. Here, we underscore key developments involved in characterizing mRNA, proteins, metabolites, and fluxes (see also Table 1). Many powerful techniques, which we do not report, are also being developed for the analysis of genomes, protein–protein interactions, protein–DNA interactions, and protein location.

Genome-wide transcriptional analysis is the most mature omic analytical method. The basic principle behind DNA microarray technology is based on the hybridization of labeled RNA or DNA prepared from extracted cellular mRNA to highly ordered nucleotide sequences attached to a solid matrix (48–50). Although many array technologies exist, spotted DNA microarrays and high-density oligonucleotide microarrays (commercially available from Affymetrix™) are the most widely used (51,52). Despite similar statistical treatments to classify significantly changed genes, these approaches fundamentally differ in the experimental setup. Namely, in spotted arrays, the two samples under comparison are labeled with different fluorescent dyes and co-hybridized on the same DNA microarray to obtain relative abundances of specific gene products. High-density oligonucleotide microarrays measure absolute abundance levels because each sample for comparison is labeled with the same dye and hybridized individually.

Techniques for quantification of cell-wide protein content have emerged as complementary tools to transcriptome analysis (53). Relative to genome-wide transcriptional profiling, proteome elucidation is based on the analysis of biopolymers made up of a larger library of molecules. Because of the increase in building block complexity (proteins being comprised of 20 amino acids, whereas mRNA has 4 different nucleotides), methods to probe the proteome are slightly less established than DNA microarrays. Moreover, protein coverage is an issue. Four primary methods have been used to globally examine proteome status. The

Table 1. Summary of Pros and Cons of Common Omics Methods and Technologies.

Method	Advantages	Disadvantages
Transcriptome		
Spotted microarrays	—Minimizes problems with noise and background —Prepared in-house	—Measures relative abundances —Comparison between arrays/experiments is complicated
High density oligonucleotide microarrays	—Measures absolute abundances —Manufactured commercially —Standardized probes and protocols	—Concerns about background adjustments and normalization
Proteome		
2DE-MS	—Well suited to large-scale separation and identification of numerous proteins in one sample —Comparison of protein quantity from 2–3 gels is possible with new staining techniques	—Amount of material needed —Reproducibility —Limited sensitivity —Excludes many large, hydrophobic and basic proteins
Protein arrays	—High specificity —Currently beneficial for analysis of specific classes of proteins —Enormous potential	—Nonmature technology —Currently limited by proteome coverage
ICAT	—Good sensitivity —Relative protein abundances for 2 samples are determined in one experiment	—Limited to a binary set of reactions —Limited to proteins containing cysteine —Not amenable to post-translationally modified proteins
iTRAQ	—Good sensitivity and consistency —Relative protein abundances of up to 4 samples are determined in one experiment —Post-translational modifications can be identified —Higher confidence identification than ICAT	—Samples must be prepared according to strict guidelines —MS time is increased due to increased number of peptides

Table 1. *Continued*

Method	Advantages	Disadvantages
Metabolome (Adapted from Villas-Boas et al., 2005 (59))		
GC-MS	—High separation efficiency —Easy interface between GC and MS —Simultaneously resolves different classes of metabolites —Reproducible	—Unable to analyze thermo-labile metabolites —Requires derivatization of nonvolatile metabolites —Difficult to identify unknown compounds after derivatization
LC-MS	—High sensitivity —Enables analysis of thermo-labile metabolites —Average to high chromatographic resolution	—Matrix effects —Restrictions on LC eluents due to interface issues from LC to MS —De-salting may be necessary
CE-MS	—Uses small volumes —High resolution —Fast and efficient separation of charged and uncharged species	—Difficult to interface CE with MS —Complex methodology and quantification —Least developed
MS	—Allows for rapid screening of metabolites (2–3 min per sample) —High sensitivity —Negligible sample clean-up for profiling	—Identification of metabolites generally requires tandem MS —Matrix effects —Requires elegant data deconvolution methods
Fluxome		
NMR analysis	—Requires minimal sample preparation —High reproducibility —Nondestructive to the sample	—Higher cost and lower throughput relative to MS analysis —Requires complicated data deconvolution and statistical fitting procedures
MS analysis	—High sensitivity —Rapid throughput —More resolvable metabolites	—Requires complicated data deconvolution and statistical fitting procedures

workhorse of this analysis has been separating and quantifying proteins by two-dimensional electrophoresis (2DE) and identifying them by mass spectrometry (MS) (54). More recently, protein arrays, similar in concept to DNA arrays, but based on highly specific protein interactions, are also being developed (55). However, the most rapid advances in the proteomic field have originated in techniques that make use of labeling proteins with either isotope or isobaric tags (56–59). These elegant strategies use detection of peptide fragments by MS for the identification of relative protein abundances and comparison between samples. The two most popular methods are the "isotope/coded affinity tag" (ICAT) strategy (57) and the "isobaric tags for relative and abundance quantification" (iTRAQ) method (59). Although both approaches have advantages, it has been argued that the iTRAQ method offers more comprehensive proteome coverage because of its labeling scheme and ability to analyze up to 4 samples simultaneously.

Although transcriptome and proteome analysis have been developed most extensively over the past decade, the tools necessary for quantitative high-throughput metabolome analysis are just now emerging. As the intermediates of biochemical reactions, metabolites represent the amplification and integration of signals from upstream molecular players within the cell (i.e., mRNAs and proteins). The highly diverse chemical nature of metabolite structures makes quantifying all metabolites of the cellular system impossible in practice. However, strategies attempting to cover a wide range of metabolites in a single step continue to evolve. Typical quantitative approaches couple an analytical separation technique (e.g., capillary electrophoresis [CE], liquid chromatography [LC], and gas chromatography [GC]) with MS or NMR based detection (60–62). GC-MS is the most common method for global metabolite quantification. In one recent illustration, GC-MS was applied for the integrated analysis of approximately 80 metabolites (intra- and extracellular) involved in amino acid and central carbon metabolism from *S. cerevisiae* (63). Because most naturally occurring metabolites are not volatile, GC-MS is limited by the requirement for sample volatility. Efficient derivatization methods are available, but alternative approaches for quantitative profiling also enable analysis of a large number of metabolites. To illustrate, LC-MS has been shown to uniquely combine sensitivity and specificity to study the intermediates of the glycolytic pathway (64). In addition to dynamic developments in refined analytical techniques and MS sensitivity, advances in internal standardization, one of the main challenges in quantitative metabolome analysis, are also paving the way for more robust measurements. Mashego and co-workers have developed an approach that uses extracts from ^{13}C-saturated microbial cultivations to provide an internal standard for all intracellular metabolites to be quantified (65). This work has created a platform independent of ion suppression effects and holds significant promise for unifying quantitative metabolome analysis. Although not the focus here, qualitative scanning of metabolites is also an integral part of metabolome analysis (60,61).

Although cell-wide quantitative measurements of mRNAs, proteins, and metabolites yield insights to the molecular status of the cell, experimentally determining the flow of material through metabolic networks provides detailed information on the actual functional operations that determine cellular phenotype. Quantitative metabolic flux analysis, fluxomics, relies on the use of ^{13}C-labeled substrates followed by determination of characteristic patterning found in intracellular metabolites (11,66–68). Both proteinogenic amino acids and direct metabolites have been used for determination of labeling patterns (68). When possible, direct measurements of metabolites are preferred, as they enable direct monitoring of flux distributions and they are able to detail transient phenomena (69), but it is important to consider the rapid dynamics of exchange between the metabolites and amino acids incorporated into cellular proteins (70). Typically, NMR or MS analyses have been utilized with mathematical data integration methods to identify the metabolic flow of carbon through the cellular network; however, Sauer has suggested that because of its sensitivity and rapid pace, MS sample analysis is poised to have the greatest impact in high-throughput flux analysis (68,71). To this end, several reports have now demonstrated the power of MS analysis in the large-scale determination of *in vivo* fluxes (69,71–73).

4. Models for Metabolic Engineering in the Systems Biology Era

4.1. Making Sense of the Omics

In the previous sections, we described some of the omics tools and exemplified how omics data are being used in metabolic engineering, within a systems biology framework. Several methods were referred to for data analysis, interpretation of observations, and quantitative prediction of cellular behavior. These methods include models for comparative analysis of omics profiling (e.g., statistical tests and reduction of dimensionality methods), models for integrative analysis (e.g., graph theory–based models), and predictive models (e.g., based on linear and quadratic optimizations). Selection of the appropriate type of model(s) to deal with a given problem plays an important role in extraction of knowledge. Furthermore, experimental design should anticipate the focal point of data analysis and clearly define the biological question/hypothesis, while assuring statistical treatment of the results.

Analysis of omics data represents an important step toward understanding of cellular response. Analysis typically starts through assessment of statistical significance, followed by methods to group genomic features by profiles or co-regulation patterns. Interpretation of results coming from these methods is often a challenge by itself, therefore being useful to focus on a particular biological question. An overview on classical and integrative methods for data analysis is given in section 4.2.

Predictive models are an attempt to put together (the increasing number of) biological insights into a coherent whole (74). Stoichiometric metabolic models, derived from fundamental principles of conservation of mass and energy, have been particularly successful in exploring the relationship between genotype and phenotype and in predicting product yields and growth rates under changing environmental and genetic conditions, at steady state. Omics data have brought increasing predictive capabilities to these types of models. Section 4.3 deals with the predictive capabilities of stoichiometric metabolic models.

4.2. Data Analysis

4.2.1. Classical Methods

Traditionally, analysis of quantified molecular components is made by comparing their relative occurrence (e.g., fold change) between two or more conditions. However, when dealing with high-throughput technologies, technical noise is an important source of variation (75). Therefore, methods for statistical assessment of variance are often used to identify statistically significant biological changes among conditions.

Omics analysis generates vast amounts of data. For example, genome-wide gene expression data generates tens of thousands of data, with at least as many variables as measured transcript abundances. This high-dimensionality of omics data makes it difficult to visualize relationships between variables and to group experiments by similarity of profiles. Methods such as clustering (76), PCA (77), independent component analysis (78), and singular value decomposition (79) have been applied to reduce dimensionality of omics data, facilitating its visualization, allowing an overall characterization of the structure of the data and contributing to the separation of biologically meaningful information from noise.

4.2.1.1. Statistical Significance Analysis: When quantifying a few molecular components such as mRNA or protein levels using classic Southern or Western blots, analysis of results typically focuses on how the fold ratio changes. The introduction of high-throughput arrays for genome-wide measurements of mRNA levels in the late 1990s brought new dimensions to data analysis. Initially, because of high costs, only one chip was being used per experiment and conclusions were made based on fold changes (49,80,81). However, further studies have shown that fold changes are not always an indicator of significance because there is an intrinsic biological variability (even in biological replicates) and also because technical reproducibility of microarrays is not very strong (75). Therefore, an alternative way to assess biological significance of differential expression is to run replicate experiments (biological replicates) and to apply a statistical test to identify significant changes (75,82). For statistical reasons, these replicates should be in a minimum of three, although it has been noticed that the number of false negatives among the significant changing genes decreases when the number of biological replicates increases (83,84).

The main idea behind most of the statistical significance tests is to evaluate whether two different groups of numbers are similar. The most common test for a pair-wise comparison is the Student's *t*-test, which assumes that each set of numbers follows a normal distribution and tests how similar the two distributions are, e.g., if they have the same mean. Other methods have been developed specifically for microarray data to improve the assessment of significance, including significance analysis of microarrays (85) and variability and error assessment (86). When comparing more than two datasets simultaneously, statistical tests like analysis of variance or multivariate analysis of variance are often used. In general, statistical significance tests assign a probability value (*p*-value) to each feature (gene, protein, metabolite, etc.) under analysis. The *p*-value indicates the likelihood that the observed differential expression occurs by chance, i.e., the lower the *p*-value, the more significant the change.

After applying a statistical test and generating a list of *p*-values for all features under analysis, it is necessary to define "what is significant." This can be done by establishing a cut-off and defining that all cases with a *p*-value below the cut-off should be called "significant." In engineering sciences, a *p*-value cut-off of 0.05 (95% confidence) is usually accepted as a threshold for significance. However, it is often argued that when testing for thousands of features the threshold for significance has to be chosen so that the probability of having any false-positive among all features tested should be ≤0.05 (87), and this calls for the use of methods that account for multiple testing, such as the Bonferroni correction or the Benjamini–Hochberg correction (88). However, these corrections produce very strict cut-offs, and may lead to the exclusion of many true positives. Storey and Tibshirani (87) reported a good alternative method for choosing a cut-off based on false discovery rate, by defining a measure of statistical significance called *q*-value. Their approach leads to a less stringent cut-off, while keeping a good balance between false-positives and false-negatives.

4.2.1.2. PCA: Principal component analysis is a method for reduction of dimensionality that allows the visualization of high-dimensional data in a low-dimensional space, projecting the omics data onto a plane in such a way that similar variables (e.g., mRNA transcripts) or experiments will be located close to each other. PCA decomposes the original space in a low-dimensional space of dimension *n*, where *n* is the number of principal components. PCA identifies the direction in space that captures most of the variance, and this direction corresponds to the first principal component (PC1). The second principal component (PC2) is determined as being a vector orthogonal to PC1 that captures most of the remaining variance (and this process can be continued to find other principal components). Data can then be projected onto this low-dimensional space, whose axes are the principal components. If PC1 and PC2 retain most of the variance of the data, it is possible to have a good 2D visualization of the relationships between variables and experiments.

When performing PCA, it may be convenient to mean-center and scale the data; i.e., transform each variable vector so it has mean 0 and standard deviation 1. Moreover, it may be convenient to perform PCA only on significant changing genes/proteins/metabolites. Although PCA decomposition "filters" for features with high variance, it is known that this variance can be caused by either technical noise or biological changes. Therefore, nonbiological variance should, at best, be subtracted before PCA is performed.

A PCA bi-plot depicts both loadings and scorings, that is, the projection of both variables and samples in the principal component space. Loadings contain information on how variables relate to each other, whereas scorings refer to how samples are related. Analysis of loadings tells us how the variance of certain features (e.g., a gene) is explained by that principal component. Loadings weight should be read in the principal component axis, and genes with high absolute values are the ones that contribute more for that component. The distribution of the scorings tells us how the samples can be explained by the loadings. For example, a sample standing in the upper right quadrant of a bi-plot is positively influenced by the variables also standing in the upper right quadrant, and is negatively influenced by variables in the lower left quadrant.

4.2.1.3. Clustering: Clustering was one of the first methods proposed for analysis of transcriptome data, and it is probably the most wide spread method for grouping profiles of omics data. In the context of gene expression data, the basic idea consists in grouping genes based on their similarity profile (76). Genes sharing a common profile throughout a series of experiments will cluster together, and can be further analyzed as having a common profile. Therefore, clustering is also referred to as a method for reduction of dimensionality.

An important concept when dealing with clustering is the notion of "similarity of genes." Different metrics can be used to assess gene similarity, based on either distances or correlation factors. Metrics frequently used are Euclidian distance and the Pearson correlation. The latter is often preferred because it measures the similarity of the directions of two gene expression vectors, and it is insensitive to the amplitude. More considerations of metrics selection can be found in other studies (76,89,90). Once the appropriate metric is selected, a distance matrix can be calculated for all pair-wise distances among all genes. Genes can then be joined based on similarity using one of the different clustering methods, such as hierarchical clustering, K-means clustering, or self-organized maps (see Jewett et al. [29] and Kaminski and Friedman [91] for reviews).

Once genes are grouped into clusters, cluster analysis is often based on the concept of guilt by association, suggesting that genes belonging to the same cluster are involved in related processes. This has been used to assign a function to genes with unknown function (92), to search for common binding motifs in the upstream region of co-regulated genes (93,94) or to analyze the expression profile of a particular biological process, such as a pathway (95,96).

4.2.2. Integrative Methods

In our quest to answer questions like "how is control of fluxes exerted at different levels of cellular regulation?" we need to understand the principles and architecture of the regulatory machinery at its different levels. Because transcription is hierarchically the "first layer" of cellular regulation, and transcriptome data is currently the most widely quantified omics, much effort has been put into methods for dissecting the transcriptional regulatory machinery. In particular, methods combining transcriptome data with known biological interactions have brought new insights into cellular transcriptional programs.

Classic methods, such as clustering and PCA are data driven, i.e., they attempt to search for hidden correlations in the data by using data alone. The main hypothesis behind the interpretation of results derived from these methods is that co-regulated genes show similar expression patterns in the underlying experiment(s). However, these methods implicitly assume that there may be an all–all interaction amongst the genes being analyzed. This high degree of freedom makes data-driven methods sensitive to noise in the data. Consequently, relatively weak, but biologically significant, correlations may be overshadowed by stronger, but biologically insignificant, correlations. One way to overcome this issue is to reduce the degree of freedom during data analysis by integrating omics data with known biological interactions occurring in the cell, such as protein–protein interaction networks or metabolic networks. Several integrative methods have been reported and applied with different purposes.

Several studies have focused on elucidating local network architecture and functionality by showing that mRNA transcriptional patterns of genes belonging to a group of interacting genes (or gene products) are significantly more similar than in a random set of genes. This correlation was shown to exist for genes that belong to certain metabolic pathways (96–98), for genes belonging to a particular functional class as defined by gene ontology (99), and for genes belonging to a cluster of interacting proteins (100). For example, Ihmels et al. (96) combined gene expression data with metabolic pathway topology of *S. cerevisiae* to analyze how the coordinated expression of enzymes shapes the metabolic network of yeast. They observed that genes belonging to a particular metabolic pathway show higher coexpression than a random control, and further conclude that transcriptional regulation biases metabolic flow toward linear pathways and that isozymes are often regulated separately, thus preventing cross-talk between pathways. Moreover, from their integrative analysis, they also observed that transcriptional regulation of metabolic pathways obeys a hierarchical regulation.

Ideker et al. (33) introduced an integrative approach to search for highly transcriptionally co-regulated subnetworks in molecular interaction networks, aiming to uncover modules of cellular transcriptional regulation in response to a genetic or environmental perturbation. As described in section 2.3.2, they combined the topology of protein–protein and protein–DNA interaction networks with gene expression data to discover regulatory and signaling circuits. They represented the network as a graph and applied graph theory and optimization algorithms to search for high-scoring subnetworks, which correspond to groups of interacting proteins or genes that change their transcriptional response the most. Remarkably, these subnetworks may contain genes without large expression changes, but are still required to connect to other differentially expressed genes. This may be the case of regulatory proteins that are constitutively expressed, being mainly regulated at the posttranscriptional level, but playing a key role in the transcriptional response of other genes.

A similar graph-theory approach was followed by Patil and Nielsen (15) to uncover transcriptional regulation of metabolism through the integration of gene expression data within the metabolic network. In this case, the metabolic network was represented as a graph where enzymes sharing common metabolites are connected. The resulting high-scoring subnetwork corresponds to the most highly correlated (and connected) enzymes and reflects the propagation effects of a genetic/environmental perturbation on metabolism, at the transcriptional level. Notably, the reported method describes an interesting attempt at looking into the metabolic network as a whole, highly connected structure (14,101), instead of looking into "textbook-defined" pathways, which are small and isolated entities. Another novelty of the work by Patil and Nielsen (15) lies in the definition and identification of "reporter metabolites," which are metabolites that might be functionally related to the perturbation factor (gene deletion or change in environmental condition). This represents one of the first attempts to infer the global role of a metabolite based on mRNA expression and metabolic network topology without direct measurement of metabolite concentration.

Liao and co-workers introduced network component analysis (NCA), a method for reduction of dimensionality that incorporates network connectivity information and that can be used to reconstruct regulatory signals (102,103). In their work, they show that NCA can be applied to gene expression data to determine transcription factor activities given the mRNA transcript levels and the transcription factor's connectivity matrix (matrix with 1s if a transcription factor binds a certain gene; with 0s otherwise). NCA is a powerful data decomposition method that may be further used with other types of connectivity information derived, for instance, from protein–protein or metabolic networks.

4.3. Predictive Models

A wide range of models is used for the simulation and prediction of cellular function. Many models are based on the description of the kinetics of several individual processes, e.g., enzyme-catalyzed reactions or protein–protein interactions, and integration of the kinetics expression into dynamic mass balances. This results in a set of coupled differential equations that can be used to simulate operation of the system at both steady state and dynamic conditions. Because of the requirement for a large number of fitted parameters, most of these models are currently limited to describe only relatively small systems, i.e., a given signal transduction pathway or a dedicated metabolic pathway (104–106).

As an alternative to kinetic models, stoichiometric models have proven to have good predictive capabilities for assessment of gene lethality and essentiality during growth under different carbon sources (39,40,72,107), for determination of product yields (41,42,108), and for analysis of flexibility and robustness of the metabolic network (12,18). This makes them valuable tools to evaluate whole metabolic function and a major aid for identification of metabolic engineering targets for obtaining a desirable phenotype. Furthermore, these models are characterized by having only few fitted parameters; hence, it is possible to extend the concept to a genome scale (109).

Flux balance analysis has been extensively used to explore the capabilities of large metabolic networks and for *in silico* prediction of fluxes in metabolic mutants (2,110,111). FBA is a linear programming–posed problem where constraints are defined by stoichiometry (derived from mass balances around each metabolite) and by physiological/thermodynamic limitations, and the objective function is defined as the optimization of a certain flux of interest, e.g., the flux toward formation of biomass. Maximization of biomass production has been shown to allow description of overall metabolic behavior in several cases, probably because most cells have evolved, under laboratory conditions, toward the maximization of their growth performance (16). More recently, another approach has been proposed for dealing with the effect of gene deletions in the prediction of flux distributions, based on quadratic programming (112). The so-called minimization of metabolic adjustment (MOMA) relies on the assumption that optimal growth may initially not be true for mutants generated artificially in the laboratory, as those mutants usually have not yet undergone evolution toward optimal growth.

Several methods to improve predictions from FBA and MOMA have been reported. Namely, the use of constraints that account for metabolic regulation have been shown to improve the predictive capability of models whenever regulation plays an important role (46,113). In particular, the availability of genome-wide gene expression data has offered the opportunity to integrate transcriptional level regulation into these models (44,47), as described in section 2.3.4. A significant improvement for the design of knockout strategies for metabolic engineering was the introduction of OptKnock, which is a bilevel optimization algorithm coupling biomass formation with chemical production (41). Later, the same authors introduced OptStrain (43), which is an extension of OptKnock that allows the addition of heterologous reactions, opening opportunities for the design of hosts for production of heterologous products. A fundamental problem with OptKnock (and its extension) is that it is computationally demanding if several different mutations are allowed. Patil and co-workers found a solution to this problem by developing a genetic algorithm that allows for rapid identification of a relatively large number of mutations that results in a desirable phenotype (114).

Another mathematical framework that holds interesting predictive capabilities is elementary flux mode (EFM) analysis (115). Elementary flux modes are defined as the smallest subnetworks enabling the metabolic system to operate at steady state (116), i.e., any nondecomposable routes connecting two defined metabolites in the network. Like FBA, EFM analysis has been applied to the quantification of maximal conversion yields and to the determination of gene essentiality (12). Furthermore, Stelling et al. (12) have used EFM analysis to show that metabolic structure by itself can be used to derive regulation. In their work, they used a stoichiometric metabolic model of the central carbon metabolism of *E. coli*, and, by introducing a parameter called "control-effective flux," they successfully predicted the expression ratios of many genes involved in central carbon metabolism. Similar results have been reported for the metabolic network of *S. cerevisiae* (18).

5. Conclusions

Omics technologies have generated vast amounts of system-level data. Understanding the state, including location, of all genes, mRNAs, proteins, and metabolites does not, however, explicitly reveal phenotype, and it also does not enable trivial prediction of relations between genotype and phenotype. It is necessary to have quantitative information mapping these molecules under various perturbations to comprehend the orchestrated web of complex interactions that propagate from the genetic architecture through the metabolic network.

In this chapter, we illustrated how systems biology approaches are being used in metabolic engineering to relate genotype with phenotype. Several analysis tools and predictive models were described for exploring properties and capabilities of microbial systems, uncovering hidden regulatory structures and designing new strategies to redirect fluxes. Because analytical techniques for quantification of mRNA levels are the most widespread and well-established genome-wide high-throughput technologies, particular focus has been given to transcriptome data analysis and its application in designing enhanced microbial cell factories. Nevertheless, transcriptional data by itself has a limited capability in predicting phenotype, so identification and quantification of other molecular components and integration of different molecular levels into common models will help to reveal the different levels of cellular regulation and flux control.

References

1. Hood L, Heath JR, Phelps ME, et al. Systems biology and new technologies enable predictive and preventative medicine. *Science* 2004;306(5696): 640–643.
2. Patil KR, Akesson M, Nielsen J. Use of genome-scale microbial models for metabolic engineering. *Curr Opin Biotechnol* 2004;15(1):64–69.
3. Kitano H. Systems biology: a brief overview. *Science* 2002;295(5560): 1662–1664.
4. Ideker T, Galitski T, Hood L. A new approach to decoding life: systems biology. *Annu Rev Genomics Hum Genet* 2001;2:343–372.
5. Penalva MA, Rowlands RT, Turner G. The optimization of penicillin biosynthesis in fungi. *Trends Biotechnol* 1998;16(11):483–489.
6. Bailey JE. Toward a science of metabolic engineering. *Science* 1991; 252(5013):1668–1675.
7. Stephanopoulos G, Aristidou A, Nielsen J. Metabolic Engineering: Principles and Methodologies. San Diego: Academic Press; 1998.
8. Nielsen J. Metabolic engineering. *Appl Microbiol Biotechnol* 2001;55(3): 263–283.
9. Fraenkel DG. The top genes: on the distance from transcript to function in yeast glycolysis. *Curr Opin Microbiol* 2003;6(2):198–201.
10. Christensen B, Nielsen J. Metabolic network analysis. A powerful tool in metabolic engineering. *Adv Biochem Eng Biotechnol* 2000;66:209–231.
11. Wiechert W. 13C metabolic flux analysis. *Metab Eng* 2001;3(3):195–206.
12. Stelling J, Klamt S, Bettenbrock K, et al. Metabolic network structure determines key aspects of functionality and regulation. *Nature* 2002; 420(6912):190–193.
13. Papin JA, Stelling J, Price ND, et al. Comparison of network-based pathway analysis methods. *Trends Biotechnol* 2004;22(8):400–405.
14. Jeong H, Tombor B, Albert R, et al. The large-scale organization of metabolic networks. *Nature* 2000;407(6804):651–654.

15. Patil KR, Nielsen J. Uncovering transcriptional regulation of metabolism by using metabolic network topology. *Proc Natl Acad Sci USA* 2005;102(8): 2685–2689.
16. Edwards JS, Ibarra RU, Palsson BO. In silico predictions of *Escherichia coli* metabolic capabilities are consistent with experimental data. *Nat Biotechnol* 2001;19(2):125–130.
17. Famili I, Forster J, Nielsen J, et al. *Saccharomyces cerevisiae* phenotypes can be predicted by using constraint-based analysis of a genome-scale reconstructed metabolic network. *Proc Natl Acad Sci USA* 2003;100(23): 13134–13139
18. Cakir T, Kirdar B, Ulgen KO. Metabolic pathway analysis of yeast strengthens the bridge between transcriptomics and metabolic networks. *Biotechnol Bioeng* 2004;86(3):251–260.
19. Albert R, Jeong H, Barabasi AL. Error and attack tolerance of complex networks. *Nature* 2000;406(6794):378–382.
20. Hatzimanikatis V, Bailey JE. MCA has more to say. *J Theor Biol* 1996;182(3):233–242.
21. Wang L, Birol I, Hatzimanikatis V. Metabolic control analysis under uncertainty: framework development and case studies. *Biophys J* 2004;87(6): 3750–3763.
22. Nielsen J, Olsson L. An expanded role for microbial physiology in metabolic engineering and functional genomics: moving towards systems biology. *FEMS Yeast Res* 2002;2(2):175–181.
23. Ostergaard S, Olsson L, Johnston M, et al. Increasing galactose consumption by *Saccharomyces cerevisiae* through metabolic engineering of the GAL gene regulatory network. *Nat Biotechnol* 2000;18(12):1283–1286.
24. ter Kuile BH, Westerhoff HV. Transcriptome meets metabolome: hierarchical and metabolic regulation of the glycolytic pathway. *FEBS Lett* 2001; 500(3):169–171.
25. Gygi SP, Rochon Y, Franza BR, et al. Correlation between protein and mRNA abundance in yeast. *Mol Cell Biol* 1999;19(3):1720–1730.
26. Futcher B, Latter GI, Monardo P, et al. A sampling of the yeast proteome. *Mol Cell Biol* 1999;19(11):7357–7368.
27. Daran-Lapujade P, Jansen ML, Daran JM, et al. Role of transcriptional regulation in controlling fluxes in central carbon metabolism of *Saccharomyces cerevisiae*. A chemostat culture study. *J Biol Chem* 2004;279(10):9125–9138.
28. Bro C, Nielsen J. Impact of 'ome' analyses on inverse metabolic engineering. *Metab Eng* 2004;6(3):204–211.
29. Jewett MC, Oliveira AP, Patil KR, et al. The role of high-throughput transcriptome analysis in metabolic engineering. *Biotechnol Bioprocess Eng* 2005;10(5):385–399.
30. Bro C, Knudsen S, Regenberg B, et al. Improvement of galactose uptake in *Saccharomyces cerevisiae* through overexpression of phosphoglucomutase: example of transcript analysis as a tool in inverse metabolic engineering. *Appl Environ Microbiol* 2005;71(11):6465–6472.
31. Askenazi M, Driggers EM, Holtzman DA, et al. Integrating transcriptional and metabolite profiles to direct the engineering of lovastatin-producing fungal strains. *Nat Biotechnol* 2003;21(2):150–156.
32. Ideker T, Thorsson V, Ranish JA, et al. Integrated genomic and proteomic analyses of a systematically perturbed metabolic network. *Science* 2001; 292(5518):929–934.
33. Ideker T, Ozier O, Schwikowski B, et al. Discovering regulatory and signalling circuits in molecular interaction networks. *Bioinformatics* 2002;18 Suppl 1:S233–S240.

34. Steffen M, Petti A, Aach J, et al. Automated modelling of signal transduction networks. *BMC Bioinformatics* 2002;3:34.
35. Yeang CH, Ideker T, Jaakkola T. Physical network models. *J Comput Biol* 2004;11(2–3):243–262.
36. Lee TI, Rinaldi NJ, Robert F et al. Transcriptional regulatory networks in *Saccharomyces cerevisiae*. *Science* 2002;298(5594):799–804.
37. Bro C, Regenberg B, Nielsen J. Genome-wide transcriptional response of a *Saccharomyces cerevisiae* strain with an altered redox metabolism. *Biotechnol Bioeng* 2004;85(3):269–276.
38. Nissen TL, Kielland-Brandt MC, Nielsen J, et al. Optimization of ethanol production in *Saccharomyces cerevisiae* by metabolic engineering of the ammonium assimilation. *Metab Eng* 2000;2:69–77.
39. Edwards JS, Palsson BO. The *Escherichia coli* MG1655 in silico metabolic genotype: its definition, characteristics, and capabilities. *Proc Natl Acad Sci USA* 2000;97(10):5528–5533.
40. Forster J, Famili I, Palsson BO, et al. Large-scale evaluation of in silico gene deletions in *Saccharomyces cerevisiae*. *OMICS* 2003;7(2):193–202.
41. Burgard AP, Pharkya P, Maranas CD. OptKnock: a bilevel programming framework for identifying gene knockout strategies for microbial strain optimization. *Biotechnol Bioeng* 2003;84(6):647–657.
42. Pharkya P, Burgard AP, Maranas CD. Exploring the overproduction of amino acids using the bilevel optimization framework OptKnock. *Biotechnol Bioeng* 2003;84(7):887–899.
43. Pharkya P, Burgard AP, Maranas CD. OptStrain: A computational framework for redesign of microbial production systems. *Genome Res* 2004;14(11): 2367–2376.
44. Akesson M, Forster J, Nielsen J. Integration of gene expression data into genome-scale metabolic models. *Metab Eng* 2004;6(4):285–293.
45. Covert MW, Schilling CH, Palsson B. Regulation of gene expression in flux balance models of metabolism. *J Theor Biol* 2001;213(1):73–88.
46. Covert MW, Palsson BO. Constraints-based models: regulation of gene expression reduces the steady-state solution space. *J Theor Biol* 2003;221(3): 309–325.
47. Covert MW, Knight EM, Reed JL, Herrgard MJ, Palsson BO. Integrating high-throughput and computational data elucidates bacterial networks. *Nature* 2004;429(6987):92–96.
48. Lipshutz RJ, Fodor SP, Gingeras TR, et al. High density synthetic oligonucleotide arrays. *Nat Genet* 1999;21(1 Suppl):20–24.
49. DeRisi JL, Iyer VR, Brown PO. Exploring the metabolic and genetic control of gene expression on a genomic scale. *Science* 1997;278(5338):680–686.
50. Schena M, Shalon D, Davis RW, et al. Quantitative monitoring of gene expression patterns with a complementary DNA microarray. *Science* 1995;270(5235):467–470.
51. Harrington CA, Rosenow C, Retief J. Monitoring gene expression using DNA microarrays. *Curr Opin Microbiol* 2000;3(3):285–291.
52. Lockhart DJ, Winzeler EA. Genomics, gene expression and DNA arrays. *Nature* 2000;405(6788):827–836.
53. Griffin TJ, Gygi SP, Ideker T, et al. Complementary profiling of gene expression at the transcriptome and proteome levels in *Saccharomyces cerevisiae*. *Mol Cell Proteomics* 2002;1(4):323–333.
54. Aebersold R, Goodlett DR. Mass spectrometry in proteomics. *Chem Rev* 2001;101(2):269–295.

55. Jenkins RE, Pennington SR. Arrays for protein expression profiling: towards a viable alternative to two-dimensional gel electrophoresis? *Proteomics* 2001;1(1):13–29.
56. Tao WA, Aebersold R. Advances in quantitative proteomics via stable isotope tagging and mass spectrometry. *Curr Opin Biotechnol* 2003;14(1):110–118.
57. Gygi SP, Rist B, Gerber SA, et al. Quantitative analysis of complex protein mixtures using isotope-coded affinity tags. *Nat Biotechnol* 1999;17(10): 994–999.
58. Ong SE, Blagoev B, Kratchmarova I, et al. Stable isotope labeling by amino acids in cell culture, SILAC, as a simple and accurate approach to expression proteomics. *Mol Cell Proteomics* 2002;1(5):376–386.
59. Ross PL, Huang YN, Marchese JN, et al. Multiplexed protein quantitation in *Saccharomyces cerevisiae* using amine-reactive isobaric tagging reagents. *Mol Cell Proteomics* 2004;3(12):1154–1169.
60. Goodacre R, Vaidyanathan S, Dunn WB, et al. Metabolomics by numbers: acquiring and understanding global metabolite data. *Trends Biotechnol* 2004;22(5):245–252.
61. Villas-Boas SG, Mas S, Akesson M, et al. Mass spectrometry in metabolome analysis. *Mass Spectrom Rev* 2005;24(5):613–646.
62. Kell DB. Metabolomics and systems biology: making sense of the soup. *Curr Opin Microbiol* 2004;7(3):296–307.
63. Villas-Boas SG, Moxley JF, Akesson M, et al. High-throughput metabolic state analysis: the missing link in integrated functional genomics of yeasts. *Biochem J* 2005;388(Pt 2):669–677.
64. van Dam JC, Eman MR, Frank J, et al. Analysis of glycolytic intermediates in *Saccharomyces cerevisiae* using anion exchange chromatography and electrospray ionization with tandem mass spectrometric detection. *Analytica Chimica Acta* 2002;460(2):209–218.
65. Mashego MR, Wu L, Van Dam JC et al. MIRACLE: mass isotopomer ratio analysis of U-13C-labeled extracts. A new method for accurate quantification of changes in concentrations of intracellular metabolites. *Biotechnol Bioeng* 2004;85(6):620–628.
66. Gombert AK, Moreira dos SM, Christensen B, et al. Network identification and flux quantification in the central metabolism of *Saccharomyces cerevisiae* under different conditions of glucose repression. *J Bacteriol* 2001;183(4): 1441–1451.
67. Sauer U, Lasko DR, Fiaux J, et al. Metabolic flux ratio analysis of genetic and environmental modulations of *Escherichia coli* central carbon metabolism. *J Bacteriol* 1999;181(21):6679–6688.
68. Sauer U. High-throughput phenomics: experimental methods for mapping fluxomes. *Curr Opin Biotechnol* 2004;15(1):58–63.
69. Van Winden WA, Van Dam JC, Ras C, et al. Metabolic-flux analysis of *Saccharomyces cerevisiae* CEN.PK113-7D based on mass isotopomer measurements of (13)C-labeled primary metabolites. *FEMS Yeast Res* 2005; 5(6–7):559–568.
70. Grotkjaer T, Akesson M, Christensen B, et al. Impact of transamination reactions and protein turnover on labeling dynamics in (13)C-labeling experiments. *Biotechnol Bioeng* 2004;86(2):209–216.
71. Fischer E, Zamboni N, Sauer U. High-throughput metabolic flux analysis based on gas chromatography-mass spectrometry derived 13C constraints. *Anal Biochem* 2004;325(2):308–316.
72. Blank LM, Kuepfer L, Sauer U. Large-scale 13C-flux analysis reveals mechanistic principles of metabolic network robustness to null mutations in yeast. *Genome Biol* 2005;6(6):R49.

73. Fischer E, Sauer U. Large-scale in vivo flux analysis shows rigidity and suboptimal performance of Bacillus subtilis metabolism. *Nat Genet* 2005; 37(6):636–640.
74. Bailey JE. Mathematical modeling and analysis in biochemical engineering: past accomplishments and future opportunities. *Biotechnol Prog* 1998;14(1): 8–20.
75. Churchill GA. Fundamentals of experimental design for cDNA microarrays. *Nat Genet* 2002;32 Suppl:490–495.
76. Eisen MB, Spellman PT, Brown PO, et al. Cluster analysis and display of genome-wide expression patterns. *Proc Natl Acad Sci USA* 1998;95(25): 14863–14868.
77. Yeung KY, Ruzzo WL. Principal component analysis for clustering gene expression data. *Bioinformatics* 2001;17(9):763–774.
78. Scholz M, Gatzek S, Sterling A, et al. Metabolite fingerprinting: detecting biological features by independent component analysis. *Bioinformatics* 2004.
79. Alter O, Brown PO, Botstein D. Singular value decomposition for genome-wide expression data processing and modeling. *Proc Natl Acad Sci USA* 2000;97(18):10101–10106.
80. Spellman PT, Sherlock G, Zhang MQ, et al. Comprehensive identification of cell cycle-regulated genes of the yeast *Saccharomyces cerevisiae* by microarray hybridization. *Mol Biol Cell* 1998;9(12):3273–3297.
81. Chu S, DeRisi J, Eisen M, et al. The transcriptional program of sporulation in budding yeast. *Science* 1998;282(5389):699–705.
82. Quackenbush J. Microarray data normalization and transformation. *Nat Genet* 2002;32 Suppl:496–501.
83. Knudsen S. Guide to analysis of DNA microarray data. 2nd ed. New York: John Wiley & Sons, Inc;2004.
84. Piper MD, Daran-Lapujade P, Bro C, et al. Reproducibility of oligonucleotide microarray transcriptome analyses. An interlaboratory comparison using chemostat cultures of *Saccharomyces cerevisiae*. *J Biol Chem* 2002; 277(40):37001–37008.
85. Tusher VG, Tibshirani R, Chu G. Significance analysis of microarrays applied to the ionizing radiation response. *Proc Natl Acad Sci USA* 2001;98(9): 5116–5121.
86. Ideker T, Thorsson V, Siegel AF, Hood LE. Testing for differentially-expressed genes by maximum-likelihood analysis of microarray data. *J Comput Biol* 2000;7(6):805–817.
87. Storey JD, Tibshirani R. Statistical significance for genomewide studies. *Proc Natl Acad Sci USA* 2003;100(16):9440–9445.
88. Benjamini Y. Controlling the false discovery rate: a practical and powerful approach to multiple testing. *J Royal Stat Soc* 1995;57(1):289–300.
89. Cherepinsky V, Feng J, Rejali M, et al. Shrinkage-based similarity metric for cluster analysis of microarray data. *Proc Natl Acad Sci USA* 2003;100(17): 9668–9673.
90. Sturn A, Quackenbush J, Trajanoski Z. Genesis: cluster analysis of microarray data. *Bioinformatics* 2002;18(1):207–208.
91. Kaminski N, Friedman N. Practical approaches to analyzing results of microarray experiments. *Am J Respir Cell Mol Biol* 2002;27(2):125–132.
92. Brown MP, Grundy WN, Lin D, et al. Knowledge-based analysis of microarray gene expression data by using support vector machines. *Proc Natl Acad Sci USA* 2000;97(1):262–267.

93. Gasch AP, Spellman PT, Kao CM, et al. Genomic expression programs in the response of yeast cells to environmental changes. *Mol Biol Cell* 2000; 11(12):4241–4257.
94. Zhu J, Zhang MQ. Cluster, function and promoter: analysis of yeast expression array. *Pac Symp Biocomput* 2000;479–490.
95. Ihmels J, Friedlander G, Bergmann S, et al. Revealing modular organization in the yeast transcriptional network. *Nat Genet* 2002;31(4):370–377.
96. Ihmels J, Levy R, Barkai N. Principles of transcriptional control in the metabolic network of *Saccharomyces cerevisiae*. *Nat Biotechnol* 2004;22(1): 86 92.
97. Zien A, Kuffner R, Zimmer R, Lengauer T. Analysis of gene expression data with pathway scores. *Proc Int Conf Intell Syst Mol Biol* 2000;8:407–417.
98. Pavlidis P, Lewis DP, Noble WS. Exploring gene expression data with class scores. *Pac Symp Biocomput* 2002;474–485.
99. Breitling R, Amtmann A, Herzyk P. Graph-based iterative Group Analysis enhances microarray interpretation. *BMC Bioinformatics* 2004;5:100.
100. Jansen R, Greenbaum D, Gerstein M. Relating whole-genome expression data with protein-protein interactions. *Genome Res* 2002;12(1):37–46.
101. Fell DA, Wagner A. The small world of metabolism. *Nat Biotechnol* 2000; 18(11):1121–1122.
102. Liao JC, Boscolo R, Yang YL, et al. Network component analysis: reconstruction of regulatory signals in biological systems. *Proc Natl Acad Sci USA* 2003;100(26):15522–15527.
103. Kao KC, Yang YL, Boscolo R, et al. Transcriptome-based determination of multiple transcription regulator activities in *Escherichia coli* by using network component analysis. *Proc Natl Acad Sci USA* 2004;101(2):641–646.
104. Klipp E, Nordlander B, Kruger R, et al. Integrative model of the response of yeast to osmotic shock. *Nat Biotechnol* 2005;23(8):975–982.
105. Teusink B, Passarge J, Reijenga CA, et al. Can yeast glycolysis be understood in terms of in vitro kinetics of the constituent enzymes? Testing biochemistry. *Eur J Biochem* 2000;267(17):5313–5329.
106. Heinrich R, Neel BG, Rapoport TA. Mathematical models of protein kinase signal transduction. *Mol Cell* 2002;9(5):957–970.
107. Papp B, Pal C, Hurst LD. Metabolic network analysis of the causes and evolution of enzyme dispensability in yeast. *Nature* 2004;429(6992):661–664.
108. Alper H, Miyaoku K, Stephanopoulos G. Construction of lycopene-overproducing *E. coli* strains by combining systematic and combinatorial gene knockout targets. *Nat Biotechnol* 2005;23(5):612–616.
109. Bailey JE. Complex biology with no parameters. *Nat Biotechnol* 2001; 19(6):503–504.
110. Covert MW, Schilling CH, Famili I et al. Metabolic modeling of microbial strains in silico. *Trends Biochem Sci* 2001;26(3):179–186.
111. Palsson B. In silico biology through "omics". *Nat Biotechnol* 2002;20(7): 649–650.
112. Segre D, Vitkup D, Church GM. Analysis of optimality in natural and perturbed metabolic networks. *Proc Natl Acad Sci USA* 2002;99(23): 15112–15117.
113. Herrgard MJ, Covert MW, Palsson BO. Reconciling gene expression data with known genome-scale regulatory network structures. *Genome Res* 2003;13(11):2423–2434.
114. Patil KR, Rocha I, Forster J, et al. Evolutionary programming as a platform for *in silico* metabolic engineering. *BMC Bioinformatics* 2005;6:308.

115. Schuster S, Fell DA, Dandekar T. A general definition of metabolic pathways useful for systematic organization and analysis of complex metabolic networks. *Nat Biotechnol* 2000;18(3):326–332.
116. Schuster S, Dandekar T, Fell DA. Detection of elementary flux modes in biochemical networks: a promising tool for pathway analysis and metabolic engineering. *Trends Biotechnol* 1999;17(2):53–60.

Part II

Experimental Techniques for Systems Biology

4

Handling and Interpreting Gene Groups

Nils Blüthgen, Szymon M. Kielbasa, Dieter Beule

Summary

Systems biologists often have to deal with large gene groups obtained from high-throughput experiments, genome-wide predictions, and literature searches. Handling and functional interpretation of these gene groups is rather challenging. Problems arise from redundancies in databases, where a gene is given several names or identifiers, and from falsely assigned genes in the list. Moreover, genes in gene groups obtained by different methods are often represented by different types of identifiers, or are even genes from other model organisms. Thus, research in systems biology requires software tools that help to handle and interpret gene groups.

This chapter will review tools to store and compare gene groups represented by various identifiers. We introduce software that uses Gene Ontology (GO) annotations to infer biological processes associated with the gene groups. Additionally, we review approaches to further analyze gene groups regarding their transcriptional regulation by retrieving and analyzing their putative promoter regions.

Key Words: Gene groups; homology; promoter analysis; GO; redundancy; functional interpretation.

1. Introduction

Many modern experimental techniques in molecular biology produce large gene lists. Microarray studies, for example, result in groups of interesting genes, like genes that are differentially expressed in different tissues or in normal versus transformed cells. Yeast two-hybrid screens provide groups of hundreds of interacting proteins. Genome-wide screens for transcription factor binding sites yield lists of genes potentially regulated by the transcription factor under study.

Thus, when analyzing and designing experiments and building models, systems biologists often have to deal with large gene groups. They face the problem of handling, comparing, and analyzing gene lists from diverse data sources, like databases, previous experiments, and literature searches. One hard nut to crack is that each gene can have several names, and several types of accession numbers can be assigned to one gene. Especially if one needs to handle large gene lists, comparing and analyzing these lists must be done in an automatic way.

In this chapter, several types of accession numbers are briefly reviewed. Subsequently, algorithms to convert accession numbers and to compare gene groups are introduced, with a focus on the Web service HomGL. Functional analysis and interpretation of these gene groups is another difficult task. Instead of looking at gene groups and finding the most familiar gene names, a more unbiased automation of this task is facilitated by the systematic annotation provided by the GO. This chapter shall discuss tools and statistical issues of doing so. Finally, the extraction and analysis of sequence information for gene groups will be discussed.

2. Accession Numbers

Individual researchers may submit sequences of (for example) sequenced genes, expressed sequence tags (EST), or even sequences of entire genomes to GenBank (1). To each submitted sequence, a unique identifier, the so-called accession number, is assigned. Under this accession number, one can access the sequence and its annotation, for example, via Entrez (2). Other databases, like SwissProt, follow a similar procedure (3). Currently, GenBank is growing exponentially. As of August 2005, GenBank contained more than 50 billion nucleotide bases from 47 million individual sequences for more than 165,000 named organisms (1).

Given a GenBank accession number, researchers can easily access the sequence information and some annotation at the National Center for Biotechnology Information (NCBI) Website (http://www.ncbi.nlm.nih.gov/). However, full-length mRNA sequences of a gene might by stored under several different accession numbers, as it might have been submitted by different groups of researchers. Additionally, several subsequences from EST projects might be stored in GenBank. Also, transcripts corresponding to different splice variants or various transcription start sites are stored under different accession numbers. Thus, the multiplicity of sequences in the public databases for genes, transcripts, and proteins make it challenging for researchers to compare two lists of GenBank accession numbers, even though each identifier points uniquely to one entry in GenBank.

UniGene is one of the first approaches to address this problem (2). UniGene is a system for automatic partitioning of GenBank sequences, including ESTs, into a nonredundant set of gene-oriented clusters. For each organism in GenBank with sufficient sequences (currently 16 animals and 13 plants) UniGene clusters are created. UniGene clusters should represent a unique gene. In the human UniGene November 2005 release (build 187), over 5 million human ESTs in GenBank have been

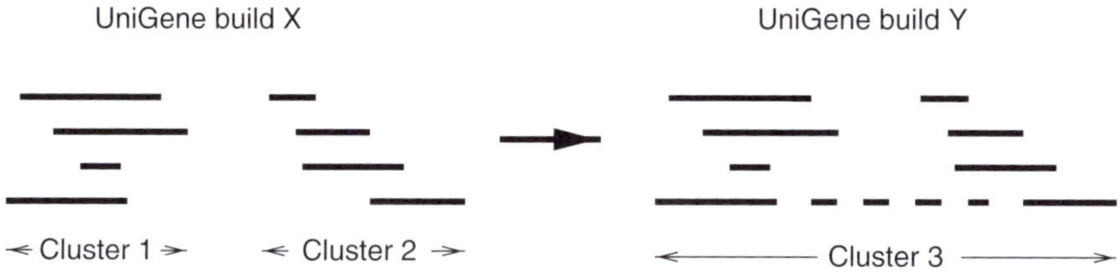

Figure 1. UniGene clusters are not stable. If a new sequence enters GenBank, UniGene clusters may join. Here, clusters 1 and 2 join and become cluster 3 in the next release, as the new (dashed) sequence joins the two clusters.

reduced 100-fold in number to 54,576 sequence clusters. UniGene has been used extensively as unique sequences for microarrays. UniGene databases are updated weekly with new EST sequences, and bimonthly with newly characterized sequences.

However, automatic clustering does not guarantee that each gene becomes represented by only a single cluster. Thus, in each new UniGene build, clusters may split or join to form new clusters when new sequence information is provided. Therefore, the cluster accession numbers are subject to change, and a gene might be referenced by different cluster numbers in different versions of UniGene (Figures 1 and 2).

Approaches overcoming this problem include RefSeq and LocusLink, and their successor Entrez Gene (4). They are manually curated sequence collections that use whole-genome information to align and cluster sequences based on their genomic locus. Entrez Gene is constructed such that the identifiers are stable. That is, once an identifier is assigned to a gene, it will be identified by the same identifier in future database releases. The relations between the databases are exemplified in Figure 3.

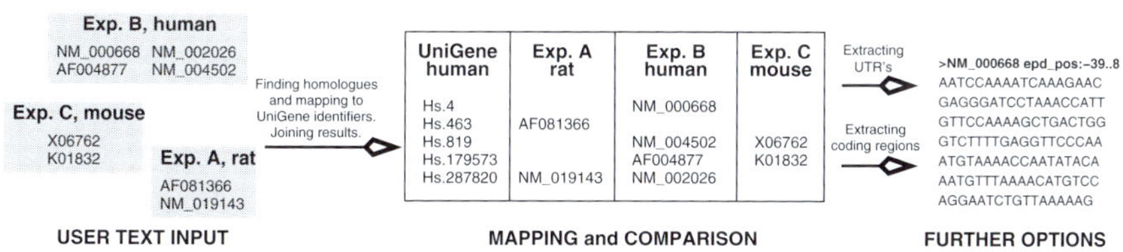

Figure 2. Example of the usage of HomGL to compare three different accession numbers. Assume that an experiment A with rat cell lines, experiment B with human cell lines, and experiment C with mice generated three lists of interesting genes. These lists of genes represented by different accession numbers can be uploaded to HomGL. All accession numbers are matched to the corresponding UniGene cluster numbers of their organism. Then, using HomoloGene, these UniGene clusters are mapped to human UniGene clusters, after which the groups can be compared. Furthermore, HomGL allows one to download the sequences for the genes by linking to EnsEMBL and GO.

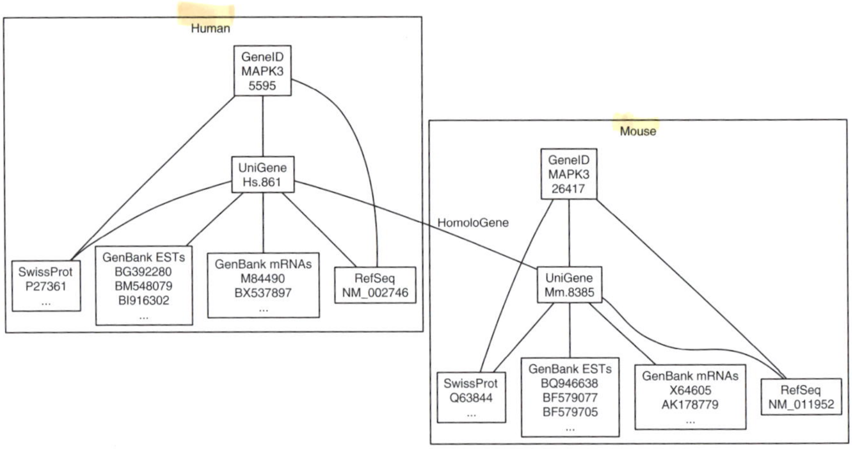

Figure 3. Relations between databases and accession numbers in HomGL.

3. Handling and Comparing Gene Groups

To compare gene groups from different sources, one needs to map the genes to one nonredundant set of identifiers, where each identifier points to one gene. Entrez Gene ID is getting closer to become such an identifier, and it allows nearly complete coverage of the genome (e.g., with 32,853 identifiers). However, most current microarrays are built upon UniGene clusters (for example, the Affymetrix U chips), such that UniGene is still the choice for comparing chips. Nevertheless, UniGene identifiers from different UniGene builds (i.e., versions) do not allow comparison, as UniGene identifiers are not stable (*see* previous section). Therefore, a possible strategy to compare gene groups would be to represent each gene by a stable GenBank or chip-specific identifier, and then link these identifiers to one (preferably recent) build of UniGene, and compare these UniGene identifiers. Doing this manually is rather time-consuming, but there are tools to help facilitate this task (Table 1). In the following sections, we introduce HomGL, a Web-based tool developed by the authors (5).

Table 1. Tools to convert accession numbers.

Application	Type of identifier
HomGL[1]	Human, mouse, and rat identifier
Resourcerer[2]	Human, mouse, and rat microarray identifier (e.g., from Research Genetics, Operon, Affymetrix, RZPD)
KARMA[3]	Several array platforms, multiple organisms
ProbematchDB[4]	Match two human/mouse/rat microarrays

[1] http://homgl.gene-groups.net/
[2] http://www.tigr.org/tigr-scripts/magic/r1.pl
[3] http://biryani.med.yale.edu/karma/cgi-bin/mysql/karma.pl
[4] http://brainarray.mhri.med.umich.edu/brainarray/

HomGL uses UniGene cluster numbers as a common nonredundant identifier. It provides mapping for GenBank, SwissProt, RefSeq, LocusLink/GeneID, and Affymetrix probe set identifiers to UniGene cluster numbers for human, mouse, and rat sequences (Figure 3). Additionally, it utilizes HomoloGene, which is a database of homologous genes, to map between UniGene clusters to homologous genes in different organisms. This way, gene groups resulting from experiments performed in different organisms may be compared. HomGL provides a user interface to facilitate the upload, storage, mapping, and comparison of gene groups. It can be accessed via http://homgl.gene groups.net/ or installed locally. In brief, HomGL imports diverse flat-files from databases like UniGene, SwissProt, GenBank, and HomoloGene and extracts accession number/UniGene cluster number pairs.

To illustrate the use of HomGL, we analyze gene groups of platelet-derived growth factor (PDGF)-induced genes from a study by Tullai et al. (6). Here, we started with 79 identifiers of probes that showed induction after treatment with the growth factor PDGF. These identifiers were uploaded to HomGL. They matched with 59 unique UniGene identifier (Figure 4 for a screenshot and part of the list of identifiers). We find that not 79, but 59, genes are induced, and HomGL helps to identify these redundancies in the database.

HomGL makes use of the homology database HomoloGene, which is available from NCBI (2). HomoloGene is a database of both curated and calculated gene orthologs and homologues, and it covers 21

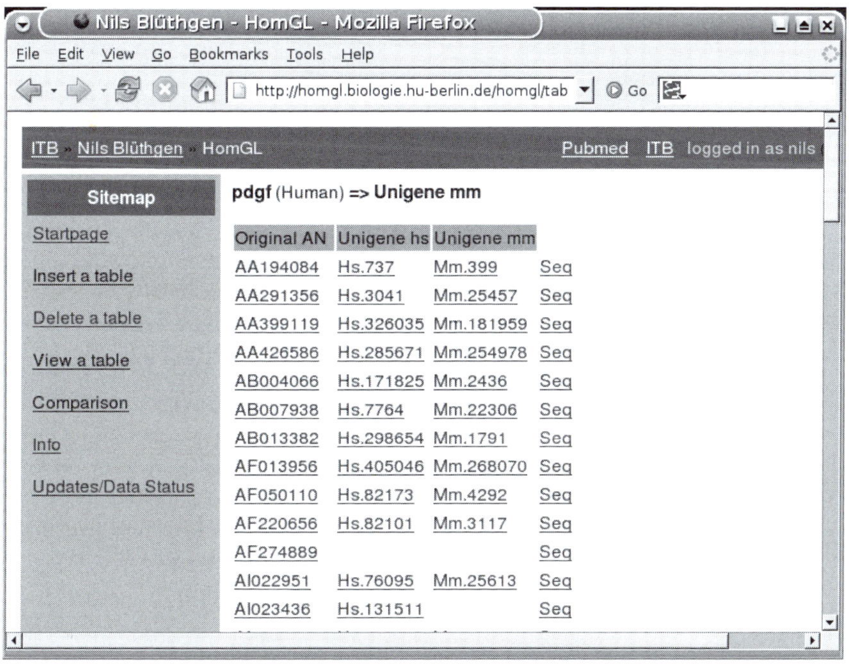

Figure 4. A HomGL screenshot, where accession numbers from an experiment performed in human cell lines are mapped to mouse UniGene cluster numbers.

organisms. Curated orthologs include gene pairs from the Mouse Genome Database, at The Jackson Laboratory, and from published reports. Computed homologies, which are considered putative, are identified by BLAST nucleotide sequence comparisons between all UniGene clusters for each pair of organisms. HomGL imports this database and constructs a map between UniGene cluster numbers of human, mouse, and rat clusters. This way, for example, 15,622 human and mouse UniGene clusters can be mapped as homologues (HomGL release August 2005). By use of this mapping between homologous UniGene clusters, the 59 genes discussed above can be mapped to 56 unique mouse UniGene identifiers, which facilitates comparison with other experiments performed in mice, or to design a customized chip for immediate-early genes in mouse cells.

There are other tools and approaches to compare and link different identifiers. These include KARMA (7), SRS (8), Resourcerer (9), and ProbematchDB (10). The availability of these tools is shown in Table 1.

4. Functional Interpretation of Gene Groups

Once the genes are annotated with suitable identifiers, the next step of analysis is typically the identification of the biological processes and functions associated with the gene groups. This interpretation task, called functional profiling, is laborious and complex. Ontologies that provide a systematic representation of existing knowledge are a suitable starting point for an automation of this step. An ontology specifies a controlled vocabulary and the relations between the terms within the vocabulary. For molecular biology the GO Consortium provides the GO as an international standard (11). The terms of GO describe molecular functions, biological processes, or cellular locations of genes and gene products. The relations between the terms form a directed acyclic graph (DAG), where the nodes represent the terms (12) (Figure 5). A variety of tools are available for browsing and editing the ontology, as well as for using it in statistical and biological analysis of data, e.g., http://www.geneontology.org/GO.tools.shtml. Many chip manufacturers provide GO terms for the genes covered by their chips, e.g., Affymetrix (http://www.affymetrix.com/support/); otherwise, gene groups can be annotated with GO terms using tools like HomGL, as discussed in the previous section.

4.1. GO-Based Functional Profiling

In the following section, we discuss how to utilize the knowledge represented by GO to automatically test if any term is significantly associated with a studied group of genes. To profile gene groups, we require four data sources: a test group of genes (e.g., up-regulated genes), a reference group (e.g., all significantly expressed genes), GO annotations for these genes, and the GO. For each term in the ontology, we ask whether this particular term is enriched in the test group as compared to the reference group. The null hypothesis is that the test group is sampled randomly from the reference group.

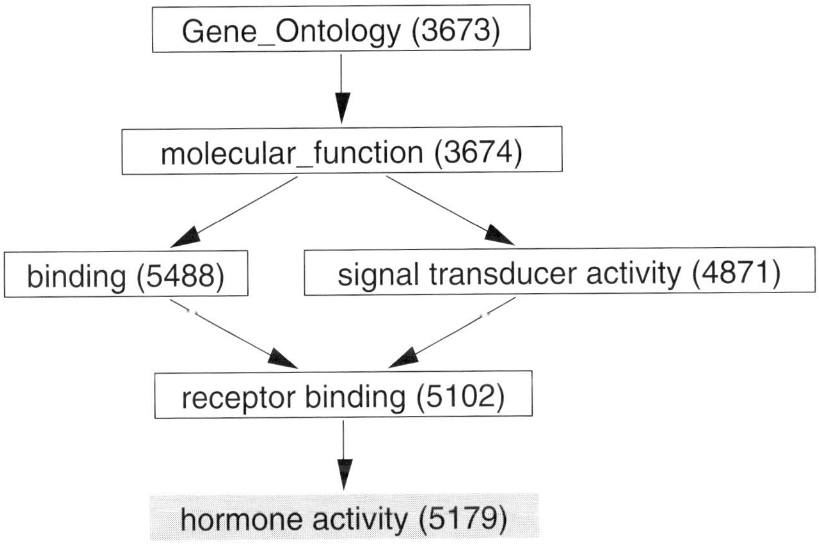

Figure 5. Part of the DAG representing the GO. Annotations are usually given as terms in different parts of the DAG, e.g., term 5179 *hormone activity*, implying a series of more general terms (identifiers 5102, 5488, 4871, 3674, 3673).

To test for this association, we categorize each gene in two ways: first, whether it is annotated with the term under consideration or not, and second, whether it belongs to the test group or not. Based on these categories we build a 2×2 contingency table of gene frequencies for each term. Figure 6 shows the structure of such a contingency table. Using

		Test Group Yes	Test Group No	Reference Group
Annotated	Yes	20	417	Zi=437
Annotated	No	65	13978	
		T=85		N=14480

Figure 6. Contingency table of gene frequency that is calculated for each term. Each gene is categorized in two ways: whether it belongs to the test group and whether it is annotated with the term under consideration. In total, 14,480 genes are in the reference group. 437 genes are annotated with this specific term, and 20 of them are in the test gene group. 14,043 are not annotated with this term. Out of this, 65 are in the test group. The number of genes in the test group is 85.

Fisher's exact test http://home.clara.net/sisa/fishrhlp.htm, we compute p-values that allow us to detect and quantify associations between the two categorizations. Fisher's exact test is based on the hypergeometric distribution, and works in a similar manner as the -test for independence. The -test provides only an estimate of the true probability values, and it is not accurate if the marginal distribution is very unbalanced or if we expect small frequencies (less than five) in one of the cells of the contingency table. Both situations are typical for the task and data under consideration. Fisher's exact test can, in principle, quantify the reduction of a term in respect to the reference group, but a reduction is unlikely to be detected in typical data sets.

4.2. Multiple Testing

The definition of statistical significance is a major challenge due to the large number of terms that need to be tested. The use of single-test p-values is justified only if we test whether a single term is associated with a specific gene group. However, in genome-wide screening experiments, the situation is fundamentally different; the current GO includes approximately 19,000 terms, out of which typically several thousand terms appear in the annotation of an investigated gene group and have to be tested. If one performs that many tests, problems arising from multiple testing cannot be ignored. Namely, even when we apply a very conservative threshold, like $p < 0.001$, a few terms will be reported to be associated with the test group by sheer chance. This phenomenon is known as a false-positive or type-I error. The standard solution for this problem is to calculate adjusted p-values. These adjusted p-values control the number of false discoveries in the entire list and can be used similarly to normal p-values for single tests (13).

To control the number of false discoveries, these methods determine adjusted p-values to control the false discovery rate (FDR) that quantifies the expected portion of false discoveries within the positives. If there is no prior expectation about an association between the gene list and any biological process, one might favor the family-wise error rate (FWER). However, the typical case in profiling gene lists is that one expects some terms to be enriched. In this case, the FDR is an adequate measure of false discoveries. Both rates can be reliably estimated by resampling simulations, but this method suffers from very long runtime, even on modern computers. Alternatively, several approaches exist to estimate the FDR from the single-test p-values (e.g., Benjamini-Hochberg and Benjamini-Yekutieli [13]). These methods are designed to cope with general problems, but turn out to be not particularly suitable for the specific problem considered here. For the specific problem of profiling gene groups, the expected FDR can also be determined exactly by an analytical expression (14). It was shown that the resulting profile does not depend critically on the precise composition of the test group. This is an important characteristic, as the extraction of gene group from high-throughput experiments always has to deal with the trade-off between specificity and sensitivity. Therefore, it is often not clear which genes to include or exclude from a certain gene group, e.g., by choosing a specific threshold.

4.3. Software

There are several software implementations available to profile gene groups using GO, including Onto-Express (15), EASE+David (16,17), GoSurfer (18), GoMiner (19), GeneMerge (20), FatiGO (21), GOstat (22), and Bingo (23). Most of the software packages listed before have some multiple testing correction. Results from applications that do not use multiple testing corrections are hard to interpret because false-positive predictions dominate. The same applies to the Benjamini-Hochberg estimate used in Bingo. On the other hand, the packages using appropriate standard multiple testing corrections (Bonferroni in Gene-Merge, Benjamini-Yekutieli in FatiGo, GoStat, and GOSurfer do give control of the number of false discoveries, but are too conservative, and therefore have less power. EASE uses a jackknife procedure that is similar to resampling to correct for multiple testing, which can give more robust scores, although they cannot be interpreted as adjusted *p*-values. GoMiner uses resampling to determine the multiple testing correction, and Gossip (14) (http://gossip.gene-groups.net/) uses the exact analytical results for the FDR determination and provides profiling gene group visualizations of the results (Figure 7). The exact analytical expression and the resampling method perform significantly better then the

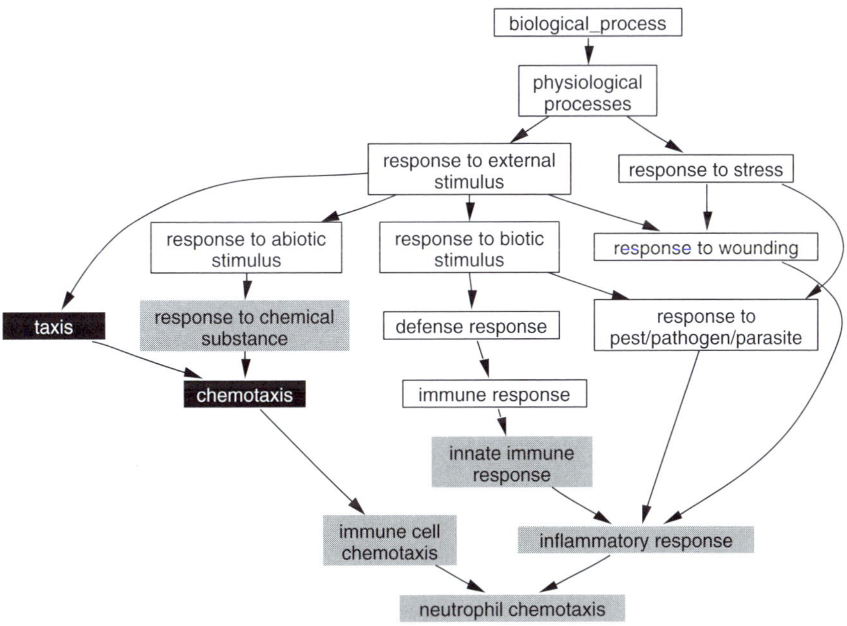

Figure 7. The processes significantly (black $p < 0.01$; gray $p < 0.05$) associated with genes induced by the growth factor PDGF via the kinase Erk (6). The gene group was first mapped to UniGene identifiers and annotated using HomGL, and then analyzed with Gossip.

Table 2. Selection of Web and desktop applications for functional profiling, and their approach to the multiple testing issues.

Application	Multiple Testing	Comments
Gossip (14)	FDR, Analytical exact FWER, Approx.	Desktop Appl. and Integrated into (24,26)
GoMiner (19)	Resampling	Queuing System, Web
EASE (16,17)	Jackknife	Desktop Appl.
Bingo (23)	Bonferroni (FWER) Benjamini–Hochberg (FDR)	Desktop Appl. Cytoscape Plugin
FatiGo (21)	Benjamini–Yekutieli	Web Appl.
GoStat (22)	Holm, Benjamini–Yekutieli	Web Appl.
GOSurfer (18)	Benjamini–Yekutieli	Desktop Appl.
Gene-Merge (20)	Bonferroni	Web & Desktop

For an up-to-date list of available tools refer to http://www.geneontology.org/GO.tools.shtml.

estimation methods (14). The analytical expression also allows a very fast computation, and thus opens the door to combining the automatic functional profiling with other bioinformatics methods, like sequence alignment and motif search, to build powerful and efficient perdition tools (24–26). Table 2 summarizes information on the tools available for the functional profiling of gene groups and their multiple testing corrections.

5. Retrieval and Analysis of Sequences

Although the abundance of proteins can be controlled through different processes, alteration of gene transcriptional rates is the most directly utilized cellular mechanism (27). Therefore, a standard next step of analysis of a co-regulated gene group is the prediction of transcription factors controlling expression of the genes. In this process, nucleotide sequences expected to contain regulatory regions of the genes (promoters) are predicted, extracted, and searched for putative transcription factor binding sites.

The motifs that are recognized by the transcription factors are similar to their surroundings and, consequently, difficult to detect by computational methods. Because it is believed that the majority of the regulatory sites are located close to transcription start sites (TSSs), usually only short fragments (1–10 kbp) upstream of the transcripts are selected for further analysis. It has been shown that a single gene may have multiple TSSs and, consequently, multiple promoter regions. The following collections of experimentally verified promoters may be useful for sequence extraction. The eukaryotic promoter database EPD (28) (http://www.epd.isb-sib.ch) provides an annotated nonredundant collection of eukaryotic POL II promoters based on underlying experimental evidence obtained from full-length cDNA clones. The Functional Annotation of the Mouse consortium (29) (http://fantom3.gsc.riken.jp/) annotated the mouse genome with variations in transcripts arising from alternative promoter usage and splicing. Another library, DBTSS (30) (http://dbtss.hgc.jp/), defines putative promoter groups by clustering TSSs within a

500-base interval. Finally, the EnsEMBL (31) application programming interface provides a convenient framework to implement the extracting subroutines in the PERL language.

There are two fundamental approaches for searching putative binding sites in the chosen promoter regions (32). Based on the assumption that genes of the group are co-regulated by the same transcription factor, one may search for a pattern recognized by the factor, which should occur repeatedly in the sequences. Gibbs sampler (33), AlignACE (34), or GLAM (35) detect such motifs by aligning short fragments of the input promoter sequences based on the statistical method of iterative sampling. MEME (36) uses expectation maximization and artificial intelligence heuristics to construct an alignment. Approaches by van Helden et al. (37) and Kielbasa et al. (38) go enumeratively through the space of possible motifs and then select those that are overrepresented in the promoter sequences with respect to randomized sequences.

Alternatively, the promoters may be scanned for patterns similar to those extracted from libraries containing sequences experimentally found to be bound by transcription factors. Such libraries (Jaspar [39] and Transfac [40,41,42]) are prepared out of binding sites, for which there exists clear and direct evidence for function and identity of the bound transcription factor. Many tools offering different scoring strategies predict new candidate sites on the basis of similarity to profiles retrieved from such precompiled libraries (43,44). Clover (available at http://zlab.bu.edu/clover/) allows one to assess which, if any, of the motifs are statistically over- or underrepresented in the sequences (45). The method developed by Rahmann et al. (46) provides a statistically well-founded method to calculate score thresholds optimized for detection of true positive binding sites.

The rate of transcription factor binding site predictions varies for different binding models and their parameters, but typically, a candidate site is reported every 500–5,000 bp (47) for each studied transcription factor. Consequently, concepts utilizing various properties of regulatory mechanisms have been developed to improve specificity of the predictions. Bussemaker et al. (48) and Caselle et al. (49) correlate the presence of binding sites with gene expression levels. Wagner et al. (50) proposes to detect closely spaced binding sites of the same transcription factor, based on the observation that many transcription factors show cooperativity in transcriptional activation. The algorithms of Pilpel et al. (51), Frith et al. (52,53), and Murakami et al. (54) score overrepresented close occurrences of binding sites recognized by different transcription factors. Experimental evidence for such interactions is also collected in a dedicated database (55). Moreover, phylogenetic footprinting, which is preferential conservation of functional sequences over the course of evolution by selective pressure, results in a striking enrichment of regulatory sites among the conserved regions (56,57). These lines were followed to combine the knowledge of co-regulation among different genes and conservation among orthologous genes to improve the identification of binding sites (58,59). Another method reducing the number of false-positive predictions associates cooperative binding of transcription factors with biological functions of the corresponding genes using GO and Gossip (25,26). Finally, it should be noted that the profiles recognized by transcription factors that are available in the libraries are often similar to each other, leading to overlapping binding site predictions. Therefore, the predictions are easier to interpret when only independent profiles are selected for searching (60,61).

6. Conclusions

This chapter discusses several tools that help systems biologists to handle and analyze gene groups from different high-throughput experiments and databases. Figure 8 shows a typical analysis pipeline that can be set up to generate hypotheses from given gene groups. First, the gene groups represented by diverse identifiers have to be matched to identifiers that uniquely represent one gene. We discussed in the first part of this chapter that UniGene and Entrez Gene are such identifiers. We also introduced tools like HomGL (5) to perform this mapping. These tools also facilitate the comparison of different gene groups, allowing us to find common genes.

These gene groups can then be further analyzed. With the systematic functional annotation provided by the GO, it is possible to perform automated functional profiling of gene groups. We discussed statistical frameworks that test whether functions, processes, or locations defined

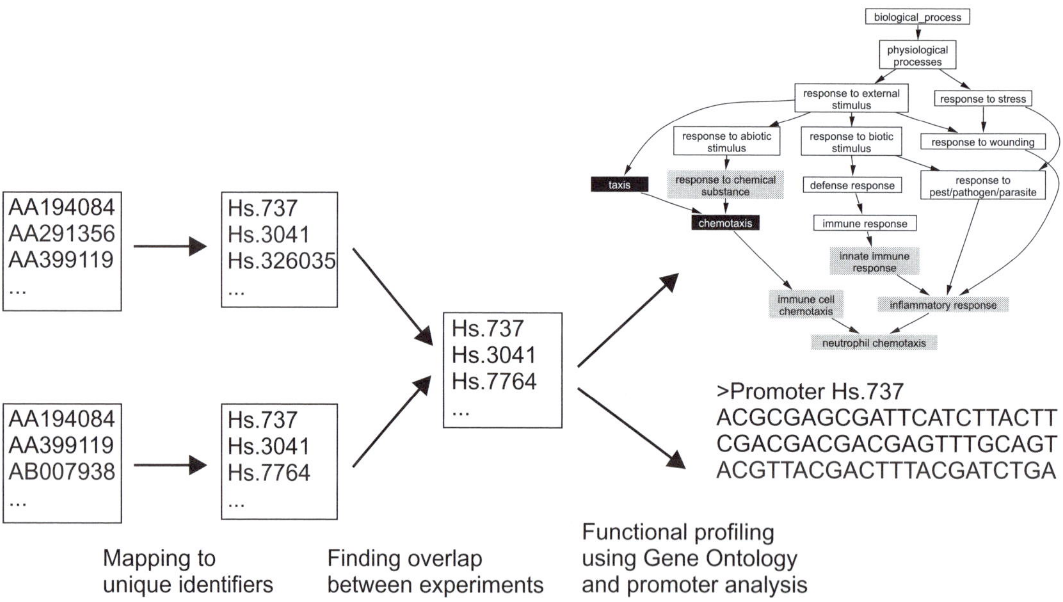

Figure 8. Suggested pipeline to analyze gene groups from high-throughput experiments. Gene groups from different experiments are to be mapped to unique identifiers, such as UniGene or Entrez Gene. Subsequently, these lists can be compared to find common/consistent entries. These lists can then be further analyzed functionally using GO, and their promoter sequences can be analyzed.

by the GO are significantly enriched within a gene group when compared to a reference group. To avoid misleading results, multiple testing issues must be taken into account. We discussed tools for performing automatic functional analysis, and hinted at the pitfalls and useful combinations of this method with other bioinformatics methods like sequence alignment and promoter analysis.

Finally, promoters of coexpressed genes might be studied to identify transcription factors and their binding sites, which might explain gene co-regulation. Although several methods are available to extract promoter sequences and perform a search for transcription factor binding sites, the results are still very error prone and dominated by false predictions. Thus, although many tools exist to analyze promoter regions, the refinement of the methods is still an active research topic. Although the result should be interpreted with care, the functional analysis of gene groups and the analysis of promoter sequences may collectively facilitate the generation of new hypotheses from genome-wide experiments.

References

1. Benson D, Karsch-Mizrachi I, Lipman D, et al. GenBank. *Nucleic Acids Res* 2005;33:D34–38.
2. Wheeler D, Barrett T, Benson D, et al. Database resources of the National Center for Biotechnology Information. *Nucleic Acids Res* 2005;33:D39–45.
3. Boeckmann B, Bairoch A, Apweiler R, et al. The SWISS-PROT protein knowledge base and its supplement TrEMBL in 2003. *Nucleic Acids Res* 2003;31:365–370.
4. Maglott D, Ostell J, Pruitt K, Tatusova T. Entrez Gene: gene-centered information at NCBI. *Nucleic Acids Res* 2005;33:D54–58.
5. Blüthgen N, Kielbasa SM, Cajavec B, Herzel H. HOMGL-comparing genelists across species and with different accession numbers. *Bioinformatics* 2004;20:125–126.
6. Tullai JW, Schaffer ME, Mullenbrock S, et al. Identification of transcription factor binding sites upstream of human genes regulated by the phosphatidylinositol 3-kinase and MEK/ERK signaling pathways. *J Biol Chem* 2004;279:20167–20177.
7. Cheung K, Hager J, Pan D, et al. KARMA: a web server application for comparing and annotating heterogeneous microarray platforms. *Nucleic Acids Res* 2004;32:W441–444.
8. Veldhoven A, de Lange D, Smid M, et al. Storing, linking, and mining microarray databases using SRS. BMC *Bioinformatics* 2005;6:192.
9. Tsai J, Sultana R, Lee Y, et al. RESOURCERER: a database for annotating and linking microarray resources within and across species. *Genome Biol* 2001;2:SOFTWARE0002.
10. Wang P, Ding F, Chiang H, et al. ProbeMatchDB—a web database for finding equivalent probes across microarray platforms and species. *Bioinformatics* 2002;18:488–489.
11. Ashburner M, Ball CA, Blake JA, et al. Gene Ontology: tool for the unification of biology. The Gene Ontology Consortium. *Nat Genet* 2000;25: 25–29.
12. Bard JB, Rhee SY. Ontologies in biology: design, applications and future challenges. *Nat Rev Genet* 2004;5:213–222.
13. Dudoit S, Shaffer JP, Boldrick JC. Multiple hypothesis testing in microarray experiments. *Stat Sci* 2003;18:71–103.

14. Blüthgen N, Brand K, Cajavec B, et al. Biological profiling of gene groups utilizing gene ontology. *Genome Inform* 2005;16:106–115.
15. Draghici S, Khatri P, Martins RP, et al. Global functional profiling of gene expression. *Genomics* 2003;81:98–104.
16. Hosack DA, Dennis G Jr, Sherman BT, et al. Identifying biological themes within lists of genes with EASE. *Genome Biol* 2003;4:R70.
17. Dennis G, Sherman BT, Hosack DA, et al. DAVID: Database for Annotation, Visualization, and Integrated Discovery. *Genome Biol* 2003;4:P3.
18. Zhong S, Li C, Wong WH. ChipInfo: Software for extracting gene annotation and gene ontology information for microarray analysis. *Nucleic Acids Res* 2003;31:3483–3486.
19. Feng W, Wang G, Zeeberg B, et al. Development of gene ontology tool for biological interpretation of genomic and proteomic data. *AMIA Annu Symp Proc* 2003;839.
20. Castillo-Davis CI, Hartl DL. GeneMerge–post-genomic analysis, data mining, and hypothesis testing. *Bioinformatics* 2003;19:891–892.
21. Al-Shahrour F, Diaz-Uriarte R, Dopazo J. FatiGO: a web tool for finding significant associations of Gene Ontology terms with groups of genes. *Bioinformatics* 2004;20:578–580.
22. Beissbarth T, Speed TP. GOstat: find statistically overrepresented Gene Ontologies within a group of genes. *Bioinformatics* 2004;20:1464–1465.
23. Maere S, Heymans K, Kuiper M. BiNGO: a Cytoscape plug-in to assess overrepresentation of gene ontology categories in biological networks. *Bioinformatics* 2005;21:3448–3449.
24. Conesa A, Gotz S, Garcia-Gomez J, et al. Blast2go: A universal tool for annotation, visualization and analysis in functional genomics research. *Bioinformatics* 2005;21:3674–3676.
25. Kielbasa S, Blüthgen N, Herzel H. Genome-wide analysis of functions regulated by sets of transcription factors. Proceedings of the German Conference on Bioinformatics. 2004;105–113.
26. Blüthgen N, Kielbasa S, Herzel H. Inferring combinatorial regulation of transcription in silico. *Nucleic Acids Res* 2005;33:272–279.
27. Wasserman W, Fickett J. Identification of regulatory regions which confer muscle-specific gene expression. *J Mol Biol* 1998;278:167–181.
28. Schmid C, Praz V, Delorenzi M, et al. The Eukaryotic Promoter Database EPD: the impact of in silico primer extension. *Nucleic Acids Res* 2004;32:D82–85.
29. Carninci P, Kasukawa T, Katayama S, et al. The transcriptional landscape of the mammalian genome. *Science* 2005;309:1559–1563.
30. Suzuki Y, Yamashita R, Sugano S, Nakai K. DBTSS, DataBase of Transcriptional Start Sites: progress report 2004. *Nucleic Acids Res* 2004;32:D78–81.
31. Birney E, Andrews D, Bevan P, et al. Ensembl 2004. *Nucleic Acids Res* 2004;32 Database issue:D468–D470.
32. Stormo G. DNA binding sites: representation and discovery. *Bioinformatics* 2000;16:16–23.
33. Lawrence CE, Altschul SF, Boguski MS, et al. Detecting subtle sequence signals: a Gibbs sampling strategy for multiple alignment. *Science* 1993;262:208–214.
34. Roth FR, Hughes JD, Estep PE, Church GM. Finding DNA regulatory motifs within unaligned noncoding sequences clustered by whole-genome mRNA quantitation. *Nature Biotechnol* 1998;16:939–945.
35. Frith M, Hansen U, Spouge J, Weng Z. Finding functional sequence elements by multiple local alignment. *Nucleic Acids Res* 2004;32:189–200.

36. Bailey TL, Elkan C. Fitting a mixture model by expectation maximisation to discover motifs in biopolymers. In: Proceedings of the International Conference on Intelligence Systems for Molecular Biology. AAAI Press; 1994: 28–36.
37. van Helden J, André B, Collado-Vides J. Extracting regulatory sites from the upstream region of yeast genes by computational analysis of oligonucleotide frequencies. *J Mol Biol* 1998;281:827–842.
38. Kielbasa S, Korbel J, Beule D, et al. Combining frequency and positional information to predict transcription factor binding sites. *Bioinformatics* 2001;17:1019–1026.
39. Sandelin A, Alkema W, Engstrom P, et al. JASPAR: an open-access database for eukaryotic transcription factor binding profiles. *Nucleic Acids Res* 2004;32 Database issue:D91–D94.
40. Wingender E, Dietze P, Karas H, Knuppel R. TRANSFAC: a database on transcription factors and their DNA binding sites. *Nucleic Acids Res* 1996;24: 238–241.
41. Wingender E, Chen X, Hehl R, et al. TRANSFAC: an integrated system for gene expression regulation. *Nucleic Acids Res* 2000;28:316–319.
42. Matys V, Fricke E, Geffers R, et al. TRANSFAC: transcriptional regulation, from patterns to profiles. *Nucleic Acids Res* 2003;31:374–378.
43. Quandt K, Frech K, Karas H, et al. MatInd and MatInspector: new fast and versatile tools for detection of consensus matches in nucleotide sequence data. *Nucleic Acids Res* 1995;23:4878–4884.
44. Kel A, Gossling E, Reuter I, et al. MATCH: A tool for searching transcription factor binding sites in DNA sequences. *Nucleic Acids Res* 2003;31:3576–3579.
45. Frith M, Fu Y, Yu L, et al. Detection of functional DNA motifs via statistical over-representation. *Nucleic Acids Res* 2004;32:1372–1381.
46. Rahmann S, Müller T, Vingron M. On the power of profiles for transcription factor binding site detection. *Stat Appl Genet Mol Biol* 2003;2:7.
47. Wasserman W, Sandelin A. Applied bioinformatics for the identification of regulatory elements. *Nat Rev Genet* 2004;5:276–287.
48. Bussemaker H, Li H, Siggia E. Regulatory element detection using correlation with expression. *Nat Genet* 2001;27:167–171.
49. Caselle M, Di Cunto F, Provero P. Correlating overrepresented upstream motifs to gene expression: a computational approach to regulatory element discovery in eukaryotes. *BMC Bioinformatics* 2002;3:7.
50. Wagner A. A computational genomics approach to the identification of gene networks. *Nucleic Acids Res* 1997;25:3594–3604.
51. Pilpel Y, Sudarsanam P, Church G. Identifying regulatory networks by combinatorial analysis of promoter elements. *Nat Genet* 2001;29:153–159.
52. Frith M, Spouge J, Hansen U, Weng Z. Statistical significance of clusters of motifs represented by position specific scoring matrices in nucleotide sequences. *Nucleic Acids Res* 2002;30:3214–3224.
53. Frith M, Li M, Weng Z. Cluster-Buster: Finding dense clusters of motifs in DNA sequences. *Nucleic Acids Res* 2003;31:3666–3668.
54. Murakami K, Kojima T, Sakaki Y. Assessment of clusters of transcription factor binding sites in relationship to human promoter, CpG islands and gene expression. *BMC Genomics* 2004;5:16.
55. Kel-Margoulis O, Romashchenko A, Kolchanov N, et al. COMPEL: a database on composite regulatory elements providing combinatorial transcriptional regulation. *Nucleic Acids Res* 2000;28:311–315.
56. Dieterich C, Cusack B, Wang H, et al. Annotating regulatory DNA based on man-mouse genomic comparison. *Bioinformatics* 2002;18 Suppl 2 S84–S90.

57. Wasserman W, Palumbo M, Thompson W, et al. Human-mouse genome comparisons to locate regulatory sites. *Nat Genet* 2000;26:225–228.
58. Wang T, Stormo G. Combining phylogenetic data with co-regulated genes to identify regulatory motifs. *Bioinformatics* 2003;19:2369–2380.
59. Lenhard B, Sandelin A, Mendoza L, et al. Identification of conserved regulatory elements by comparative genome analysis. *J Biol* 2003;2:13.
60. Roepcke S, Grossmann S, Rahmann S, Vingron M. T-Reg Comparator: an analysis tool for the comparison of position weight matrices. *Nucleic Acids Res* 2005;33:W438–441.
61. Kielbasa S, Gonze D, Herzel H. Measuring similarities between transcription factor binding sites. *BMC Bioinformatics* 2005;6:237.

5

The Dynamic Transcriptome of Mice

Yuki Hasegawa and Yoshihide Hayashizaki

Summary

The mouse transcriptome was comprehensively analyzed by the collection of mouse full-length cDNA clones. With the development of several technologies, the international Functional Annotation Of Mouse (FANTOM) sketched the outlines of the transcriptional framework, finding an unexpectedly large number of noncoding and sense/antisense (S/AS)-pairing RNAs, which was called a "novel RNA continent." In this chapter, the mouse encyclopedia project and the complexity of the mouse transcriptome will be introduced to give you a new view of the transcriptome world.

Key Words: FANTOM; transcriptome; ncRNA; promoter; sense/antisense; full-length cDNA (FL-cDNA).

1. Introduction

In the past decade, several mammalian genomes have been sequenced, including the genomes of human and mouse (1–3). In 2004, the Encyclopedia of DNA Elements (ENCODE) Consortium announced that only 2% of the total genome encodes proteins, which are the essential building blocks for constructing a human body, thereby implying that the genomic sequence in itself cannot explain the complexity of gene functions and mechanisms (4). Turning to the transcriptome for answers, the transcriptome consisting not only of transcripts heading for translation, but also of a large number of RNA transcripts with biological activity of their own, the first step is to find all transcripts. Transcriptome analysis is much more demanding than genome sequencing because thousands of RNA molecules must be isolated and sequenced from all tissues and from all developmental stages. To explore the functions attached to these transcripts, the mouse transcriptome has been comprehensively analyzed after creating a large collection of full-length cDNA (FL-cDNA) clones (5–7). After developing multiple technologies, including the construction of full-length mouse cDNA libraries, cDNA microarrays, and transcriptional starting site detection, the framework of the transcriptome was tentatively analyzed by the international FANTOM consortium (5–7). Previously, the FANTOM1 and FANTOM2 meetings launched an approach based on comprehensive full-length cDNA isolation, which allowed the identification of noncoding RNA (ncRNA) expression and showed that ncRNAs are more widespread than believed (5–7). After FANTOM3, which focused on transcription start-site identification, transcriptional units (TUs), and framework clustering analysis, larger numbers of ncRNAs were identified, among which more than 30,000 may function as coexpressed S/AS regulatory pairs. In September 2005, the Institute of Physical and Chemical Research (RIKEN) group, together with collaborators from 45 institutes in 11 countries, published two milestone papers to transform our understanding of the information content in the mammalian genome (5,8). In this chapter, the history of the Riken Mouse Encyclopedia will be introduced, and the discovery of a novel RNA continent through the comprehensive analysis of mammalian transcriptome will give you a new view of the genomic world.

2. Mouse Encyclopedia Project

2.1. The Choice of Mouse

To collect maximum RNA samples to construct libraries, mouse was chosen over human, due to the fact that mice have a very short life cycle and are easy for crossbreeding; various tissues from different developmental stages can be extracted easily. Because mouse and human are said to share approximately 80% of genes (2), collection of mouse full-length cDNA will provide us with an amazing view of a mammalian transcriptome close to our own.

2.2. The Transcriptome Dataset

The characterization of all transcripts present in a cell or tissue at a given time, and the mechanisms driving their expression, is the aim of transcriptome research (9). To describe the mammalian transcriptome in the Riken Mouse Encyclopedia project, we combined isolated full-length cDNA clone sequences with the new cap analysis gene expression (CAGE) data (10), the gene identification signature (GIS) (11), and gene signature cloning (GSC) ditag technologies for the identification of transcription initiation and termination sites (12). Each technology and the new landscape of the mammalian transcription will be described in the following chapters.

3. Technology Used for the Mouse cDNA Encyclopedia

3.1. Full-Length cDNA Library Construction

The mouse full-length cDNA (FL-cDNA) library has been constructed from 263 RNA samples, including samples from various developmental stages and it contains a complete copy of the mRNA population, including splice variants (13). FL-cDNAs are the starting material for the construction of the RIKEN FL-cDNA encyclopedia. Cloning of FL-cDNA inserts has previously been hampered by problems related to both the preparation and the cloning of long cDNA inserts. The first difficulty was the selection of full-length cDNAs, for several reasons; one of them is the RNA's ability to form secondary structures (Figure 1). To solve these problems, four key technologies were developed: an mRNA elongation procedure, a new selection method (13), the construction of a new vector

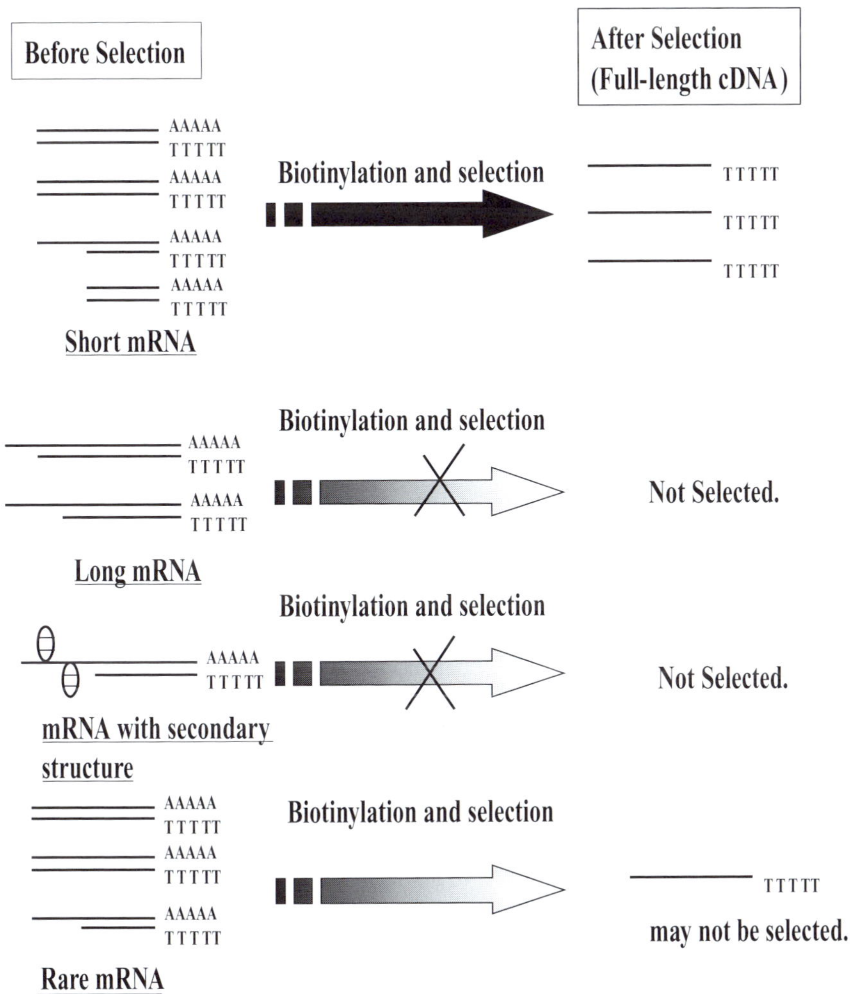

Figure 1. Commonly encountered problems in creating and selecting full-length cDNAs. Of the four groups of mRNAs in the figure, only short mRNA is efficiently converted to full-length cDNA in conventional reverse transcription reactions. Long mRNA and mRNAs rich in secondary structures are transcribed predominantly into truncated first-strand cDNAs that are discarded during the selection process. Rare mRNAs, which tend to be longer in length than highly expressed mRNAs, face the same difficulties. Most of these difficulties can be overcome by using RT together with trehalose and sorbitol.

family (14), and a combined normalization and subtraction procedure (15).

3.2. mRNA Elongation Strategies

The first difficulty faced was an inefficient synthesis of first-strand cDNA. The major obstacle to preparing high-quality cDNA libraries has been the low efficiency of reverse transcriptase (RT) to synthesize full-length cDNA because full-length cDNA tend to form strong mRNA secondary structures. To inhibit the formation of these structures, a higher reaction temperature is needed, which also requires more thermostable enzymes. So far, an isolation of thermostable enzymes has been restricted to enzymes existing in thermophilic organisms. To overcome this limited availability, we explored a new method to confer thermal stability to enzymes by using disaccharide trehalose and sorbitol, which can be used as reaction additives to stabilize or stimulate enzymatic activity at unusually high temperatures. The trehalose can also be used for thermosensitive enzymes, as though they would be thermostable. In fact, even thermoactivated RT, which displays full activity at 60°C instead of the standard 42°C (13), became powerful enough to synthesize full-length cDNA without early termination. At an increased temperature, the stability of secondary structures within single-stranded mRNA templates is decreased, allowing easier passage of RT through the knotty structures typically found in the 5′-untranslated regions of mRNAs. Thermoactivation of RT results in longer cDNAs, a higher representation of FL-cDNAs in the library, and an overall higher recovery of FL-cDNA in the subsequent cap-trapper protocol.

3.3. Avoidance of Internal Cleavage

The RT reaction contains 5-methyl-dCTPs instead of the standard dCTP, causing the first strand of cDNA to be methylated. Therefore, synthesis of the second strand in the presence of standard 4 dNTPs results in hemimethylated double-stranded cDNAs. Internal sequences of hemimethylated double-stranded cDNA are resistant to cleavage by methylation-sensitive restriction enzymes. In the cloning step, only the restriction sites located in the unmethylated primer-adapter sequences at the ends of the double stranded cDNA will be cleaved. In this technology, SstI restriction enzyme was used because it never cuts hemimethylated cDNA.

3.4. Selection of FL-cDNAs

Another difficulty was to collect only the completed sequences from the elongation procedure. The selection of FL-cDNAs is achieved by cap-trapping (13,16,17) (Figure 2). After the completed first strand cDNA synthesis, a biotin residue is chemically linked to the diol group of the cap structure located at the 5′ ends of the mammalian mRNAs, "trapping the cap." RNase I (a single-strand–specific nuclease that cleaves the phosphodiester bond between any two ribonucleotides in exposed

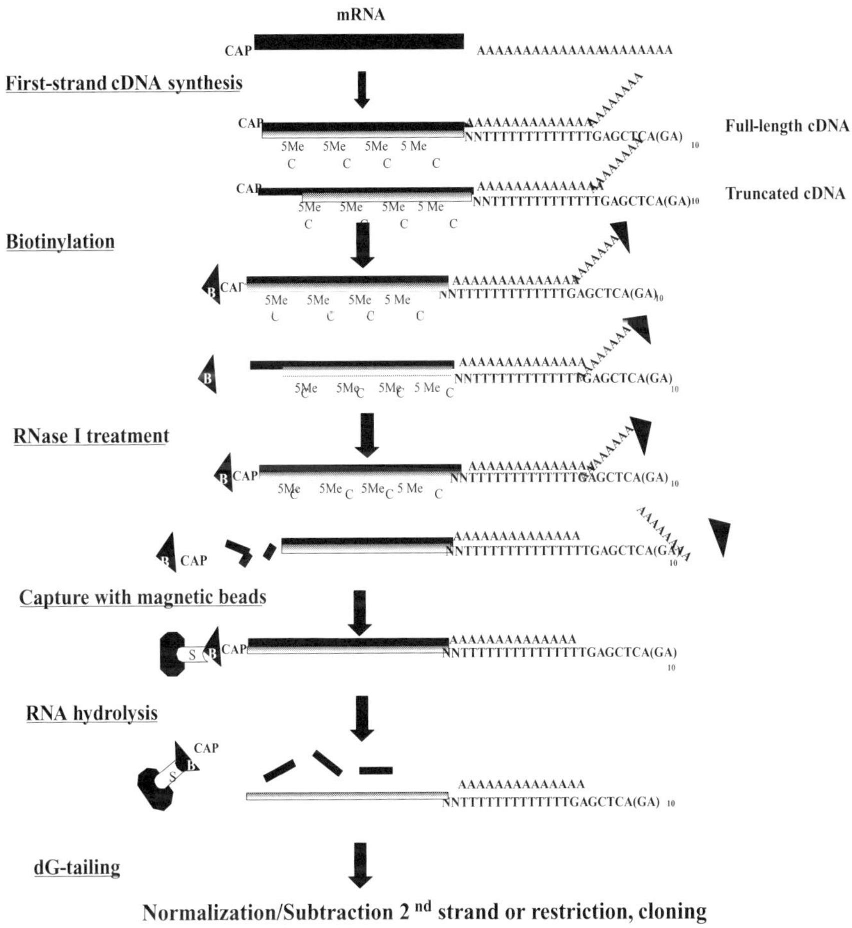

Figure 2. CAP trapper technology. After single-stranded cDNA synthesis, CAP and polyA sites are labeled. RNase I is used to cleave remaining single-stranded, uncompleted RNA, while DNA-RNA hybrids survive due to their protective hemimethylation. Single-stranded RNA is cleaved by RNase I at 5′-sites and at polyA sites. The non-polyT complementary sequence is designed as a primer for the first cDNA. The biotinylated site of mRNA is preserved from RNase I cleavage by hybridization of mRNA with FL-cDNA. As a result, only mRNA hybridized with full-length single-stranded cDNA, which can be trapped by streptavidin beads, can be gained. The first-strand cDNA is then isolated and subjected to the cloning process. (Reproduced with permission from Carninci et al.)

single-stranded RNA) is then used to eliminate the biotinylated cap from incompletely synthesized cDNA–mRNA hybrids, which will be degraded into a mixture of mono- and oligonucleotides. The RNA moiety in the hybridized segment of mRNA in cDNA–mRNA hybrids is protected from digestion by hemimethylation.

The full-length cDNA–mRNA hybrids retain their caps and are then recovered by binding to streptavidin-coated magnetic beads.

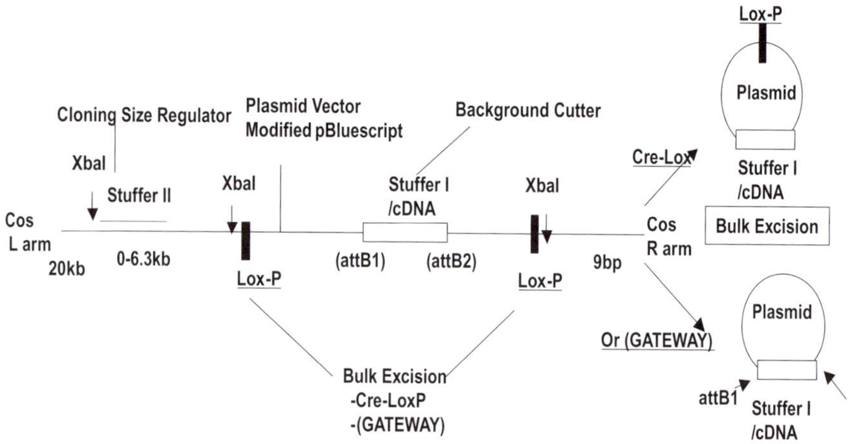

Figure 3. General scheme of the λ-FLC family of functional elements. The functional elements of the vectors are the left and right arms, the cloning size regulator (stuffer II); a plasmid derivative of pBluescript, the bulk excision elements (loxP or Gateway sequence); and the stuffer, including background-reducing and -monitoring sequences. The plasmid is excised using Cre recombinase, or only the cDNA inserts are transferred with the Gateway system. (Reproduced with permission from Carninci et al.)

3.5. Construction of a New Vector

A selection of adequate cloning vectors is a key component for a successful production of FL-cDNA library clones. FL-cDNA library often exhibits problems with cloning biases compared with normal library constructions because of the following reasons:

1. Longer mRNA is more susceptible to damage.
2. Shorter FL-cDNA is easier to synthesize than longer cDNAs, even under higher temperature with trehalose and sorbitol.
3. Commercially available vectors prefer the cloning of shorter cDNA, especially in plasmid vectors.

To overcome these problems, a new class of cloning vectors was developed: λ-full-length cDNA (λ-FLC) cloning vectors (14) (Figure 3). These vectors can be bulk-excised for the preparing of FL-cDNA libraries in which a high proportion of the plasmids carry large inserts that can be transferred into other vectors. By using λ vectors, long cDNA clones can be cloned without size bias.

3.6. Subtraction and Normalization Technology

Many of the problems in discovering new genes stem from the differences in expression levels between mRNAs in different mammalian cells. Largely abundant mRNA disguises the presence of rare transcripts. By enhancing the efficiency of single-pass sequencing (minimum number of sequencing/maximum number of different transcripts), by reducing the prevalence of superabundant and intermediately expressed cDNAs, the rare transcripts can be trapped. The normalization and subtraction technologies allow the construction of a library without any bias toward abundant transcripts inherent in the original mRNA population (15). Further improvement can be achieved by subtracting unwanted cDNAs from the final library. Both normalization and subtraction are accomplished by carefully controlled hybridization of first-strand cDNA to RNA drivers. For normalization, the driver is a biotinylated aliquot of the RNA initially used as a template for cDNA synthesis; for subtraction, the driver consists of a biotinylated RNA population prepared by in vitro transcription of well-characterized sets of cDNA clones. The hybrids formed between the abundant and unwanted cDNAs are then easily subtracted with another probe. The normalization procedure almost doubled the gene discovery compared with nonnormalized cDNA libraries, and this is further improved for the libraries created with a combination of normalization and subtraction technologies.

3.7. High-Throughput Sequence Analysis System: Riken Integrated Sequencing Analysis

To accomplish the huge number of sequencing required for the Mouse cDNA Encyclopedia Project, a high-throughput sequencing system was developed. The Riken Integrated Sequencing Analysis (RISA) system (18) was constructed, which consists of 384-multicapillary auto sequencer, a 384-multicapillary array assembler, and a 384-multicapillary casting device. The RISA sequencer is capable of simultaneously analyzing 384 independent sequencing products. RISA system can be used with any fluorescent-labeled sequencing reaction. RISA could read 99.2% of the mouse FL-cDNA sequences successfully and an average 654.4-bp could be read with more than 99% accuracy. RISA also contains an incubator, a plasmid preparator, a RISA filtrator, a densitometer, and a high-throughput thermal cycler. To complete the sequencing of all cDNAs from various tissues in a few years, sequencing speed had to be accelerated. With 16 RISA sequencers, it has become possible to process 50,000 DNA samples per day.

3.8. New Distribution Method for Transcriptome Resources: The DNA Book

As described in the previous section, the full-length cDNA clone bank and the database will be the major platform resource for postgenomic and posttranscriptome analysis. However, because the storage and the delivery system will be a huge logistical problem for a total of 2 million clones, new technology was developed to handle this number of physical clone samples; the "DNABook" (19,20). The DNABook is basically a book where the physical clones are printed on paper sheets, thereby enabling shipping in the form of a DNA book. The DNA printing was tested in various conditions, including humidity, temperature, and resistance to touching and/or scratching. Once a user receives the encyclopedia DNABook, clones can be amplified by PCR by simply punching out DNA spots on the paper, then dissolving the paper in the reaction mixture, and the clones can be used for further analysis in 2 hours. This

DNA book will be the new distribution and storage method for clone resources produced by large-scale transcriptome analysis.

3.9. Full-Length cDNA Microarrays

The mouse full-length cDNA microarray was developed for validation of the reproducible expression profile of mouse full-length cDNA encyclopedia. At the FANTOM1 meeting, microarrays with 18,816 mouse cDNAs were used for expression profiling analysis for 49 adult and embryonic mouse tissues (21). Target DNAs were collected from RIKEN mouse cDNA libraries (which were constructed as described above). The experimental setup consisted of a Stanford-type arrayer with 16 tips giving a spot diameter of 100–150 µm, producing slides with 14,000 spots. In comparison to the earlier experimental time of 40 hours, the process was speeded up 6-fold by the use of the 48 pins double-headed arrayer.

cDNA microarray is a powerful tool for high-throughput analysis for expression profiling; however, it also has limitations because it can be applicable only for genes or transcripts with determined sequences.

3.10. CAGE Technology

Approaches in transcriptome research have focused on large collections of a "representative clone" for each gene. These approaches addressed neither the dynamics of transcriptional regulation nor regulatory principles like alternative promoter usage or splicing (22,23). Partial identification of the promoter sites has been provided by gene discovery programs based on the sequencing of FL-cDNA libraries (24,25). By sequencing 5′ ends from FL-cDNA libraries and mapping the sequences to the genome, several thousand promoters can be determined and corresponding coding and regulatory regions can be identified. Although the analysis can produce statistics on transcriptional start sites derived from large numbers of 5′ end sequences, these methods require a massive number of comprehensive sequencing, which is prohibitively expensive. The serial analysis of gene expression (SAGE) is based on the cloning and subsequent sequencing of concatamers of short cDNA fragments derived from 3′ end of multiple mRNAs (http://cgap.nci.nih.gov/SAGE) (26–29). Although SAGE is a relatively cheap high-throughput digital data collection technology, and it gives the information for counting transcription units (TUs), it cannot be used for identifying promoters in FL-cDNA cloning. To solve these problems, CAGE tag technology (10), which allows high-throughput identification of sequence tags corresponding to 5′-ends of mRNA at the cap sites and the identification of the transcription start site (TSS), was developed. It also allows promoter usage analysis, gene expression analysis, and obtains absolute expression values for expressed genes in a common reference sample.

CAGE libraries are based on a preparation of tags that contain the initial 20 nucleotides from mRNA 5′ ends, which are concatenated and sequenced (Figures 4 and 5). The method uses cap-trapper FL-cDNAs, to which linkers are attached in the 5′ end. This is followed by the cleav-

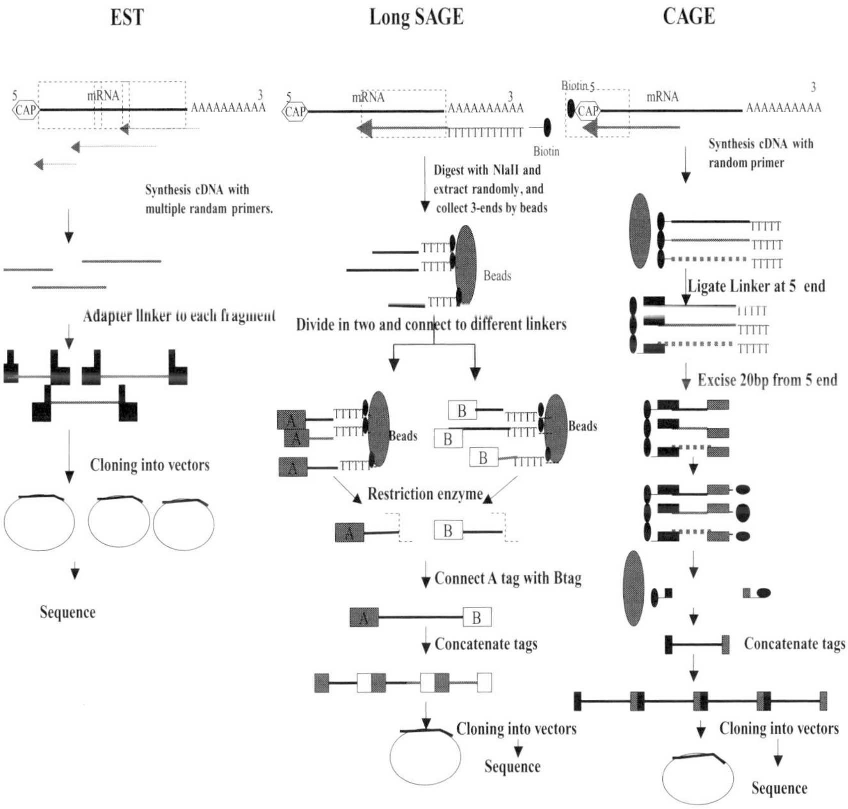

Figure 4. Comparison between EST, SAGE, and CAGE. EST technology consists of cloning and sequencing of each tag derived from random segments of mRNA into each vector, followed by a sequencing step. SAGE is based on the cloning and subsequent sequencing of concatamers of short DNA fragments derived from 3′-ends of multiple mRNAs. CAGE uses cap-trapper FL-cDNAs, to which linkers are attached in the 5′-ends. This is followed by the cleavage of 20-bp-long segments by class II restriction enzymes, PCR, concatamerization, and cloning of the CAGE tags. (Reproduced with permission from Shiraki et al.)

age of a 20-bp-long segment by class II restriction enzymes, PCR, concatamerization, and cloning of the CAGE tags. CAGE tags derived by sequencing these libraries are mapped to the genome and used for TSS and expression analysis. In addition, it can be used for the determination of the 5′ end borders of new TUs. Approximately 70% of the CAGE tags could be mapped by BLAST alignment program and a mapping strategy to the genome (5). After mapping CAGE tags, plenty of unclassifiable clones appeared. These unclassifiable tags could be confirmed by RACE experiments. The 5′-end–specific tags will become one of the most important tools in system biology in being fast, cheap, specific and by giving us information about TSS.

3.11. GIS and GSC Technologies

At the same time as the development of CAGE technology, Genomic Institute of Singapore also announced another method for promoter

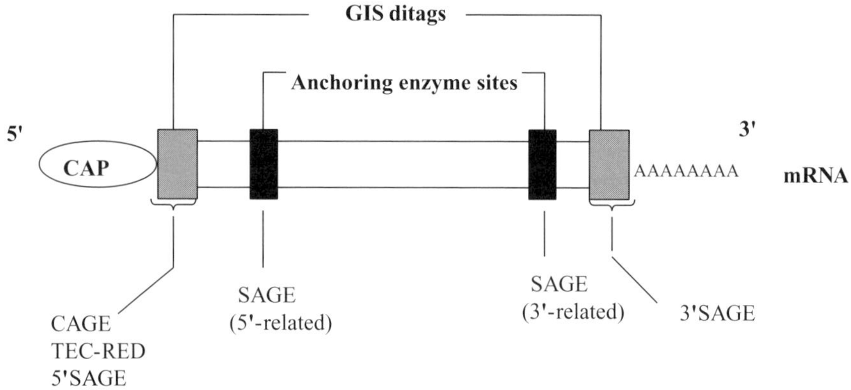

Figure 5. Usage of sequencing tags in transcript identification. Any transcript with a known sequence can be identified by a short signature sequence or tag, depending on the approach; such tags can be obtained from different regions. For SAGE, the location of the tags is determined by the recognition sequence of an anchoring enzyme; in most cases, tags close to the 3′-ends are isolated. Alternatively, SAGE tags can also be isolated from regions close to the 5′-end. New approaches aim to isolate the true 5′- (CAGE, TEC-RED, 5′SAGE) or 3′-ends (3′SAGE), including the option to clone 5′- and 3′-ends into ditags (GIS)(as indicated in light gray). (Reproduced with permission from Harbers et al.)

identification analysis, GIS (11). GIS contains not only 5′ end tags, but also contains 3′ end tags. GIS and GSC combine their 3′ and 5′ tags into ditags. By combining both ends of a cDNA into one, ditags could overcome the limitations of the separate preparation of 5′-SAGE and 3′-SAGE libraries from the same sample. However, one of the limitations in GIS mapping is that it cannot allow for the direct identification of trans-spliced transcripts in the initial mapping. Additional experiments are needed for the amplification of the corresponding cDNA inserts required for the annotation of unmapped ditags. New class IIs restriction enzymes will greatly help in the preparation of longer tags. The only difference between GIS and GSC is that the latter use subtracted libraries. Subtraction libraries can provide more data on rare transcripts and a combination of 5′-end cDNA arrays and CAGE tags will provide us with new views of expression profiling.

4. FANTOM

4.1. FANTOM1 and 2

After the collection of FL-cDNA data resources was started, the data was clustered into groups with 5′ end and 3′ end sequence homology. The full-length sequenced clones were analyzed according to motif and domain, and annotated to reveal the transcriptome complexity. This annotation process needs not only computational power but also human effort for the curation process. FANTOM was the name for the project in which molecular biologists from various countries and institutes were called on to annotate cDNAs with the use of various databases. The

FL-cDNA libraries were integrated into a database, where clustering of the sequenced full-length cDNAs into a nonredundant and comprehensive set provided a platform for functional analyses of the transcriptome and proteome. However, manual curation to identify truncated transcripts and inappropriate clustering of closely related sequences was still required. The Representative Transcript and Protein Sets (RTPS) pipeline (Figure 6) was designed (30) and introduced to automate and make the comprehensive analysis faster. On the FANTOM2, the RTPS provided the framework of global transcriptome analysis on mouse genome, based on available sequences information, such as reference sequences (refseqs) and expressed sequence tags (ESTs) at a single point at one time (31). However, it was not enough to reflect the comprehensive mouse transcriptome, such as tissue differences or developmental stages, and the manual curation was not fast enough to catch up with the frequent data update. After the FANTOM2, the RTPS pipelines were redesigned into a fully automated search engine, including other species, such as human and rat (30).

Before the discovery of the novel RNA continent in 2005, there were three major activities organized by the FANTOM consortium. In

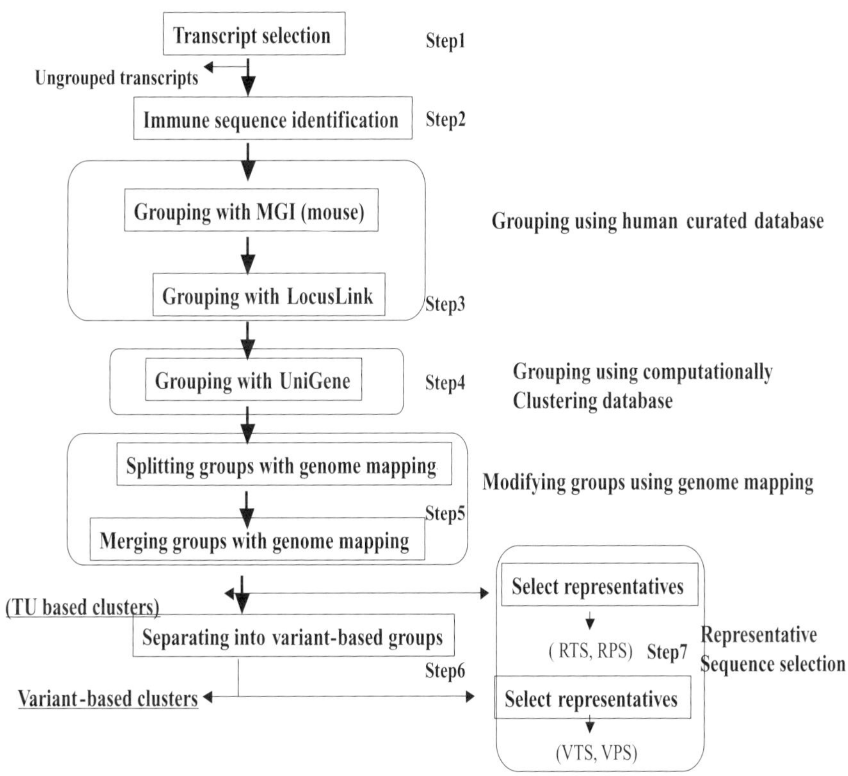

Figure 6. RTPS pipeline. Representative sequences are selected from the RTPS pipeline by a series of steps including transcript selection, clustering based on genome mapping and splicing patterns and are then merged based on grouping by curated databases, computational clustered databases, and genome mapping. (Reproduced with permission from Kasukawa et al.)

FANTOM1 in 2000, approximately 12,000 of FL-cDNAs were annotated, and the number of genes in mice was estimated. The FL-cDNA collection continued after FANTOM1, and the second meeting took place in 2002. In FANTOM2, more than 60,000 FL-cDNAs were analyzed, and 40,000 newly collected clones were annotated. FANTOM2 also contributed data on the existence of ncRNA and a large number of alternative splicing events.

4.2. FANTOM 3

After FANTOM2, an additional 40,000 FL-cDNAs were sequenced, and the number of whole FL-cDNAs (in which 5′- and 3′-end–sequenced cDNAs were included) reached 2 million. In FANTOM3, the functional analysis was accelerated because CAGE data were added. The CAGE tags provided TSS sequences, and GIS and GSC contributed with the transcriptional framework. More than 180,000 CAGE tags were mapped as transcription starting sites, of which 91 were confirmed to indicate true TSS. All data from the FANTOM3 can be accessed at http://fantom3.gsc.riken.jp/db/ and http://www.ddbj.nig.ac.jp (Table 1).

4.2.1. The Novel RNA Continent: New Definitions of Vocabulary

After comprehensive analyzing of the transcriptome, the definition of "gene" needs to be changed. The following three words have been newly defined (Figure 7).

- Transcriptional Forest (TF): A genomic region in which either strand will be the target of transcription as an mRNA precursor.
- Transcriptional Desert (TD): A genomic region in which none of the strands will be the object for transcription.
- Transcriptional Unit (TU): The complex of exon regions in which exons overlap more than 1 bp on the same strand as a group. This definition is closest to the original concept of a gene.
- Transcriptional Framework (TK): The grouping of transcripts that share common expressed regions, as well as splicing events, termination events, or TSS.

Table 1. DATA set resources.

Data Sources	TOTAL	Number of Libraries	Safely mapped
Riken full-length cDNAs	102,801	237	100,303
Public(non-RIKEN) mRNAs	56,009		52,119
CAGE tags (mouse)	11,567,973	145	7,151,511
CAGE tags (human)	5,992,395	24	3,106,472
GIS ditags	385,797	4	118,594
GSC ditags	2,079,652	4	968,201
RIKEN 5 ESTs	722,642	266	607,462
RIKEN 3 ESTs	1,578,610	265	907,007
5/3 pairs of RIKEN cDNA	448,956	264	277,702

This table represents the summary of the total data sources used for FANTOM3. Reproduced with permission from Reference 5.

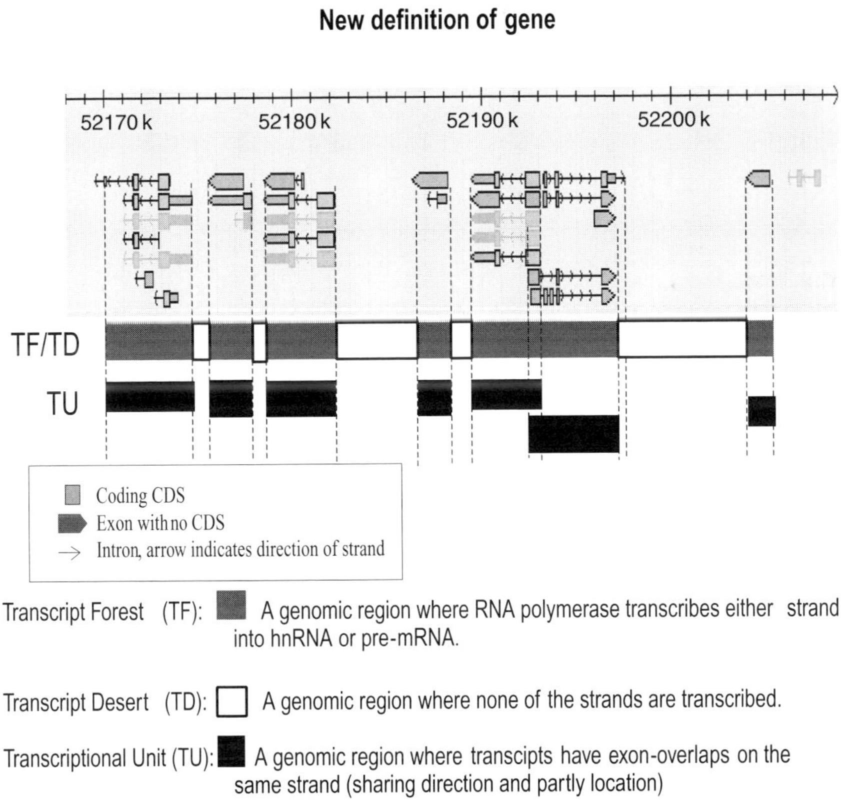

Figure 7. New definition of "gene." The upper part of the figure shown here is a part of genomic element viewer in FANTOM DB. The numbers at the top of the figure indicate the base position in the genome. The squares indicate coding sequences (CDS) and the arrow shapes indicate exons, which does not have CDS. The black arrow lines indicate introns, the direction of the arrow indicates the direction of the strand. At the lower part of the figure, the gray boxes indicate TF, the light gray boxes between TFs indicate TD, and the black boxes indicate TUs.

Analyses with this new vocabulary have disclosed that unexpected large regions of the genome are transcribed into RNA; approximately 44,000 TUs, which is more than 70% of the genome. Of these TUs, more than 20,000 were found as ncRNA (5).

4.2.2. Decrease in TU Numbers

The CAGE, GIS, and GSC analysis showed many cases of TU fusion, in which unrelated and differently annotated transcripts can be joined to make a TK, and the TK can be clustered together into TFs. The total number of TFs (with GSC data) is 18,461, and they encompass 62.5% of the mouse genome. By this grouping, the number of TUs experienced a decline; however, subsequent analysis showed that on the contrary, an increase (65%) in TU numbers caused by alternative promoters, polyadenylation sites, and a higher frequency of alternative splicing occurred.

4.2.3. Functional RNA Research

Numerous reports of functional analysis on RNAs have been published in 2005, including studies stating that micro RNAs have important regulatory functions in malignant alterations and/or inhibition of translational initiation (32–36). Exploring the role and diversity of these numerous ncRNAs now constitutes a main challenge in transcription research.

4.2.3.1. ncRNA: One of the most amazing discoveries made after the genome wide transcriptome analysis, is the finding of large numbers of ncRNA. It had been said that there were only approximately 100 ncRNAs in mice; however, more than 23,000 ncRNAs (which is approximately 43% of total genes) were found. These ncRNAs may have critical roles in gene regulation. Interestingly, approximately 60% of these ncRNAs form S/AS pairs (Figure 8). One notable observation was the conservation of promoter sequence between human and mice in these

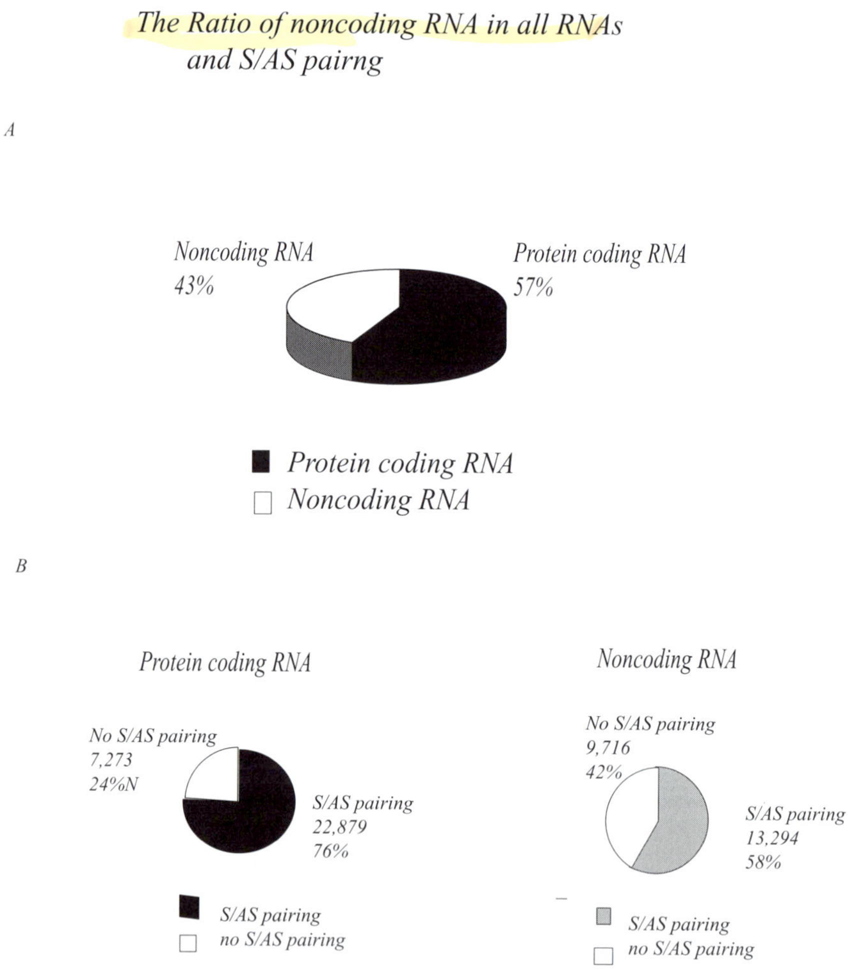

Figure 8. The ratio of ncRNA in all RNA and the ratio of S/AS pairing. (A) In all RNA, except tRNA and hnRNA, protein-coding RNA consists of 53%, and ncRNA consists of 47%. (B) S/AS pair forming RNA consists of 76% in protein-coding RNA and 58% in ncRNA.

ncRNAs, although exons show no homology. This implies that ncRNA may have mechanism mediated by double-stranded RNA from S/AS pairings, and it is more important to know when and where they are expressed than the exon sequences. This data will be the basis of high-throughput comparison analysis for transcription regulation in the evolution and differentiation in mammals.

4.2.3.2. S/AS RNA: An additionally striking discovery was the fact that almost 70% of all RNA forms S/AS pairs as shown in Table 2. Antisense transcription, which is a transcription from the opposite strand to the protein-coding strand, has captured scientists' attention in biology because it may have a role in gene regulation involving RNA interference (RNAi) and gene silencing at the chromatin level by hybridization to the sense strand of DNA. Previously, mammalian transcriptome analyses suggested that as much as approximately 20% of all transcripts may contribute to S/AS pairs (37,38), and the imprinted loci are generally known to display numerous S/AS transcripts, which are selectively expressed depending on parental chromosomal origin, such as the *genus* locus (39). However, global genome-wide analysis in FANTOM3 demonstrated that paired S/AS expression is not restricted to imprinted loci. This S/AS pairing phenomenon occurs universally in the whole genome, especially in genes that function in cell cycling, transporters, cell death, interleukin, cell structure, adhesion, phosphatase, and ubiquitination. S/AS pairs were found unevenly distributed across the genome; some chromosomes (4,17) showed a lower overall S/AS pair density. S/AS hybrids can also provide the templates for transcript cleavage involving the enzyme Dicer, which forms the molecular basis for RNAi. However, knockout or overexpression analysis showed that RNAi mechanism is not enough to explain the regulation of S/AS pairs. In FANTOM3, CAGE tag frequency data represents a *de facto* expression-profiling approach, together with microarray analysis and S/AS pairs. That S/AS pairs were found in randomly primed CAGE libraries rather than in oligo-dT–primed CAGE libraries, suggests that some polyadenylate (polyA) minus RNA transcripts, or very long ncRNA transcripts are involved in S/AS pairs. Whether concordant or discordant RNAi regulation reflects common or divergent regulation, it will require more detailed analysis in the near future.

Table 2. Number of individual TUs showing S/AS overlap.

TU	total no. of TUs	Overlapping cDNA, Tagor tag pair. Single or multiple evidence	Multiple evidence	TUs with overlapping cDNA evidence
Coding TU	**20,714**	**18,021 (87.0%)**	**13,711 (66.2%)**	**7,223 (34.9%)**
Noncoding TU	**22,839**	**13,401 (58.7%)**	**8,593 (37.6%)**	**5,296 (23.2%)**
Total	**43,553**	**31,422 (72.1%)**	**22,304 (51.2%)**	**12,519 (28.7%)**

"Single or multiple evidence" means that at least one type of evidence was used for classification. "Multiple evidence" that at least two independent transcripts were detected. "Overlapping cDNA" indicates overlap using only the cDNA data set. Noncoding TUs do not have any coding cDNA in the cluster. Coding TUs may contain noncoding variants of coding transcripts.

Source: Reproduced with permission from Reference 8.

4.2.4. Summary of FANTOM3

In the mammalian genome, there are twice as many protein-coding genes as compared with *Drosophila* melanogaster. In September 2005, the FANTOM consortium announced that 56,722 cDNAs with new cDNA sequences were discovered, in which more than 23,000 ncRNAs were included, except ribosomal RNA and transfer RNA. This discovery of an unexpectedly high number of ncRNAs dramatically changed the current knowledge of gene transcription, which is that protein is the final functional substance coded by the genome. Together with the discovery of a novel RNA continent, this is the first step into uncharted territory. These ncRNAs, which were considered as "junk," have now been shown to have important roles in the regulation of gene expression, and to implicitly contribute to the variance and rapid evolutionary change needed to establish the variation between species. If this is true, we will have to reconsider how genetic information is stored in the genome, and how this genetic information would be treated to regulate complicated mammalian developmental stages.

5. Future Prospects

5.1. New Technology Combinations

With the discoveries of ncRNAs and S/AS pairs transcripts, the underlying concept of a gene needs to be revised, from the previous concept that genes are scattered in the genome like oases in a desert, to the new concept that "junk" RNA regions actually have functions. To understand the network of molecules connecting genes and phenotypes, the ncRNAs and S/AS pairing will give us critical information, and new research on new RNA mechanisms in gene regulation at various stages will be started. For exploring more complex transcriptomes, new types of microarrays and sequencing devices will be required. New genome science and postgenome science technologies are required to be high-throughput, but also low cost and faster. Two new powerful tools for future genome-wide transcriptome analysis will be introduced in the following sections.

5.1.1. Tiling Arrays

Affymetrix, Inc. pioneered whole-genome tiling array technology to interrogate the genome at resolutions approaching every nucleotide. Tiling arrays have high-density oligonucleotide probes spanning the entire genome, and are a new type of microarray that gives large-scale data on transcriptional binding sites and splicing variants (40–43).

Uses of genome tiling arrays are:

1. Mapping transcription
2. Detection of transcription factors
3. Detection of transcripts bound by RNA-binding proteins
4. Detection of chromatin modification sites
5. Detection of DNA methylation sites
6. Detection of chromosomal origins of replications

In 2004, the binding sites for several transcription factors were determined by using the approach of chromatin immunoprecipitation with tiling arrays (41,42). However, the tiling array itself has several shortcomings, such as its inability to show precise gene structures or detect splicing junctions. On the other hand, FL-cDNA provides clear connections of how exons form a transcript; the combination of these techniques with additional knowledge will provide a more dynamic view of the transcriptome.

5.1.2. $1,000 Genome Technologies

In addition to the modification of microarrays, there are many more new high-throughput sequencing technologies. In 2004, the National Institutes of Health (NIH) published a request for applications for grants to develop low-cost genome-sequencing technologies. Sequencing an entire mammalian-sized genome currently costs between $10 million and $50 million. However, in the next 10 years, this number can be reduced by four orders of magnitude, with the ultimate goal being a $1000 genome. To achieve this goal, nanopore or microchannel approaches need to be developed. Companies like Helicos Bioscience Corporation (http://www.helicosbio.com/), VisiGen Biotechnologies (http://www.visigenbio.com) and 454 Life Sciences (http://www.454.com/), have developed revolutionizing single-molecule sequencing devices to reach the $1000 human genome goal. Helicos is currently developing an instrument for individual molecule sequencing of DNA or RNA. The new technology uses modified bases, which are attached to fluorescent chemicals (fluorophores) to visualize the synthesis process in sequencing reactions on a glass chip. When the laser light of a specific color falls upon a fluorophore, it will emit an intense light of another color, which can be detected by using a sensitive digital camera connected to a microscope. VisiGen is engineering DNA polymerases and nucleotide triphosphates to function as direct molecular sensors of DNA base identity. With this technology, VisiGen can read 1 Mbp/s, enabling 3,600 Mbp to be read in 1 hour. Both technologies have neither PCR amplification nor a cloning step, which allows analyses with higher density, low cost, and high-throughput. 454 Life Sciences technology has developed a sequencing system that involves a pyrosequencing method that has the potential to perform sequencing 100 times faster than conventional sequencing machines (44). This system can produce up to 20 million sequences per day, which means a whole genome can be read per run. Although these three technologies contain one shortage, which is that they cannot have 1:1 correspondence between substrate clones and the sequence trace because they are based on shotgun sequencing methods, the throughput and efficiency is comparably high, and even larger genomes can be sequenced in a few days with a minimum of personnel. These new technologies will greatly contribute to high-throughput transcriptome analysis in the future.

5.2. Achieving Systematic Genome Network Analysis

The understanding of the genome network is the first goal of molecular biology; this will be a great challenge for all life science researchers.

Large-scale scientific projects comprehensively cover each molecule or functional target in all biological fields, and can thus be termed "horizontal basic science." So far, research accomplished by small groups focusing on individual biological phenomena can be termed "vertical point science." The establishment of a genome structure and function database will drastically accelerate vertical point science; connecting phenotypes and genes will become much easier and faster. The horizontal basic scientific projects have enabled vertical point research groups to avoid tedious and time consuming experiments, and instead allowed them to choose candidate genes for their target molecules from genome structure data and functional databases. The first milestone in integrating horizontal and vertical research is the comprehensive molecular understanding of the relationship between genes and phenotypes. Such relationships have included those between a disease and its causative genes, a drug's effect and its target molecule. This complicated and extensive genome network needs to be analyzed. Vertical point research groups will work on the mechanisms of biological phenomena using the information contained in the database created by horizontal basic science research groups. The results of this vertical point research will then be unified to form a genome network database. A systematic collaborative framework for combining vertical and horizontal research will facilitate genome network research. Groups analyzing each biological phenomenon will have access to genome-wide screening; it will be as if a genome center were located within each small group. The key issue for life science research in the 21st century will be the establishment of this framework of collaborative research. Such collaboration will utilize and expand both the genome structure and function database and the genome network database (45).

6. Conclusions

The Mouse cDNA encyclopedia has revealed a novel RNA continent and showed a greater complexity than we expected. The result of more than 100,000 clones derived from transcriptome analysis opened the gate for a new RNA functional and proteome analysis research field. New technologies and databases will give us the keys to the mammalian mystery world.

Acknowledgments: We thank the Laboratory for Genome Exploration Research Group and all FANTOM consortium members for data production and analysis. We also thank Ann Karlsson for critical reviewing on the writing. This research was mainly supported by Research Grant for the Genome Network Project from the Ministry of Education, Culture, Sports, Science and Technology of Japan (MEXT), the RIKEN Genome Exploration Research Project from MEXT (to Yoshihide Hayashizaki), Advanced and Innovational Research Program in Life Science (to Yoshihide Hayashizaki), National Project on Protein Structural and Functional Analysis from MECT (to Yoshihide Hayashizaki), and a grant for Intersystem Collaboration of RIKEN (to Yoshihide

Hayashizaki). All the sequences (CAGE and cDNAs) are available through the DNA Data Bank of Japan (DDBJ) and other public databases. The FANTOM3 cDNAs are available through Yoshihide Hayashizaki (e-mail: yosihide@gsc.riken.jp)

References

1. Lander ES, Linton LM, Birren B, et al. Initial sequencing and analysis of the human genome. *Nature* 2001;409(6822):860–921.
2. Venter JC, Adams MD, Myers EW, et al. The sequence of the human genome. *Science* 2001;291(5507):1304–1351.
3. Waterston RH, Lindblad-Toh K, Birney E, et al. Initial sequencing and comparative analysis of the mouse genome. *Nature* 2002;420(6915):520–562.
4. Consortium TEP. The ENCODE (ENCyclopedia Of DNA Elements) Project. *Science* 2004;306(5696):636–640.
5. Carninci P, Kasukawa T, Katayama S, et al. The transcriptional landscape of the mammalian genome. *Science* 2005;309(5740):1559–1563.
6. Kawai J, Shinagawa A, Shibata K, et al. Functional annotation of a full-length mouse cDNA collection. *Nature* 2001;409(6821):685–690.
7. Okazaki Y, Furuno M, Kasukawa T, et al. Analysis of the mouse transcriptome based on functional annotation of 60,770 full-length cDNAs. *Nature* 2002;420(6915):563–573.
8. Katayama S, Tomaru Y, Kasukawa T, et al. Antisense transcription in the mammalian transcriptome. *Science* 2005;309(5740):1564–1566.
9. Ruan Y, Le Ber P, Ng HH, Liu ET. Interrogating the transcriptome. *Trends Biotechnol* 2004;22(1):23–30.
10. Shiraki T, Kondo S, Katayama S, et al. Cap analysis gene expression for high-throughput analysis of transcriptional starting point and identification of promoter usage. *Proc Natl Acad Sci USA* 2003;100(26):15776–15781.
11. Ng P, Wei CL, Sung WK, et al. Gene identification signature (GIS) analysis for transcriptome characterization and genome annotation. *Nat Methods* 2005;2(2):105–111.
12. Harbers M, Carninci P. Tag-based approaches for transcriptome research and genome annotation. *Nat Methods* 2005;2(7):495–502.
13. Carninci P, Westover A, Nishiyama Y, et al. High efficiency selection of full-length cDNA by improved biotinylated cap trapper. *DNA Res* 1997;4(1):61–66.
14. Carninci P, Shibata Y, Hayatsu N, et al. Balanced-size and long size cloning of full-length, cap-trapped cDNAs into vectors of the novel λ-FLC family allows enhanced gene discovery rate and functional analysis. *Genomics* 2001;77(1–2):79–90.
15. Carninci P, Shibata Y, Hayatsu N, et al. Normalization and subtraction of cap-trapper-selected cDNAs to prepare full-length cDNA libraries for rapid discovery of new genes. *Genome Res* 2000;10(10):1617–1630.
16. Carninci P, Shiraki T, Mizuno Y, Muramatsu M, Hayashizaki Y. Extra-long first-strand cDNA synthesis. *Biotechniques* 2002;32(5):984–985.
17. Shibata Y, Carninci P, Watahiki A, et al. Cloning full-length, cap-trapper-selected cDNAs by using the single-strand linker ligation method. *Biotechniques* 2001;30(6):1250–1254.
18. Shibata K, Itoh M, Aizawa K, et al. RIKEN integrated sequence analysis (RISA) system–384-format sequencing pipeline with 384 multicapillary sequencer. *Genome Res* 2000;10(11):1757–1771.
19. Hayashizaki Y, Kawai J. A new approach to the distribution and storage of genetic resources. *Nat Rev Genet* 2004;5(3):223–228.

20. Kawai J, Hayashizaki Y. DNA book. *Genome Res* 2003;13(6B):1488–1495.
21. Miki R, Kadota K, Bono H, et al. Delineating developmental and metabolic pathways in vivo by expression profiling using the RIKEN set of 18,816 full-length enriched mouse cDNA arrays. *Proc Natl Acad Sci USA* 2001;98(5):2199–2204.
22. Black DL. Mechanisms of alternative pre-messenger RNA splicing. *Annu Rev Biochem* 2003;72:291–336.
23. Landry JR, Mager DL, Wilhelm BT. Complex controls: the role of alternative promoters in mammalian genomes. *Trends Genet* 2003;19(11):640–648.
24. Suzuki Y, Tsunoda T, Sese J, et al. Identification and characterization of the potential promoter regions of 1031 kinds of human genes. *Genome Res* 2001;11(5):677–684.
25. Ueda HR, Chen W, Adachi A, et al. A transcription factor response element for gene expression during circadian night. *Nature* 2002;418(6897):534–539.
26. Hwang BJ, Muller HM, Sternberg PW. Genome annotation by high-throughput 5′ RNA end determination. *Proc Natl Acad Sci USA* 2004;101(6):1650–1655.
27. Man MZ, Wang X, Wang Y. POWER_SAGE: comparing statistical tests for SAGE experiments. *Bioinformatics* 2000;16(11):953–959.
28. Matsumura H, Ito A, Saitoh H, et al. SuperSAGE. *Cell Microbiol* 2005;7(1):11–18.
29. Velculescu VE, Zhang L, Vogelstein B, Kinzler KW. Serial analysis of gene expression. *Science* 1995;270(5235):484–487.
30. Kasukawa T, Katayama S, Kawaji H, Suzuki H, Hume DA, Hayashizaki Y. Construction of representative transcript and protein sets of human, mouse, and rat as a platform for their transcriptome and proteome analysis. *Genomics* 2004;84(6):913–921.
31. Kasukawa T, Furuno M, Nikaido I, et al. Development and evaluation of an automated annotation pipeline and cDNA annotation system. *Genome Res* 2003;13(6B):1542–1551.
32. He L, Thomson JM, Hemann MT, et al. A microRNA polycistron as a potential human oncogene. *Nature* 2005;435(7043):828–833.
33. Lu J, Getz G, Miska EA, et al. MicroRNA expression profiles classify human cancers. *Nature* 2005;435(7043):834–838.
34. O'Donnell KA, Wentzel EA, Zeller KI, Dang CV, Mendell JT. c-Myc-regulated microRNAs modulate E2F1 expression. *Nature* 2005;435(7043):839–843.
35. Pillai RS, Bhattacharyya SN, Artus CG, et al. Inhibition of translational initiation by Let-7 MicroRNA in human cells. *Science* 2005;309(5740):1573–1576.
36. Ting AH, Schuebel KE, Herman JG, Baylin SB. Short double-stranded RNA induces transcriptional gene silencing in human cancer cells in the absence of DNA methylation. *Nat Genet* 2005;37(8):906–910.
37. Kiyosawa H, Yamanaka I, Osato N, Kondo S, Hayashizaki Y. Antisense transcripts with FANTOM2 clone set and their implications for gene regulation. *Genome Res* 2003;13(6B):1324–1334.
38. Yelin R, Dahary D, Sorek R, et al. Widespread occurrence of antisense transcription in the human genome. *Nat Biotechnol* 2003;21(4):379–386.
39. Holmes R, Williamson C, Peters J, Denny P, Wells C. A comprehensive transcript map of the mouse Gnas imprinted complex. *Genome Res* 2003;13(6B):1410–1415.
40. Bertone P, Stolc V, Royce TE, et al. Global identification of human transcribed sequences with genome tiling arrays. *Science* 2004;306(5705):2242–2246.

41. Cawley S, Bekiranov S, Ng HH, et al. Unbiased mapping of transcription factor binding sites along human chromosomes 21 and 22 points to widespread regulation of noncoding RNAs. *Cell* 2004;116(4):499–509.
42. Cheng J, Kapranov P, Drenkow J, et al. Transcriptional maps of 10 human chromosomes at 5-nucleotide resolution. *Science* 2005;308(5725):1149–1154.
43. Kampa D, Cheng J, Kapranov P, et al. Novel RNAs identified from an in-depth analysis of the transcriptome of human chromosomes 21 and 22. *Genome Res* 2004;14(3):331–342.
44. Margulies M, Egholm M, Altman WE, et al. Genome sequencing in microfabricated high-density picolitre reactors. *Nature* 2005;437(7057):376–380.
45. Hayashizaki Y, Kanamori M. Dynamic transcriptome of mice. *Trends Biotechnol* 2004;22(4):161–167.

6

Dissecting Transcriptional Control Networks

Vijayalakshmi H. Nagaraj and Anirvan M. Sengupta

Summary

Reconstructing how transcriptional networks function involves figuring out which promoters are affected by which transcription factors. Searching for functional regulatory sites bound by particular transcription factors in a genome is therefore of great importance. The chapter discusses efforts at building classifiers that separate promoters targeted by particular transcription factors from those that are not. We start with simple sequence classifiers based on Support Vector Machines and go on to discuss how to integrate different kind of data into the analysis.

Key Words: Transcription; regulatory elements; motifs; support vector machines; probabilistic models; DNA sequence; evolution.

1. Introduction

Much of systems biology is dedicated to deciphering cellular regulatory circuits. Because transcriptional regulation often plays a crucial role in these circuits, analyzing genetic interactions representing transcriptional control is a fundamentally important enterprise. Although control of transcription is one of the main steps of regulation and is the focus of this chapter, it should be kept in mind that gene expression is controlled at many different levels. A recent surge of activity on siRNA and microRNA emphasizes this point.

Transcription initiation is often affected by the binding of transcription factors (TFs) to regulatory sites on the DNA in a sequence-specific manner (1). The important problem of locating the binding sites for specific TFs, and thus identifying the genes they regulate, has attracted much attention from the bioinformatics community (2,3). Different methods have been employed for abstracting patterns, or "motifs," from the sequences that bind particular TFs, and predictions have been made for likely binding sites in the genome of the organism under study. Factors regulating multiple genes often have binding motifs that are low in information content (4), making the task of prediction harder. Low information content might arise from the binding site being too short (4–6 bases) or from allowing too much variability in the long sequence (10–25 bases).

Examples of such highly pleiotropic and functionally important proteins range from global regulators in prokaryotes (e.g., CRP, LRP, FIS, IHF, H-NS, HU, and σ factors (5) in *E. coli*) to many eukaryotic transcription factors important in metazoan development (e.g., Hox proteins [6]). Improving the ability of bioinformatic methods to dissect regulation of gene expression requires, among other things, enhancing our ability to identify regulatory sites on DNA and deepening our understanding of context-dependent interactions that govern functionality of the sites *in vivo*. In this chapter, we will discuss the development of classifiers of promoters that originate from our understanding of mechanistic models, as well as the development of the evolutionary process.

Ideally, analysis of transcriptional control networks could be reduced to a completely experimental question. Experimental approaches to locating binding sites on DNA (7,8), have uncovered numerous binding sites for various factors. Several databases have been devoted to such regulatory sites, like DPInteract (9) and RegulonDB (10) for *E. coli*, SCPD for yeast (11), and TRANSFAC for many higher eukaryotic organisms (12). However, looking at these databases, it is obvious that for most pleiotropic TFs targeting a large number (100–1,000) of genes, the number of known sites is still a small fraction of all the functional sites. A high-throughput version of the chromatin immunoprecipitation method, commonly known as the "ChIP on chip," has been introduced recently (13,14,15,16). In principle, this method locates binding sites throughout the genome. However, the resolution is limited to just several hundred bases.

One could use SELEX (17), as an *in vitro* method, to find the strongest binding sites (sequences close to the consensus) from a library consisting of randomly generated oligonucleotides. However, a TF can often function at binding sites that are far weaker than the consensus, as we will see (18). Therefore, to characterize the binding preferences of a TF, we need to identify many of these potential weak binding sites and to estimate the parameters describing the statistical distribution of these sequences. The appropriate modification of the SELEX procedure, which is needed to achieve this goal, has been suggested (19), but has not yet been widely adopted by the community. Indeed, proper application of this procedure does improve our ability to describe motifs more accurately (20).

Although much progress has been made in the experimental elucidation of transcriptional control, the subject still seems to require much from computational analysis. As different kinds of data pile up, we need more and more clever ways of analyzing them, drawing upon our ability to combine different kinds of evidence.

The computational methods for binding site location can be broadly classified into supervised and unsupervised methods. Supervised algorithms are trained on a set of binding sites identified directly by experimental measurements (9,21,22). Unsupervised algorithms identify regulatory binding sites on the basis of statistical significance, i.e., apparent overrepresentation of certain short sequences (23–30). We start our discussion with supervised learning tools and their performance.

2. Information Theoretic Weight Matrix and the Problem of Thresholds

Although it is common among biologists to refer to regulatory sequences by the best binding sequence (e.g., E-box or CACGTG being the sequence that binds Myc/Max), it is often noted that some of the bases in the sequence could depart from the "ideal" sequence and still give rise to functional regulatory elements. In these cases, one is often tempted to use more complex notation (CCNNNWWRGG for Mcm1 in yeast, with N = any nucleotide, W = A or T, R = A or G) trying to encompass the variability. However, for highly pleiotropic factors with "degenerate" (read extremely variable) binding motifs, it quickly becomes obvious that one needs a more quantitative way of dealing with the problem.

The widely used bioinformatic tool for quantitatively describing such motifs is the weight matrix method (23,31–36). The weight matrix is given by $w_{i\alpha} = \ln\left(\frac{f_{i\alpha}}{p_\alpha}\right)$, where $f_{i\alpha}$ is the frequency of base α appearing at position i in the example sequences, and p_α is the background probability of finding the base α. It is used to define the "information score" (31,32,33)

$$Z(S) = \sum_{i=1}^{L} \sum_{\alpha=1}^{4} w_{i\alpha} S_{i\alpha}$$

of any sequence S (where $S_{i\alpha}$ characterizes the sequence $S_{i\alpha} = 1$, if the ith base is α and $S_{i\alpha} = 0$ otherwise). The weights $w_{i\alpha}$ are given by

$$w_{i\alpha} = \ln \frac{\hat{p}_{i\alpha}}{p^0_\alpha},$$

where $\hat{p}_{i\alpha}$ is the observed fraction of the base α at site i of the sites in the training set, and p^0_α is the fraction of the base α in the genomic background.

To use weight matrix method for locating putative binding sites in a genome, one needs to set a threshold, so that sequences with a score higher than the threshold are identified as candidate regulatory sites. However, optimal setting of the threshold is a nontrivial problem, which could be settled if one thinks carefully about the physical interpretation of the bioinformatic scores.

Information score $Z(S)$ is often interpreted in terms of Gibbs free energy of TF binding to sequence S (35). Indeed, definition of w in terms of example sequences can be derived from maximization of the probability of "observing" the set of examples $S^{(a)}$, $(a = 1, \ldots, n_s)$ provided that the probability of each sequence $p(S^{(v)})$ is proportional to $\exp(Z(S^{(a)}))$, which may be identified as a Boltzmann factor so that $Z(S^{(a)}) = -E(S^{(a)})/\kappa_B T$. The binding (free) energy[1] $E(S)$ is then given by Equation (2) and the elements of the weight matrix $w_{i\alpha}$ are interpreted as (the negative of) the interaction energy contributed by base α at site i (in units of $\kappa_B T$) (34,35).

The issue of finding a principled way of defining a score threshold for site classification gets resolved if we note that the correct expression for the probability of sequence S to be bound to a protein in thermodynamic equilibrium is given by

$$p(S) = \frac{n_{tf}}{n_{tf} + Ke^{E(S)/\kappa_B T}} = \frac{1}{e^{(E(S)-\mu)/\kappa_B T} + 1} = f(E(S) - \mu), \qquad (1)$$

where μ is the chemical potential set by factor concentration $\mu = k_B T \ln(n_{tf}/K)$, n_{tf} being the concentration of the TF and K being the equilibrium constant of binding for a reference sequence for which E is set to zero. A sequence with binding energy well below the chemical potential is almost always bound to a factor. On the other hand, the sequence with energy well above the chemical potential is rarely bound and $p(S)$ is approximated by $\exp(-(E(S) - \mu)/\kappa_B T)$. The weight matrix procedure (34) assumes Boltzmann distribution of binding probability, and hence, corresponds to the latter limit. In contrast, Equation (1) provides a general description that correctly includes the saturation effect and introduces the binding threshold μ physically set by TF concentration (see Figure 1). This expression for $p(S)$, which is correct from the physi-

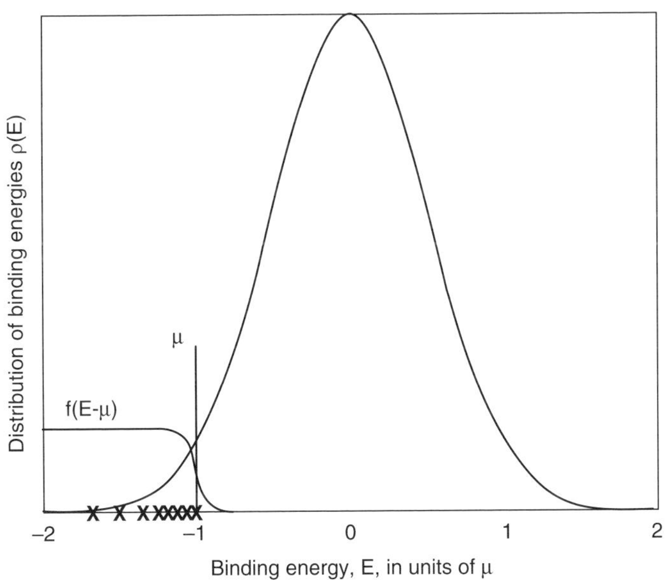

Figure 1. The distribution of binding energy, $\rho(E)$, and the binding probability, $f(E - \mu)$.

[1] Occasionally, we will refer to the free energy of binding simply as "binding energy" for the sake of brevity. In biophysical literature, the commonly used notation for this quantity would be $\Delta G(S)$ rather than $E(S)$.

cochemical point of view, can be used in a maximum likelihood framework to obtain a more rigorous estimate for the binding energy and to provide a practically useful solution to the problem of classifier threshold choice as follows.

2.1. One-Class SVM out of Maximum Likelihood Method

Let us begin by briefly recapitulating the maximum likelihood problem that leads to an algorithm, with some additional assumptions, which not only provides a way of scoring candidate sequences but also gives us a natural threshold. To formulate a probabilty model for the the training set of binding sequences, consider a hypothetical SELEX experiment: a library of randomly generated sequences of length L is mixed into a solution with a known concentration of a given transcription factor. After reaching thermodynamic equilibrium, some of the DNA sequences bound to the factor are extracted from the solution. This gives us a set O containing n_s sequences. The probability of observing the sequences comprising set O, but not other sequences is given by

$$\begin{aligned} e^{\mathcal{L}} &= \prod_{S\in O}[\gamma P_s f(E(S)-\mu)]\prod_{S'\notin O}[1-\gamma P_{S'} f(E(S')-\mu)] \\ &\approx \prod_{S\in O}[\gamma P_s f(E(S)-\mu)]e^{-\gamma\int dE\rho\varepsilon(E)f(E-\mu)}, \end{aligned} \tag{2}$$

where P_S is the probability of a sequence in the random library being S, and γ is the (yet unknown) probability of a sequence bound to the factor being extracted and sequenced. The likelihood function $\mathcal{L}$ depends on ε through $E(S)=\varepsilon\cdot S\equiv\sum_{i=1}^{L}\sum_{\alpha=1}^{4}\varepsilon_{i\alpha}S_{i\alpha}$. We want to choose all the parameters in such a way as to maximize $\mathcal{L}$.

In the case where the chemical potential μ is so low that the probability of any sequence being bound is small, maximization of $\mathcal{L}$ reduces to the weight matrix construction (34). However, it is more appropriate to retain the saturation effect in the *physical* probability of binding and use μ as a natural binding threshold.

In the limit, where the variation of the sequence-dependent part of the binding free energy is far larger than the temperature, the binding probability for a sequence S is either one or zero for most sequences, depending upon whether $E(S)$ is less than or greater than μ. As a result, the maximum likelihood method reduces to minimizing $\Sigma_{S'}P_{S'}f(E(S')-\mu)=\int dE\rho\varepsilon(E)\theta(\mu-E)=4^L\nu(\mu)$ (where $\nu(\mu)$ is the probability that a randomly chosen sequence has free energy below μ), subject to the constraints $E(S)\leq\mu$ (for all $S\in O$).

Note that by minimizing $\nu(\mu)$ we chose the threshold in such a way as to minimize expected number of false-positives without misclassifying the examples.

Provided that μ is not too low, $\nu(\mu)$ may be approximated by the integral of a Gaussian probability distribution with variance $\chi^2\equiv\sum_{i=1}^{L}\sum_{\alpha}p_\alpha(\varepsilon_{i\alpha}-\bar{\varepsilon}_i)^2$, where $\bar{\varepsilon}_i=\sum_{\alpha}p_\alpha\varepsilon_{i\alpha}$ (4,9). Our problem, then, reduces to minimization of χ^2 (because $v(\mu) \approx \mathrm{erf}(\mu/\chi)$), which is subject to constraints

$$E(S) = S \cdot \varepsilon \equiv \sum_{i=1}^{L} \sum_{\alpha=1}^{4} \varepsilon_{i\alpha} S_{i\alpha} \leq \mu = -1 \tag{3}$$

for every $S \in O$ (4,37). The overall energy scale here is arbitrary, and we have set the fixed value of $|\mu|$ to 1; hence we are determining ε in units of μ (which, more precisely, is the difference between the chemical potential and the average energy). Minimizing a nonnegative definite quadratic form, subject to linear inequalities, is a well-developed technique known as quadratic programming (37). We call this the quadratic programming method for energy matrix estimation (QPMEME) (36).

This method is very similar to support vector machines (SVMs) (38). If we think of sequences S as vectors in a vector space V, and $H = \{x = V | \varepsilon \cdot x = 1\}$ being a hyperplane separating the binding sequences from the nonbinding ones, the main difference is that in conventional applications of SVMs, one is trying to separate between positive examples and negative examples with a separator surface of largest margin. In our case, we do not have particular nonbinding sequences, and, instead, we are trying to minimize the probability that any random sequence is identified as a binding sequence, while still correctly classifying all of the positive examples. QPMEME turns out to be a one-class SVM (40–42), with $K(S, S') = \Sigma_{i\alpha} \hat{S}_{i\alpha} p_{\alpha}^{-1} \hat{S}'_{i\alpha}$, as the kernel that comes naturally from a maximum likelihood formulation and the assumption of approximately Gaussian distribution of the scores.

Conventional SVMs, with positive and negative training data, have been used in many problems (Figure 2) in bioinformatics (43–48). For systematic kernels for strings and other discrete structures, see the unpublished preprints by Haussler and by Watkins (Haussler D, Convolution Kernels on Discrete Stuctures, preprint UCSC-CRL-9910, 1999, and Watkins C, Dynamic Alignment Kernels, technical report CSD-TR-98-11, 1999.). Some of these developments are described in a recently published book (49). One-class SVMs have been occasionally used in gene expression analysis (50). However, our derivation of a one-class SVM for the binding site classification problem is novel.

The classifying hyperplane in QPMEME is affected by only the marginal binding sites. The orientation of the hyperplane, as obtained from the weight matrix, depends on all the sequences. If the orientation is slightly wrong, then trying to include all the training sets among the predicted positives produce too many false-positives for the method. Making the threshold too tight, on the other hand, leaves out many biologically verified sites. ROC analysis (51) of the trade-off between false-positives and false-negatives shows this effect. We have compared the false-negative fraction versus the number of candidate sites found within non-ORF fraction of the *E. coli* K12 genome for the weight matrix and QPMEME methods and found QPMEME to be superior by ROC analysis (36).

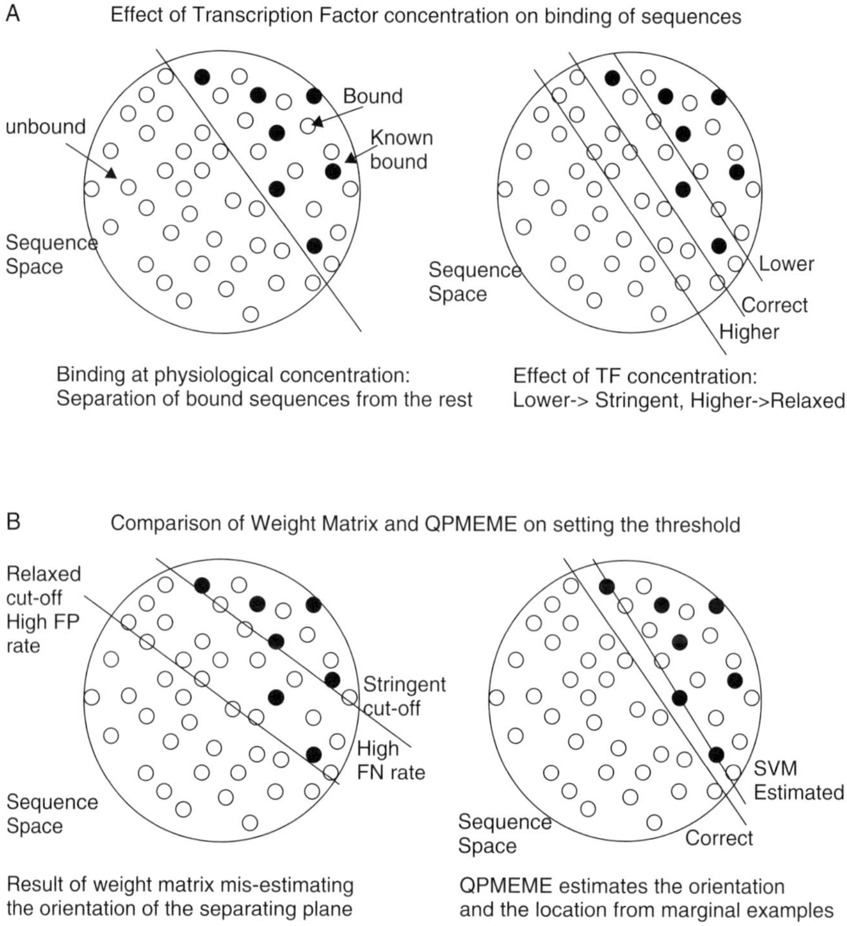

Figure 2. Geometric interpretation of the classification problem. The dark circles are the true negatives. The light circles are the true positives. The filled light circles make the training set.

2.2. Extended QPMEME with Dinucleotide Terms

Emphasizing the relation of the scoring function used in devising a classifier (for site location) with the physical properties of protein–DNA interaction is very useful because the relationship provides clues to improving the classifiers. For example, the linear classifier corresponds to the "independent nucleotide approximation," where different positions i contribute additively to the binding energy. More generally, the sequence specific interaction $E(S)$ can be parametrized by

$$E(S) = \sum_{i=1}^{L}\sum_{\alpha=1}^{4} \varepsilon_{i\alpha i} S_{i\alpha} + \sum_{i,j=1}^{L}\sum_{\alpha,\beta=1}^{4} J_{ij,\alpha\beta} S_{i\alpha} S_{j\beta} + \ldots, \qquad (4)$$

where S_i^α characterizes the sequence $S_i^\alpha = 1$, if the i-th base is α and $S_i^\alpha = 0$ otherwise. $\varepsilon_{i\alpha}$ is the interaction energy with the nucleotide α at position $i = 1, \ldots, L$ of the DNA string (35), and $J_{ij,\alpha\beta}$ is the pair dependent (α at position i and β at j) correction. The independent nucleotide approximation $E(S) = \sum_{i=1}^{L} \sum_{\alpha=1}^{4} \varepsilon_{i\alpha i} S_{i\alpha}$, which truncates Equation 4 at the first term appears to be adequate for some, but not for all, TFs (35,52,53). The nearest neighbor dinucleotide terms ($J_{ij,\alpha\beta}$ with $j = i + 1$) are expected to be particularly important whenever there is strong DNA deformation due to protein binding.

We can, once again, formulate the problem in terms of minimizing variance, which now has an additional contribution from $J_{ii+1,\alpha\beta}$,

$$x^2 \equiv \sum_{i=1}^{L} \sum_{\alpha} p_\alpha \varepsilon_{i\alpha}^2 + \sum_{i=1}^{L-1} \sum_{\alpha\beta} p_\alpha p_\beta \left(J_{ii+1,\alpha\beta}\right)^2, \tag{5}$$

which is subject to linear inequalities $E(S^{(a)}) \leq \mu < 0$ for each $a, \ldots, n_s$ and linear equalities $\Sigma_\alpha p_\alpha \varepsilon_{i\alpha} = \Sigma_\alpha p_\alpha J_{ii+1,\alpha\beta} = \Sigma_\beta p_\beta (J_{ii+1,\alpha\beta}) = 0$ for each i. This is, once more, a solvable quadratic programming problem that leads to a generalized kernel of the form $K(S,S') = \sum_{i=1}^{L} \sum_{\alpha} \hat{S}_{i\alpha} p_\alpha^{-1} \hat{S}'_{i\alpha} + \sum_{i=1}^{L-1} \sum_{\alpha\beta} \hat{S}_{i\alpha} p_\alpha^{-1} \hat{S}'_{i\alpha} \hat{S}_{i+1\beta} p_\beta^{-1} \hat{S}'_{i+1\beta}$.

This method was tested (54) on data generated from all-atom model-based calculations (55) for several transcription factors, some of which deform DNA seriously[2] upon binding. We compared results from models assuming independent base approximation as well as from those with dinucleotide terms. They were trained with 200 randomly chosen sequences with binding free energy within 5 kcal/mol of that of the consensus. Extended QPMEME clearly outperformed other methods (54).

This dinucleotide model has, formally, about four times as many parameters as the model based on independent bases. Estimating these parameters requires a larger number of training sequences. Currently, for many of the DNA-deforming transcription factors, the number of known sites available is too few so that we risk overfitting (see the book by Hastie et al. (55) for discussion of general statistical isssues related to overfitting). Clearly, we need to experimentally generate many more binding sequences, and it will be useful if our study on synthetic data could give us some idea about the number of sites required. We have, therefore, studied how the performance of extended QPMEME depends on the size of the training dataset. For example, we estimate that approximately 60–70 TBP-binding sites would be good enough to build a reasonable dinucleotide model. This is because only few of the dinucleotide terms are important in practice.

The SVMs and probabilistic approaches for classifying promoters could be developed further in several directions. One could design classifiers based on better estimation of the fraction of binding sites among random sequences, by going beyond the Gaussian approximation. There are several ways to do this, the two most promising being large deviation methods and some efficient exact evaluation based on dynamic

[2] If you are wondering why we do not test our method on real data, wait until the next paragraph.

programming. Another desirable feature would be to handle gracefully a small number of false-positives included in the training set. To have a procedure that is robust to such contamination, we have to go beyond the "all-or-nothing" approximation. We could add a linear penalty function $\max(0, E(S) - \mu)$ for each known example S to the quadratic function to be optimized, which could be solved within the quadratic programming framework (38). Alternatively, we could generalize logistic regression (57) for one-class problems.

3. Combining Heterogeneous Data: Going Beyond Sequence

Predicting regulatory binding sites in eukaryotes is complicated. Many effects other than the inherent affinity of a protein for a particular DNA sequence affects the occupancy, as well as functionality, of a site. The use of additional information often would reduce the false-positive rates of conventional computational methods. For example, analysis of gene expression using DNA microarrays provides genome-wide profiles of the genes controlled by the presence or absence of a specific transcription factor. However, whether a change in the level of transcription of a specific gene is caused by the transcription factor acting directly at the promoter of the gene, or through regulation of other transcription factors working at the promoter, could not be determined solely from steady-state microarray data. Combining microarray expression data and site preference data overcomes limitations of predictions based solely on either kind of information.

As mentioned before, tools that combine different types of data in a principled fashion to predict regulatory interactions are gaining importance, given the limitations of drawing conclusions from a single form of data. Traditionally, one often sets up filters for each kind of data, generates sets of candidates, and then examines the overlap of the sets to find out the most likely candidates. This procedure, although simple, suffers from two problems. The first is that it is too conservative and likely to miss many genuine candidates. The second is that the stringency of each of the filters is somewhat arbitrary. The first problem is sometimes dealt with by designing metaclassifiers, which are classifiers that combine the output of multiple classifiers to make a decision (58). The strategies for combining data could be as simple as the majority decision or could be a complicated rule that is learned based on some online training. However, even with a sophisticated meta-classifier, one is still stuck with the arbitrariness of the underlying classifiers.

Here is a simple procedure for combining variables that could be used to rank genes. If we have n variables, x_{ig}, $i = 1 \ldots n$, each of which scores a gene g for belonging to a category C (say regulated in a particular way). For simplicity, let us say that large values of x_{ig} indicates the gene is likely to be in that particular category. We calculate the product of cumulative probabilities $p_g = \Pi_i Prob_i(x > x_i g)$. The number p_g being small is to be considered an indication of the gene being a good candidate for belonging to C. How small is small enough? We could see how small p_g gets if we use scrambled data. Let us permute the index of each variable, take products, and generate ordered sets of p_g value. We take the gene that has the lowest significant p_g values for the unscrambled dataset. The kth p_g value (in ascending order) is considered significant if it is lower than the kth p_g value generated by a certain number of randomly scrambled datasets. The significance level is set by the number of scrambled datasets used.

Some possible problems with this method are as follows. Is case one of the probability models is wrong, and for some reason gives extremely small probabilities for certain genes, then this variable will dominate everything. One way out of this is to use empirical probabilities based on ranking: number of genes with x_i larger than x_ig divided by total number of genes. One still needs to be careful. If, for some reason, x_i has very little predictive value, then the procedure throws in random noise. We will need some method of feature selection for deciding which variable not to use.

We faced this problem while trying to identify the targets of the a1/α2 repressor complex involved in repressing haploid specific genes (59) in the yeast genome. In the analysis of expression profile using microarrays, when comparing the presence or absence of a specific transcription factor, one is forced to ask whether a change in the level of transcription of a specific gene is caused by the transcription factor acting directly at the promoter of the gene or through regulation of other transcription factors working at the promoter. To address this problem we devised a computational method that combines microarray expression and site preference data (60). We utilized the microarray data obtained in Galitski et al. (61) and the mutational study in Jin et al. (62). The mutational study allows us to score sequences away from the consensus, in terms of their ability to bind the repressor. We compute a binding p-value for promoters of genes, based on what fraction of genes have promoters with stronger sites than the best site in the particular promoter. We also defined a score that combines two components: one penalizing difference of expression between a and α, the other rewarding overexpression in haploids over diploids. An expression p-value of a gene represents the fraction of genes with better or equal score compared with that associated with the particular gene.

Instead of setting separate, and somewhat arbitrary, cutoffs on the binding and the expression p-values, and then considering the genes that look significant by either criterion, we decided to look for correlated significance in the way described above. We calculated a combined p-value for a gene by taking the product of the two p-values, which gives us the fraction of random unregulated genes that would be better on both counts compared with this gene. If this number is low, then we might say the gene is a good candidate for being a direct regulatory target. The appropriate threshold on the combined p-value is drawn by permuted combination of binding p-values and expression p-values, as previously discussed (60). For the sake of completeness, we point out parallel work on the mating-type system from the Johnson lab (63).

The early days of microarray data analysis were heavily influenced by clustering. It's no wonder that some of the early papers that integrated sequence analysis with expression data relied crucially on clustering. For example, in a pioneering paper, Tavazoie et al. clustered expression data and used multiple local sequence alignment algorithms on the promoter regions of the coclustered genes to discover regulatory motifs (24). This approach has been further refined by using Bayesian networks to incorporate additional constraints regarding relative positions and the orientations of the motifs (63). Another approach has been to break the genes into modules and perform module assignments and motif searches at the same time via an expectation maximization algorithm (as opposed to clustering first and finding motifs later) (65). Another method that does not utilize clustering is a regression model-based analysis to locate "words" in the promoter that correlates with modulation of expression (27). Most of these approaches attack the difficult problem of what to do when relatively little is known about the regulatory system and sequence recognition by the TFs. Consequently, these investigators develop pattern recognition algorithms that are essentially unsupervised. On the other hand, the approach discussed in the chapter takes advantage of knowledge about the biological system, and uses that information, combined with expression analysis, to identify potential target sites. The loss of generality resulting from such an approach is more than compensated for by the improved predictive power.

4. Combining Phylogenetic Footprinting and Motif Search

Comparison across species to discover regulatory elements is a powerful tool (28,29,66–68) and commonly goes by the name of phylogenetic footprinting (*see* the review by Martha Bulyk (69)). It is turning out to be a popular bioinformatic method for finding or verifying binding sites of transcription factors. The power of this procedure was tested by comparing the genomes of several *Saccharomyces* species involved in the study by Cliften et al. (70) and that by Kellis et al. (70). In its extreme version, phylogenetic footprinting demands perfect conservation of words of a certain length (say, hexamers) among multiple species (70). However, some of the functional sites are not so well preserved and occasionally may even be lost in some of the related species. We would therefore need some less stringent way of measuring conservation.

4.1. Evolutionary Comparison of the HO Promoter

One of the targets of a1/α2, the *HO* gene, codes for an endonuclease involved in mating-type switch. In contrast to most other yeast promoters, the *HO* promoter is very complex, with multiple binding sites for different TFs that is reminiscent of developmental promoters in metazoans. The *HO* promoter has 10 putative a1/α2 binding sites (72), which we will call HO(1–10). Several of these predicted a1/α2 binding sites have low affinity for the repressor complex. These sites, on their own, show very weak repression of transcription in a diploid. This raises the question of whether these sites have a functional role in the context of the *HO* promoter. One approach to address this question is to determine if these sites are conserved in related yeast species. Significant

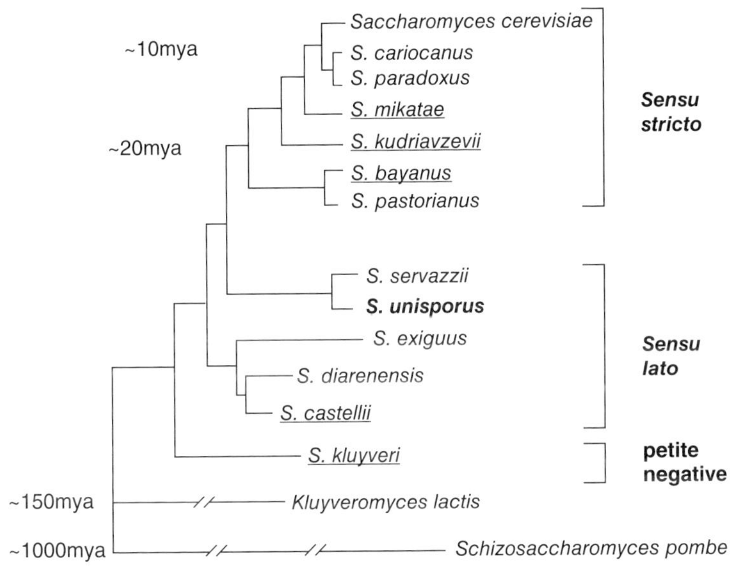

Figure 3. *Saccharomyces* phylogenetic tree. The species names that are underlined correspond to those sequenced by the group at Washington University, St. Louis, and were used in a study (69). From http://www.genome.wustl.edu/projects/yeast/.

conservation of these sequences would suggest that they have a functional role in the cell.

We obtained the *HO* promoters from *S. cerevisiae*, *S. bayanus*, *S. mikatae*, *S. paradoxus* and *S. kudriavzevii* (70,71) and performed a segmented multiple sequence alignment (Figure 3) using the DIALIGN 2 (73–75). A measure of the significance for the alignment of the sites was then constructed by taking the alignment scores, provided by DIALIGN, and averaging over each base pair in the site. The background probability of the alignment scores was calculated by a similar analysis on two other neighboring regions of the genome (18).

The background distribution of average scores fits an exponential function, as can be seen in Figure 4. The average scores of many of the *HO* sites are in the tail of the distribution, and 5 of the 10 a1/α2 sites appeared to be well conserved (as judged by 1% significance level) among the different species. As might be expected, several of the strong repressor sites, such as HO(3), HO(4), and HO(10), are highly conserved among the different species. In contrast, HO(1), HO(2), and HO(5), which were weak repressor sites on their own, showed significantly less sequence conservation. Interestingly, the two other weak repressor sites, HO(7) and HO(8), are strongly conserved among the different yeasts. In fact, these sites appear to be more highly conserved than HO(4), HO(6), and HO(9), which are all strong a1/α2 repressor sites. The fact that HO(7) and HO(8) are highly conserved strongly suggests that they have a functional role in regulating the *HO* promoter, even though they function only very weakly on their own. For examples of alignments of the sites from these categories, see Figure 5. Mutational analysis of these sites

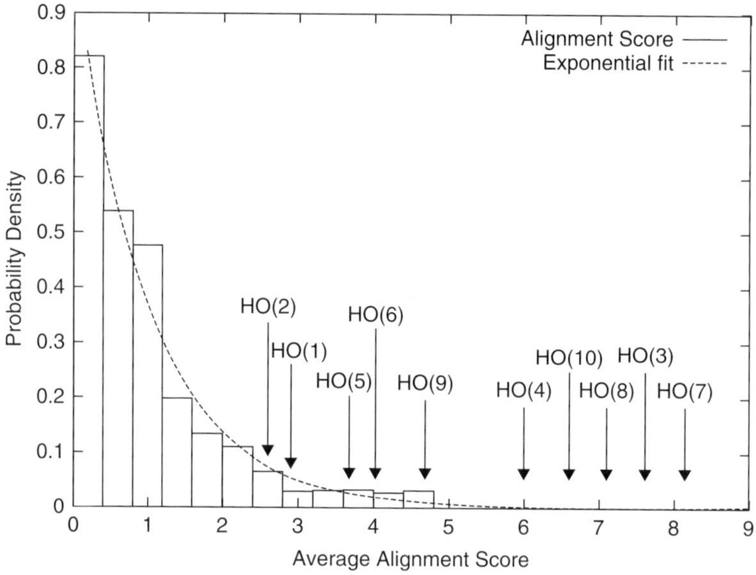

Figure 4. Conservation of the a1/$\alpha 2$ sites in the *HO* promoter among four species of yeast (*Saccharomyces sensu stricto*). The graph shows the empirical background distribution of several alignment scores for 20 base pair segments. The scores for the 10 putative a1/$\alpha 2$ sites are indicated with arrows.

```
HO(10) Strong site, highly conserved

Scer --------GTTTTGCCGCGTTAAAACCTACATC-AAAAAAGG-GGGATCA
Spar gtcaaTACGTTTTGCCGCGTTAAAACCTACATC-AAAAAAAGGCGGATCA
Smik CAAt-----TTTTACCGCGTTAAAACATACATCgAAAAAAGGGCGGATCA
Skud -----TACGTTTTACCGCGTTAAAACTTACATC-AAAAAAGGGCGGATCA
Sbay AAAgtTACATTTTACCGCGTTAAAACCTACATC-AAAAAAGGGCGGATCA
     000001112666656666667777776777777077777655455555555

HO(7) Medium strength site, highly conserved

Scer CCCAAAGGGGTATCAAATAATCGATGTGCTTTTTCACTCTACGAATGATC
Spar CCCAAAGGGGTATGAAATAATCGATGTGCTTTTTCACTCTACGAATGATC
Smik CCCAAAAGGGTATGAAATAATCGATGTGCTTTTTCACTCTACGAATGATC
Skud CTGAAAGGGGTATCGAATAATCGATGTGCTTTTTCACTCTACGAATGATC
Sbay CTGAAAGGG--ATCAAGTAACTGATTTGTCTTTTCTCTCTGCGGATGATC
     44444444444554466666666666777777777777777888988888

HO(2) Medium strength site, not well conserved

Scer AATTCA-TGTCAT-GTCCACATTAACATCATTG-CAGAGCAACAATTCAT
Spar AATTCA-TGTAAATGTTTACATTAACATCACTTGCAGGAGAACGGCTCGT
Smik AACcttaTGCGAAcGTTTACATTACTATCACTCACAGGAAAATAATAAAT
Skud AAagaA-TTTATTTGTTTACATCAACATCTCTTGTAGAGGAACAATGCAT
Sbay AACTGA-TGTAATTGTTTACATCAATATCTTCG-CAGAAGAGCAATCCAT
     22100102221111222222222222223321111122222222111122
```

Figure 5. Conservation of the a1/$\alpha 2$ sites in the *HO* promoter among four species of yeast (*Saccharomyces sensu stricto*). Aligned sequences for HO(10), HO(7), and HO(2). The numbers below indicate the degree of conservation according to DIALIGN2.

in the context of the *HO* promoter shows that these sites do have a role in regulating transcription. These experiments indicate that apparently weak sites, which may not be identified by conventional algorithms, can have an important role in transcriptional regulation (18).

The scoring system for alignment used in this study, based on the program DIALIGN, starts with a null model of independent sequences. This is not very accurate, given the structure of the phylogenetic tree. A more appropriate model would treat, say, a certain degree of similarity between *S. cerevisiae* and *S. bayanus* as more significant than the same degree of similarity between *S. cerevisiae* and *S. paradoxus*. It is, indeed, possible to have a probability model that integrates phylogeny into motif detection. However, discussion of this topic would take us too far afield.

5. Conclusion

In summary, for locating regulatory elements in genomes, it is not enough to rank the candidate sites; we need to classify them. We therefore need ways to set cutoffs in a principled manner. This chapter discusses computational approaches for that purpose. We specifically focus on combining sequence analysis, phylogenetic comparison, and microarray data analysis in a principled fashion in terms of cogent probability models. In traditional machine learning, one often puts all the domain-specific intuition into the choice of the feature space, but utilizes very standard classifiers on the feature space. We, on the other hand, intend to use biophysics and evolutionary modeling to construct appropriate classifying surfaces in these feature spaces (Figure 6). In biological problems, large numbers of verified negatives are often unavailable, making one-class classifiers (meaning those that could be trained on positive examples only) very useful. The grand challenge is to patch together the lessons learned from these studies to construct more detailed models of important submodules of the transcriptional network.

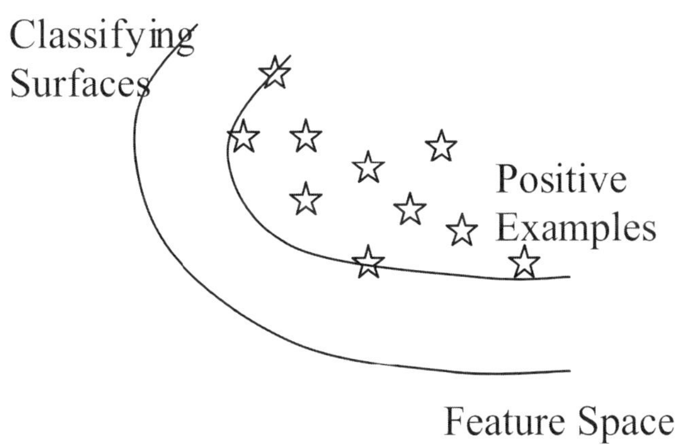

Figure 6. One-class classifiers in the general feature spaces of combined biological data.

Acknowledgments: We thank our collaborators, A.R. Bruning, M. Djordjevic, S.E. Hanlon, R. Lavery, J.R. Mathias, R.A. O'Flanagan, G. Paillard, B.I. Shraiman, and A.K. Vershon. The work was partially supported by NHGRI grant R01HG03470.

This work was finished during a visit to KITP program, supported in part by the National Science Foundation under grant PHY99-07949.

References

1. Lewin B. Genes VII. New York: Oxford University Press; 2000.
2. Fickett JW, Wasserman WW. Discovery and modeling of transcriptional regulatory regions. *Curr Opin Biotechnol* 2000;1:19–24.
3. Stormo GD, Tan K. Mining genome databases to identify and understand new gene regulatory systems. *Curr Opin Microbiol* 2002;5:149–153.
4. Sengupta AM, Djordjevic M, Shraiman BI. Specificity and robustness of transcription control networks. *Proc Natl Acad Sci USA* 2002;99:2072–2077.
5. Wagner R. Transcription Regulation in Prokaryotes. Oxford: Oxford University Press; 2000.
6. Gilbert SF. Developmental Biology, 6th edition. Sunderland: Sinauer; 2000.
7. Docherty K. Gene Transcription, DNA Binding Proteins. New York: John Wiley & Sons Ltd.; 1997.
8. Travers AA, Buckle M. DNA-Protein Interactions: A Practical Approach. Oxford: Oxford University Press; 2000.
9. Robison K, McGuire AM, Church GM. A comprehensive library of DNA-binding site matrices for 55 proteins applied to the complete *Escherichia coli* K-12 genome. *J Mol Biol* 1998;284:241–254. Available at http://arep.med.harvard.edu/dpinteract/
10. Salgado H, Santos A, Garza-Ramos U, et al. RegulonDB (version 2.0): a database on transcriptional regulation in *Escherichia coli. Nucleic Acids Res* 1999;27:59–60. http://www.cifn.unam.mx/ComputationalGenomics/regulonDB
11. Zhu J, Zhang MQ. SCPD: A Promoter Database of Yeast *Saccharomyces cerevisiae. Bioinformatics* 1999;15:607–611. Available at http://cgsigma.cshl.org/jian/
12. Wingender E, Chen X, Hehl R, et al. TRANSFAC: an integrated system for gene expression regulation. *Nucleic Acids Res* 2000;28:316–319. Available at http://transfac,gdb.de/TRANSFAC
13. Ren B, Robert F, Wyrick JJ, et al. Genome-wide location and function of DNA binding proteins. *Science* 2000;290: 2306–2309.
14. Iyer VR, Horak CE, Scafe CS, et al. Genomic binding sites of the yeast cell-cycle transcription factors SBF and MBF. *Nature* 2001;409:533–538.
15. 15. Lee TI, Rinaldi NJ, Robert F, Odom DT, et al. Transcriptional regulatory networks in *Saccharomyces cerevisiae. Science* 2002;298:799–804.
16. Harbison CT, Gordon DB, Lee TI, et al. Transcriptional regulatory code of a eukaryotic genome. *Nature* 2004;431:99–104.
17. Tuerk C, Gold L. Systematic evolution of ligands by exponential enrichment: RNA ligands to bacteriophage-T4 DNA-polymerase. *Science* 1990;249: 505–510.
18. Mathias JR, Hanlon SE, O'Flanagan RA, et al. Repression of the yeast *HO* gene by the MATα2 and MATa1 homeodomain proteins. *Nucleic Acids Res* 2004;32:6469–6478.
19. Roulet E, Busso S, Camargo AA, et al. High-throughput SELEX SAGE method for quantitative modeling of transcription factor binding sites. *Nat Biotechnol* 2002;20:831–835.

20. Nagaraj VH, O'Flanagan RA, Shraiman BI, Sengupta AM, manuscript in preparation.
21. Chen QK, Hertz GZ, Stormo GD. MATRIX SEARCH 1.0: a computer program that scans DNA sequences for transcriptional elements using a database of weight matrices. *Comput Appl Biosci* 1995;11:563–566.
22. Gralla J, Collado-Vides J. Organization and function of transcription regulatory elements. In: Neidhart FC, Ingraham F, eds. *Escherichia coli* and *Samonella typhimurium:* Cellular and Molecular Biology, Washington DC: ASM Press, 1996:1232–1245.
23. Stormo GD, Hartzell GW, 3rd. Identifying protein-binding sites from unaligned DNA fragments. *Proc Natl Acad Sci USA* 1989;86:1183–1197.
24. Tavazoie S, Hughes JD, Campbell MJ, et al. Systematic determination of genetic network architecture. *Nat Genet* 1999;22:281–285.
25. Hughes JD, Estep PW, Tavazoie S, Church GM. Computational identification of cis-regulatory elements associated with groups of functionally related genes in *Saccharomyces cerevisiae. J Mol Biol* 2000;296:1205–1214.
26. Bussemaker HJ, Li H, Siggia ED. Building a dictionary for genomes: identification of presumptive regulatory sites by statistical analysis. *Proc Natl Acad Sci USA* 2000;97:10096–10100.
27. Bussemaker HJ, Li H, Siggia ED. Regulatory element detection using correlation with expression. *Nat Genet* 2001;27:167–171.
28. McCue L, Thompson W, Carmack C, et al. Phylogenetic footprinting of transcription factor binding sites in proteobacterial genomes. *Nucleic Acids Res* 2001;29:774–782.
29. Rajewsky N, Socci ND, Zapotocky M, Siggia ED. The evolution of DNA regulatory regions for proteo-gamma bacteria by interspecies comparisons. *Genome Res* 2002;12:298–308.
30. Liu XS, Brutlag DL, Liu JS. An algorithm for finding protein-DNA binding sites with applications to chromatin-immunoprecipitation microarray experiments. *Nat Biotechnol* 2002;20:835–839.
31. Staden R. Computer methods to locate signals in nucleic acid sequences. *Nucleic Acids Res* 1984;12:505–519.
32. Schneider TD, Stormo GD, Gold L, Ehrenfeucht A. Information content of binding sites on nucleotide sequences. *J Mol Biol* 1986;188:415–431.
33. Stormo GD, Schneider TD, Gold L. Quantitative analysis of the relationship between nucleotide sequence and functional activity. *Nucleic Acids Res* 1986;14:6661–6679.
34. Berg OG, von Hippel PH. Selection of DNA binding sites by regulatory proteins: statistical-mechanical theory and application to operators and promoters. *J Mol Biol* 1987;193:723–750.
35. Stormo GD, Fields DS. Specificity, free energy and information content in protein-DNA interactions. *Trends Biochem Sci* 1998;3:109–113.
36. Stormo GD. DNA binding sites: representation and discovery. *Bioinformatics* 2000;1:16–23.
37. Djordjevic M, Sengupta AM, Shraiman BI. A biophysical approach to transcription factor binding site discovery. *Genome Res* 2003;13:2381–2390.
38. Fletcher R. Practical Methods of Optimization. New York: Wiley; 1987.
39. Cristianini N, Shawe-Taylor J. Introduction to support vector machines. Cambridge: Cambridge University Press; 2001.
40. Schölkopf B, Platt J, Shawe-Taylor J, et al. Estimating the support of a high-dimensional distribution. *Neural Comput* 2001;13:1443–1471.
41. Manevitz LM, Yousef M. One-class SVMs for document classification. *J Mach Learn Res* 2001;2:139–154.

42. Tax DMJ, Duin RPW. Uniform object generation for optimizing one-class classifiers. *J Mach Learn Res* 2002;2:155–173.
43. Jaakkola T, Diekhans M, Haussler D. Using the Fisher kernel method to detect remote protein homologies. In: Lengauer T, Schneider R, Bork P, Brutlad D, Glasgow J, Mewes H, Zimmer R editors. ISMB 99. Proceedings Seventh International Conference on Intelligent Systems for Molecular Biology; 1999 Aug 6–11; Heidelberg, Germany. Menlo Park: AAAI Press; 1999:149–158.
44. Jaakkola T, Diekhans M, Haussler D. A discriminative framework for detecting remote protein homologies. *J Comput Biol* 2000;7:95–114.
45. Furey TS, Cristianini N, Duffy N, et al. Support vector machine classification and validation of cancer tissue samples using microarray expression data. *Bioinformatics* 2000;16:906–914.
46. Brown MP, Grundy WN, Lin D, et al. Knowledge-based analysis of microarray gene expression data by using support vector machines. *Proc Natl Acad Sci USA* 2000;97:262–267.
47. Pavlidis P, Furey TS, Liberto M, Haussler D, Grundy WN. Promoter region-based classification of genes. In: Altman RB, Dunker AK, Hunter L, Lauderdale K, Klein TE editors. BIOCOMPUTING 2001. Proceedings of the Pacific Symposium; 2001 Jan 3–7; Mauna Lani, Hawaii, USA. Singapore: World Scientific; 2000:151–163.
48. Vert JP. Support vector machine prediction of signal peptide cleavage site using a new class of kernels for strings. In: Altman RB, Dunker AK, Hunter L, Lauderdale K, Klein TE editors. BIOCOMPUTING 2002. Proceedings of the Pacific Symposium; 2002 Jan 3–7; Kauai, Hawaii, USA. Singapore: World Scientific; 2001:649–660.
49. Schölkopf B, Tsuda K, Vert JP. Kernel Methods in Computational Biology. Cambridge: The MIT Press; 2004.
50. Kowalczyk A, Raskutti B. One class SVM for yeast regulation prediction, ACM SIGKDD Explorations Newsletter 2002;4:99–100.
51. Egan JP. Signal Detection Theory and ROC Analysis. New York: Academic Press, 1975.
52. Bulyk ML, Johnson PL, Church GM. Nucleotides of transcription factor binding sites exert interdependent effects on the binding affinities of transcription factors. *Nucleic Acids Res* 2002;30:1255–1261.
53. Benos PV, Bulyk ML, Stormo GD. Additivity in protein-DNA interactions: how good an approximation is it? *Nucleic Acids Res* 2002;30:4442–4451.
54. O'Flanagan RA, Paillard G, Lavery R, Sengupta AM. Non-additivity in protein-DNA binding. *Bioinformatics* 2005;21:2254–2263.
55. Paillard G, Lavery R. Analyzing protein-DNA recognition mechanisms. *Structure* 2004;12:113–122.
56. Hastie T, Tibshirani R, Friedman J. The Elements of Statistical Learning. New York: Springer; 2001.
57. Hosmer DW, Lemeshow S. Applied Logistic Regression. New York: Wiley; 2000.
58. Dietterich TG. Machine learning research: four current directions. *AI Magazine* 1997;18:97–136.
59. Johnson A. A combinatorial regulatory circuit in budding yeast. In: McKnight SL, Yamamoto KR, editors. Transcriptional Regulation. Cold Spring Harbor: Cold Spring Harbor Laboratory Press; 1992.
60. Nagaraj VH, O'Flanagan RA, Bruning AR, et al. Combined analysis of expression data and transcription factor binding sites in the yeast genome. *BMC Genomics* 2004;5:59.

61. Galitski T, Saldanha AJ, Styles CA, et al. Ploidy regulation of gene expression. *Science* 1999;285:251–254.
62. Jin Y, Zhong H, Vershon AK. The yeast a1 and alpha2 homeodomain proteins do not contribute equally to heterodimeric DNA binding. *Mol Cell Biol* 1999;19,585–593.
63. Galgoczy DJ, Cassidy-Stone A, Llinas M, et al. Genomic dissection of the cell-type-specification circuit in *Saccharomyces cerevisiae*. *Proc Natl Acad Sci USA* 2004;101:18069–18074.
64. Beer MA, Tavazoie S. Predicting gene expression from sequence. *Cell* 2004;117:185–198.
65. Segal E, Shapira M, Regev A, Pe'er D, Botstein D, Koller D, Friedman N. Module networks: identifying regulatory modules and their condition-specific regulators from gene expression data. *Nat Genet* 2003;34:166–176.
66. McGuire AM, Hughes JD, Church GM. Conservation of DNA regulatory motifs and discovery of new motifs in microbial genomes. Genome Res 2000;10:744–757.
67. Pennacchio LA, Rubin EM. Genomic strategies to identify mammalian regulatory sequences. *Nat Rev Genet* 2001;2:100–109.
68. Blanchette M, Tompa M. Discovery of regulatory elements by a computational method for phylogenetic footprinting. *Genome Res* 2002;12:739–748.
69. Bulyk ML. Computational prediction of transcription-factor binding site locations. *Genome Biol* 2003;5:201.
70. Cliften P, Sudarsanam P, Desikan A, et al. Finding functional features in *Saccharomyces* genomes by phylogenetic footprinting. *Science* 2003;301: 71–76.
71. Kellis M, Patterson N, Endrizzi M, et al. Sequencing and comparison of yeast species to identify genes and regulatory elements. *Nature* 2003;423:241–254.
72. Miller AM, MacKay VL, Nasmyth, KA. Identification and comparison of two sequence elements that confer cell-type specific transcription in yeast. *Nature* 1985;314:598–603.
73. Morgenstern B, Dress A, Werner T. Multiple DNA and protein sequence alignment based on segment-to-segment comparison. *Proc Natl Acad Sci USA* 1996;93:12098–12103.
74. Morgenstern B, Frech K, Dress A, Werner T. DIALIGN: finding local similarities by multiple sequence alignment. *Bioinformatics* 1998;14:290–294.
75. Morgenstern B. DIALIGN 2: improvement of the segment-to-segment approach to multiple sequence alignment. *Bioinformatics* 1999;15, 211–218.

7

Reconstruction and Structural Analysis of Metabolic and Regulatory Networks

Hong-wu Ma, Marcio Rosa da Silva, Ji-Bin Sun, Bharani Kumar, and An-Ping Zeng

Summary

Networks of interacting cellular components carry out the essential functions in living cells. Therefore, understanding the evolution and design principles of such complex networks is a central issue of systems biology. In recent years, structural analysis methods based on graph theory have revealed several intriguing features of such networks. In this chapter, we describe some of these structural analysis methods and show their application in analysis of biological networks, specifically metabolic and transcriptional regulatory networks (TRNs). We first explain the methods used for reconstruction of biological networks, and then compare the pros and cons of the different methods. It will be shown how graph theory–based methods can help to find the organization principle(s) of the networks, such as the power law degree distribution, the bow-tie connectivity structure, etc. Furthermore, we present an integrated network that includes the metabolite–protein (transcription factor) interaction to link the regulatory network with the metabolic network. This integrated network can provide more insights into the interaction patterns of cellular regulation.

Key Words: Metabolic network; regulatory network; network reconstruction; Scale-free network; bow tie; network centrality; systems biology; graph theory.

1. Introduction

It is recognized that the interactions between cellular components rather than the components themselves determine the behavior of a complex biological system. Therefore, understanding the complex interactions in various cellular processes in terms of large-scale biological networks is the central issue in systems biology (1–11). The rapid development in genome sequencing and other high-throughput experimental technologies makes it possible to reconstruct such networks at the whole-system level (2,4,5,12,13). Structural analysis of these biological networks can help us find certain general organization principles of biological systems. In this chapter, the methods for reconstruction of biological networks and several structural features of these network are described, with emphasis on metabolic networks and TRNs for which relatively reliable and complete data are available. We first illustrate the major approaches and available databases for the reconstruction of genome-scale metabolic networks and regulatory networks. We then introduce the methods based on graph theory for the structural and functional analysis of these networks. The different structures of the two types of networks are discussed, and an integrated network that includes the metabolite–protein interaction, as well as the metabolic reaction and the transcriptional regulation, is presented. This integrated network is more suitable for studying the functional organization of the biological system.

2. Methods for Reconstruction of Metabolic Networks and Regulatory Networks

2.1. Genome-Based Metabolic Network Reconstruction

It is well known that different organisms may have very different metabolic ability in uptaking various substrates or synthesizing diverse metabolic products because of the existence of different enzymes (pathways) in their metabolic networks. The availability of the fully sequenced genomes and the sequence similarity–based gene function annotation methods make it possible to reconstruct organism-specific metabolic networks based on the genome (14–16). The reconstruction method is depicted in Figure 1. First, all the open reading frames (ORFs) in the

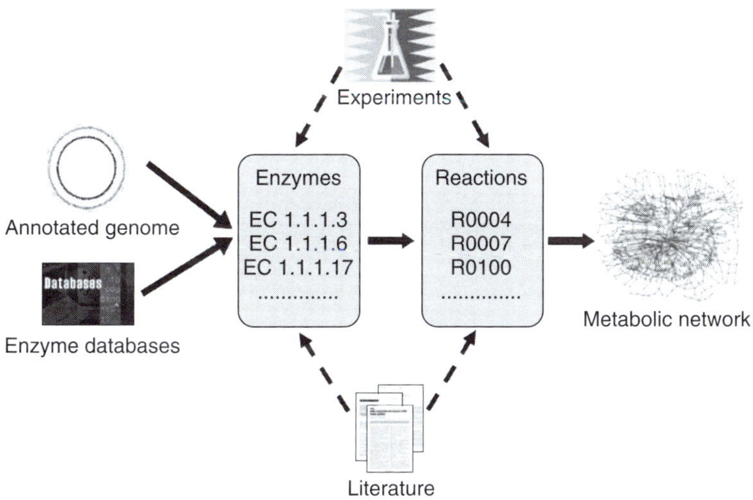

Figure 1. Processes for the reconstruction of metabolic networks. The high-throughput method (shown by the solid arrows) directly reconstructs an organism-specific metabolic network from enzyme or genome databases. New enzymes or reactions from biological experiments or literature (shown by the dashed arrows) can be added to obtain high-quality metabolic networks.

newly sequenced genome are identified, and sequence similarity searches are carried out to find the similar genes in the gene or protein databases. If an ORF is found to be similar to an enzyme gene in another organism, it may also be annotated as that enzyme. This sequence similarity–based knowledge transfer between different organisms allows us to obtain a list of enzymes existing in the metabolic network of a newly sequenced organism directly from the genome sequence, even though few or no biochemical experiments have been done to investigate the metabolic enzymes in that organism.

Modern sequencing technologies are generating a large quantity of DNA sequences, first in unfinished form or with low genome coverage because of the time-consuming, and thus limiting, steps of finishing and annotation. As of March 2006, only one third of the genome sequencing projects documented by the National Center for Biotechnology Information had been completed. Moreover, the available methods for genome annotation often encounter difficulties when dealing with unfinished, error-containing genomic sequences. To accelerate the use of genomic data from a large number of ongoing sequencing projects for studying cellular metabolism, Sun and Zeng (17) recently developed a homology-based algorithm called IdentiCS to identify protein-coding sequences, and thus to reconstruct a strain-specific metabolic network directly from unfinished raw genomic data. Compared with the conventional method, which requires more than 8-fold coverage of genome sequences, this method needs only a 3-4-fold coverage. The method was originally developed to analyze unfinished bacteria genomes, and has now been extended to eukaryotic genomes.

Currently, many enzyme nomenclature databases such as KEGG (14), BRENDA (18), and ExPASy (19) have included all the corresponding genes in different organisms for each enzyme in these databases. This provides a high-throughput way for metabolic network reconstruction. The lists of enzymes for all the organisms included in an enzyme database can be obtained at one time. It should be noted that the enzyme–gene relationships in these enzyme databases are also directly or indirectly from the sequence similarity–based gene annotation.

After obtaining an enzyme list for a specific organism, the next step would be to find all the reactions catalyzed by these enzymes. The enzyme–reaction relationships are often not simple one–one relationships. One enzyme may catalyze several different reactions. For example, the enzyme fatty-acid synthase (Enzyme Commission [EC] Number 2.3.1.85) catalyzes approximately 30 reactions in the fatty acid synthesis pathway. Unfortunately, in most enzyme databases only the main reaction catalyzed is listed for each enzyme; for some enzymes with wide substrate activity, a general compound name such as "an aldehyde" is used. Therefore, some reactions that occur in reality may not be included in the reconstructed metabolic network. As far as we know, the KEGG LIGAND database is the most complete metabolic reaction database (20), including more than 6,000 enzyme-catalyzed or non–enzyme-catalyzed biochemical reactions. Most of the known reactions catalyzed by an enzyme are listed, thus allowing for reconstruction of more complete metabolic networks.

2.2. High-Quality Metabolic Network Reconstruction

The aforementioned sequence similarity–based, high-throughput reconstruction method is necessary for comparative analysis of large-scale metabolic networks because it allows an automatic approach to reconstructing networks for many organisms at the same time. However, there is a trade-off between high productivity and the high quality. For example, the networks reconstructed in such a high-throughput manner may be not complete, for the following reasons:

1. There are some non–enzyme-catalyzed reactions that occur spontaneously in a metabolic network. These reactions should be added to the metabolic network to avoid artificial missing links in the reconstructed metabolic network.
2. Enzyme Commission numbers are often used in linking an annotated gene with one or more metabolic reactions. However, only chemically well-characterized enzymes are given an EC number by the International Union of Biochemistry and Molecular Biology (IUBMB). For this reason, many enzymes are often found to have an incomplete EC number (e.g., 1.2.-.-) in the genome annotation database. It is necessary to develop a set of new IDs, or to just use enzyme names, for these unclear enzymes to correctly map a reaction to a gene.
3. Many enzymes for which the reactions catalyzed have been experimentally determined are not found in any fully sequenced genomes. Among the 4,223 enzymes in KEGG database, 2,572 are not found to be coded by any gene in any fully sequenced organism. The reason for this may be that the functions of a large part of the genes in a genome are unknown. For this reason, Karp (21) recently called for an Enzyme Genomics Initiative to find coding sequences for these enzymes.

The poor quality of the reconstructed network is made evident by comparing the metabolic networks for the same organism from a variety of databases. Even for *Escherichia coli*, one of the best-studied organisms, there are discrepancies between the networks from different databases such as KEGG, ExPASy, and EcoCyc (22). Therefore, human curation to improve the quality of the reconstructed network is necessary for any further biological function analysis. New reactions with biochemical and physiological evidences should be added to the network. Literature survey is needed to verify or correct the gene–enzyme relationship obtained from sequence similarity search. Reactions catalyzed by unclear enzymes should be clarified and added to the network. Unfortunately, these network quality improvement processes are relatively time consuming. Thus, for only a small number of organisms, such as *E. coli* and *Saccharomyces cerevisiae* (13,23–25), are the high quality metabolic networks available.

The quality of the reconstructed network can also be improved by comparative genomics. The IdentiCS tool offered a direct visualization of the differences among the metabolic network (pathways) of closely related organisms (17). Figure 2 shows a comparison of citrate acid cycle (TCA cycle) among five *Aspergillus* species. The differences among

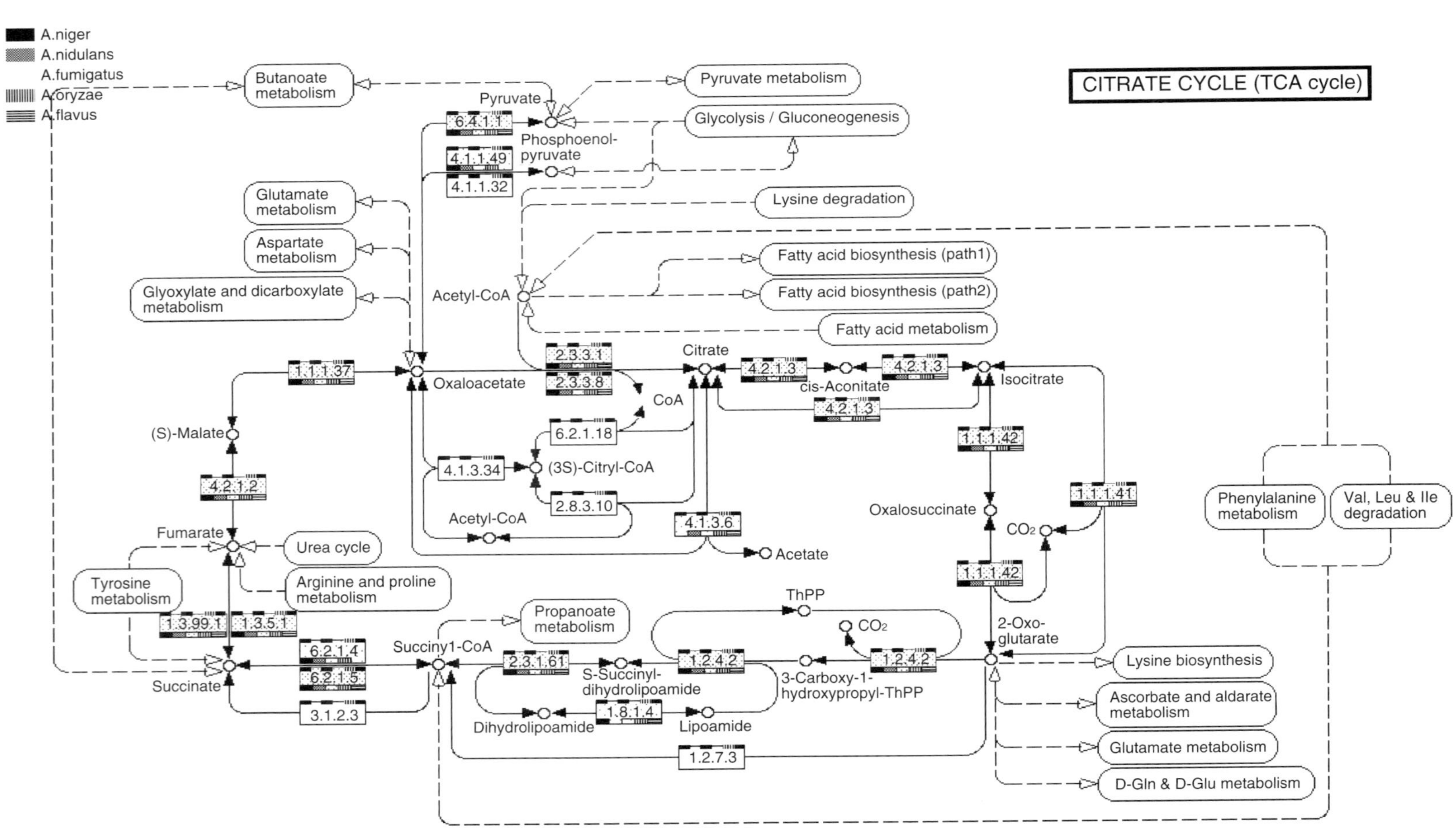

Figure 2. Visual comparison of metabolic pathways constructed from unfinished genomic sequence (*Aspergillus niger*), cDNA annotation (*A. flavus*), and genome annotation (the remaining three *Aspergilli*). Small boxes under the EC numbers indicate the presence of this enzyme in the corresponding organism, and the small rectangles above the EC numbers represent hyperlinks to online databases, such as KEGG, EcoCyc, IUBMB, BRENDA, ExPASy, or Google, for fast access to relevant information. Densely dotted background of EC numbers means that all compared organisms have such enzymes, whereas a loosely dotted one means not all, but at least one, of the compared organisms have such enzymes.

closely related organisms should be further evaluated by bioinformatics analysis or literature review to confirm the genomic annotation.

Genomic annotation efforts generally cannot annotate more than 50% of all coding sequences into a functional category. This is partially because of our limited knowledge on some metabolic pathways. Incompleteness of annotation leads to an incomplete reconstruction of metabolic network. Sometimes, a physiological phenotype (for instance, that a bacteria can live exclusively on a carbon source or produce some metabolites) may be known, but the genetic nature of the reac-tions is only partially known. In such cases, tools from comparative genomics may give useful indications. STRING from the European Molecular Biology Laboratory (http://string.embl.de) (26) is an excellent site that integrates many comparative genomics analysis tools, such as genomic neighborhood, phylogenetic/cooccurrence pattern, gene fusion events, coexpression pattern, etc., in a unified scoring system; thus, it is very useful to study the function of unknown genes and to find the potential missing links for a better reconstruction of the metabolic network.

2.3. Reconstruction of the TRN

The reconstructed metabolic network of a specific organism represents a static picture of the possible reactions in it. However, not all of these reactions occur in the cell at the same time. For example, the enzymes in a pathway for uptaking a specific substrate are induced only when the substrate is present in the growth media. In prokaryotes, the presence or the amount of an enzyme in the cell is mainly controlled by the TRN. The interactions in the TRN are between transcription factors and target genes. Responding to environmental changes, a transcription factor can bind to or dissociate from the binding site at the upstream region of its target genes, thus activating or repressing the expression of the target genes. Therefore, it would be beneficial to reconstruct the TRN for better understanding the dynamic regulation of metabolic networks.

The reconstruction of genome-scale TRN is not as easy a task as the sequence similarity–based, high-throughput metabolic network reconstruction (27). As shown in Figure 3, the regulatory relationship B1-A1 in organism 1 often cannot be directly transferred to organism 2 (B2-A2) because the short sequence of the binding site at the upstream of gene A2, rather than the sequence of A2 itself, determines if B2 can bind and regulate it. Although A2 has high similarity to A1, it may have very different binding site, and thus cannot be bound by B2. Because of this, the genome-scale TRNs are often reconstructed by systematically collecting regulatory interaction information from literature. Therefore, genome-scale TRNs are available only for certain well-studied organisms, such as *E. coli*, *Bacillus subtilis*, and *S. cerevisiae* (28–31). Several databases have been developed for storage and management of the curated regulatory interaction information from literature. Among them, RegulonDB (28) is the most prominent database for *E. coli* regulatory network, DBTBS (32) contains most of the known regulatory interactions in *B. subtilis*, and Prodoric (33) aims to include the regulatory interactions for prokaryotes; however, at this stage, the information in the database is mainly

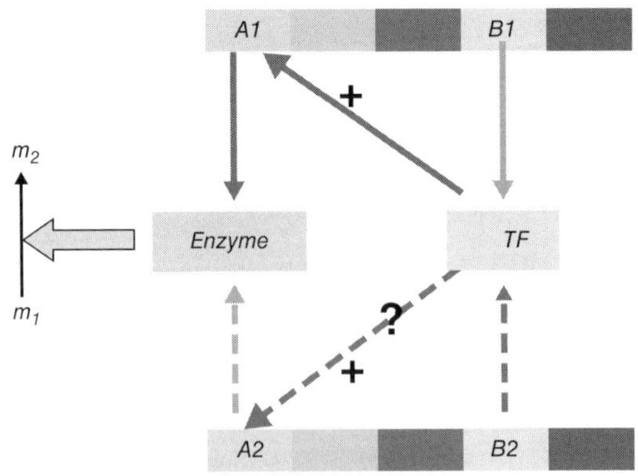

Figure 3. Comparison of the reconstruction of the metabolic network and regulatory networks. A1–A2 and B1–B2 are orthologous gene pairs between the two organisms. A1 is known to code for an enzyme that catalyzes a metabolic reaction m1 → m2. This enzyme function can be transferred to A2 from the orthologous relationship. B1 codes for a TF that activates A1. From this orthologous relationship, B2 may also be annotated as the same TF. However, it is not clear whether the regulatory interaction between A2 and B2 exists in organism 2.

for *E. coli*, *B. subtilis*, and *Pseudomonas aeruginosa*. TRANSFAC (34) is a database on eukaryotic (mainly yeast) transcription factors and their genomic binding sites. From these databases, we can extract the regulatory relationship between genes, and thus reconstruct the regulatory network. However, this process is often not straightforward because the databases often try to keep all the related information by using different files in different formats. For example, to obtain the regulatory relationships between genes from RegulonDB we have to extract information from six different files: product_table.dat (relationships between a gene and its coded polypeptide product), polyp_prot_link.dat (relationships between polypeptides and proteins), conformation_table.dat (the modified protein conformation), regulatory_interaction.dat (which promoter is regulated by which activated protein), transcription_unit.dat (relationships between transcription units and promoters), and trans_gene_link.dat (genes in a transcription unit). Therefore, to obtain the regulatory network in term of gene–gene regulation, we need to start from the interaction data between proteins and promoters in the file "regulatory_interaction.dat," and, based on the information in other files, to find the regulatory genes and regulated genes.

Because of the human curation feature of these databases, the regulatory information in one database is often not very complete. Therefore, integrating information from different databases is important to obtain more complete networks. Recently, Ma et al. (35) have produced an extended regulatory network of *E. coli* based on the information from RegulonDB, EcoCyc, and the work from Shen-Orr et al. (36). Certain new regulatory interactions are also added to the network directly from recently published literature. The resulting network includes 2,724 regulatory interactions among 1,278 genes. Surprisingly, only approximately one third of the interactions are common in all three data sources. Based on this extended network, both RegulonDB and EcoCyc have updated their databases (37,38).

2.4. The Integrated Network

In the aforementioned regulatory network of *E. coli*, approximately one third of the regulated genes code for metabolic enzymes. It is well known that changing gene expression level by transcriptional regulation is an important way to control the metabolic fluxes. On the other hand, the effect of a transcription factor on its target gene is also affected by the concentration of certain metabolites, which can bind to it. Therefore, for better understanding of cellular regulation, specifically the metabolic regulation, it is necessary to integrate the different interactions to obtain the whole picture. Recently, several papers have been published on integrating various cellular interactions for network structure or functional analysis (39,40). An ongoing study is carried out in our group to collect all the available interactions in *E. coli*, including transcriptional regulation (between transcription factors and their target genes), metabolic reactions (between metabolites), metabolite–protein interaction (between metabolites and transcription factors), and protein–protein interactions (including signal transduction, mainly the two-component systems). Currently, the partially finished integrated network (using only genes as nodes) consists of 6,497 interactions between 1,549 genes. This integrated network allows us to obtain new findings, which cannot be revealed by analysis of any individual network. An example of revealed feedback regulations through metabolite–protein interactions will be discussed later in section 4.4.

3. Graph Representation of Biological Networks

There are two components in metabolic networks: reactions and metabolites. Therefore, a direct graph representation of a metabolic network should be a hypergraph including two types of nodes, which represent reactions and metabolites, respectively (also called a two-node network) (11). However, to facilitate structural analysis, a two-node network is often converted to two types of one-node network: metabolite graph and reaction graph. In a metabolite graph, the nodes represent metabolites, and the links are reactions. Correspondingly, in a reaction graph the nodes are reactions, and two reactions are linked if a metabolite is the substrate of one reaction and the product of another. There are reversible reactions and irreversible reactions in metabolic networks. Correspondingly, the links in the graph can be directed (called arcs in graph theory) or undirected (called edges). In most cases, the metabolic network should be regarded as a directed graph in structural analysis. It should be mentioned that graph representation is a simplified way to represent the metabolic network. Some information in the reaction equations may be missing in the graph. A reaction often has several links in the graph (sometimes in very different parts) because most reactions have multiple substrates and products. On the other hand, one link in the graph may represent several different reactions. For example, reaction $A + B = C$ will be represented as two links, A–C and B–C, in the metabolite graph. Two reactions, $A = C$ and $B = C$, will also be represented by the same two links. Therefore, in the metabolite graph, we cannot distinguish which reactions lead to the links. Therefore, a reverse step to map a link to its corresponding reaction(s) is required when providing biological interpretation for the results from graph analysis of a metabolic network.

An important issue in graph representation of a metabolic network is how to deal with the currency metabolites, such as H_2O, CO_2, ATP, etc. (5). Currency metabolites are normally used as carriers for transferring electrons and certain functional groups (phosphate group, amino group, one carbon unit, methyl group, etc.). When considering the connections through currency metabolites, structural analysis often produces biologically meaningless results. For example, in the glycolysis pathway, the path length (number of reaction steps in the pathway) from glucose to pyruvate should be 9 in terms of biochemistry. However, if ATP and ADP are considered as nodes in the network, then the path length between glucose and pyruvate becomes only 2 (the first reaction uses glucose and produces ADP, whereas the last reaction consumes ADP and produces pyruvate). This calculation of path length is obviously biologically not meaningful. Different approaches have been proposed to address this problem. A simple way is to exclude the top-ranked metabolites based on their connection degree (number of links connected with a metabolite) (41). The problem is that certain primary metabolites, such as pyruvate, may also have high degrees of connection. Moreover, currency metabolites cannot be defined, per se, by compounds, but should be defined according to the reaction. For example, glutamate (GLU) and 2-oxoglutarate (AKG) are currency metabolites for transferring amino groups in many reactions, but they are primary metabolites in the following reaction:

$$\text{AKG} + \text{NH}_3 + \text{NADPH} = \text{GLU} + \text{NADP}^+ + \text{H}_2\text{O}$$

The connections through them should be considered. The same situations are for NADH, NAD^+, ATP, etc. Another problem is for reactions such as this:

$$\text{AcORN(Acetyl-ornithine)} + \text{GLU} = \\ \text{ORN(Ornithine)} + \text{AcGLU(Acetylglutamate)}.$$

The acetyl group is transferred between GLU and ORN in this reaction. Only the connections AcORN–ORN and GLU–AcGLU should be included; AcORN–AcGLU and GLU–ORN should be excluded. Otherwise the path length from GLU to ORN will be 1, and this is not in accordance with the pathway in real biochemistry.

From the previous discussion, we can see that it is difficult to remove the connections through currency metabolites automatically using a program. Therefore, we manually checked the reactions that appear in the KEGG metabolic maps and added corresponding connections one by one (5). In this way, the reaction–connection relationships can be more accurately obtained and used to generate metabolite graphs from the lists of reactions of different organisms. As an example, Figure 4 depicts the two graphs (with and without connections through currency metabolites) for the reconstructed metabolic network of *Streptococcus pneumonia*. It can be seen that the one without currency metabolites is more realistic and more amenable for analysis. In contrast, the true

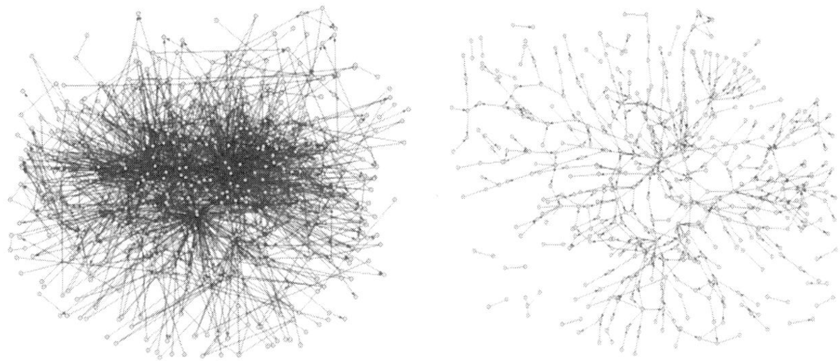

Figure 4. The metabolite graph representation of metabolic networks of *Streptococcus pneumonia*. The left network includes the connections through currency metabolites, and the right one does not. Links with arrow represent irreversible reactions, and those without arrow represent reversible reactions.

network structure in the graph with currency metabolites is masked by the large number of links through currency metabolites. Therefore, the removal of connections through currency metabolites is an essential step in drawing biologically meaningful conclusions from graph analysis of metabolic networks.

In a recent study, Croes et al. (42) presented a weighted graph approach to find biologically meaningful pathways using graph analysis. In the weighted graph, each compound in the network is assigned a weight equal to the number of reactions in which it participates. Then path finding is performed by searching for one or more paths with the lowest weight. They found that the correspondence between the computed and biologically annotated pathways approached 85% in the weighted graph. The main advantage of this method is that it can be performed automatically by a program based on the reaction list. Therefore, the very time-consuming process of manual examination of currency metabolites is not necessary.

Arita (43,44) proposed a different approach, called atomic reconstruction of metabolism, for graph representation of metabolic networks. In this approach, the atomic flow in a metabolic reaction is traced, and a substrate is connected only to the product(s) that contains at least one atom from it. An example is shown here for the following reaction:

ATP + D-Glucose = ADP + D-Glucose 6-phosphate

In this reaction, the link from D-glucose to ADP is not included in the graph because there is no atomic flow between these two metabolites. However, the other three links (ATP–ADP, ATP–D-glucose 6-phosphate, and D-glucose–D-glucose 6-phosphate) are all included in the resulting graph. Therefore, although this approach can avoid certain connections through currency metabolites, there are still biologically meaningless connections in the graph.

The graph representation of TRNs is relatively simple because there is only one type of node: genes. All the links in the TRN are directed because all the regulatory interactions are from a regulatory gene to a regulated gene. However, autoregulatory loops exist in TRN because

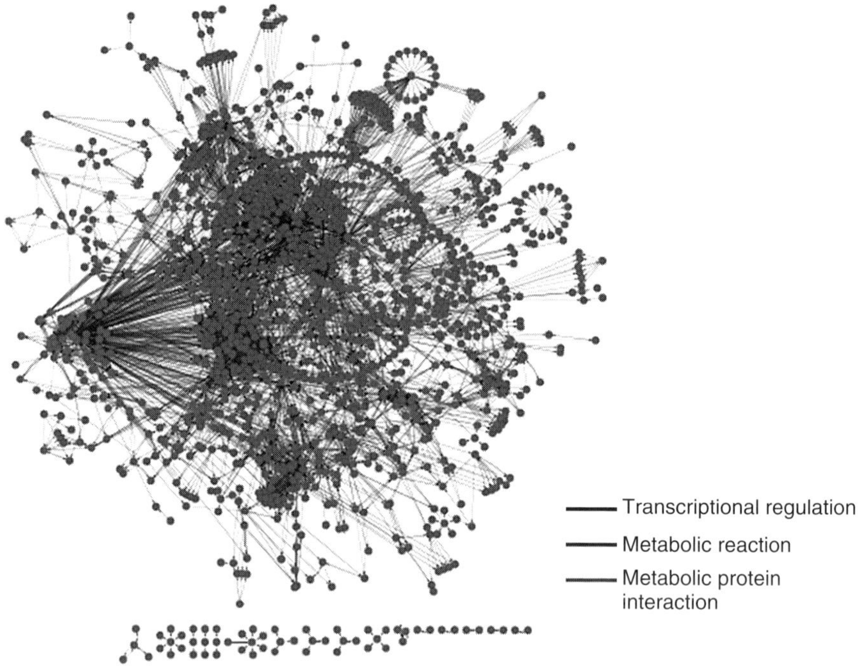

Figure 5. An integrated molecular network of *E. coli* comprising transcriptional regulation (dark gray edges), metabolic reaction (medium gray edges), and metabolic protein interaction (light gray).

many transcriptional factors (TFs) regulate their own gene expression. In TRN, a regulatory link can be active, repressed, or dual. In certain situations, such as network motif analysis, the different types of interactions should be considered.

For the integrated network, the graph representation is quite complex because it comprises different types of interactions involving different cellular components, such as genes, proteins, and metabolites. To reduce the complexity of graph representation, we have developed a way to represent the integrated network by a simple graph in which the nodes represent the genes and the multicolor edges represent various interactions. All the links in the network are directed, and an edge between two nodes *x* and *y* (genes) represents the interactions: transcription regulation if gene *x* codes for a transcription factor that regulates gene *y*, metabolic reaction if the genes *x* and *y* encode for enzymes catalyzing successive metabolic reactions, and metabolite protein interaction if the gene *x* encodes for an enzyme that catalyzes a reaction producing *metabolite m* as a co-factor for a transcription factor coded by gene *y*. The resulting graph for the integrated network of *E. coli* is illustrated in Figure 5.

4. Structural Analysis of Biological Networks

4.1. Degree Distribution and Average Path Length

The connection degree of a node is defined as the number of links connected with it. Several studies have shown that the degree distribution among all the nodes follow a power law relationship in metabolic networks and regulatory networks, as well as many other complex networks (11,45–50). This kind of network is called a scale-free network (48). The few high-degree nodes dominate the network structure and are called the hubs of the network. Most of the nodes are connected through the hubs by a relatively short path. The hub metabolites in the metabolic network include several metabolites in glycolysis pathway (glycerate-3-phosphate, pyruvate, D-fructose-6-phosphate, and D-glyceraldehyde-3-phosphate), pentose phosphate pathway (D-ribose-5-phosphate and D-xylulose-5-phosphate), and TCA cycle derivatives (acetyl-CoA, L-glutamate, and L-aspartate) (5). In regulatory networks, the hubs include several important global regulators such as CRP, RpoS, FNR, IHF, ArcA, etc (35,51). These results indicate that structurally important nodes also play functionally central roles.

In a directed network, the path length from node A to node B is defined as the number of steps in the shortest paths from A to B. Average path length (APL) of a network is the average of the path lengths for all connected pairs of nodes in the network (52). After excluding the currency metabolites, the APLs calculated for metabolic networks of various organisms are shown in Figure 6. Generally, APL tends to increase with the network scale. More interestingly, quantitative differences were found among the three domains of organisms; namely, that the metabolic networks of eukaryotes and archaea generally have a longer APL than those of bacteria. The average APL values for networks of these three domains of organisms are 9.57, 8.50, and 7.22, respectively. This result indicates that there are true structure differences between the metabolic networks of different organisms. This is in opposition to the result of Jeong et al. (11), who found that the APLs

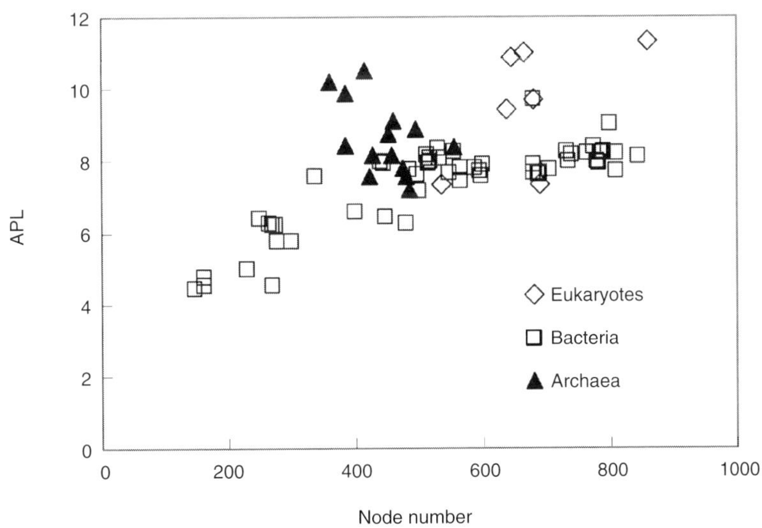

Figure 6. The calculated APL for the metabolic networks of fully sequenced organisms. A clear difference between path lengths of metabolic networks of the three domains of organisms can be seen.

for the metabolic networks of the 43 organisms studied were constant when the connections through currency metabolites were included.

The network structural differences are the result of a long evolutionary process. To explore this, we constructed evolutionary trees based on the reaction contents of metabolic networks for 82 fully sequenced organisms (53). We found that the major results from phylogenetic trees based on metabolic networks are surprisingly in good agreement with the tree based on 16S recombinant RNAs, despite the prevalence of horizontal transfer of metabolic genes among organisms, confirming the three-domain classification and the close relationship between eukaryotes and archaea at the level of metabolic networks. This indicates that the gene transfer events are constrained by some system-level organization principle(s).

4.2. Network Centrality

A method to analyze networks is to evaluate the location of the nodes in the network. Measuring node location in the network is finding the centrality of this node. The measurement of centrality helps determine the importance of a node in the network. Three different centrality measurements have been widely used in network analysis: degree centrality, closeness centrality, and betweenness centrality (47,54,55). The degree centrality of a node is defined as the fraction of nodes that are connected to each node. So:

$$C_D(n) = \frac{d_n}{N-1}, \qquad (1)$$

where d_i is the number of nodes connected to node n and N is the number of nodes in the network. Therefore in degree centrality, only the directly linked nodes are considered. In contrast, the closeness centrality of a node considers not only the directly connected nodes, but also the nodes connected with it through other nodes. The term "closeness centrality" was first introduced by Sabidussi (56). The closeness centrality of node x ($C(x)$) is defined as follows:

$$C(n) = \frac{N-1}{\sum_{m \in U, m \neq n} d(x, y)} = \frac{1}{\bar{d}}, \qquad (2)$$

where d(x,y) is the distance between node x and node y; U is the set of all nodes; $\bar{d}$ is the average distance between x and the other nodes. Based on closeness centrality, the most central nodes are the ones with the shortest paths to other nodes in the network. They form the core part of the whole network, whereas the periphery nodes have long paths to other nodes. Therefore it is better than degree centrality to show the overall location of the nodes in the whole network because some nodes in the periphery part of the network may also have a high degree centrality. The difference between the two measurements can be seen by the simple network in Figure 7. This is a typical social network called *kite network*, in which nodes represent people and links mean that the two people know each other. The centralities for all the nodes in this network

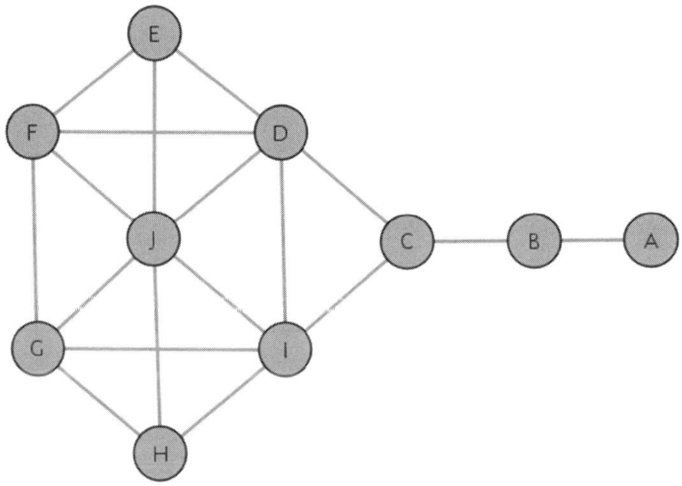

Figure 7. Kite network to show the different measures of centrality.

are listed in Table 1. It can be seen that **J** is the node with the highest degree centrality, whereas **D** and **I** have the highest closeness centrality, although they have fewer connections than **J**.

Metabolic networks are directed networks. Therefore, the average distance from a node to all other connected nodes is different from that of other nodes to that node. To address this, we can define output closeness centrality when $d(x,y)$ in eq. (2) is considered as the path length from x to y and input closeness centrality when $d(x,y)$ is the path length from y to x. The overall location of a node in the network can be described by the overall closeness centrality, which is defined as the reciprocal of the average of the mean input distance and the mean output distance. The 10 most central metabolites in the *E. coli* metabolic network based on these different centrality measures are listed in Table 2. Pyruvate is both the input and output center of the network. Eight of these central metabolites (pyruvate, acetyl-CoA, phosphoenolpyruvate [PEP], glyceraldehyde 3-phosphate [G3P], 2-dehydro-3-deoxy-6-phospho-D-gluconate [KDPG], malate, fumarate, and citrate) are in the central

Table 1. Different centrality measures for the nodes in the *Kite* network in Figure 7.

Betweenness centrality		Closeness centrality		Degree centrality	
C	**0.388889**	**D**	**0.642857**	**J**	**0.666667**
D	**0.231481**	**I**	**0.642857**	**D**	**0.555556**
I	**0.231481**	**J**	**0.6**	**I**	**0.555556**
B	**0.222222**	**C**	**0.6**	**F**	**0.444444**
J	**0.101852**	**F**	**0.529412**	**G**	**0.444444**
F	**0.023148**	**G**	**0.529412**	**H**	**0.333333**
G	**0.023148**	**H**	**0.5**	**E**	**0.333333**
H	**0**	**E**	**0.5**	**C**	**0.333333**
E	**0**	**B**	**0.428571**	**B**	**0.222222**
A	**0**	**A**	**0.310345**	**A**	**0.111111**

Table 2. The most central metabolites in the metabolic network of *E. coli*.

Output center		Input center		Overall Center	
Metabolite	**Mean distance**	**Metabolite**	**Mean distance**	**Metabolite**	**Mean distance**
Pyruvate	4.2198	Pyruvate	4.663	Pyruvate	4.4414
KDPG	4.6007	Acetyl-CoA	4.9011	Acetyl-CoA	4.7582
Acetyl-CoA	4.6154	Malate	4.9011	Malate	4.8864
G3P	4.696	Acetate	4.9194	KDPG	4.9286
Serine	4.7473	Formate	4.9853	Acetate	4.978
Acetaldehyde	4.7729	Fumarate	5.1978	Acetaldehyde	5.0311
DR5P	4.8608	KDPG	5.2564	G3P	5.0641
Cystine	4.8645	Citrate	5.2821	PEP	5.2106
Malate	4.8718	Acetaldehyde	5.2894	HOAKG	5.2491
PEP	4.8938	Methylglyoxal	5.3516	Methylglyoxal	5.2766

Abbreviations: DR5P, 2-Deoxy-D-ribose 5-phosphate; G3P, Glyceraldehyde 3-phosphate; HOAKG, D-4-Hydroxy-2-oxoglutarate.

metabolism, namely, the glycolysis and TCA cycle pathway. All other central metabolites are directly connected with one or more of these eight metabolites. For example, serine and cysteine can be directly converted to pyruvate by irreversible reactions; thus, they are output centers, but not input centers. It should also be mentioned that not all of these central metabolites have a higher connection degree. For example, KDPG is an important metabolic intermediate in the Entner–Doudoroff pathway, and it is identified as a central metabolite by all the closeness centrality measures. However, because it links with only two metabolites (pyruvate and G3P), it is not recognized as a central metabolite by degree centrality.

The betweenness centrality of a node is defined as the fraction of the number of the shortest paths that go through the node. The betweenness centralities for the nodes in the kite network are also shown in Table 1. Interestingly, node C is the one with the highest betweenness centrality, even though it has few direct connections (less than the average in the network). In many ways, node C has one of the best locations in the network—it is between two important constituencies. It plays a "broker" role in the network. Without it, **A** and **B** would be cut off from the rest of the network. Actually betweenness centrality has been used for decomposition of metabolic network into small subnetworks (57).

4.3. Network Global Connectivity: The "Bow-Tie" Structure

The scale-free property revealed by the power-law connection degree distribution is a local property of network. It cannot tell us anything about the global network structure. For example, both networks in Figure 8 show a power-law degree distribution. However, the left network is a fully connected one, whereas the right one consists of several disconnected subgraphs. To investigate the global network connectivity, new method(s) and parameter(s) are needed. In graph theory, two concepts have been widely used to describe the network connectivity: strongly connected component and weakly connected component (52). A subset of vertices in a network is called a strongly connected component if from every vertex of the subset we can reach every other vertex belonging to the same subset through a directed pathway. If the direction of links is

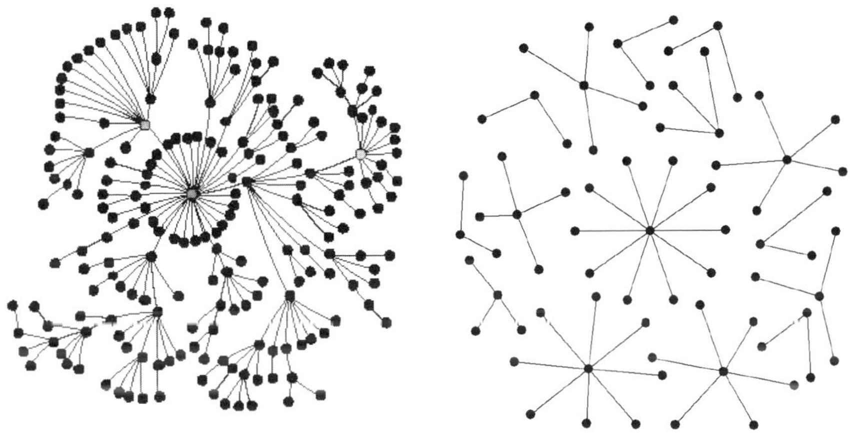

Figure 8. Two simple network examples to show the limitation of connection degree distribution. Both networks show power law degree distributions, but have apparently different network connectivity.

not important (consider the network to be undirected), such a subset is called a weakly connected component (52). Because metabolic networks are directed networks, we first calculate the strongly connected components in the network. The size distribution of these components in the metabolic network of *E. coli* is shown in Figure 9. It can be seen that the size distribution is very uneven. Most components are very small, whereas the largest component is very big (includes 10 times more nodes than the second largest component). This largest component is called the "giant strong component" (GSC). Next, we analyze the connectivity between the GSC and other parts of the network. A subset in which all the metabolites can be converted to metabolites in the GSC and a subset

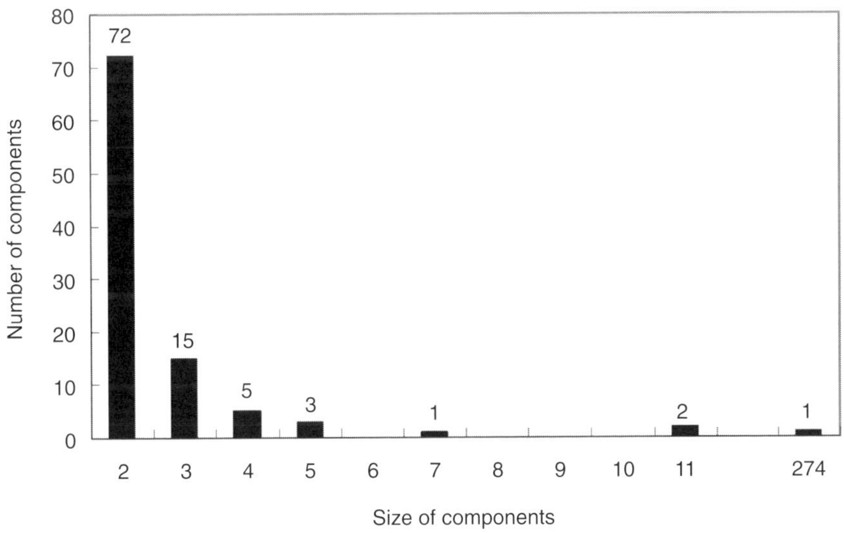

Figure 9. The size distribution of the strongly connected components in the metabolic network of *E. coli*.

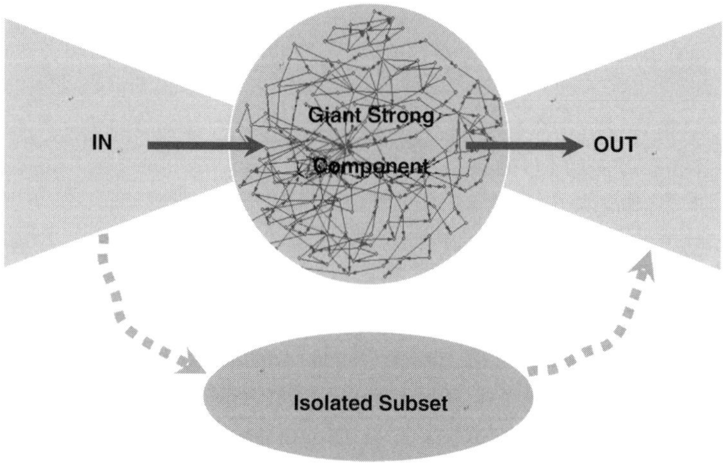

Figure 10. A cartoon show of the bow-tie connectivity structure of metabolic networks.

in which all of the metabolites can be produced from metabolites in the GSC were identified. The two subsets are called the IN and OUT subsets. All the other metabolites that are not connected with metabolites in the GSC form an isolated subset. In this way, we obtained the global connectivity structure of metabolic networks, as shown in Figure 10. This connectivity structure was also found in the metabolic networks of all other organisms studied. A similar connectivity structure has also been found by Broder et al. (58) in the Web page graph in which Web pages represent nodes and hyperlinks represent links. They called it a "bow-tie" connectivity structure. The discovery of the bow-tie structure in different kinds of networks implies that it is a common structure in large-scale networks. Organization as a bow tie may be important for the complex system to be robust under variable and undetermined environments (59,60).

The GSC is the most complex, and the core part of a metabolic network. We found that the GSC follows a similar power-law connection degree distribution to the whole network. Furthermore, the APL of the whole network was found to have a linear relationship with that of the GSC. This implies that the APL of the entire network is mainly determined by that of the GSC. Because of the large scale, it is often difficult to achieve a comprehensive understanding of the biological features of genome-based metabolic networks. A way to reduce the whole network is desired to make the network more amenable to functional analysis (7). The bow-tie connectivity structure of the metabolic network represents a step forward in this direction. For example, understanding and manipulating the distribution and control of metabolic fluxes over the metabolic network are key steps in metabolic engineering of organisms and therapy of certain metabolic diseases. However, for large-scale metabolic networks, the estimation of metabolic flux and control can be very difficult or even impossible. However, the most important part of the network,

GSC, normally contains less than one third of the nodes of the entire network, although it preserves the main features of the whole network. Large-scale metabolic networks are more feasible for analysis of flux distribution and identification of all the possible elementary flux modes or extreme pathways. The distribution of metabolic fluxes is mainly controlled by regulating the flux ratio at branch points. Most of the branch points are in the GSC. Therefore, one may focus on the GSC when studying the flux distribution and its regulation in metabolic network. This can largely simplify the analysis process.

4.4. Regulatory Network: The Multilayer Acyclic Structure

As described in the previous section, the metabolic networks are connected through a bow-tie structure. It would be interesting to investigate whether this structure exists in other biological networks. Surprisingly, we found that there are no strongly connected components in the extended regulatory network of *E. coli* (35). This means that there are no regulatory cycles (e.g., gene A regulates gene B and gene B also regulates gene A through another path) in the TRN of *E. coli*. Or, in other words, the flow of the regulatory signal in the network is one-way only, and there is no feedback. This implies an acyclic structure of the *E. coli* TRN in which the nodes can be placed in different layers according to their depth in the regulatory cascade. To identify such a structure, we rearranged the nodes in the following way:

1. Nodes that do not regulate other nodes (output connection degree is 0 when the autoregulatory loops are not considered) were assigned to layer 1 (the lowest layer).
2. We removed all the nodes already assigned to layer 1, and from the remaining network identified a set of nodes whose output connection degree is 0 and assigned them to layer 2.
3. We repeated step 2 to remove nodes that have been already assigned to a layer and identified the nodes with 0 output degree to make a new layer, until all the nodes were assigned to different layers.

Using this method, a 9-layer hierarchical structure of the *E. coli* regulatory network was uncovered, as shown in Figure 11. All the regulatory links in this graph are downward, and there is no link between operons in the same layer (except the autoregulatory loops). The genes at the bottom layers are the target genes, which are regulated only by other genes. Most of the global regulator genes such as *crp*, *rpoS*, *ihf*, *cspA*, *hns*, *rpoN*, *fis*, and *rpoE* are at the top layers of the hierarchy (36,61). However, this does not mean a global regulator requires more steps (through other, more specific regulators) to regulate a gene at the bottom layer. Actually, there are many shortcuts between the top global regulators and the genes at the bottom. The APL of the whole network is only 1.85, meaning most of the regulatory signal can be transferred to a target gene in less than 2 steps. This is important for the cell to respond to environmental perturbations in a fast and efficient way. In many cases, a global regulator regulates a target gene in the bottom layer together with a specific regulator, which is also regulated by the global regulator, forming a

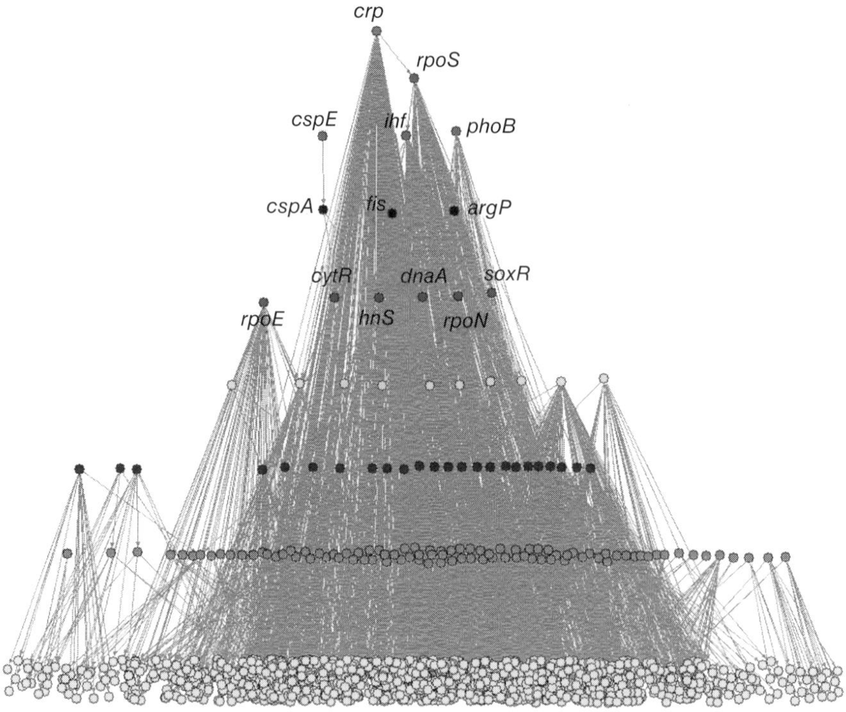

Figure 11. The multilayer hierarchical structure of the extended *E. coli* transcriptional regulatory network.

feed-forward loop, which is the most important network motif in the regulatory network (36).

The multilayer hierarchical structure of the *E. coli* TRN implies that no feedback regulation exists at transcription level. This raises the question of why the TRNs of these organisms possess such an acyclic hierarchical structure. A possible biological explanation for the existence of this hierarchical structure is that the interactions in TRN are between proteins and genes. Only after a regulating gene has been transcribed, translated, and, eventually, further modified by cofactors or other proteins, can it regulate the target gene. A feedback from the regulated gene at transcriptional level may delay the process for the target gene to access a desired expression level in a new environment. Feedback control may be mainly through other interactions (e.g., metabolite and protein interaction) at posttranscriptional level, rather than through transcriptional interactions between proteins and genes. For example, a gene at the bottom layer may code for a metabolic enzyme, the product of which can bind to a regulator, which in turn regulates its expression. In this case, the feedback is through metabolite–protein interaction to change the activity of the transcription factor, and then to affect the expression of the regulated gene. With the help of the integrated network, we really identified many feedbacks through metabolite–protein interaction. For example, the transcription factor TreR regulates two genes, *treB* and *treC*, in the trehalose-degradation pathway. One of the regulated genes, *treB*, located at the lowest layer in the multilayer hierarchical structure, codes for an enzyme in trehalose PTS transport system that catalyzes the following reaction:

PEP + trehalose = Trehalose-6-phosphate + Pyruvate

The metabolic product trehalose-6-phosphate (T6P) is a cofactor for TreR. Thus, through the interaction between T6P and TreR, we obtain a feedback link from *treB* to TreR.

5. Conclusions

One of the goals of systems biology is to develop theoretical models to describe and predict cellular behavior at the whole-system level. The structural and functional analysis of genome-based metabolic networks described in this chapter represents one step toward this goal. The macroscopic structures of the biological networks (scale-free, bow-tie, multilayer hierarchy), which were uncovered by analysis of the network as a whole, represent certain system-level principles governing the organization of interacting cellular components. Although these structural properties still give only a static picture of the whole system, they can serve as a basis or blueprint for analyzing the dynamic behavior of the network (e.g., information and material flows), which is the next necessary and more demanding step in network analysis.

References

1. Kitano H. Computational systems biology. *Nature* 2002;420(6912):206–210.
2. Papin JA, Hunter T, Palsson BO, et al. Reconstruction of cellular signalling networks and analysis of their properties. *Nat Rev Mol Cell Biol* 2005; 6(2):99–111.
3. Barabasi AL, Oltvai ZN. Network biology: understanding the cell's functional organization. *Nat Rev Genet* 2004;5(2):101–113.
4. Herrgard MJ, Covert MW, Palsson BO. Reconstruction of microbial transcriptional regulatory networks. *Curr Opin Biotechnol* 2004;15(1):70–77.
5. Ma HW, Zeng AP. Reconstruction of metabolic networks from genome data and analysis of their global structure for various organisms. *Bioinformatics* 2003;19(2):270–277.
6. Alon U. Biological networks: the tinkerer as an engineer. *Science* 2003;301(5641):1866–1867.
7. Bray D. Molecular networks: the top-down view. *Science* 2003;301(5641): 1864–1865.
8. Stelling J, Klamt S, Bettenbrock K, et al. Metabolic network structure determines key aspects of functionality and regulation. *Nature* 2002;420(6912): 190–193.
9. Milo R, Shen-Orr S, Itzkovitz S, et al. Network motifs: simple building blocks of complex networks. *Science* 2002;298(5594):824–827.
10. Wagner A, Fell DA. The small world inside large metabolic networks. *Proc R Soc Lond B Biol Sci* 2001;268(1478):1803–1810.
11. Jeong H, Tombor B, Albert R, et al. The large-scale organization of metabolic networks. *Nature* 2000;407(6804):651–654.
12. Palsson BO. In silico biotechnology. Era of reconstruction and interrogation. *Curr Opin Biotechnol* 2004;15(1):50–51.

13. Forster J, Famili I, Fu P, Palsson BO, et al. Genome-scale reconstruction of the *Saccharomyces cerevisiae* metabolic network. *Genome Res* 2003;13(2): 244–253.
14. Kanehisa M, Goto S, Kawashima S, et al. The KEGG resource for deciphering the genome. *Nucl Acids Res* 2004;32(90001):D277–D280.
15. Karp PD, Riley M, Saier M, et al. The EcoCyc and MetaCyc databases. *Nucleic Acids Res* 2000;28(1):56–59.
16. Overbeek R, Larsen N, Pusch GD, et al. WIT: integrated system for high-throughput genome sequence analysis and metabolic reconstruction. *Nucleic Acids Res* 2000;28(1):123–125.
17. Sun J, Zeng AP. IdentiCS–identification of coding sequence and in silico reconstruction of the metabolic network directly from unannotated low-coverage bacterial genome sequence. *BMC Bioinformatics* 2004;5(1):112.
18. Schomburg I, Chang A, Ebeling C, et al. BRENDA, the enzyme database: updates and major new developments. *Nucleic Acids Res* 2004;32(Database issue):D431–D433.
19. Gasteiger E, Gattiker A, Hoogland C, et al. ExPASy: the proteomics server for in-depth protein knowledge and analysis. *Nucl Acids Res* 2003;31(13): 3784–3788.
20. Goto S, Okuno Y, Hattori M, et al. LIGAND: database of chemical compounds and reactions in biological pathways. *Nucleic Acids Res* 2002;30(1): 402–404.
21. Karp P. Call for an enzyme genomics initiative. *Genome Biol* 2004;5(8):401.
22. Keseler IM, Collado-Vides J, Gama-Castro S, et al. EcoCyc: a comprehensive database resource for *Escherichia coli*. *Nucleic Acids Res* 2005;33(Database issue):D334–D337.
23. Becker SA, Palsson BO. Genome-scale reconstruction of the metabolic network in *Staphylococcus aureus* N315: an initial draft to the two-dimensional annotation. *BMC Microbiol* 2005;5(1):8.
24. Edwards JS, Palsson BO. The *Escherichea coli* MG1655 in silico metabolic genotype: its definition, characteristics, and capabilities. *Proc Natl Acad Sci USA* 2000;97(10):5528–5533.
25. Thiele I, Vo TD, Price ND, et al. Expanded metabolic reconstruction of *Helicobacter pylori* (iIT341 GSM/GPR): an in silico genome-scale characterization of single- and double-deletion mutants. *J Bacteriol* 2005;187(16): 5818–5830.
26. von Mering C, Huynen M, Jaeggi D, et al. STRING: a database of predicted functional associations between proteins. *Nucleic Acids Res* 2003;31(1): 258–261.
27. Yu H, Luscombe NM, Lu HX, et al. Annotation transfer between genomes: protein-protein interologs and protein-DNA regulogs. *Genome Res* 2004; 14(6):1107–1118.
28. Salgado H, Gama-Castro S, Martinez-Antonio A, et al. RegulonDB (version 4.0): transcriptional regulation, operon organization and growth conditions in *Escherichia coli* K-12. *Nucleic Acids Res* 2004;32 Database issue: D303–D306.
29. Ishii T, Yoshida K, Terai G, et al. DBTBS: a database of Bacillus subtilis promoters and transcription factors. *Nucleic Acids Res* 2001;29(1):278–280.
30. Luscombe NM, Babu MM, Yu H, et al. Genomic analysis of regulatory network dynamics reveals large topological changes. *Nature* 2004;431(7006): 308–312.
31. Guelzim N, Bottani S, Bourgine P, et al. Topological and causal structure of the yeast transcriptional regulatory network. *Nat Genet* 2002;31(1): 60–63.

32. Makita Y, Nakao M, Ogasawara N, et al. DBTBS: database of transcriptional regulation in Bacillus subtilis and its contribution to comparative genomics. *Nucleic Acids Res* 2004;32(Database issue):D75–D77.
33. Munch R, Hiller K, Barg H, et al. PRODORIC: prokaryotic database of gene regulation. *Nucleic Acids Res* 2003;31(1):266–269.
34. Matys V, Fricke E, Geffers R, et al. TRANSFAC: transcriptional regulation, from patterns to profiles. *Nucleic Acids Res* 2003;31(1):374–378.
35. Ma HW, Kumar B, Ditges U, et al. An extended transcriptional regulatory network of *Escherichia coli* and analysis of its hierarchical structure and network motifs. *Nucleic Acids Res* 2004;32(22):6643–6649.
36. Shen-Orr SS, Milo R, Mangan S, et al. Network motifs in the transcriptional regulation network of *Escherichia coli*. *Nat Genet* 2002;31(1):64–68.
37. Salgado H, Gama-Castro S, Peralta-Gil M, et al. RegulonDB (version 5.0): *Escherichia coli* K-12 transcriptional regulatory network, operon organization, and growth conditions. *Nucleic Acids Res* 2006;34(Database issue): D394–D397.
38. Salgado H, Santos-Zavaleta A, Gama-Castro S, et al. The comprehensive updated regulatory network of *Escherichia coli* K-12. *BMC Bioinformatics* 2006;7(1):5.
39. Herrgard MJ, Lee BS, Portnoy V, et al. Integrated analysis of regulatory and metabolic networks reveals novel regulatory mechanisms in *Saccharomyces cerevisiae*. *Genome Res* 2006;16(5):627–635.
40. Yeger-Lotem E, Sattath S, Kashtan N, et al. Network motifs in integrated cellular networks of transcription-regulation and protein-protein interaction. *Proc Natl Acad Sci USA* 2004;20;101(16):5934–5939.
41. Croes D, Couche F, Wodak SJ, et al. Metabolic PathFinding: inferring relevant pathways in biochemical networks. *Nucleic Acids Res* 2005;33(Web Server issue):W326–W330.
42. Croes D, Couche F, Wodak SJ, et al. Inferring meaningful pathways in weighted metabolic networks. *J Mol Biol* 2006;356(1):222–236.
43. Arita M. In silico atomic tracing by substrate-product relationships in *Escherichia coli* intermediary metabolism. *Genome Res* 2003;13(11):2455–2466.
44. Arita M. The metabolic world of *Escherichia coli* is not small. *Proc Natl Acad Sci U S A* 2004;101(6):1543–1547.
45. Barrett CL, Price ND, Palsson BO. Network-level analysis of metabolic regulation in the human red blood cell using random sampling and singular value decomposition. *BMC Bioinformatics* 2006;7:132.
46. Wolf YI, Karev G, Koonin EV. Scale-free networks in biology: new insights into the fundamentals of evolution? *Bioessays* 2002;24(2):105–109.
47. Jeong H, Mason SP, Barabasi AL, et al. Lethality and centrality in protein networks. *Nature* 2001;411(6833):41–42.
48. Strogatz SH. Exploring complex networks. *Nature* 2001;410(6825):268–276.
49. Albert R, Barabasi AL. Topology of evolving networks: local events and universality. *Phys Rev Lett* 2000;85(24):5234–5237.
50. Van N, V, Snel B, Huynen MA. The yeast coexpression network has a small-world, scale-free architecture and can be explained by a simple model. *EMBO Rep* 2004;5(3):280–284.
51. Ma HW, Buer J, Zeng AP. Hierarchical structure and modular organisation in the *Escherichia coli* transcriptional regulatory network. *BMC Bioinformatics* 2004;2004.
52. Batagelj V, Mrvar A. Pajek. Program for Large Network Analysis. *Connections* 1998;21(2):47–57.
53. Ma HW, Zeng AP. Phylogenetic comparasion of metabolic capacities of organisms at genome level. *Mol Phylogenet Evol* 2004;31(1):204–213.

54. Ma HW, Zeng AP. The connectivity structure, giant strong component and centrality of metabolic networks. *Bioinformatics* 2003;19(11):1423–1430.
55. Freeman LC. Centrality in social networks: Conceptual clarification. *Social Networks* 1979;1:215–239.
56. Sabidussi G. The centrality index of a graph. *Psychometrika* 1966;31:58–603.
57. Holme P, Huss M, Jeong H. Subnetwork hierarchies of biochemical pathways. *Bioinformatics* 2003;19(4):532–538.
58. Broder A, Kumar R, Maghoul F, et al. Graph structure in the Web. *Comp Networks* 2000;33(1–6):309–320.
59. Csete M, Doyle J. Bow ties, metabolism and disease. *Trends Biotechnol* 2004;22(9):446–450.
60. Kitano H. Biological robustness. *Nat Rev Genet* 2004;5(11):826–837.
61. Martinez-Antonio A, Collado-Vides J. Identifying global regulators in transcriptional regulatory networks in bacteria. *Curr Opin Microbiol* 2003;6(5): 482–489.

8

Cross-Species Comparison Using Expression Data

Gaëlle Lelandais and Stéphane Le Crom

Summary

Molecular evolution, which is classically assessed by the comparison of individual proteins or genes between species, can now be studied by comparing coexpressed functional groups of genes. This approach, which better reflects the functional constraints on the evolution of organisms, can exploit the large amount of data generated by overall, genome-wide expression analyses. To optimize cross-species comparison, particular caution must be used in the selection of expression data, using, for example, the most related experimental conditions between species. In addition, defining gene pairs having interspecies correspondence is also a critical step that can create misleading relations between genes and that could benefit from international annotation efforts like the Gene Ontology (GO) Consortium.

In this chapter, we describe methodologies based on global approaches or gene-centered methods that can be used to answer precise biological questions. Finally, through a set of examples, we show that expression profile comparison between species can help to discover functional annotation for unknown genes and improve orthology links between organisms.

Key Words: Microarrays; transcriptome; cross-species comparison; gene ontology; orthologs; paralogs.

1. Introduction

Comparing genomic properties of different species at varying evolutionary distances is a powerful approach to studying biological and evolutionary principles. Because entire genome sequences are available for a large number of organisms (1), gene and protein sequences have received the highest attention as the basis for investigating evolutionary changes (2). It has been valuable to develop methodologies based on comparative analyses for identifying coding and functional noncoding sequences, as well as sequences that are unique for a given organism. But evolution involves biological variations at many levels, and one of the next major steps is to understand how the genes interact to perform particular biological processes. High-throughput genomic technologies, particularly DNA microarray methods (3,4), monitor gene expression levels on a genomic scale. An increasing number of studies use DNA microarrays for comprehensive investigations of genetic network architecture, and this approach lends itself to comparative analysis of two or more model organisms (5). Distinguishing the similar from the dissimilar in large-scale data sets promises to improve fundamental understanding of both the universality and the specialization of molecular biological mechanisms.

2. Chapter Outline

This chapter is organized as follows. In section 3, we describe the minimal information required to compare expression data between species. In section 4, available methodologies are presented. This part focuses mainly on the yMGV and MiCoViTo tools developed respectively by Marc et al. (6) and Lelandais et al. (7). Finally in section 5, several applications that illustrate the potential of cross-species comparisons using expression data are discussed.

3. Information Required to Compare Expression Between Species

3.1. Choosing Expression Data

The accumulation of microarray data from multiple species provides new opportunities to i) discover how the genes interact to perform specific biological process and ii) study the evolution of properties of expression networks. Recent works have initiated comparative analyses of expression profiles from different organisms (*see* [5,8–12] for review). Results presented in these studies provide a lot of evolutionary information that substantially depends on the choice of expression data from the many available. In this section, an overview of two different approaches is presented: comparison via a compendium of expression profiles and comparison of a specific biological process.

3.1.1. Cross-Species Comparison Via a Compendium of Expression Profiles

A compendium of expression profiles is an expression matrix composed of a large number of DNA microarray experiments (generally more than 100). Introduced by Hughes et al. (13), the compendium approach has been used to assign potential functions to previously unknown genes by comparing their expression profiles to those of genes with known functions. Indeed, genes that encode proteins that participate in the same pathway or are part of the same protein complex often exhibit expression profiles that are correlated under a large number of various conditions in DNA microarray experiments. However, similar expression patterns do not necessary imply that genes are functionally related. For instance, apparent coexpression can occur by chance, as a result of the noisiness of microarray data. The statistical justification for the compendium approach is that with a thousand data points, it is highly unlikely to observe a significant correlation between two expression patterns by chance (14). In that respect, several studies have attempted to consolidate the compendium approach by identifying gene pairs exhibiting coexpression in multiple species and across a large number of arrays in each species (8–10). These gene pairs are likely to be functionally related, thanks to the evolutionary conservation of their coexpression.

3.1.2. Cross Species Comparison of a Specific Biological Process

In spite of very interesting results that demonstrate the potential of cross-species comparisons using expression data, the global approach that consists of the integration of large sets of unrelated microarray data prevents a precise comparison of the sets of genes involved in a particular cellular process. In this context, cross-species comparisons based on selected experiments that are as close as possible between species appear to be a more promising approach. Comparing expression results in various organisms that are subject to the same environmental changes is an alternative way to bring the most interesting answers on how conserved regulatory networks are. This principle was first pursued by Alter et al. (12), who compared time points during the cell cycle between yeast and human. Cross-species comparison of a specific biological process allows the separation of the expression profiles into those common to both species, as well as those specific for one or the other dataset (11). But a critical assessment of the obtained results relies on the evaluation of background noise associated to microarray data (15). When working with microarray data coming from different sources, a challenging part of the work is to be sure that the comparison is feasible. Until 2005, doubts persisted on how reproducible and comparable microarray results were, and studies coming from various laboratories and platforms found low correlation between expression data (*see* [16,17] for review). Three papers published in the May 2005 issue of *Nature Methods* (18–20) show that there is a high laboratory effect that takes place when analyzing microarray data that comes from various sources. It stresses the importance of the quality of the interspecies biological data if one wishes to carry out comparative transcriptomic analyses. The standardization of microarray protocols is of fundamental importance.

3.1.3. Standards for Microarray Data

To reduce the experimental variability coming from various collaborators, it is necessary to use the same standard protocols for sample preparation, RNA extraction, labeling, and hybridization. Concerning the raw data pretreatment, it is also essential to apply the same procedure for each experiment's results. To ensure the best comparative analysis of gene expression data coming from different laboratories, all associated parameters (protocols, pretreatment steps, statistical analyses, etc.) have to be clearly available, along with the measured expression values. In this context, a great international effort has been carried out by the Microarray Gene Expression Data Society supporting a standard, called Minimal Information About a Microarray Experiment, to enclose all descriptions of an experiment necessary to understand how data have been processed (21). Therefore, the standardization of microarray protocols, which is now accessible to numerous research groups, and the DNA sequence data from closely related species (22,23), led to a rapid accumulation of expression data that was directly comparable between organisms.

3.2. Defining Gene Pairs Having Interspecies Correspondence

Once expression data have been chosen for each species to be compared, coherent pairs of related genes, one in each organism, have to be defined. These gene pairs are needed to connect expression results between species, and thus compare properties of the transcriptional programs. To define gene pairs having interspecies correspondence, several types of information can be used. In particular, we can use sequence conservation or coherent functional gene annotation.

3.2.1. Sequence Conservation

Sequence comparisons provide insights into evolutionary relationships between species (2). These comparisons allow the detection of sequence conservation and give an overview of the components potentially conserved or species-specific in an organism. In particular, candidate orthologous gene pairs are now routinely identified between entirely sequenced genomes (24). Orthology defines the relationship between genes in different species that originate from a single gene in the last common ancestor of these species (25). Such gene pairs are most likely to share the same function and are good starting points to compare expression data across species.

3.2.1.1. How to Find Orthologous Gene Pairs: Detection of orthologous gene pairs is an important, but challenging, problem. Because orthologs, by definition, are related through evolutionary history, they should ideally be identified using phylogenetic methods (26). But construction of phylogenetic trees needs several steps (search for similar sequences, multiple alignments, etc.), which are poorly automated and require large resources of computing power to be systematically applied to entire genomes. Therefore, several automated ortholog detection methods have been developed and are now available (INPARANOID, OrthoMCL, etc.) (26,27). They are based on initial candidate ortholog identification using all-versus-all sequence comparisons between two genomes. To summarize, these algorithms start the detection of orthologs with calculation of all pairwise similarity scores between the complete sets of protein sequences from the two genomes. This is generally done with the BLAST program (28). Then, different approaches can be used. One of the most popular consists in detecting sequence pairs with mutually best hits (29). The idea is that if the sequences are orthologs, they should score higher with each other than with any other sequences. These sequences are called "reciprocal best hit" and their identification is finally followed by different methods (clustering and resolution of overlapping groups) to refine the output list of orthologs.

3.2.1.2. Orthology, Paralogy, and Homology: The concept of homologs is strictly defined as genes coming from a common evolutionary ancestor

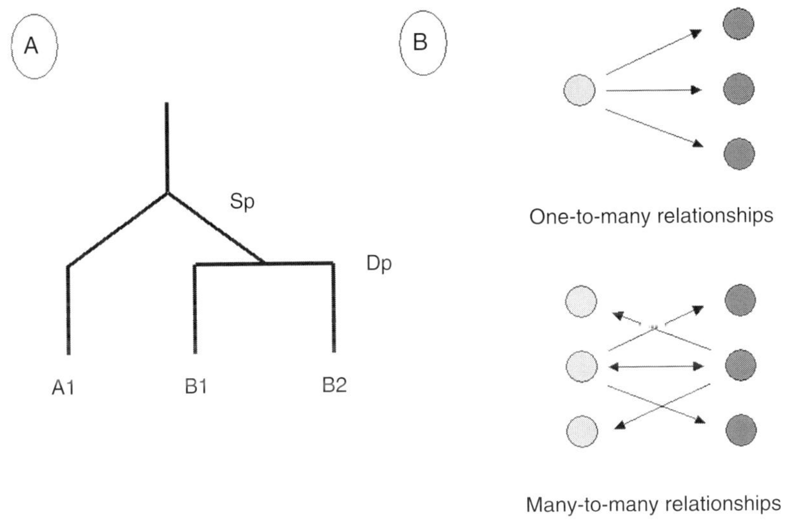

Figure 1. Orthology and paralogy definitions. Orthologs and paralogs are two types of homologous sequences. Orthology describes the relationship genes in different species that derive from a common ancestor. Paralogy describes the relationships between homologous genes within a single species that diverged by gene duplication. (A) After a speciation event (Sp), characters A and B were fixed in the two different species, creating orthologous relationships between genes A1 and B1/B2. Next, a duplication event (Dp) created two paralogous genes B1 and B2 in the second species. (B) Often, genetic rearrangements mask phylogenetic links between genes, leading to complex relationships between orthologous genes. For instance, we can observe one-to-many relationships when one gene in the first organism (light gray circle) gave multiple orthologs in the second organism (dark gray circles), or many-to-many relationships when multiple links exist between orthologous genes.

(25) (Figure 1A). Homologs are further classified as orthologs (two genes that predate speciation and that code functionally equivalent proteins that arise from evolution; *see* previous paragraph) and paralogs (genes which have arisen by duplication events and whose function generally have diverted from the original ancestor). Thus, two genes similar to each other at the sequence level can be related by different evolutionary history, but in addition to orthology and paralogy relationships, they can also result from convergent evolution, where two genes, previously unrelated, became similarly independently. This illustrates the complexity of orthology analyses (30), and that is the reason why ortholog detection methods often identify more than one putative ortholog in one or both species. In this case, the interspecies-related genes have one-to-many or many-to-many relationships (26) (Figure 1B). Defining interspecies gene pairs using sequence information could sometimes lead to complex situations, especially when species are separated by a long evolutionary distance.

3.2.2. Coherent Functional Gene Annotation

Another way to define gene pair associations is to use functional gene annotation. It could help to solve complex similarity links between genes by separating neofunctionalization from subfunctionalization phenomenan (31). In this context, the GO Consortium (32) has made a great effort to standardize gene annotation across species. GO is a structural network consisting of defined terms and relationships between them that describe three attributes of gene products: molecular function, biological process, and cellular components (33). Functional annotation is therefore a promising tool, but is still facing some shortcomings that could prevent its use to define precise interorganism gene pairs. In particular, functional annotations are comparatively incomplete because of the way they are built (manually curated) and of mass data gathering on a very small number of model organisms and genes (34). Annotation bias is then found in databases according to researcher interests for some scientific domains. Nevertheless, since the beginning of the GO project, the number of organism groups participating in the consortium, as well as the number of genes precisely annotated, has prodigiously grown every year (35). This should lead to the definition of more and more relevant interorganism gene associations.

4. Available Tools and Methodologies

In the previous sections, we have presented how to choose expression data and how to define gene pairs to compare gene expression between species. In the following, we will take a closer look at the methodologies available. There are two main approaches to perform gene expression comparison between species. The first group of methodologies is based on mostly using a large dataset without any *a priori* on the genes to be found, which is in contrast to the second group of tools, which is based on a "gene-centered" concept.

4.1. Global Approaches

To ensure interspecies comparison using expression data, one of the first methods to be used was to work with large datasets using various experimental conditions. As introduced in Section 3, the first compendium dataset (13) was done in *Saccharomyces cerevisiae* using a set of 300 microarray experiments. This compendium was available in a database, which allowed one to gather experimental results; for example, to compare expression profiles to one obtained in another organism. This methodology was reinforced by the work done by the laboratory of Jürgh Bähler on gene expression in the fission yeast *Schizosaccharomyces pombe*. Using microarrays, a set of experiments was obtained on cell cycle regulation, sexual differentiation, and response to stress and environment (36,37). These experiments were conducted in the same experimental conditions as those used with *S. cerevisiae* (38,39), and allowed scientists to make interspecies comparisons between these two related, but distant, yeast species and to draw correlations between gene expression and gene conservation (40). If one wants to compare the expression of a new organism, the best way to do so is to follow the same experimental conditions as those already available.

Some help can also be found using tools linking expression results and GO-functional annotation. As an example, the GoMiner tool (41) allows browsing expression results according to GO annotation to search for functional correlations. This can be done for microarray experiment results coming from various organisms, and correlation can be found searching for common motifs in the di-acyclic graph (DAG) outputs obtained with significantly expressed genes. In addition, the integration of "context" related to experimental conditions can increase the power of association made using GO annotation. Currently, when multiple annotations are found for one gene, searching in DAG can rapidly increase the complexity of the information. Merging experimental information from other experimental procedures, or reducing the complexity by focusing on specific experimental conditions, helps in finding relevant correlations.

4.2. Gene-Centered Approaches

Contrary to the global methodologies, gene-centered approaches aimed at discovering correlations between expression profiles across different species, focusing on one gene. Indeed, one of the most reliable ways to go deeper into the data to capture interesting trends is to be an expert in the field. In that respect, tools allowing the biologist to analyze a subset of genes related to his area of expertise were highly desirable. That is the reason why we developed the yMGV (42) and MiCoViTo (7) tools to help researchers mining expression data from this gene point of view.

4.2.1. yMGV (yeast Microarray Global Viewer)

The initial philosophy of yMGV was to empower biologists with a data-mining interface, and generate easily interpretable and mostly graphical outputs (6). Recently, intraspecies data analyses carried out by yMGV have been extended to incorporate data allowing comparisons of gene expression between orthologs (42). To facilitate these comparisons, a *S. cerevisiae* to *S. pombe* orthology table based on sequence similarity has been stored in the database (43). The Web interface allows users to retrieve genes based on expression levels in specified experiments. When used with discrimination, this tool should help the fission yeast community to easily take advantage of the huge amount of available information on the budding yeast transcriptome. In addition, as it has been shown that standard clustering methods are usually less effective when applied to large numbers of data sets (compendium) that are biologically unrelated (44), the microarray experiments in yMGV are hand-curated and classified into 17 biologically coherent categories. A module is available that lists genes that are significantly coexpressed in respect to a user-selected reference gene according to one of the 17 biological categories. This proved to be very efficient for isolating genes co-regulated only in specific conditions.

4.2.2. MiCoViTo (Microarray Comparison Visualization Tool)

With the MiCoViTo tool, users can identify and visualize groups of genes having similar expression in two sets of microarray experiments representing two distinct transcriptome states. This tool allows the biologist to mine microarray results to find expression modifications in a subset of genes related to his area of expertise. Such "gene-centric" clustering analysis, which distinguishes differentially expressed genes in specific parts of the transcriptome, overcomes some drawbacks of global analysis approaches. With MiCoViTo, a given transcriptome state can be represented as a network where genes are joined pairwise by a weighted link proportional to their corresponding expression profiles. The basic idea is therefore to compare the immediate transcriptome neighborhood of a given gene (seed gene) in two sets of microarray experiments describing two distinct transcriptome states. Using this approach to compare transcriptomes from different organisms captured in the same state will be possible by incorporating functional annotation in the same format for two organisms and defining pairs of seeds having interorganism correspondence. This will be part of future development of the MiCoViTo tool.

5. Using Expression Data Provides New Insight into Cross-Species Comparisons

The availability of genome sequences and genome-wide biological data provides a large amount of information that can be analyzed to enhance fundamental understanding of the universality, as well as the specialization, of molecular biological mechanisms. Comparative analysis of expression data between two or more model organisms provides new insight into cross-species comparisons. In this last section, several applications for cross-species comparisons are discussed. Note that for the sake of clarity, particular examples are also presented. They have been extracted from a cross-species comparison of the yeasts *S. cerevisiae* and *S. pombe* (45). These two model organisms have been chosen because two very similar sets of microarray time-course experiments (one for each organism) were available. Two different laboratories have used DNA microarray to study the transcriptional program that drives the developmental process of sporulation (36,38), in which diploid cells undergo meiosis to produce haploid germ cells.

5.1. Application 1: Improving Functional Gene Annotations

With the rapid increase in the number of sequenced genomes (1), one of the major ambitions of the postgenomic era is the functional elucidation of newly sequenced ORFs (46). Functional gene annotations of new ORFs are often predicted based on sequence similarity with genes of known functions. Despite the success of this approach, the absence of a direct relationship between sequence similarity and functional similarity of two proteins is a well-recognized limitation. An ORF can have several close homologs, all involved in different functions. In this case, gene expression analysis can provide functional information, complementary to that from sequence data. Indeed, homologous genes whose function has been conserved are expected to be expressed in a similar way. For that reason, finding an evolutionary conservation of expression patterns between species can help to identify genes that are genuinely function-

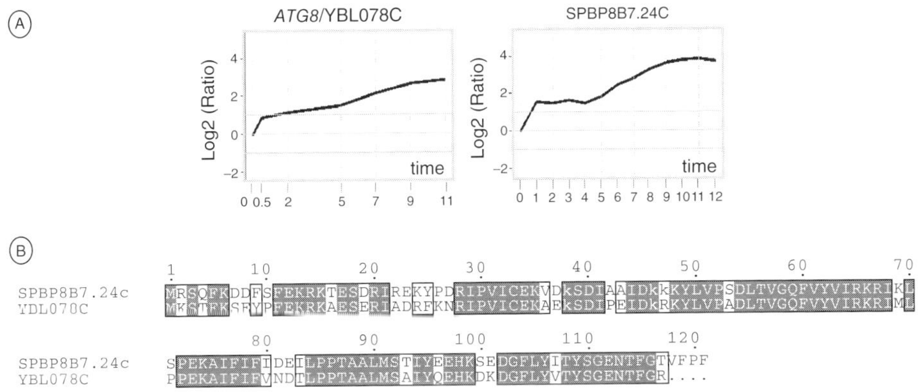

Figure 2. Conservation of expression between organisms can be used for improving functional gene annotation. (A) Expression profile of *S. cerevisiae* gene *ATG8* during the time course experiments described in Chu et al. (38) and expression profile of its *S. pombe* ortholog gene SPBP8B7.24C during the time course experiments described in Mata et al. (36). (B) Sequence alignment between the amino acid sequences of the genes *ATG8* and SPBP8B7.24C. Alignment and colored version of the result were generated using the Tcoffee Web server (50).

ally related. Several studies have integrated cross-species expression and sequence comparisons to systematically infer gene functions (8,9). In Stuart et al. (8), the authors compared the correlated patterns of gene expression in more than 3,000 DNA microarrays from humans, flies, worms, and yeasts. They identified genes that were coexpressed across multiple organisms and demonstrated that multiple-species analyses tend to retain coexpression links between functionally related genes, whereas it discards spurious gene-expression links.

As an illustration, we can consider the *S. cerevisiae* gene ATG8 (YBL078C) and its *S. pombe* ortholog SPBP8B7.24C. In this case, sequence homology and expression profiles during sporulation are conserved between the two yeasts (Figure 2). The SGD database (47) contains experimental data and a precise description of the gene ATG8 (48); it encodes a protein that mediates attachment of autophagosomes to microtubules. Whereas in the GeneDB database (49), the only information (as of October 2005) about the gene SPBP8B7.24C (a "predicted autophagy-related microtubule-associated protein") is inferred from sequence homology. However, the demonstration of conservation of expression substantially strengthens the functional gene annotation (Figure 2). The evolutionary conservation of expression patterns between species provides functional information, complementary to that from sequence data, and helps identify genes that are functionally related.

5.2. Application 2: Refining Orthologous Links Between Organisms

We saw in section 3 that the identification of genes that are orthologs between species is an important point in cross-species comparisons. But we also pointed out a major difficulty; using automated

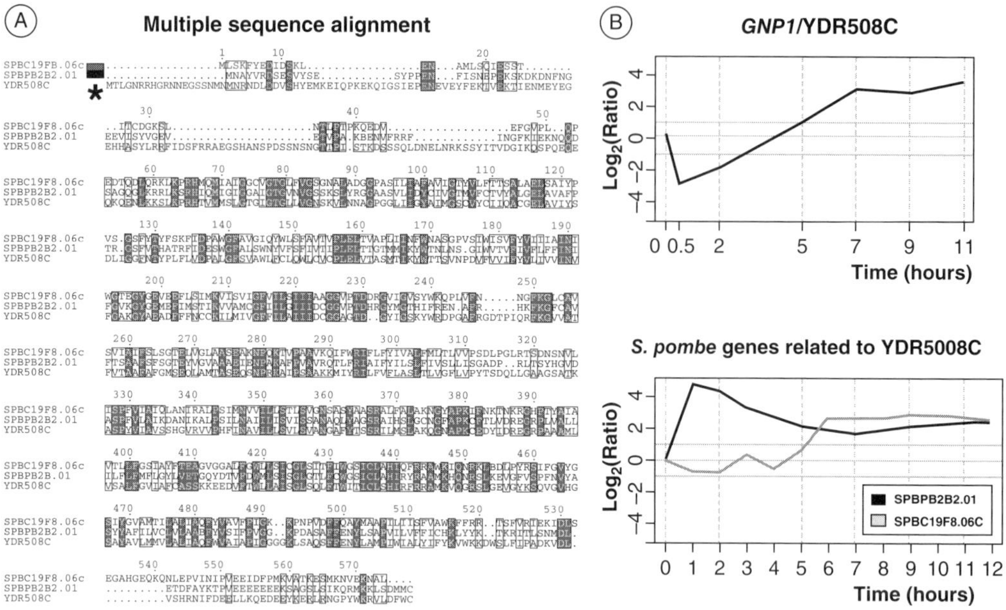

Figure 3. Coexpression can be used for refining orthologous links between organisms. (A) Multiple sequence alignment between the amino acid sequences of the three genes, whose expression profiles are represented in B. Boxes at the top allow for discrimination between sequences (SPBC19F8.06C/meu22, light gray; SPBPB2B2.01, black; GPN1/YDR508C, black asterisk). Alignment and colored version of the alignment were generated using the Tcoffee Web server (50). Gray residues correspond to highly reliable portions of the multiple alignments. (B) Expression profile of GPN1 during the time course experiments as described in Chu et al. (38), and expression profiles of the two *S. pombe* genes related to GPN1 with high similarity scores (SPBPB2B2.01, black; SPBC19F8.06C/meu22, light gray) during the time course experiments described in Mata et al. (36).

orthology-detection methods, it is sometimes not possible to determine a unique relationship between two organisms for some amino acid sequences. Again, the combination of sequence and expression data allows the discrimination of homolog genes, which cannot be distinguished by sequence comparison alone.

As a simple illustration, we can consider the *S. cerevisiae* gene *GPN1* (YDR508C). Using the INPARANOID algorithm (29), two orthologs have been detected in the *S. pombe* genome: SPBPB2B2.01 and SPBC19F8.06C/meu22 (Figure 3A). This situation may be the consequence of lineage-specific gene duplications generating multiple paralogs in one species (in this case *S. pombe*), or deletion events resulting in the loss of the "true ortholog" of a gene in *S. cerevisiae*. Nevertheless, in such a case it is nontrivial to determine which of the genes is functionally equivalent to the ortholog in the other species. But, as two laboratories have used DNA microarrays to study the transcriptional program that drives the sporulation process, it has been possible to plot the corresponding expression profiles (Figure 3B). Interestingly, of the two *S. pombe* expression profiles, that of meu22 (SPBC19F8.06C, light gray) is clearly different from the expression profile of SPBPB2B2.01, but very similar to that of GPN1. Such an observation suggests that, *in fine*, only the orthologous link between GPN1 and meu22 is reliable.

6. Conclusion

Comparing genomic properties of different organisms is of fundamental importance. It is now clear that a large fraction of the genes specifying the core biological function is shared by all eukaryotes. In this sense, comparative functional genomics is a powerful approach to distinguishing the similar from the species-specific features of biological process. Yet, the challenge of systematically comparing expression data between organisms is only starting to be addressed, but provide unprecedented opportunities to understand the evolution of biological systems.

We can hope in the future that cross-species comparison will benefit from the improvements made on microarray data quality and gene annotation. Indeed, we can expect that standardization process will create more reliable microarray datasets that will be comparable between each other. In the meantime, increasing knowledge on gene annotation will enrich GO databases. With this information, powerful cross-species comparison could be performed using, for example, a complex graph-based algorithm approach. We can also hope that a real demand will come from the biologist community for easy to use and reliable tools to mine these cross-species data.

References

1. Bernal A, Ear U, Kyrpides N. Genomes OnLine Database (GOLD): a monitor of genome projects world-wide. *Nucleic Acids Res* 2001;29:126–127.
2. Frazer KA, Elnitski L, Church DM, et al. Cross-species sequence comparisons: a review of methods and available resources. *Genome Res* 2003;13: 1–12.
3. Schena M, Shalon D, Davis RW, et al. Quantitative monitoring of gene expression patterns with a complementary DNA microarray. *Science* 1995; 270:467–470.
4. Eisen MB, Brown PO. DNA arrays for analysis of gene expression. *Methods Enzymol* 1999;303:179–205.
5. Zhou XJ, Gibson G. Cross-species comparison of genome-wide expression patterns. *Genome Biol* 2004;5:232.
6. Marc P, Devaux F, Jacq C. yMGV: a database for visualization and data mining of published genome-wide yeast expression data. *Nucleic Acids Res* 2001;29:E63–3.
7. Lelandais G, Marc P, Vincens P, et al. MiCoViTo: a tool for gene-centric comparison and visualization of yeast transcriptome states. *BMC Bioinformatics* 2004;5:20.
8. Stuart JM, Segal E, Koller D, et al. A gene-coexpression network for global discovery of conserved genetic modules. *Science* 2003;302:249–255.
9. Bergmann S, Ihmels J, Barkai N. Similarities and differences in genome-wide expression data of six organisms. *PLoS Biol* 2004;2:E9.
10. Lefebvre C, Aude JC, Glemet E, et al. Balancing protein similarity and gene co-expression reveals new links between genetic conservation and developmental diversity in invertebrates. *Bioinformatics* 2005;21:1550–1558.

11. McCarroll SA, Murphy CT, Zou S, et al. Comparing genomic expression patterns across species identifies shared transcriptional profile in aging. *Nat Genet* 2004;36:197–204.
12. Alter O, Brown PO, Botstein D. Generalized singular value decomposition for comparative analysis of genome-scale expression data sets of two different organisms. *Proc Natl Acad Sci USA* 2003;100:3351–3356.
13. Hughes TR, Marton MJ, Jones AR et al. Functional discovery via a compendium of expression profiles. *Cell* 2000;102:109–126.
14. Kim SK, Lund J, Kiraly M, et al. A gene expression map for *Caenorhabditis elegans*. *Science* 2001;293:2087–2092.
15. Carter DE, Robinson JF, Allister EM, et al. Quality assessment of microarray experiments. *Clin Biochem* 2005;38:639–642.
16. Jordan BR. How consistent are expression chip platforms? *Bioessays* 2004; 26:1236–1242.
17. Marshall E. Getting the noise out of gene arrays. *Science* 2004;306:630–631.
18. Bammler T, Beyer RP, Bhattacharya S, et al. Standardizing global gene expression analysis between laboratories and across platforms. *Nat Methods* 2005;2:351–356.
19. Irizarry RA, Warren D, Spencer F, et al. Multiple-laboratory comparison of microarray platforms. *Nat Methods* 2005;2:345–350.
20. Larkin JE, Frank BC, Gavras H, Sultana R, Quackenbush J. Independence and reproducibility across microarray platforms. *Nat Methods* 2005;2:337–344.
21. Brazma A, Hingamp P, Quackenbush J, et al. Minimum information about a microarray experiment (MIAME)-toward standards for microarray data. *Nat Genet* 2001;29:365–371.
22. Kellis M, Patterson N, Endrizzi M, et al. Sequencing and comparison of yeast species to identify genes and regulatory elements. *Nature* 2003;423:241–254.
23. Dujon B, Sherman D, Fischer G, et al. Genome evolution in yeasts. *Nature* 2004;430:35–44.
24. Remm M, Storm CE, Sonnhammer EL. Automatic clustering of orthologs and in-paralogs from pairwise species comparisons. *J Mol Biol* 2001;314: 1041–1052.
25. Fitch WM. Homology a personal view on some of the problems. *Trends Genet* 2000;16:227–231.
26. Storm CE, Sonnhammer EL. Automated ortholog inference from phylogenetic trees and calculation of orthology reliability. *Bioinformatics* 2002;18: 92–99.
27. Li L, Stoeckert CJ, Jr., Roos DS. OrthoMCL: identification of ortholog groups for eukaryotic genomes. *Genome Res* 2003;13:2178–2189.
28. Altschul SF, Gish W, Miller W, Myers EW, Lipman DJ. Basic local alignment search tool. *J Mol Biol* 1990;215:403–410.
29. O'Brien KP, Remm M, Sonnhammer EL. Inparanoid: a comprehensive database of eukaryotic orthologs. *Nucleic Acids Res* 2005;33 Database Issue: D476–D480.
30. Sonnhammer EL, Koonin EV. Orthology, paralogy and proposed classification for paralog subtypes. *Trends Genet* 2002;18:619–620.
31. He X, Zhang J. Gene complexity and gene duplicability. *Curr Biol* 2005; 15:1016–1021.
32. Ashburner M, Ball CA, Blake JA, et al. Gene ontology: tool for the unification of biology. The Gene Ontology Consortium. *Nat Genet* 2000;25: 25–29.
33. Harris MA, Clark J, Ireland A, et al. The Gene Ontology (GO) database and informatics resource. *Nucleic Acids Res* 2004;32:D258–261.

34. Ogren PV, Cohen KB, Hunter L. Implications of compositionality in the gene ontology for its curation and usage. *Pac Symp Biocomput* 2005:174–185.
35. Lewis SE. Gene ontology: looking backwards and forwards. *Genome Biol* 2005;6:103.
36. Mata J, Lyne R, Burns G, et al. The transcriptional program of meiosis and sporulation in fission yeast. *Nat Genet* 2002;32:143–147.
37. Chen D, Toone WM, Mata J, et al. Global transcriptional responses of fission yeast to environmental stress. *Mol Biol Cell* 2003;14:214–229.
38. Chu S, DeRisi J, Eisen M, et al. The transcriptional program of sporulation in budding yeast. *Science* 1998;282:699–705.
39. Gasch AP, Spellman PT, Kao CM, et al. Genomic expression programs in the response of yeast cells to environmental changes. *Mol Biol Cell* 2000;11: 4241–4257.
40. Mata J, Bahler J. Correlations between gene expression and gene conservation in fission yeast. *Genome Res* 2003;13:2686–2690.
41. Zeeberg BR, Feng W, Wang G, et al. GoMiner: a resource for biological interpretation of genomic and proteomic data. *Genome Biol* 2003;4:R28.
42. Lelandais G, Le Crom S, Devaux F, et al. yMGV: a cross-species expression data mining tool. *Nucleic Acids Res* 2004;32 Database issue:D323–D325.
43. Wood V. Schizosaccharomyces pombe comparative genomics: from sequence to systems. In: Comparative genomics using fungi as models (Sunnerhagen P, Piskur J, eds.), vol. 15, pp. 233–285. New York: Springer; 2006.
44. Gasch AP, Eisen MB. Exploring the conditional coregulation of yeast gene expression through fuzzy k-means clustering. *Genome Biol* 2002;3: RESEARCH0059.
45. Lelandais G, Vincens A, Badel-Chagnon S, et al. Comparing gene expression networks in a multi-dimensional space to extract similarities and differences between organisms. *Bioinformatics* 2006;22(11):1359–1366.
46. Enault F, Suhre K, Claverie JM. Phydbac "Gene Function Predictor": a gene annotation tool based on genomic context analysis. *BMC Bioinformatics* 2005;6:247.
47. Christie KR, Weng S, Balakrishnan R, et al. *Saccharomyces* Genome Database (SGD) provides tools to identify and analyze sequences from *Saccharomyces cerevisiae* and related sequences from other organisms. *Nucleic Acids Res* 2004;32 Database issue:D311–D314.
48. Lang T, Schaeffeler E, Bernreuther D, et al. Aut2p and Aut7p, two novel microtubule-associated proteins are essential for delivery of autophagic vesicles to the vacuole. *EMBO J* 1998;17:3597–3607.
49. Hertz-Fowler C, Peacock CS, Wood V, et al. GeneDB: a resource for prokaryotic and eukaryotic organisms. *Nucleic Acids Res* 2004;32 Database issue: D339–D343.
50. Poirot O, O'Toole E, Notredame C. Tcoffee@igs: A Web server for computing, evaluating and combining multiple sequence alignments. *Nucleic Acids Res* 2003;31:3503–3506.

9

Methods for Protein–Protein Interaction Analysis

Keiji Kito and Takashi Ito

Summary

Protein–protein interactions (PPIs) play key roles in various aspects of cellular regulation. Accordingly, PPI analysis is crucial for systems-level understanding of any biological process. This chapter describes two major methods for PPI analysis, namely, mass spectrometry (MS)–based approaches and yeast two-hybrid (Y2H)–based ones. Both methods have been the major driving forces of interactome mapping to reveal the characteristics of global PPI networks. Furthermore, MS-based methods enable a quantitative analysis of interaction, thereby providing insight into system dynamics. Alternatively, Y2H-based methods provide interaction-defective and separation-of-function alleles, which are useful for system perturbation. These two methods would thus contribute not only to system identification but also to system analysis, thereby serving as invaluable tools for systems biologists.

Key Words: Mass spectrometry; affinity tag purification; tandem affinity purification tag (TAP tag); stable isotope labeling; yeast two-hybrid system; reverse two-hybrid system; interactome.

1. MS-Based Approaches for PPIs

1.1. Identification of Proteins with MS

In MS-based proteomics, two major methods, namely, peptide mass fingerprinting (PMF) and tandem MS (MS/MS) analysis, are widely used for protein identification (1–3). In either method, proteins were digested with proteases into a set of fragments, and the resultant peptides are ionized and introduced into a mass spectrometer.

1.1.1. Protein Identification with PMF

In PMF, masses of proteolytic peptides derived from a protein are measured simultaneously (or as a set) with MS (Figure 1). Trypsin is the most popular enzyme to cleave the main chain of protein at the C-terminal side of arginine and lysine. It is possible to calculate a set of masses for

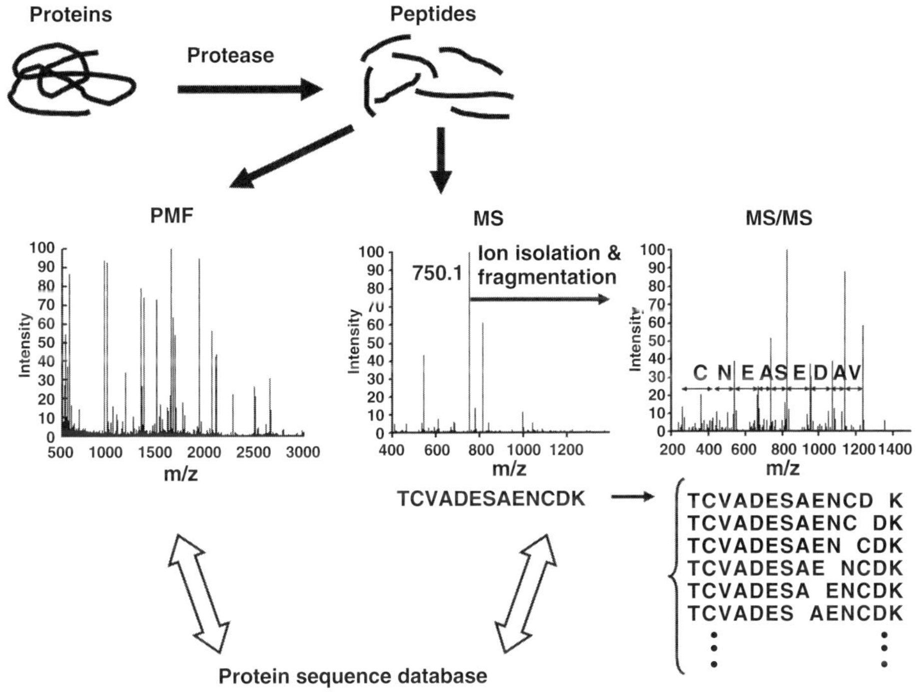

Figure 1. Protein identification with PMF and MS/MS analysis.

the peptides produced by trypsin digestion of each protein in the database. An experimentally measured set of peptide masses can be compared with each theoretically calculated set of masses using a search engine (Figure 1). If sufficient matches are found between the experimental data and a theoretical set of peptide masses from a protein X in the database, the protein of interest is identified as protein X.

In PMF, mass spectra are usually obtained using matrix-assisted laser desorption/ionization (MALDI)–time-of-flight (TOF)–MS, in which the analytes are ionized via MALDI, a method for soft ionization, and the masses of ions are measured by a TOF mass analyzer. Samples crystallized with matrix are sublimated and ionized by laser pulse, and the ionized analytes are separated according to the difference in duration of flight from ion source to mass detector in TOF-mass analyzer, which depends on the mass-to-charge ratio (*m/z*).

PMF is a simple and easy method for protein identification. MALDI-TOF-MS can detect a wide range of masses to achieve high coverage of peptides from a simple sample that contains only a few proteins. On the other hand, PMF is not suitable for the analysis of complex samples composed of multiple protein species because of the difficulty in assignment of the complicated spectra derived from many proteins.

1.1.2. Protein Identification with MS/MS

In MS/MS analysis, fragment ions of individual peptides generated in MS equipment are measured to obtain sequence information. In the first step, MS spectrum is recorded for ionized peptides introduced into the instrument. In the second step, a peptide ion to be analyzed, such as the most intense one, is isolated and fragmented, and MS/MS spectrum is obtained for the resultant fragment ions (Figure 1). Peptides dissociate most frequently at the peptide bonds via a low-energy, collision-induced dissociation (CID), which is a method for ion fragmentation by collision with inert gas molecules and is frequently used for proteomics analysis.

If every fragment ion generated from the analyzed peptide can be detected, it is possible to read out the amino acid sequence (*de novo* sequence). However, perfect MS/MS spectra including every fragment ion are rarely recorded, because of uneven breakage of individual peptide bonds. Usually, masses of the isolated peptide and its fragment ions are compared with those theoretically calculated from the protein sequence database (Figure 1). A score, which indicates the reliability of peptide identification, is assigned to each MS/MS analysis according to the degree of matches between the experimentally measured masses and theoretically predicted ones.

In this manner, peptides can be identified by MS/MS analysis through the patterns of fragment ions. Because we know which entry in the database contains each identified peptide sequence, proteins in the analyzed samples can be specified by the identified peptides. MS/MS analysis is highly sensitive and suitable for analyzing complex samples, because target ion is isolated from the others, including background noise, before the analysis and because a single highly reliable assignment of MS/MS spectrum to a peptide may be enough to identify the protein. Of note, the MS/MS approach provides not only the mass of a peptide but also its amino acid sequence, thereby making protein identification much more reliable than the PMF approach, which depends solely on the masses. As described in the following sections, MS/MS analysis is amenable to be coupled with peptide separation steps such as liquid chromatography (LC) before ion introduction into MS equipment, providing a technical platform for high-throughput analysis of complex protein mixtures.

Two types of MS instrument, the ion-trap analyzer and hybrid type of the quadrupole-TOF combination, are most widely used in MS/MS analysis for proteomics, coupled with electrospray ionization (ESI) as an ion source. In the ESI method, which is one of the soft ionization techniques, the analytes in a solvent are ionized by application of high voltage. Ionization of the analytes in solution makes it easy to couple the ESI with LC, thereby allowing the implementation of online separation steps before ionization.

1.1.3. Identification of Multiple Proteins in Complex Samples

To identify multiple proteins from a complex mixture sample, separation steps are required before MS analysis. In PMF approach, proteins are usually resolved with sodium-dodecyl sulfate (SDS)-polyacrylamide gel electrophoresis (PAGE) or two-dimensional (isoelectric focusing [IEF] and SDS-PAGE) gel electrophoresis (2DE), and individual bands or spots are excised and subjected to in-gel digestion with a sequence-specific protease (1,2). A set of peptide masses were measured with MALDI-TOF-MS as described in previous sections.

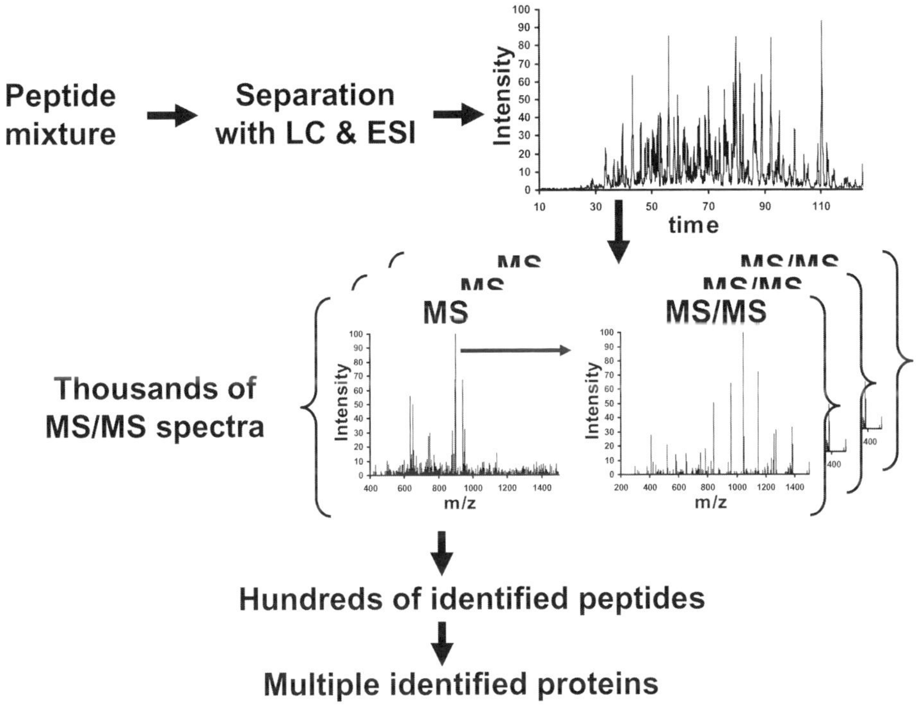

Figure 2. Identification of multiple proteins with LC-MS/MS analysis.

In MS/MS analysis, one or two-dimensional electrophoresis is also used as the separation step for proteins to be analyzed (1,2). In contrast with PMF, MS/MS relies on peptide sequence and the mass of each peptide, but not the combination of peptide masses. Accordingly, there is no need for simultaneous detection of a set of peptide ions derived from a protein; each peptide can be separated and introduced individually (or at different times) into the MS instrument. In the case of ESI, it is possible to connect LC directly with the ion source, thereby minimizing the loss of samples to achieve higher sensitivity and throughput (2,3). In such LC-ESI-MS/MS analysis, peptides separated using reverse-phase LC are directly introduced into the MS instrument via ESI, and a multitude of MS/MS spectra are obtained with an automated data acquisition system. This system is widely used in proteomics and has a significant impact on the analysis of complex samples containing a great number of proteins (Figure 2). In particular, multidimensional protein identification technology (MudPIT) using 2-dimensional LC, which combines strong cation-exchange and reverse-phase chromatographies, enables high-throughput identification of more than 1,000 proteins at once (4). Thus, a MS-based protein identification system serves as a powerful technical platform for large-scale analysis of protein interactions, in combination with the technology for purification of protein complexes.

1.2. Isolation of Protein Complexes with Affinity Tag Purification

1.2.1. Purification of Protein Complexes with Affinity Tags Introduced into the Target Proteins

Protein complexes were traditionally purified by biochemical methods, such as gel filtration chromatography and density gradient centrifugation. These approaches, which require optimization of purification procedures in a case-by-case manner, are labor-intensive and time-consuming. Affinity purification, which uses an affinity probe for the target protein (bait), is a convenient method that can be applied to a variety of protein complexes using a unified procedure. Although antibodies against bait protein would be the best affinity probes to purify the protein complex, a good antibody is not always available for the protein of interest. It is thus general to attach an affinity tag specifically recognized by antibodies or ligands to the bait protein (1–3,5).

In affinity purification, a bait protein fused with an affinity tag and its associated proteins are captured on an appropriate affinity resin to be separated from unbound proteins (Figure 3). Protein complexes are subsequently eluted with reagents, to dissociate the interaction between the affinity tag and the resin. Various tags have been developed for affinity purification, including short peptides, binding domains, or proteins, which have been also used for purification of recombinant proteins (6). Because a standard purification procedure can be established for each tag, a large-scale analysis of protein complexes can be conducted using a unified protocol.

For the budding yeast, a DNA fragment coding an affinity tag can be introduced into the genomic locus of the target protein via homologous

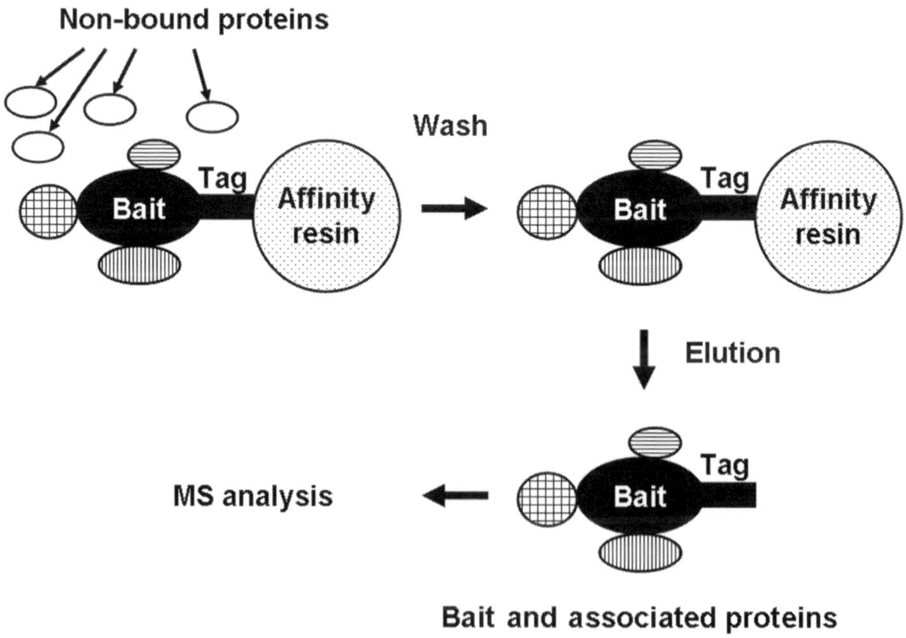

Figure 3. Purification of protein complexes with affinity tag.

recombination. This strategy allows the tagged proteins to be expressed under its own promoter or at the physiological level, thereby eliminating artifacts induced by overproduced proteins. For higher eukaryotes, such as mammals, tagged-proteins are generally expressed from a plasmid or viral vector. In any case, the expression levels should be compared between the tagged protein and its endogenous counterpart to eliminate the effects of over- and underexpression of the former protein. Although affinity tags can be attached either to the N- or to the C-terminal, it is needless to say that every possible care should be taken to keep the protein function as intact as possible.

1.2.2. TAP Tag

Affinity tags can be designed not only for single-step but also for two-step purification procedures. Single-step purification generally results in higher yield and retention of weak interactors, but tends to suffer from a higher amount of contaminants. In contrast, two-step purification, by a serial use of two different affinity tags, substantially improves the purity, but tends to result in a lower yield and dissociation of weak interactors. The most popular procedure for two-step purification would be the tandem affinity purification (TAP) method, in which the affinity tag (TAP tag) consists of calmodulin-binding peptide (CBP), tobacco etch virus (TEV) protease recognition sequence, and immunoglobulin-binding domain of protein A (PrA) (7). At the first step of TAP procedure, the TAP-tagged complex, which is composed of the TAP-tagged bait protein and its associated proteins, is captured on immunoglobulin resin via PrA. After an appropriate washing step, the captured protein complex is specifically released from the resin by cleavage with TEV protease. The TEV eluate is subjected to the second purification step, in which the TAP-tagged complex is bound to calmodulin resin via CBP. After an appropriate washing step, the complex is released by chelating Ca^{2+} with EGTA. The protein components of the purified complex can be identified by MS analysis. The TAP method has been widely used for identification of novel proteins in a particular complex and for comparative analyses of orthologous complexes in different organisms (8–12).

1.3. Large-Scale Analysis of Protein Complexes

1.3.1. Two Approaches for Large-Scale Analysis of Protein Complexes

A large-scale analysis of protein complexes was enabled via a combination of high-throughput MS-based protein identification and systematic standardization of affinity purification procedures. As pioneering works, two studies on yeast protein complexes were carried out using TAP and FLAG epitope tag for affinity purification (13,14). In the former case, TAP tag was inserted into the genomic locus by homologous recombination, producing the fused protein at a natural expression level. Protein complexes were isolated with the aforementioned two-step purification, followed by protein identification with SDS-PAGE and PMF analysis. In the latter approach, termed as HMS-PCI (high-throughput mass spectrometric protein complex identification), each FLAG-tagged bait protein was expressed from a plasmid vector using an inducible promoter. Protein purification was performed with a single-step immunoprecipitation using anti-FLAG antibody resin, followed by SDS-PAGE and LC-MS/MS analysis.

1.3.2. Comparison Between the Two Large-Scale Data Sets

The TAP and HMS-PCI studies used 589 and 600 bait proteins to successfully purify 454 and 493 complexes, respectively, each containing at least one associated protein identified by MS (15). Although 115 baits are common to both approaches, the interacting proteins (prey) of these baits showed only a marginal overlap (Figure 4): the TAP and HMS-PCI data sets contain 628 and 875 interactions from common baits, respectively, of which only 198 (15%) were shared. Among the protein complexes purified by the common 115 baits, 48 (42%) shared at least one protein, but the remaining 67 (58%) shared no proteins at all.

The small overlap may result from the differences in strategy between the two studies, because both studies reported that 70% of protein interactions were reproducibly detected (13,14). Several proteins recovered by many different bait proteins were assumed as contaminants, which presumably associated nonspecifically with bait proteins or affinity resin and were thus omitted from the analysis. HMS-PCI has a higher contaminant background (50% of total interactions) than TAP (20%). This may be attributed to the overexpression of tagged proteins and one-step purification procedure in HMS-PI (13,14). The use of a more sensitive detection system (or LC-MS/MS) may also contribute to the high incidence of contaminants in HMS-PCI. Alternatively, one-step purification could increase the chance of detecting weak and transient interactions. Indeed, HMS-PCI successfully identified protein interactions between

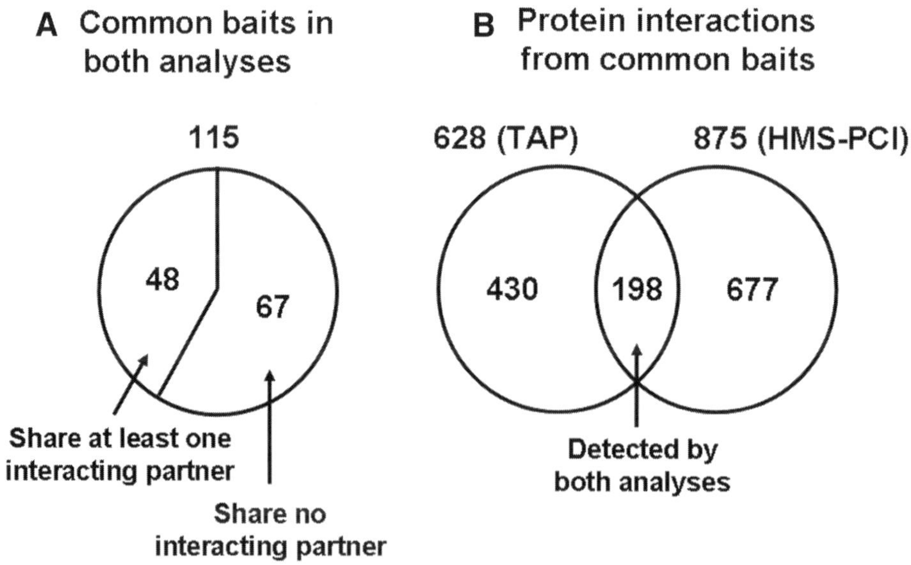

Figure 4. Comparison of two large-scale analyses of yeast protein complexes based on affinity tag purification and MS technology. The numbers indicated are from the previous report to compare these two analyses (15).

substrates and kinases, including mitogen-activated protein kinase and cyclin-dependent kinase (Cdk) (14).

The two large data sets were compared to other available information, such as functional categories of each protein and interactions previously reported in conventional studies or detected with different experimental strategies. Interacting protein pairs from TAP data set more often share the same functional categories than those from HMS-PCI (16). In addition, comparison of the interactions detected by the common baits indicates that TAP shows a higher coverage of known interactions than HMS-PCI (15). These assessments suggest that TAP strategy is more reliable than the HMS-PCI approach. On the other hand, HMS-PCI data contained more unknown proteins, which may be weak or transient interactors leading to the identification of novel interactions. Thus, in a sense, these two large-scale analyses are of complementary nature, and their integration would improve the accuracy and coverage of identified protein interactions. In terms of comparison with interactions detected by other methods, it should be noted that the two data sets show a similar small overlap with the protein interactions detected by comprehensive Y2H analyses (15).

1.3.3. Genome-Wide Analysis for Protein Complexes

After the initial large-scale analysis, the TAP strategy was further extended to a genome-wide scale: all of the open reading frames (ORFs) in the budding yeast were TAP-tagged, and 3,206 purifications were conducted to successfully purify 1,993 TAP-tagged proteins, approximately 90% of which contained at least one associated protein (17). These data clustered cellular proteins into different complexes, each of which often has several isoforms that share "core" components, but have different "attachment" components, presumably generating functional variety (Figure 5). Intriguingly, some of the proteins classified as attachments frequently co-occur in different complexes, thereby serving as a module for functional diversification (Figure 5). The analysis identified 491 complexes with 5,477 isoforms, 478 cores, and 147 modules to reveal the modular architecture of cellular machinery.

1.4. Focused Analysis on PPI Networks

Besides the global studies, MS-based PPI analysis was also performed in targeted studies to reveal a more precise interaction network in a particular cellular pathway. Targeted analysis allows more detailed, or finer, experiments, in which the proteins identified as prey in the first round of experiments are tagged to be used as bait in the second round to validate the interactions (reverse-tagging experiments); the biological relevance of the identified interactions is evaluated by integrating the results of various biochemical, genetic, and cell-based assays. Two examples are described in the following sections.

1.4.1. Analysis of Associated Proteins with Cdk

Cyclin-dependent kinases are the main regulators of cell cycle and are activated by binding to cell cycle stage–specific cyclins. Cyclin-Cdk modules phosphorylate a variety of substrates involved in the cell cycle

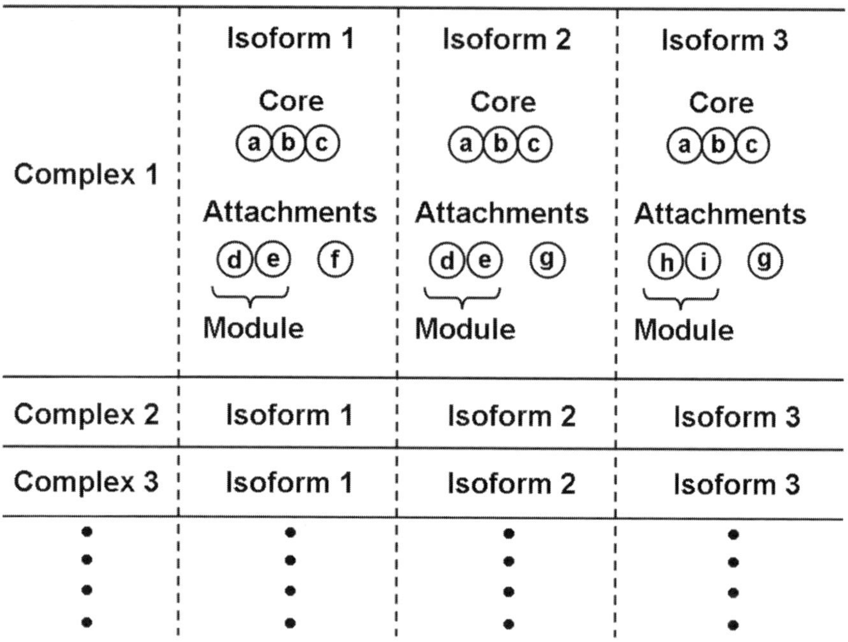

Figure 5. Schematic illustration for modularity of the yeast protein complexes. Architecture of protein complexes is illustrated, based on the report for a genome-wide analysis of yeast protein complexes (17). The letter in each circle represents a component.

progression, including transcription factors, replication machinery, chromosome segregation machinery, ubiquitin ligases, and cyclin inhibitors. Understanding of PPIs around Cyclin-Cdk is, thus, vital to uncover the mechanism by which cell cycle proceeds in a highly regulated manner.

Cyclin-associated proteins, including known interacting partners, were identified via MS analysis of the proteins copurified with each PrA-tagged cyclin, which was expressed from their genomic loci and purified by a single-step purification using immunoglobulin resin (18). Detected interactions were confirmed by reciprocal purification, in which the cyclin-associated proteins were tagged and purified for MS analysis to identify the cyclin as their binding partner. Intriguingly, the analysis revealed that some of the identified proteins were phosphorylated at Cdk consensus motifs, suggesting that they serve as novel substrates of Cyclin-Cdk.

Of note, many of the associated proteins with biological relevance confirmed in this approach had escaped detection in the large-scale TAP and HMS-PCI analyses (13,14,17). This may be because weak and transient interactions would be lost during two-step purification in TAP, and because overexpression of tagged proteins in HMS-PCI would inhibit effective identification of physiological interactions. Although the single-step purification results in the detection of many contaminants, specific interactions can be distinguished from the nonspecific ones via a quantitative analysis using stable isotope-labeling (19).

1.4.2. Protein Interactions in the Tumor Necrosis Factor (TNF)-α–induced NF-κB Signaling Pathway

NF-κB is activated by proinflammatory cytokines to induce expression of various genes playing central roles in mammalian immune response. For mapping the PPI network around this protein, 32 known or candidate components in the TNF-α-induced NF-κB signaling pathway were TAP-tagged and purified, along with associated proteins (20). To generate a reliable data set, purification was repeated at least four times for each bait protein, and the obtained data were compared with those from control purifications conducted at the same scale. To validate the involvement of detected interactors in the signaling pathway, they were knocked down by RNA interference, and the effects on the signal transduction were evaluated. This strategy succeeded in the ability to identify previously unknown components in this pathway with high confidence.

1.5. Quantification of Dynamics of PPIs with Stable Isotope-Labeling Methods

Besides protein identification, MS can be used to reveal quantitative differences between the samples, provided that either of them is labeled with stable isotope. Thus, integration of isotope-labeling techniques to the MS-based protein complex analysis allows one to obtain information on not only static but also dynamic aspects of PPIs, which would be more vital for modeling and systems-level understanding of the molecular events in the cells.

1.5.1. Methods of Protein Labeling with Stable Isotope

The intensities of ions detected by MS equipment do not directly correspond to the abundance of the peptide, as each peptide displays different ionization efficiency, depending on its intrinsic chemical property, the complexity of the sample, and the natures of coexisting peptides. Therefore, in quantitative MS analysis of proteins, intensity of each peptide ion should be compared with that of the peptide sharing the same chemical property, but having a different mass; i.e., the isotopically labeled peptide.

The labeling methods are largely classified into two categories, namely, chemical and metabolic labeling (21–24). In the former, proteins or peptides isolated from the cells are chemically labeled *in vitro* with stable isotope tags. In the latter, cells are cultivated in the presence of stable isotope-labeled essential nutrients, such as amino acids, to metabolically label the proteins *in vivo*. The chemical labeling methods can be applied to almost any type of samples and proteins. In contrast, metabolic labeling methods are more suitable for metabolically active cells and proteins turning over at a substantial rate.

1.5.1.1. Chemical Labeling Methods: The most popular chemical labeling method is the isotope-coded affinity tag (ICAT) approach, in which a compound consisting of a biotin affinity tag, a linker containing stable isotope, and a reactive part against the thiol group, is coupled to cysteine residues in the proteins (25). After the labeling of one protein sample

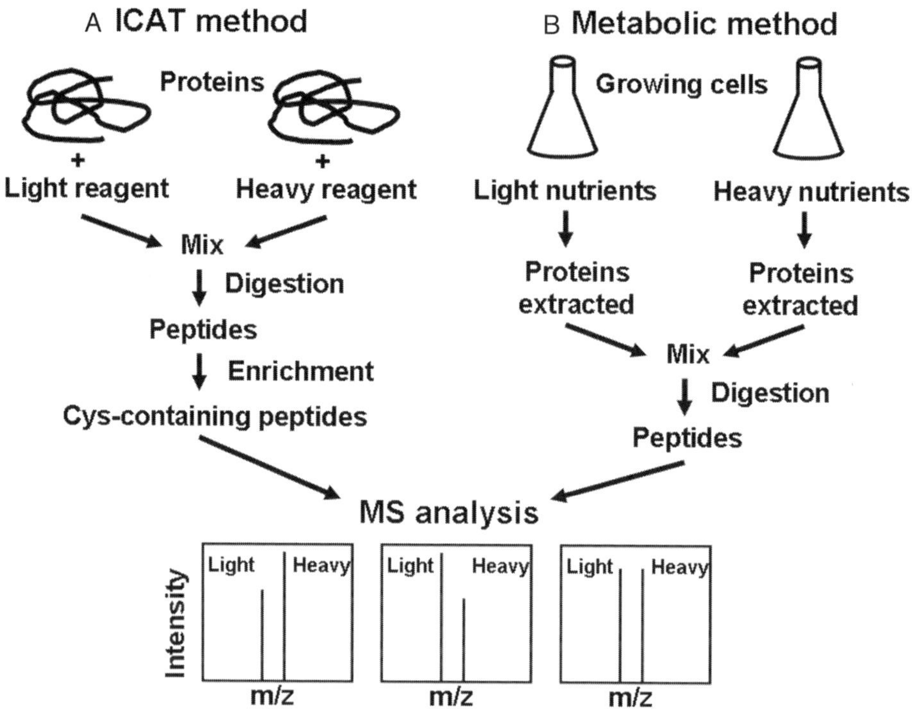

Figure 6. Labeling protocols of proteins based on (A) chemical and (B) metabolic methods. (A) In ICAT method, two protein samples labeled with light and heavy forms of the reagent are mixed and digested into peptides, followed by enrichment of Cys-containing peptides with biotin affinity tag. (B) Proteins extracted from the cells grown in a medium containing either light (native) or heavy (stable isotope-labeled) nutrient, are mixed and digested into peptides. In either method, differences in abundance of peptides are quantified with the ratio of intensities for light and heavy ions.

with light affinity tag and the other with heavy affinity tag, these two samples are combined and subjected to protease digestion, followed by enrichment of cysteine-containing peptides and MS analysis (Figure 6). Besides ICAT protocol, various methods have been developed for labeling carboxyl, amino, or thiol moieties (21–24).

1.5.1.2. Metabolic Labeling Methods: In the first attempt at metabolic labeling, yeast cells were grown in medium containing an isotope-labeled nitrogen source to generate the quantitative data for multiple proteins with MS analysis (26). Later, alternative protocols were developed, in which stable isotope-labeled amino acids were used as essential nutrients to label the proteins in the growing cells (27–29) (Figure 6). Trypsin and lysil-endopeptidase (Lys-C) are generally used for digestion of proteins to produce peptides containing at least one basic amino acid, either lysine or arginine, at their C-terminal ends. Thus, if stable isotope-labeled lysine is incorporated into the proteins, all of the peptides generated by Lys-C digestion would be quantifiable. Among other amino acids, leucine is preferentially used because it is one of the most abundant residues in the yeast proteome. For instance, approximately 70% of unique trypsin or Lys-C–digested peptides of yeast contain at least one leucine. Thus, lysine and leucine are widely used for isotope labeling of proteins.

1.5.2. Quantitative Analysis of Protein Complexes

MS-based technology, combined with affinity purification and stable isotope labeling, enables quantitative analysis of protein complexes. For instance, quantitative change in the number of proteins associated with the PrA-tagged yeast Ste12 transcription factor was examined using the ICAT method (30). Ste12 complex containing Ste12 and two known interactors was unambiguously found to increase in its amount after stimulation with mating pheromone, providing a possible model for regulation of Ste12 activity during response to this factor. Similarly, metabolic labeling was successfully used to examine the change of proteins associated with phosphorylated form of the epidermal growth factor receptor (EGFR), which was purified using a recombinant SH2 domain that specifically binds to the phosphorylated EGFR (31). A wide variety of proteins, including both previously reported interactors and novel ones never implicated in this signaling pathway, were detected as the molecules showing enhanced association with EGFR upon EGF stimulation, providing novel insights into the EGF signaling pathway.

1.5.3. Perspective for Quantitative Analysis of Protein Interactions

The aforementioned approaches can generate quantitative data on protein complexes. Although these data would be quite useful, they are relative quantification data on protein abundance in the isolated complexes, thereby still falling short of grasping the actual picture of PPIs in the cells. To obtain much more useful data for calculating the kinetic parameters and quantitative modeling, absolute quantification is ideal. The use of the known number of stable isotope-labeled standards enables absolute quantification of the proteins in the purified sample. One should, however, know and calculate the efficiency of bait recovery and the dissociation of prey during the purification. It is also an issue to be critically evaluated whether the detected PPIs in the cell extracts faithfully reflect those in the living cells. Although *in vivo* chemical cross-linking of PPIs would provide a snapshot of protein complexes in the cells (32,33), efficiency of cross-linking is, unfortunately, far less than 100%, in general. Therefore, despite remarkable progress in recent years, further technical advances, especially in the preparation of protein complexes and the evaluation of absolute quantity, are necessary for MS-based proteomics to contribute to delineation and modeling of the molecular events in the cells.

2. Two-Hybrid Approaches for Protein Interactions

2.1. Principle of the Y2H System

The Y2H was originally developed by Stanley Fields and his colleagues based on the modular architecture of the yeast transcription factor Gal4 (34). Molecular anatomy of Gal4 in the early 1980s had revealed a well-defined DNA-binding domain (DBD) and transcription activation

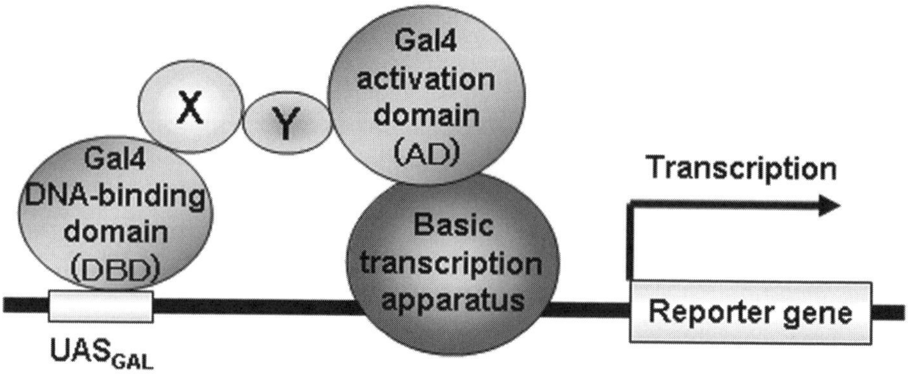

Figure 7. Principles of the Y2H system.

domain (AD). The role of the DBD can be interpreted as binding the upstream activation sequence (UAS_{GAL}) to place the AD in close proximity to target gene promoters, thereby inducing their transcription. Although the recruitment is usually achieved via the covalent bonds linking the two domains, it can be mediated by a non-covalent interaction between the two domains. Accordingly, when a pair of interacting proteins, namely, X and Y, are coexpressed as two hybrid proteins Gal4-DBD-X and Gal4-AD-Y, the former hybrid (or bait) can bind the UAS_{GAL} to recruit the latter hybrid (or prey) via the interaction between X and Y, thereby inducing the expression of the gene lying downstream from the UAS_{GAL} (i.e., the reporter gene) (Figure 7). In other words, the expression of the reporter gene indicates an interaction between the two proteins X and Y. In brief, "interaction-mediated reconstitution of Gal4 activity" was the basis of the original Y2H system.

The Y2H does not always use Gal4. Some systems use *Escherichia coli* LexA and its operator (*LexOp*) instead of Gal4 and UAS_{GAL}, respectively. Similarly, VP16 and B42 are also used instead of Gal4-AD. For the reporter gene, *E. coli* β-galactosidase gene (*lacZ*) is frequently used because its expression can be detected by simple X-gal staining and quantified by the measurement of the enzyme activity. However, for library screening to identify unknown binding partners, nutritional selections such as *HIS3*, *ADE2*, and *URA3*, are much more useful than *lacZ*, because these genes allow selection on plates lacking histidine, adenine, and uracil, respectively. For *HIS3*, it is common to use a medium not only lacking histidine but also containing 3-aminotriazole (3-AT), which is an inhibitor of imidazoleglycerol-phosphate dehydratase or His3 protein, to confer severe histidine starvation.

2.2. Pros and Cons of Y2H

The beauty of the Y2H system is that it allows one to detect PPIs under a physiological *in vivo* condition without any need to handle proteins. It can be quite sensitive, and it can be used to screen a library to identify unknown binding partners of the protein of your interest. In addition, it can be used for finer mapping of binding domains, as well as isolation of interaction-defective alleles, which help decipher a biological role for the PPI.

However, it is generally believed that Y2H tends to give false-positives (FPs). The FPs can be divided into two categories, namely, technical FP and biological FP. Technical FP is based on reporter gene activation independent of the two-hybrid interaction, which may be caused by self-activation of bait and/or promoter-specific fortuitous activation. The former can be eliminated by rigorous examination of bait constructed before library screening. The latter can be eliminated using host strains bearing multiple reporter genes underregulated by different Gal4-responsive promoters. For example, PJ69-4A and its derivatives bear *GAL1pr-HIS3*, *GAL2pr-ADE2*, and *GAL7pr-lacZ*, thereby minimizing fortuitous promoter-specific activation (35). Alternatively, biological FP is inevitable, because Y2H is an artificial system in which two proteins that would never encounter each other in the living cells may be forced to meet in the yeast nucleus. If they have some affinity in a purely physicochemical sense, they may well interact with each other to induce reporter gene expression. Such an interaction is valid as a two-hybrid interaction, but is of no biological relevance. Thus, biological FP should be eliminated by knowledge-based curation of the data or by integrating other lines of experimental evidence.

One should also bear in mind that Y2H also suffers from false-negatives. This is because some fractions of proteins inevitably fail to fold properly when fused to DBD and/or AD. In addition, because Y2H can detect only simple binary interactions by its nature, it would miss interactions requiring more than three proteins. Similarly, it would miss, in principle, interactions involving membrane proteins and posttranslational modifications.

Although Y2H has contributed to the identification of a plethora of exciting interactions, it is not a magic bullet. However, it would indeed serve as a powerful bullet for careful hunters who know its pros and cons well.

2.3. Comprehensive Y2H Analysis for Interactome Mapping

The Y2H system was initially used to analyze a particular PPI interaction (i.e., one-to-one application) and was then used in the screening of a prey library for binding partners of the protein of interest (i.e., one-to-many application). Logical extension of its application would be the screening in a "many-to-many" mode, the extreme of which would be a comprehensive analysis that examines every possible binary combination between the proteins encoded by an organism's genome. Indeed, a real proteome-wide Y2H analysis of the budding yeast was conducted independently by two groups, including one of our own (36–38).

We amplified all annotated yeast ORFs in their full-length forms by means of PCR, and then cloned each of them into two vectors, one expressing each ORF as a Gal4-DBD fusion (bait) and the other expressing each ORF as a Gal4-AD fusion (prey) (36). The bait and prey plasmids were then introduced into Y2H host strains bearing mating type **a** and α, respectively, using an optimized chemical transformation protocol

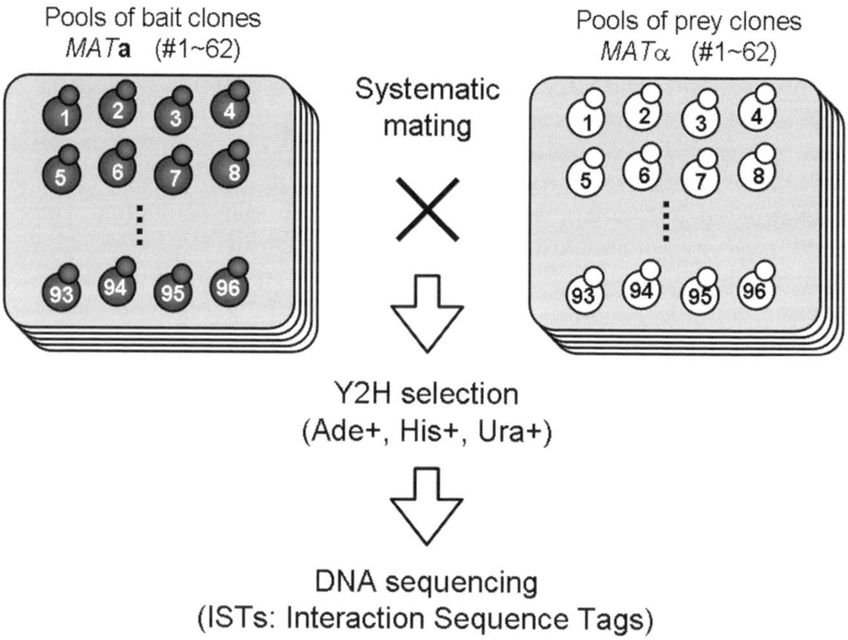

Figure 8. An IST approach in a comprehensive Y2H analysis of the budding yeast proteome.

in a 96-well plate format. Bearing opposite mating types, bait and prey clones can be crossed to form diploid cells, each of which bears a unique combination of bait and prey. If the bait and prey interact, the reporter genes are activated to allow the cells to survive the Y2H selection. Accordingly, each survivor should bear a pair of mutually interacting bait and prey, the identities of which can be revealed by tag-sequencing of the cohabiting plasmids to generate an interaction sequence tag (IST) for subsequent database search.

In an actual screening, we used a pooling strategy. Each pool contained 96 bait or prey clones and was subjected to the mating-based screening to examine all possible combinations between the bait and prey pools (Figure 8). The colonies of survivors were subjected to DNA sequencing to obtain ISTs. Consequently, 4,549 independent two-hybrid interactions were revealed in total (37). Of these, 841 were detected more than 3 times and assumed to be of high relevance (i.e., core data set) (37). Notably, more than 80% of these interactions were novel at that time. A similar IST project was conducted by CuraGen, who screened a pool of 5,331 preys with each of their 4,665 baits (39). They revealed 691 interactions in total, most of which were, again, novel.

Comparison between our core data set and theirs revealed an unexpectedly small overlap; the two data sets share 141 PPIs, which correspond to approximately 10% of the total independent PPIs (37,39). There are several plausible reasons for this small overlap. The systems used by the two groups were different; we used multicopy vectors in the host, bearing multiple reporter genes under different Gal4-regulated promoters, whereas they used single-copy vectors, but used only a single reporter gene. Because both groups' PCR amplified the ORFs, some of them would inevitably bear mutations affecting PPIs. Although both groups pooled clones, the screen does not seem to reach saturation: two-thirds of our 4,549 interactions and one-third of CuraGen's 691 interactions were detected only once. Of course, any two-hybrid screen contains false signals; accuracy of these data sets is assumed to be 50%–60% (39). Nevertheless, these data provide a wealth of implications to various aspects in yeast biology and have borne a new field of "interactome informatics."

After the pioneering works on yeast, large-scale Y2H screens were conducted on nematode, fruit fly, and human (40–43). In contrast with yeast studies, these studies integrate various methods to select reliable PPIs from raw Y2H data.

2.4. Reverse Y2H for Functional Analysis of PPIs

2.4.1. Mapping Interaction Domains by Means of Y2H

Although Y2H plays a major role in the search for new interactors and the cataloguing of PPIs, it also serves as an invaluable platform for finer analyses of verified PPIs. To further characterize a certain PPI, it is vital to know the domains mediating the interaction because the interaction domain can be overexpressed in the cell for competitive inhibition of the endogenous PPI to learn its biological role. For interaction domain mapping, Y2H is the most versatile method; the trimming at the DNA level readily converts to the examination at the protein level. It is even possible to screen a library of fragmented proteins derived from a single target protein to identify the minimal region to interact with its binding partner (44).

2.4.2. Isolation of Interaction-Defective Alleles by Reverse Y2H

Besides the mapping of interaction domains, it is also important to isolate an interaction-defective allele that encodes a protein defective in the interaction of interest. For this purpose, a smart method of "reverse" Y2H was developed (45). The most popular reverse Y2H system uses *URA3* as its reporter gene. The *URA3* reporter can be used in the selection *for* the interaction; its induction allows the cell to survive in the absence of uracil. Of note, it can also be used in the selection *against* the interaction in the presence of pro-toxin 5-fluoroorotic acid (5-FOA). Orotidine-5′-phosphate decarboxylase or Ura3 protein converts 5-FOA to 5-fluorouracil, which is a toxin that kills the yeast. In other words, *URA3* functions as a "suicide" reporter in the presence of 5-FOA. Accordingly, we can select interaction-defective mutants from a library of mutagenized prey, using the reverse Y2H system. Once an interaction-defective allele is isolated, it can be used to generate cells defective in the interaction to learn a biological role of the PPI. At the same time, sequencing of these interaction-defective alleles would help one pinpoint the site of interaction.

2.4.3. Isolation of Separation-of-Function Alleles by Dual-Bait Reverse Y2H

It should be noted that the proteins playing pivotal roles in regulation of biological systems often have several binding partners, thereby serving

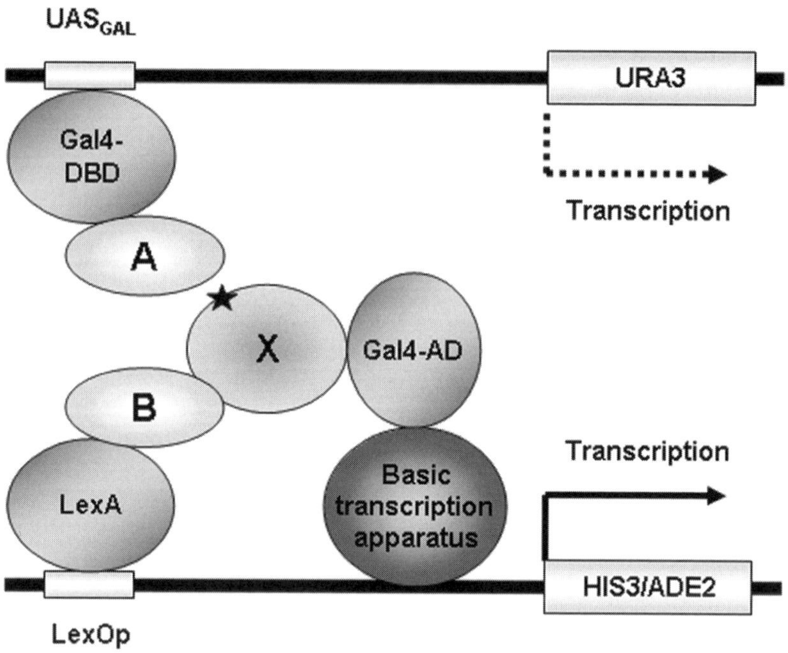

Figure 9. Dual-bait reverse Y2H to isolate separation-of-function alleles. The star indicates a mutation abolishing the binding of X to A.

as a merging or branching point in cellular signaling, or as a hub in the PPI network. To decipher the role of each interaction or path, it is vital to have a separation-of-function allele encoding a protein that is defective in one interaction, but not in the others. For effective isolation of such separation-of-function alleles, we developed a dual-bait reverse Y2H system in which Gal4- and LexA-based baits are used to induce the expression of *URA3* and *HIS3* (and/or *ADE2*) reporters in the same cell (46) (Figure 9).

Let's say protein X can bind both proteins A and B. To learn a role of X–A interaction, we have to have a mutant X that is defective in binding A, but not in binding B. For this purpose, we expressed Gal4-DBD-A and LexA-B in the strain bearing both UAS_{GAL}-*URA3* and *LexOp-HIS3* as reporter genes. This strain is then mated with another strain with the opposite mating type that bears a mutated library of Gal4-AD-X. The resultant diploid cells were selected on a medium containing uracil, 5-FOA, and 3-AT, but lacking histidine. To survive this selection, X has to bear mutations rendering it incapable of binding A so as not to induce the suicide reporter *URA3*. On the other hand, X has to retain its ability to bind B to induce *HIS3* to tolerate 3-AT–induced severe histidine starvation. Therefore, each of the survivors would have a desired allele encoding a mutant protein X that interacts with B but not with A, and that can be subsequently used to learn a role of X–A interaction.

The dual-bait reverse Y2H can be also applied to selective isolation of missense mutations leading to defective interaction. Indeed, one of the pitfalls of reverse Y2H is that it frequently picks up nonsense mutants, leading to truncation of the protein of interest. Truncated alleles cannot be used for functional analysis and should be eliminated from the screening. For this purpose, we developed a system for "guaranteed" reverse Y2H as a variation of dual-bait reverse Y2H (46). In this system, the prey Y to be mutated was fused with Gal4-AD at its N-terminal end and with the PC motif-containing region (PCCR) of Cdc24 at its C-terminal end (Gal4-AD-Y-PCCR). The bait X was expressed as Gal4-DBD-X to activate UAS_{GAL}-*URA3* for 5-FOA–mediated counterselection. In addition, the PB1 domain of Bem1, which was shown to interact specifically with the Cdc24 PCCR, is expressed as a LexA-fusion so that the PB1–PCCR interaction induces the expression of *HIS3* to confer 3-AT resistance. In this system, survivors on medium containing both 5-FOA and 3-AT have to bear mutations on Y that abolish its binding to X but maintain the PB1–PCCR interaction, thereby eliminating nonsense mutants lacking the C-terminal–attached PCCR. In other words, the PB1-PCCR interaction guarantees that the protein Y is not truncated. Indeed, this system successfully isolated Gcn1 mutants defective in interaction with Gcn2 (46).

Collectively, the Y2H system allows one to obtain various useful alleles that can be used for perturbation or elimination of the interaction of interest, thereby contributing to an understanding of the systems involving the PPI.

2.5. Three-Hybrid System for Dissection of Complex Interactions

Another interesting Y2H application is the three-hybrid system, in which a third protein, Z, is coexpressed in a nuclear-targeted form in addition to a pair of bait, X, and prey, Y (Figure 10). If the two-hybrid interaction between X and Y turns out to be dependent on the expression of Z, we can assume that Z mediates or bridges an indirect interaction between X and Y. Such information would be quite useful to decipher the topology or architecture of a multiprotein complex, which is typically identified by the aforementioned MS analysis.

Conversely, it is possible that the expression of Z suppresses or eliminates the two-hybrid interaction between X and Y. We can then assume

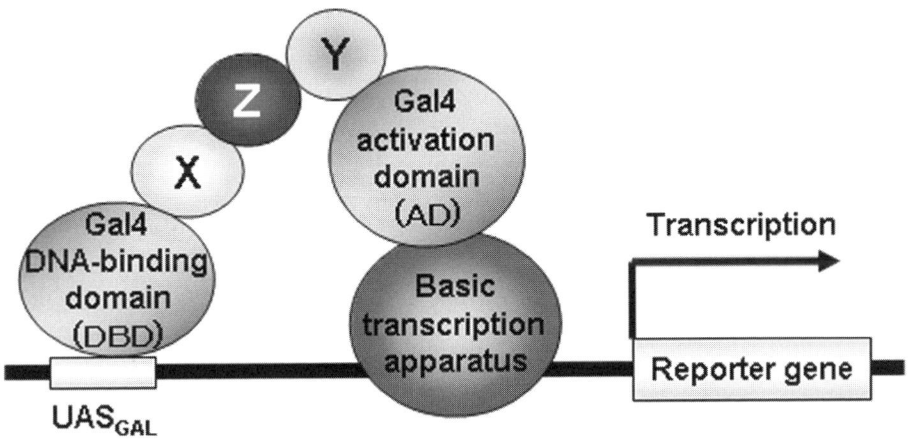

Figure 10. A three-hybrid system to analyze complex interactions.

that X (or Y) and Z bind Y (or X) in a mutually exclusive manner, suggesting that the interaction around X–Y may well constitute a switch in the network.

In a similar context, it is also possible to screen peptide inhibitors of X–Y interaction using this system. Furthermore, integration of the dual-bait system would enable one to search peptides inhibiting X–Y interaction, but not other similar interactions. Such inhibitors can be used as specific perturbagens that are useful for systems analysis.

2.6. Other Two-Hybrid Systems

Although the two-hybrid system was developed in yeast, it is possible to construct a similar system in other organisms. Indeed, two-hybrid systems in *E. coli* and mammalian cells were developed. Although the former works faster than Y2H, the folding of eukaryotic proteins, especially those bearing multiple domains, would be compromised. Conversely, the latter may be ideal for mammalian proteins to fold correctly and to be modified, but is not suitable for library screening and large-scale applications.

Similarly, although Y2H was developed using transcription factors, it is possible to develop a two-hybrid system based on the interaction-mediated reconstitution of other protein activities. For instance, an interesting method called split ubiquitin system was developed to examine PPIs, especially those involving membrane proteins (47) (Figure 11).

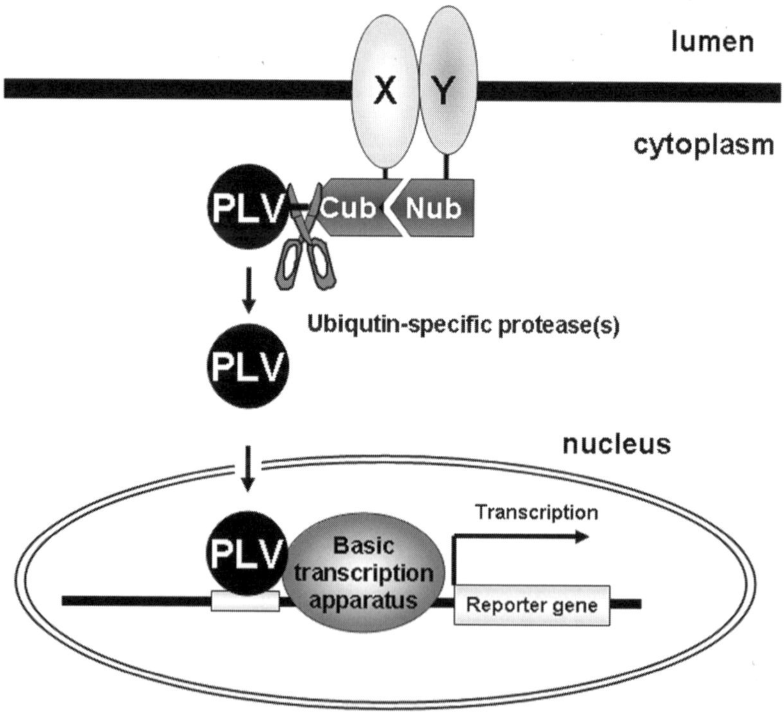

Figure 11. A split ubiquitin system to examine interactions between integral membrane proteins.

Ubiquitin is a small protein comprising 76 amino acids, and it functions as a tag for protein degradation and intracellular sorting. In the split ubiquitin system, ubiquitin was divided into its N-terminal half (N-ub) and C-terminal half (C-ub). Two membrane proteins of interest, namely, X and Y, were fused at its cytoplasmic portion with N-ub and C-ub, the latter of which was further fused with an artificial transcription factor, PrA-LexA-VP16 (PLV). If X and Y interact on membrane, the N-ub and C-ub were put in close proximity to reconstitute a ubiquitin molecule beneath the cytoplasmic surface. The reconstituted ubiquitin molecule is recognized by ubiquitin-specific proteases that cleave at the C-terminal end of ubiquitin. The cleavage liberates PLV, and it migrates into the nucleus to induce the expression of reporter gene. If X and Y do not interact, PLV is kept tethered to the membrane, failing to activate the expression of the reporter gene. The split ubiquitin system is quite useful for the detection of interactions among membrane proteins, and it was used for a large-scale analysis of interactions among yeast integral membrane proteins (48). Various modifications similar to those described for conventional Y2H would be applicable to this system, further improving its usefulness.

3. Conclusion

Systematic identification of PPIs or interactome mapping has been achieved by MS- and Y2H-based approaches. Although other potential techniques, including proteome chip (49), are being developed, only these two methods have thus far succeeded in the generation of large data sets. Besides interactome mapping, MS can be used for quantitative analysis of PPIs when combined with stable isotope labeling. Although fluorescent imaging technologies such as FERT and FCCS would provide data with higher spatiotemporal resolution (50,51), their throughput is still limited, and integration of both approaches would be desirable. On the other hand, Y2H is a genetic method suitable for isolation of mutants for interactions. Interaction-defective alleles are indispensable for the perturbation of PPIs, from which we can learn its biological role, as well as the property of the system including it. Therefore, MS- and Y2H-based methods for PPI analysis will stay in the toolbox for systems biologists.

References

1. Pandey A, Mann M. Proteomics to study genes and genomes. *Nature* 2000;405:837–846.
2. Aebersold R, Goodlett DR. MS in proteomics. *Chem Rev* 2001;101:269–295.
3. Aebersold R, Mann M. MS-based proteomics. *Nature* 2003;422:198–207.
4. Washburn MP, Wolters D, Yates JR 3rd. Large-scale analysis of the yeast proteome by multidimensional protein identification technology. *Nat Biotechnol* 2001;19:242–247.
5. Dziembowski A, Seraphin B. Recent developments in the analysis of protein complexes. *FEBS Lett* 2004;556:1–6.

6. Terpe K. Overview of tag protein fusions: from molecular and biochemical fundamentals to commercial systems. *Appl Microbiol Biotechnol* 2003;60: 523–533.
7. Rigaut G, Shevchenko A, Rutz B, et al. A generic protein purification method for protein complex characterization and proteome exploration. *Nat Biotechnol* 1999;17:1030–1032.
8. Caspary F, Shevchenko A, Wilm M, et al. Partial purification of the yeast U2 snRNP reveals a novel yeast pre-mRNA splicing factor required for pre-spliceosome assembly. *EMBO J* 1999;18:3463–3474.
9. Chen CY, Gherzi R, Ong SE, et al. AU binding proteins recruit the exosome to degrade ARE-containing mRNAs. *Cell* 2001;107:451–464.
10. Westermarck J, Weiss C, Saffrich R, et al. The DEXD/H-box RNA helicase RHII/Gu is a co-factor for c-Jun-activated transcription. *EMBO J* 2002; 21:451–460.
11. Schmitt C, von Kobbe C, Bachi A, et al. Dbp5, a DEAD-box protein required for mRNA export, is recruited to the cytoplasmic fibrils of nuclear pore complex via a conserved interaction with CAN/Nup159p. *EMBO J* 1999; 18:4332–4347.
12. Estevez AM, Kempf T, Clayton C. The exosome of *Trypanosoma brucei*. *EMBO J* 2001;20:3831–3839.
13. Gavin AC, Bosche M, Krause R, et al. Functional organization of the yeast proteome by systematic analysis of protein complexes. *Nature* 2002;415: 141–147.
14. Ho Y, Gruhler A, Heilbut A, et al. Systematic identification of protein complexes in *Saccharomyces cerevisiae* by MS. *Nature* 2002;415:180–183.
15. Bader GD, Hogue CW. Analyzing yeast protein-protein interaction data obtained from different sources. *Nat Biotechnol* 2002;20:991–997.
16. von Mering C, Krause R, Snel B, et al. Comparative assessment of large-scale data sets of protein-protein interactions. *Nature* 2002;417:399–403.
17. Gavin AC, Aloy P, Grandi P, Krause R, et al. Proteome survey reveals modularity of the yeast cell machinery. *Nature* 2006;440:631–636.
18. Archambault V, Chang EJ, Drapkin BJ, et al. Targeted proteomic study of the cyclin-Cdk module. *Mol Cell* 2004;14:699–711.
19. Tackett AJ, DeGrasse JA, Sekedat MD, et al. I-DIRT, a general method for distinguishing between specific and nonspecific protein interactions. *J Proteome Res* 2005;4:1752–1756.
20. Bouwmeester T, Bauch A, Ruffner H, et al. A physical and functional map of the human TNF-alpha/NF-kappa B signal transduction pathway. *Nat Cell Biol* 2004;6:97–105.
21. Ong SE, Foster LJ, Mann M. Mass spectrometric-based approaches in quantitative proteomics. *Methods* 2003;29:124–130.
22. Sechi S, Oda Y. Quantitative proteomics using MS. *Curr Opin Chem Biol* 2003;7:70–77.
23. Goshe MB, Smith RD. Stable isotope-coded proteomic MS. *Curr Opin Biotechnol* 2003;14:101–109.
24. Tao WA, Aebersold R. Advances in quantitative proteomics via stable isotope tagging and MS. *Curr Opin Biotechnol* 2003;14:110–118.
25. Gygi SP, Rist B, Gerber SA, et al. Quantitative analysis of complex protein mixtures using isotope-coded affinity tags. *Nat Biotechnol* 1999;17:994–999.
26. Oda Y, Huang K, Cross FR, et al. Accurate quantitation of protein expression and site-specific phosphorylation. *Proc Natl Acad Sci USA* 1999;96:6591–6596.

27. Ong SE, Blagoev B, Kratchmarova I, et al. Stable isotope labeling by amino acids in cell culture, SILAC, as a simple and accurate approach to expression proteomics. *Mol Cell Proteomics* 2002;1:376–386.
28. Berger SJ, Lee SW, Anderson GA, et al. High-throughput global peptide proteomic analysis by combining stable isotope amino acid labeling and data-dependent multiplexed-MS/MS. *Anal Chem* 2002;74:4994–5000.
29. Zhu H, Pan S, Gu S, et al. Amino acid residue specific stable isotope labeling for quantitative proteomics. *Rapid Commun Mass Spectrom* 2002;16:2115–2123.
30. Ranish JA, Yi EC, Leslie DM, et al. The study of macromolecular complexes by quantitative proteomics. *Nat Genet* 2003;33:349–355.
31. Blagoev B, Kratchmarova I, Ong SE, et al. A proteomics strategy to elucidate functional protein-protein interactions applied to EGF signaling. *Nat Biotechnol* 2003;21:315–318.
32. Vasilescu J, Guo X, Kast J. Identification of protein-protein interactions using in vivo cross-linking and MS. *Proteomics* 2004;4:3845–3854.
33. Tagwerker C, Flick K, Cui M, et al. A tandem-affinity tag for two-step purification under fully denaturing conditions: application in ubiquitin profiling and protein complex identification combined with in vivo cross-linking. *Mol Cell Proteomics* 2006;5:737–748.
34. Fields S, Song, O. A novel genetic system to detect protein-protein interactions. *Nature* 1989;340:245–246.
35. James P, Halladay J, Craig EA. Genomic libraries and a host strain designed for highly efficient two-hybrid selection in yeast. *Genetics* 1996;144:1425–1436.
36. Ito T, Tashiro K, Muta S, et al. Toward a protein-protein interaction map of the budding yeast: A comprehensive system to examine two-hybrid interactions in all possible combinations between the yeast proteins. *Proc Natl Acad Sci USA* 2000;97:1143–1147.
37. Ito T, Chiba T, Ozawa R, Yoshida M, Hattori M, Sakaki Y. A comprehensive two-hybrid analysis to explore the yeast protein interactome. *Proc Natl Acad Sci USA* 2001;98:4569–4574.
38. Uetz P, Giot L, Cagney G, et al. A comprehensive analysis of protein-protein interactions in *Saccharomyces cerevisiae*. *Nature* 2000;403:623–627.
39. Ito T, Ota K, Kubota H, Yamaguchi Y, et al. Roles for the two-hybrid system in exploration of the yeast protein interactome. *Mol Cell Proteomics* 2002; 1:561–566.
40. Li S, Armstrong CM, Bertin N, et al. A map of the interactome network of the metazoan *C. elegans*. *Science* 2004;303:540–543.
41. Giot L, Bader JS, Brouwer C, et al. A protein interaction map of *Drosophila melanogaster*. *Science* 2003;302:1727–1736.
42. Rual JF, Venkatesan K, Hao T, et al. Towards a proteome-scale map of the human protein-protein interaction network. *Nature* 2005;437:1173–1178.
43. Stelzl U, Worm U, Lalowski M, et al. A human protein-protein interaction network: a resource for annotating the proteome. *Cell* 2005;122:957–968.
44. Ikeuchi A, Sasaki Y, Kawarasaki Y, et al. Exhaustive identification of interaction domains using a high-throughput method based on two-hybrid screening and PCR-convergence: molecular dissection of a kinetochore subunit Spc34p. *Nucleic Acids Res* 2003;31:6953–6962.
45. Vidal M, Brachman RK, Fattaey A, et al. Reverse two-hybrid and one-hybrid systems to detect dissociation of protein-protein and DNA-protein interactions. *Proc Natl Acad Sci USA* 1996;93:10315–10320.

46. Kubota H, Ota K, Sakaki Y, et al. Budding yeast GCN1 binds the GI domain to activate the eIF2α kinase GCN2. *J Biol Chem* 2001;276:17591–17596.
47. Stagljar I, Korostensky C, Johnsson N, et al. A genetic system based on split-ubiquitin for the analysis of interactions between membrane proteins in vivo. *Proc Natl Acad Sci USA* 1998;95:5187–5192.
48. Miller JP, Lo RS, Ben-Hur A, et al. Large-scale identification of yeast integral membrane protein interactions. *Proc Natl Acad Sci USA* 2005;102:12123–12128.
49. Bertone P, Snyder M. Advances in functional protein microarray technology. *FEBS J* 2005;272:5400–5411.
50. Miyawaki A. Visualization of the spatial and temporal dynamics of intracellular signaling. *Dev Cell* 2003;4:295–305.
51. Bacia K, Kim SA, Schwille P. Fluorescence cross-correlation spectroscopy in living cells. *Nat Methods* 2006;3:83–89.

10

Genome-Scale Assessment of Phenotypic Changes During Adaptive Evolution

Stephen S. Fong

Summary

Adaptive evolution is a process that influences and alters all biological organisms over time. Changes involved in adaptive evolution begin with genetic mutations and can lead to large changes in phenotypic behavior. Thus, the relationship between genotype and phenotype is a central issue in studying adaptive evolution.

The whole-cell phenotype of an organism is the result of integrated functions at various levels of cellular organization. As methods have been developed and improved to study the components involved with the different levels of cellular organization in a high-throughput and genome-wide scale, it is becoming possible to establish the link between genotype and phenotype. In this chapter, different means of studying and establishing connections between genotype and phenotype in the context of adaptive evolution will be discussed.

Key Words: Adaptive evolution; phenotype; phenomics; transcriptomics; proteomics; fluxomics; metabolomics.

1. Introduction to Adaptive Evolution

Adaptive evolution is the process by which the behavior of an organism is adjusted in response to a stimulus (the surrounding environment). This process can be thought to have two main components: an adaptive component, where specific behaviors (phenotypes) impart some benefit to the organism that are positively selected, and an evolutionary component, where the beneficial behavior is maintained in the organism through a genetic change. The current paradigm in evolutionary biology is that genetic changes (mutations) naturally arise during DNA replication, thus creating a heterogeneous population. Within this population, mutations that confer a beneficial phenotype (fitness advantage) are selected and propagated in subsequent generations (Figure 1). Thus, this entire

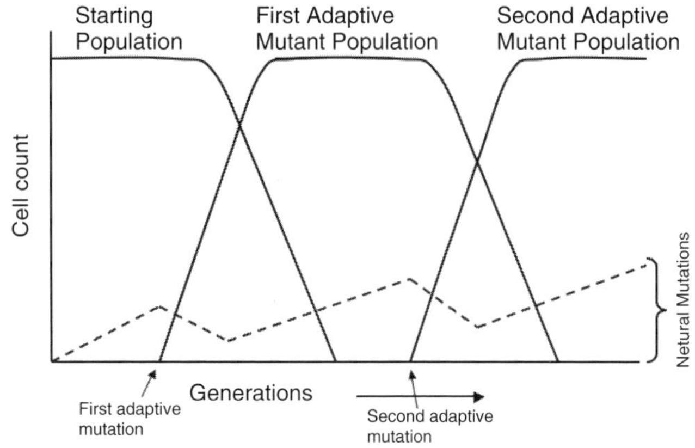

Figure 1. Schematic representation of population dynamics during adaptive evolution. As neutral mutations (dashed lines) accumulate, an adaptive mutation can occur that confers an evolutionary benefit allowing a mutant carrying the adaptive trait to overtake the population.

process is an elegant example of the interrelatedness of the genotype and phenotype.

Although the adaptive evolutionary process is common to all biological organisms, microorganisms, in particular, have been frequently utilized in experimental research to study the evolutionary process. Microorganisms have several beneficial attributes that make them conducive to experimental evolution studies (1) including the ability to carefully control growth environments and evolutionary selection pressure, fast generation times, and relatively small genome sizes. Laboratory experiments using microorganisms can thus be used to study adaptive evolution in real-time, and to attempt to determine some of the links between genotype and phenotype.

The simplest means of conducting a laboratory evolution experiment would be to grow a culture of an organism and to transfer some of the population to a new culture after the nutrients in the medium have been depleted. This process of serial passage of batch cultures is continued for the duration of the evolution experiment (2). By using this process, thousands to tens of thousands (3) of generations can be observed often resulting in large phenotype changes, such as growth rate increases of more than 100% (4,5) or adaptations that facilitate growth in stressful environments (6,7).

2. Adaptive Evolution and Systems Biology

Although the field of evolutionary biology has become defined in scope through the course of theoretical and experimental developments over history, the nascent field of systems biology is not as clearly defined. It is therefore important to begin with describing the perspective of systems biology that will be used in this section and how it relates to evolutionary biology and adaptive evolution. Thus far, systems biology has been characterized by high-throughput methods of generating data at different levels of cellular organization ("omic" datasets) and developing methods to manage, analyze, and interpret these large datasets. These, however, are only the tools implemented in systems biology, and alone they are not sufficient to define systems biology. The field of systems biology has been developed out of the belief that biological systems are complex, highly interconnected systems and that it is through the study of intact systems (and the connections in an intact system) that new insight into biological function will be obtained. The simultaneous measurement of all cellular components of a given type (mRNA transcripts, proteins, metabolites, etc.) should allow for details of the interconnectedness of these components to be studied. It is the eventual goal to not only interpret and understand the relationships between all cellular components of one type (all mRNA transcripts), but then to establish connections between different levels of cellular organization (mRNA transcripts and metabolites) to establish how a specific genotype will be manifested in terms of phenotype.

Adaptive evolution is a fascinating biological process from the perspective that large phenotypic changes can arise because of systemic changes and become permanent within a population. Given a long enough period of time with exposure to a stimulus, the behavior of an organism is refined in a coordinated manner, as combinations of cellular adjustments are evaluated and selected as a result of evolutionary selection pressure. In this manner, systematic optimization of an organism's functionality for a specific stimulus occurs. The changes occurring during adaptive evolution can potentially be anywhere within the biological system and can affect single proteins or entire regulatory, signaling, or other functional modules.

As systems biology is interested in studying how biological systems function in a coordinated manner and, ultimately, how genotypes are translated into phenotypes, adaptive evolution is a good process to study using the systems biology approach. Genetic changes in the form of mutations naturally arise and accumulate over time during evolution and these give rise to altered phenotypes. This process gives a situation where both genotype and phenotype are modified in a directed, biologically significant fashion.

Laboratory studies of adaptive evolution, where both genotype and phenotype concurrently change, are a good reflection of real biological evolution; however, it also complicates results and analysis. In most systems biology studies, the biological system being studied is defined by the known starting genotype (often specific genetic alterations are introduced) such that the topology of the functioning biological network is known. Measurements are then made to determine how an organism uses different portions of its biological capabilities to exhibit a phenotype. In the case of adaptive evolution, the problem arises that the genotype changes in an unknown manner because of random mutations. This means that measurements need to determine both how the

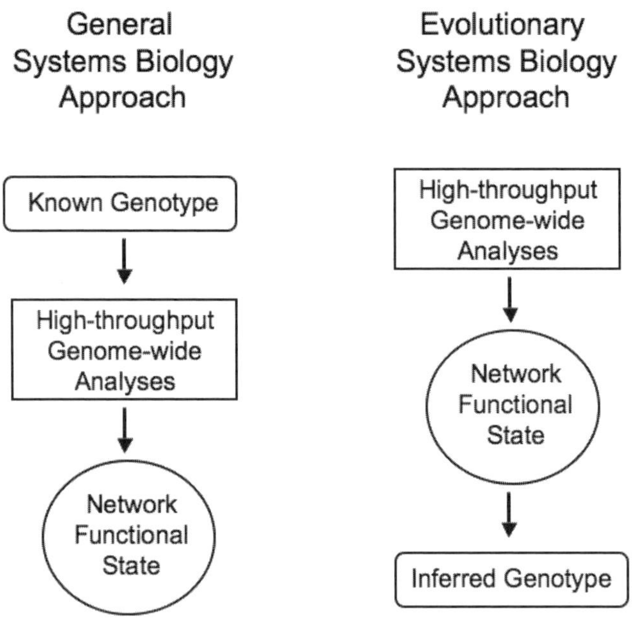

Figure 2. Schematic representation of the systems biology methodology used to study adaptive evolution.

network is used in expressing a particular phenotype and also how the network itself may have fundamentally changed in structure or function (Figure 2).

3. Genotype to Phenotype in Adaptive Evolution

Fundamental to cell biology is the central dogma of molecular biology that describes the progression from DNA to proteins that will ultimately play a role in the manifested phenotype. Understanding this progression, how the genotype relates to phenotype, is one of the largest current challenges in biology and has become a focal point of research because genomic data have become readily available for hundreds of organisms over the past several years, starting with the complete genome sequencing of *Haemophilus influenzae* (8).

The genotype of an organism can give an indication of the gene content of an organism, roughly giving the equivalent to a "parts catalogue" of what proteins are potentially present in the organism (9). This defines the potential capabilities of the organism, but gives little indication of how these parts are assembled or used. This genomics aspect of systems biology essentially establishes the biological infrastructure of an organism, giving the hard-wired structure of the organism's biological network. In this view, the amount of functional biological information that can be obtained directly from DNA sequencing is limited.

Infrastructure Analogy

The biological infrastructure established by genomics can be conceptually equated to a road map detailing all known streets. In this case, individual streets would be analogous to individual chemical reactions. As a cell undergoes certain biological processes, it can utilize combinations of different reactions to achieve a certain outcome, just as different streets can be used to get from point A to point B. Both genomics and a road map indicate possible means to reach an objective; however, neither is sufficient to select the most likely means to be used. It is necessary to obtain additional data (such as speed limits, number of traffic lights, etc.) to determine the most efficient route. One problem is that current genomic road maps are not complete. An added complication in evolution is that the biological infrastructure might change (construction of new roads, destruction of old roads, or detours), which necessitates constant reevaluation of utilized routings.

As one seeks to connect genotype to phenotype, the problem involves associating structure (genotype) to function (phenotype). Given the complexity of biological systems and the multiple layers of biological organization, the connection between genotype and cellular phenotype is indirect. The indirect connection between genotype and phenotype is in part demonstrated in evolutionary biology through the fact that genotype mutation rates and phenotype mutation rates (at the cellular level) in a given system are highly disparate (10). The indirect nature of the genotype–phenotype relationship is one of the most confounding aspects of evolutionary biology, and necessitates careful experimental design in respect to understanding limitations to the type of information that can be gained from a given data type. Thus, the questions that must be addressed are how are we going to define the phenotype of an organism, and what pieces of information will be most critical to establishing a genotype–phenotype relationship.

Genotype–Phenotype Analogy

The challenges of establishing the relationship between genotype and phenotype can be conceptually illustrated by the process of refining iron ore. If the iron ore is considered to be our starting raw material (similar to the genotype of an organism), we would like to determine how this starting material can give rise to different end products (phenotypes), such as a steel beam or a cast-iron skillet. In this scenario, knowledge of the starting point (iron ore) and end point (steel beam) are not sufficient to know how the steel beam was produced or that other types of end products could be produced from the same starting point. Just as details of the smelting and processing are critical to understanding how iron ore is transformed into an end product, so too are details of the biological progression from genotype to phenotype needed.

4. Genome-Scale Phenotype Assessment

In the most general sense, a phenotype can be viewed as any detectable characteristic of an organism for a given environment. These "detectable characteristics" can include any quantifiable biological property, from growth rates to production of cellular components. High-throughput, genome-scale measurements of biological traits or characteristics are thus used to elucidate the functional consequences of genotypic or environmental changes. This field of study has broadly been termed "phenomics" (11), and the types of measurements used can be divided into two main categories: measurement of whole-cell phenotypes and measurement of cellular components (Figure 3).

4.1. Whole-Cell Measurements

Whole-cell phenotype measurements are conducted on intact, living cells and are quantified observations of how an organism is functioning in a given environment. Characteristics such as growth rate, consumption rates, metabolite secretion, and motility can be categorized as whole-cell phenotypes. These whole-cell phenotypes represent the integrated behavior of the organism and are often subject to evolutionary selection pressures. Thus, although measurement of whole-cell phenotypes can give insight into evolutionary outcomes and selection, these measurements typically are not detailed enough to elucidate specific mechanistic changes at the molecular level.

4.1.1. Cellular Growth Rates

The evolutionary concept of natural selection stipulates that individuals that are more successful at producing progeny will out-compete those who are not as successful at producing progeny. In the microbial world, this is translated into a scheme where faster growing cells will tend to out-compete slower growing cells. This represents the most common and

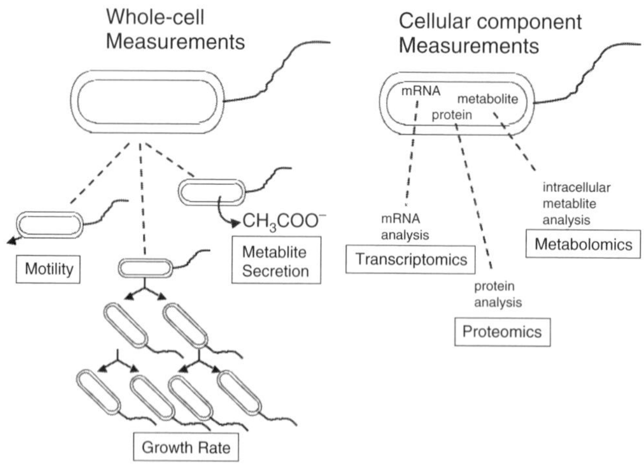

Figure 3. Illustration depicting examples of different whole-cell phenotype measurements and cellular component phenotype measurements.

widely applicable evolutionary scenario; thus, detailed quantitative measurements of microbial growth rates are often essential to studying microbial adaptive evolution. Small differences in cellular growth rates of a mixed population, when propagated over thousands of generations, can lead to one subpopulation becoming dominant and another subpopulation becoming extinct.

For microorganisms, the specific growth rate can be easily and precisely measured in a growing liquid culture by determining the change in optical density (OD) time. At low cell densities, the OD is proportional to the cell density so during exponential microbial growth, the specific growth rate is found to be the slope of log-scale plot of OD versus time. A related metric that is often useful in microbial systems is the doubling time (time necessary for the population to double in size), which is given as:

$$\text{Doubling time} = \frac{\ln(2)}{\text{Growth rate}}.$$

In evolutionary studies, quantitative growth rate measurements can be used to monitor subpopulation dynamics as the microbial population evolves in a specific growth environment. These measurements are used to determine the speed and magnitude of phenotypic changes during evolution (Figure 4A). In addition, evolving populations can be tested for growth on a panel of different nutrient conditions (Figure 4B). These measurements indicate the robustness of the evolutionary changes being retained, and may also give some leads into specific subsystems that may have changed during evolution (4,12).

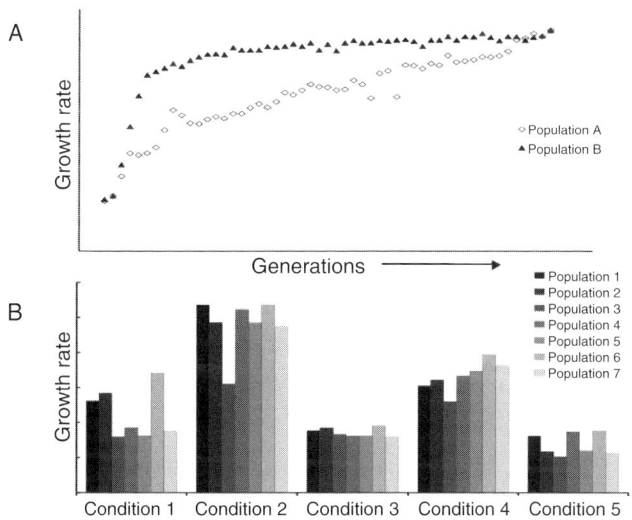

Figure 4. Sample results of high-throughput growth rate measurements for adaptive evolution. (A) Quantitative measurements of growth rates over the course of adaptive evolution showing growth rate versus evolutionary generation. (B) Growth rate testing in different growth environments where 7 different populations were tested for growth in 5 different growth conditions.

The direct measurement of the specific growth rates for microbial cultures can be conducted in a high-throughput manner using plate-reader systems. Quantitative growth rate measurements can be conducted in parallel, with hundreds of cultures being monitored simultaneously in plate-reader systems that have an optical device to detect OD, agitation for culture aeration, and temperature control for incubation (e.g., VersaMax™ system by Molecular Devices and the Bioscreen C™ system by Thermo Finnigan).

4.1.2. Cellular Respiration

A whole-cell phenotype that can be measured and is closely related to cellular growth is cellular respiration. A high-throughput means of monitoring cellular respiration has been developed by associating cellular respiration with a colorimetric change. Using this principle and proprietary chemistry, the Phenotype Microarrays™ developed by Biolog (13) can be used to test for cellular respiration under thousands of different growth conditions. Respiration is indicated by a positive colorimetric change, whereas the absence of respiration results in no colorimetric change.

A second method of monitoring cellular respiration has been developed, implementing a device that fluoresces in the absence of oxygen (Becton, Dickinson and Company). The device is placed in the bottom of a microplate well to which cells and media are added. For a dilute culture (when the culture is first inoculated), the cells will not utilize all of the available oxygen dissolved in the medium, so oxygen will reach the bottom of the well and quench the fluorescence of the insert. As the culture grows, the cells will begin to utilize all of the available oxygen, thus not allowing any oxygen to reach the device at the bottom of the well and the insert fluoresces.

High-throughput measurements of cellular respiration are reliant on secondary indicators (colorimetric or fluorescent changes) to monitor respiration, and thus are only semiquantitative in nature. Although these measurements are not as accurate as a direct measurement of cellular growth, they do provide a measure of oxygen consumption and respiration rate that are not determined in normal growth rate testing.

4.1.3. Metabolite Secretion

As cells live and grow, they consume nutrients and convert them to needed biomolecules through metabolic reactions. In all cases, the net result of the metabolic reactions is that cells produce chemical byproducts that are often released into the environment. Analysis and quantitative measurement of the produced chemicals is another important characteristic of whole-cell phenotypes.

One of the most commonly used methods to characterize chemical compounds secreted during microbial growth is to use high-performance liquid chromatography (HPLC). In this analysis, secreted chemical compounds are maintained in liquid solution and mixed with a mobile phase solution. This liquid mixture is separated in a chromatography column where different chemical compounds are delineated by their chemical properties (size, polarity, chemical affinity to column packing). After separation in the chromatography column, amounts of each chemical are quantitatively measured using a detector (typically, ultraviolet or refractive index). In this manner, many secreted chemical compounds can be quantitatively measured in a single experimental run.

Although HPLC analysis is an established and commonly used method for measurements of extracellular metabolites, it is limited in its detection capabilities. A method using mass spectrometry called "metabolic footprinting" (14) has been developed to expand the ability to detect and quantitatively measure extracellular metabolites. Although the sample preparation in metabolic footprinting is similar to that in HPLC, identification of compounds is more precisely accomplished through measurements of chemical masses. In addition, the quantitative detection range for mass spectrometry is larger than the detection range of HPLC detectors, so more chemical compounds (especially those with low concentration) are analyzed.

Quantitative analysis of secreted chemical compounds is an important aspect of whole-cell phenotypes that can be measured in a high-throughput manner. Although this aspect of the phenotype is seldom under direct evolutionary selection pressure, it can play important secondary roles in evolutionary survival, and is of primary interest in metabolic engineering applications. With the increased detection capabilities demonstrated in metabolic footprinting studies (14), it has become possible to extrapolate certain details about the internal functional state of an organism, such that properties like metabolic efficiency and pathway usage can be characterized.

4.2. Genome-Scale Cellular Component Measurements

Genome-scale measurements of cellular components are conducted by pooling populations of cells and processing them to isolate components of interest. These typically include mRNA transcripts, proteins, and intracellular metabolites. Although whole-cell phenotype measurements are primarily concerned with determining what a cellular response is in a certain environment, measurements of cellular components are more directed toward determining how a certain cellular response is manifested. In studying the genotype–phenotype relationship, genomics specifies the genotype, whole-cell phenotype measurements are the phenotype, and changes in the cellular components can be viewed as the links between the two. The main focus and challenge of studying cellular components in an evolutionary setting is to differentiate causal from noncausal changes. A causal change would directly influence the manifested whole-cell phenotype, whereas it is possible to have changes in specific cellular components that do not affect the whole-cell phenotype (noncausal).

4.2.1. mRNA Transcripts

Genome-scale measurement of messenger RNA (mRNA) transcripts was the first of the high-throughput, omic data types to be developed and implemented. Concurrent measurement of all known mRNA transcripts using spotted microarrays or synthesized oligonucleotide arrays can be used to determine the relative abundance of individual genes. In terms of the molecular biology progression, the mRNA transcripts for individual genes are the next major biological component derived from DNA, and they can yield insight into transcriptional and regulatory mechanisms.

As the most established high-throughput cellular component data type, genome-wide analysis of mRNA transcripts has been used in a handful of studies on experimental adaptive evolution (15–18). In all of these cases, the genetic mutations that occurred during evolution were unknown, so one of the overall goals of the transcriptional analysis was to determine what mechanistic changes occurred in the evolution populations. Genome-wide transcriptional analysis typically led to the implication of several specific changes at the gene expression level; however, results were also typically limited by the statistical confidence generated during data processing and analysis.

Genome-wide mRNA transcript analysis is the most well developed and widely accessible cellular component data type. This data type can give insight into the transcriptional state and regulatory network of a cell. When applied to studying adaptive evolution, the most significant problem is differentiating causal from noncausal gene expression changes. Because mRNA transcripts are closely related to DNA, they can be used to study the transcriptional process, but they are far removed from the whole-cell phenotype, so it is often difficult to directly establish that a change in mRNA transcript abundance directly affects the observed whole-cell phenotype.

4.2.2. Proteins

Genome-wide analysis of protein quantities, or proteomics, is one of the most recently developed analytical methods, and it is still being refined. In the biological setting, DNA is transcribed to RNA that is translated to proteins. It is at the protein level that biological functionality is primarily determined, as proteins act as enzymes to mediate biochemical reactions *in vivo*. Proteomic methods typically involve a derivatization step for the sample preparation and mass spectrometry for identification and quantitation of protein levels.

Quantitative measurements of protein levels is generally thought to be a more functionally important measurement than measurement of mRNA transcripts, as the proteins are the metabolically active biological component. Although an increase in mRNA transcript levels should lead to an increase in the amount of protein, this may not always be true, so a direct measurement of the proteins is more reliable.

The major concern with implementing proteomics to evolutionary studies is connected to the fact that proteins carry out enzymatic functions in a biological setting. Mutations during adaptive evolution can lead to small and subtle differences in DNA that in turn can lead to a different amino acid sequence during translation from RNA to protein. This change can greatly affect the enzymatic function of a protein. Thus, as the functional efficiency of a protein is independent from the amount of protein measurements of protein levels may not be sufficient to indicate net functional changes. It is possible to have a lower quantity of a protein, but improved enzymatic efficiency, such that the overall effect is increased throughput by that protein.

4.2.3. Fluxes

Closely related to proteomics is the genome-wide measurement of metabolic fluxes, fluxomics (19). Proteomics is concerned with measuring the quantities of proteins, but is limited in its ability to evaluate changes in protein function. Because enzymatic proteins functionally mediate chemical reactions, changes in protein function can be determined by measuring changes in the flux through the chemical reactions. This is the aim of fluxomics.

Fluxomic measurements are conducted by culturing cells in medium with carbon substrates containing mixtures of different carbon isotopes. Labeled carbon atoms are processed and incorporated into different biological molecules that can be isolated and analyzed using mass spectrometry to determine the chemical pathways that were used to synthesize the end molecule. Typically, amino acids are isolated and measured using mass spectrometry in this analysis. In this manner, intracellular functional data can be obtained in terms of the usage of specific metabolic pathways.

Fluxomic measurements have been successfully used to study several different biological systems, including adaptive evolution (20), and fluxomics has become a highly reproducible method. Currently, two main concerns limit the application of fluxomic measurements. The major current limitation with fluxomic measurements is that the level of detail that can be obtained is very limited. Organisms typically contain on the order of hundreds to thousands of possible chemical reactions, but fluxomic measurements are normally only able to distinguish flux splits for 20–30 points in the metabolic network. This means that detailed flux measurements for a large number of individual reactions is not possible. Another consideration is that some knowledge of the metabolic network being studied is necessary before metabolic flux analysis can be conducted.

4.2.4. Metabolites

In parallel with the development of proteomics and fluxomics, methods for genome-wide intracellular metabolite analysis have arisen. The common factor-facilitating progress in proteomics, fluxomics, and metabolomics has largely been the improvements to and availability of mass spectroscopy. In the case of metabolomics, the technical capabilities of the equipment are a critical component in determining the quality of the results that can be obtained. Organisms can contain well over a thousand different chemical compounds, some of which may be present in nanomolar concentrations, so the ability of the mass spectrometer to accurately and reproducibly detect low-concentration chemicals will greatly affect results.

The ability to quantitatively measure genome-wide intracellular metabolites would yield complementary data to compensate for current limitations of proteomics and fluxomics. Changes in metabolite concentrations will reflect functional changes in enzymes, as alterations in reaction rates will either deplete or build up metabolite pools. Also, if a sufficient number of metabolites are positively identified, then a high level of detailed pathway-specific information can be obtained.

Despite the promise of genome-wide metabolomics analysis, the use of metabolomics is currently hindered by several technical limitations. Although some mass spectrometers are able to detect thousands of chemical compounds in an experimental run, it is often a long and difficult process to positively identify each of the compounds. Reproducible methods for sample preparation are also a concern, as intracellular metabolites can be quickly degraded. In application, this degradation problem, along with other sample preparation steps, has often led to wide variation in the reproducibility of metabolomics measurements.

5. Summary

The process of adaptive evolution can lead to large and system-wide changes in an organism. These characteristics make the systems biology approach well suited to analyzing and understanding adaptive evolution. Evolutionary changes in genotype and whole-cell phenotype can be measured, but are so disparate in nature that it is difficult to determine how the two are connected from these measurements alone. Insight into the connection between genotype and phenotype can be obtained using high-throughput, genome-wide analyses, such as transcriptomics, proteomics, fluxomics, or metabolomics. Each of these different analyses can contribute unique information about the functional state of an organism, but each also has distinct limitations that must be considered. Often, combining the strengths of different analyses can compensate for limitations of certain data type, such that the most useful current approaches integrate different data types representing different levels of cellular organization.

References

1. Elena SF, Lenski RE. Evolution experiments with microorganisms: the dynamics and genetic bases of adaptation. *Nat Rev Genet* 2003;4(6):457–469.
2. Atwood KC, Schneider LK, Ryan FJ. Periodic selection in *Escherichia coli*. *Proc Natl Acad Sci USA* 1951;37(3):146–155.
3. Lenski RE, Travisano M. Dynamics of adaptation and diversification: a 10,000-generation experiment with bacterial populations. *Proc Natl Acad Sci USA* 1994;91(15):6808–6814.
4. Fong SS, Palsson BO. Metabolic gene-deletion strains of *Escherichia coli* evolve to computationally predicted growth phenotypes. *Nat Genet* 2004;36(10):1056–1058.
5. Ibarra RU, Edwards JS, Palsson BO. *Escherichia coli* K-12 undergoes adaptive evolution to achieve in silico predicted optimal growth. *Nature* 2002; 420(6912):186–189.
6. Riehle MM, Bennett AF, Long AD. Genetic architecture of thermal adaptation in *Escherichia coli*. *Proc Natl Acad Sci USA* 2001;98(2): 525–530.
7. Velkov VV. New insights into the molecular mechanisms of evolution: Stress increases genetic diversity. *Mol Biol* 2002;36(2):209–215.
8. Fleischmann RD, et al. Whole-genome random sequencing and assembly of Haemophilus influenzae Rd. *Science* 1995;269(5223):496–498,507–512.

9. Palsson BO. The challenges of in silico biology. *Nat Biotechnol* 2000; 18(November):1147–1150.
10. Burger RM. Willensdorfer, Nowak M.A. Why are phenotypic mutation rates much higher than genotypic mutation rates? *Genetics* 2006;172(1): 197–206.
11. Schilling CH, Edwards JS, Palsson BO. Towards metabolic phenomics: Analysis of genomic data using flux balances. *Biotechnol Progress* 1999;15(3): 288–295.
12. Fong SS, Marciniak JY, Palsson BO. Description and interpretation of adaptive evolution of *Escherichia coli* K-12 MG1655 by using a genome-scale in silico metabolic model. *J Bacteriol* 2003;185(21):6400–6408.
13. Bochner BR, Gadzinski P, Panomitros E. Phenotype microarrays for high-throughput phenotypic testing and assay of gene function. *Genome Res* 2001;11(7):1246–1255.
14. Kell DB, Brown M, Davey HM, et al. Metabolic footprinting and systems biology: the medium is the message. *Nat Rev Microbiol* 2005;3(7):557–565.
15. Riehle MM, Bennett AF, Lenski RE, et al. Evolutionary changes in heat-inducible gene expression in lines of *Escherichia coli* adapted to high temperature. *Physiol Genomics* 2003;14(1):47–58.
16. Fong SS, Joyce AR, Palsson BO. Parallel adaptive evolution cultures of *Escherichia coli* lead to convergent growth phenotypes with different gene expression states. *Genome Res* 2005;15(10):1365–1372.
17. Cooper TF, Rozen DE, Lenski RE. Parallel changes in gene expression after 20,000 generations of evolution in *Escherichia coli. Proc Natl Acad Sci USA* 2003;100(3):1072–1077.
18. Ferea TL, Botstein D, Brown PO, et al. Systematic changes in gene expression patterns following adaptive evolution in yeast. *Proc Natl Acad Sci USA* 1999;96(17):9721–9726.
19. Sauer U. High-throughput phenomics: experimental methods for mapping fluxomes. *Curr Opin Biotechnol* 2004;15(1):58–63.
20. Fong SS, Nanchen A, Palsson BO, et al. Latent pathway activation and increased pathway capacity enable *Escherichia coli* adaptation to loss of key metabolic enzymes. *J Biol Chem* 2005;281(12):8024–8033.

11

Location Proteomics

Ting Zhao, Shann-Ching Chen, and Robert F. Murphy

Summary

Location proteomics is the systematic study of subcellular locations of proteins. It seeks to provide a thorough understanding of location patterns and integrate such knowledge into systems biology studies. Progress in this field depends on the quantitative and automated analysis of images of location patterns. This chapter introduces various approaches to such analysis and summarizes successes in using them to investigate different image types and different cell types.

These approaches can be divided into two categories, feature-based analysis, and pattern modeling. In feature-based analysis, each image is converted into a feature vector and all further analysis is carried out on the features. This has enabled the automated comparison, classification, and clustering of location patterns on a large scale. An important conclusion from this work has been that, at least for certain problems, computerized analysis can perform better than visual examination. To take a further step, object-based models have been built to describe location patterns in a compact and portable form. This facilitates more complicated analysis, such as the decomposition of patterns that are themselves mixtures of more basic patterns. Moreover, generative models can be learned from collections of images so that new examples of location patterns can be synthesized from them. This provides a way to integrate location information into systems biology, by combining generative models of many proteins and using the synthesized images as initial conditions for cell behavior simulations.

Key Words: Protein subcellular location; subcellular location trees; subcellular location features; fluorescence microscopy; pattern recognition; cluster analysis; protein distribution comparison; CD tagging.

1. Introduction

In the past decade, a major focus of biological research has become the creation of comprehensive, systematic databases describing different biological phenomena. For instance, biological macromolecules have been studied extensively with respect to their sequences, structures, expression levels, and interactions. The availability of these data has enabled the growth of a new field, systems biology, which seeks to construct models of complex biological systems at many levels of organization.

Proteomics projects provide systematic information about the characteristics of an entire proteome. Proteins can support the structure and shape of the cell, they serve as transporters and receptors to mediate flow among tissues, cells, and organelles, and they catalyze most metabolic reactions in a cell.

A major focus of current research is on understanding signaling networks in cells. A network can be represented by reaction–diffusion equations relating different species within or between compartments. In some cases, good approximations can be achieved by assuming a homogeneous distribution of molecules in the equations, but information regarding the actual spatial distributions of molecules can lead to a more thorough understanding of the networks. This is especially important for proteins whose functions are highly related to their locations. For example, external stimuli or mutation can change protein location, and a mislocalized protein can produce a diseased cell (1). This illustrates the importance of integrating high-resolution subcellular location information to build an accurate model of cell pathways. In a review of signaling networks, White et al. (2) suggest that "an important next step is to develop a high resolution map of signaling networks in living cells, and the location of interacting signaling units (i.e., hubs, motifs, and modules) relative to cell structures like the plasma membrane, mitochondria, the nucleus, etc."

Toward this end, the goal of location proteomics is to generate knowledge of location patterns and organize it so that it can be easily integrated into systems biology studies. Efforts in this area can be divided into three types. The first is *knowledge-capture approaches*, which seek to collect existing unstructured information about location and place it in a systematic framework. These efforts make use of a standardized vocabulary or ontology that defines the location categories and the relationships between them. The most widely used ontology for this purpose is the cellular component ontology created by the Gene Ontology Consortium (3). Databases containing GO terms for many proteins have been assembled. However, there are three significant limitations of knowledge capture approaches for location. The first is that, of course, they are limited to proteins that have been previously studied and reported on in the literature. The second is that assignments are primarily made by human curators based on reading journal articles, leading to potential inconsistency in these assignments (both for two different curators annotating the same protein pattern and for the same curator at different times). The third is that text-based descriptors do not adequately capture the complexity of patterns displayed by proteins within cells.

The second type of approach to protein location seeks to bridge the gap between proteins whose location is known and those whose location is not known by *predicting location from sequence*. A variety of methods have been used, all of which group known proteins by location and seek rules that can assign new proteins to one of those groups (4–8). This highlights the major limitation of this approach, which is that a limited number of classes are used (typically just the major organelles). There are typically also limited numbers of well-characterized training examples.

The third, and most expensive, type of approach is the direct *determination of subcellular location*. When properly applied, this approach can avoid all of the limitations described above, and, it is important to note, can provide more and better training data for location prediction systems. The remainder of this chapter will focus on methods for determining subcellular location, especially focusing on automated methods. Our lab has worked on designing such automated systems over the past 10 years. These systems are based on combining informative numerical features with powerful machine-learning methods to recognize subcellular location patterns objectively (9) or to objectively group proteins by their location patterns (10). We are also working on building generative models to describe protein distributions. Such models are especially helpful for cellular simulation, which has been listed as one of the 10 bioinformatic challenges for the next decade (11).

2. Approaches to Determining Subcellular Location

2.1. Protein-Tagging Methods

Although some systematic studies of protein location have been carried out using cell fractionation (12–14), microscopy is the major method used for this task. With some exceptions, the major type of microscopy used has been fluorescence microscopy. In this technique, proteins are tagged with fluorescence probes that absorb light of a specific wavelength range and emit light of a different (usually higher) wavelength. The emitted light can form an image of the location pattern of interest in a microscope.

The most widely used technique to tag a protein is immunofluorescence, in which a complex of fluorescent dye molecule and antibody attaches to a specific protein as an antigen. Usually two antibodies are used. The first, or primary, antibody is specific to target the protein of interest but has no dye molecule attached. The dye is conjugated to a secondary antibody, which has high affinity for the primary antibody. The secondary antibody can often recognize all antibodies derived from a given species, so a single dye-coupled antibody is reusable for a set of primary antibodies.

The availability of antibodies can limit the utility of immunofluorescence (or immunohistochemistry, in which a chromogenic probe is used instead of a fluorescent probe) for comprehensive tagging purposes. Another disadvantage is that immunotechniques cannot be used to observe living cells.

A powerful alternative to immunofluorescence is tagging proteins by fusing their coding sequence with that of green fluorescence protein (GFP) or other fluorescent proteins. To tag a protein of interest, molecular biology techniques are used to combine the coding sequence of

GFP with the coding sequence of the protein (this approach can be used for cDNA or genomic DNA). The result is a sequence that codes for a combination of the original protein and GFP. Because the mechanism is general, it is well suited for tagging each member of a set of proteins. However, the possibility that the GFP can alter the properties of the tagged protein must be considered.

2.2. Subcellular Location Image Databases

In recent years, proteins in organisms from yeast to human have been visualized and several databases of the images have been created (Table 1). These projects are discussed below.

A subcellular location database for the yeast *Saccharomyces cerevisiae* has been created by Huh et al. (15). The proteins were tagged with GFP at the C-terminal end through homologous recombination. A total of 6,029 genes were tagged, of which 4,156 yielded significant fluorescent signal upon expression. Images of these were taken using a digital imaging-capable Nikon TE200/300 inverted microscope at 100× magnification. Based on visual inspection and co-localization experiments, the authors assigned one or more of 23 location labels to each of the 4,156 proteins.

A similar approach has been applied to characterize the location of uncharacterized coding regions in the human genome. In this project, 107 open reading frames (ORFs) were examined (16). The cDNA sequence for each ORF was fused with the GFP coding sequence in a constitutive expression vector. Monkey Vero cells were transfected with the tagged cDNAs and imaged using a Leica DM/IRBE microscope at a 63× magnification. Locations were assigned to each fusion protein by visual inspection. For proteins whose locations could not be identified visually, or proteins whose N- and C-terminal fusion localizations were not identical, the authors used predictions from their sequences to assign them.

Table 1. Protein subcellular location databases.

Project	Tag	Expression	Live/Fixed	Resolution	Analysis	Species
Yeast GFP Fusion Localization Database (yeastgfp.ucsf.edu)	GFP fusion to C terminus	Endogenous	Live	High (100×)	Visual	Yeast
GFP-cDNA Localization Project (gfp-cdna.embl.de)	GFP fusion to C terminus	Transfection	Live	High (63×)	Visual	Human (transfected into monkey cells)
CD-tagging (cdtag.bio.cmu.edu)	Internal GFP fusion	Endogenous	Live	High (60×)	Visual	Mouse
Protein atlas (www.proteinatlas.org)	Immunohisto-chemistry with monospecific antibodies	Endogenous	Fixed (Formalin)	Low (20×)	Visual	Human
PSLID (murphylab.web.cmu.edu/services/PSLID)	Immunofluorescence and GFP	Both	Both	High (60–100×)	Automated	Mouse and Human

An alternative approach to creating GFP-fusions has been used in the CD-tagging project (17,18). CD tagging creates *internal* GFP fusions rather than terminal fusions. An engineered retroviral construct is created containing the GFP coding sequence flanked by splicing acceptor and donor sites. Infection of cells by the retrovirus results in undirected (approximately random) insertion into the genome. If the insertion occurs in the proper reading frame in an intronic region, a new GFP exon is created between two exons. Stable clones that express different tagged proteins can then be isolated, and the tagged gene identified by RT-PCR (17). The advantage of this genomic-tagging approach is that endogenous regulatory sequences are preserved and thus, normal levels of expression occur. Several mouse NIH 3T3 clones were created by this approach (17), and high-resolution images were collected (19). The patterns in these images were automatically analyzed using the methods described in the following sections.

Yet another approach to analysis of protein location has been taken by the Protein Atlas project, which has focused primarily on determining protein location at the cellular level within tissues, but also provides some information on subcellular location. The Protein Atlas database contains images for more than 700 proteins in 48 normal human tissues and 20 different cancers (20). The proteins analyzed included five major types of protein families: receptors, kinases, phosphatases, transcription factors, and nuclear receptors. Proteins were tagged by immunohistochemistry using well-characterized, monospecific primary antibodies and secondary antibodies (conjugated with horseradish peroxidase) in human tissue microarrays. The microarrays were scanned using an automated slide-scanning system at 20× magnification. The resulting images were annotated by visual inspection by pathologists.

3. Automated Analysis of Subcellular Patterns

As discussed in the previous section, most interpretation of subcellular location patterns has been performed by visual examination, with or without comparison to marker proteins whose location has been characterized. Visual examination has disadvantages when considered in the context of proteome-wide location analysis. It is very time-consuming for researchers to examine thousands or tens of thousands of images. Even when this is not an issue, the results are a qualitative description using words, which limits the resolution with which determinations can be made. Last, interobserver variation can be significant.

Therefore, we have developed methods for automated analysis of subcellular patterns. These systematic methods, for the first time, yielded quantitative descriptions of subcellular location patterns, and they have enabled the field of location proteomics. The remainder of this chapter will provide a review of these methods.

3.1. Subcellular Location Features

A key to the success of automated systems for subcellular pattern analysis is the design of good numerical features to capture essential informa-

Table 2. Summary of 2D SLFs.

Features for 2D Images	Description
Zernike moment features	49 features that are calculated as the dot product between a normalized cell image and each of the Zernike polynomials up to order 12
Haralick texture features	13 features calculated as the statistics of a gray-level cooccurrence matrix formed after down-sampling the image to a specified number of gray levels (if necessary)
Morphological features	16 features that capture information about the number, size, shape, and position of fluorescent objects in an image
Wavelet features	30 features that are the average energies of the coefficients from a Daubechies-4 wavelet transform at various levels; 60 features calculated using Gabor transforms of various orientations and scales
DNA features	6 features calculated by comparison with a parallel DNA image (when available)
Edge features	5 features that measure the fraction of protein along edges and the homogeneity of edge orientation
Nonobject features	1 feature calculated as the fraction of fluorescence that is not found in objects (i.e., fluorescence in below threshold pixels)

tion from images without being overly sensitive to variations induced by cell shape or orientation. We have defined several sets of features to describe protein subcellular distributions in fluorescence microscope images. Each image can then be quantitatively represented as a point in an n-dimensional feature space (where n is the number of features used). With this mathematical description, we can apply statistical and computational methods to analyze groups of images.

Many types of features have been investigated in the field of computer vision, including features designed to characterize colors, textures, and edges. Some types of features are quite general and can be applied to many different applications, but others are more specific to each application. For example, for the purpose of computer recognition of faces, features such as the distance between the eyes or the height of the nose are often used.

For cultured cells grown on a surface, there is typically no significance to a cell's position or orientation in a field. It is therefore desirable to design features that are translationally and rotationally invariant, as well as robust to variations in cell size and shape. We have described several such features of various types, which we term subcellular location features (SLFs). Tables 2 and 3 contain brief summaries of the types of SLFs we have used for 2D and 3D images, respectively. Figure 1 shows examples of some of these features for two 2D images with different location patterns. More detailed descriptions of these features are available in recent reviews (21,22) and also at http://murphylab.web.cmu.edu/services/SLF/.

Table 3. Summary of 3D SLFs.

Features for 3D Images	Description
3D morphological features	14 features which capture information about the properties of 3D fluorescent objects in the 3D image
3D DNA features	14 features calculated by comparison with a parallel DNA image (when available)
3D texture features	26 features based on the average and range of the 13 Haralick texture features over all 13 directions in which pixels can be adjacent in 3 dimensions
3D edge features	2 features calculated as statistics of the edges in each 2D slice of a 3D image

3.2. Segmentation of Images into Single Cell Regions

The success of subcellular location pattern analysis often requires segmentation of images into single-cell regions. Because of the large variations of image patterns resulting from different imaging techniques and cell types, it is not always straightforward to get satisfactory segmentation results. When no information about cell boundaries is available, but an image of a DNA stain is available to allow nuclei to be found, Voronoi segmentation is frequently used (23). However, the resulting boundaries usually enclose full, single cells only when cell density is low and cells are well separated. To improve upon this, experiments can be designed to provide some information to help determine cell boundaries, such as staining of total protein or plasma membrane (24). When parallel images

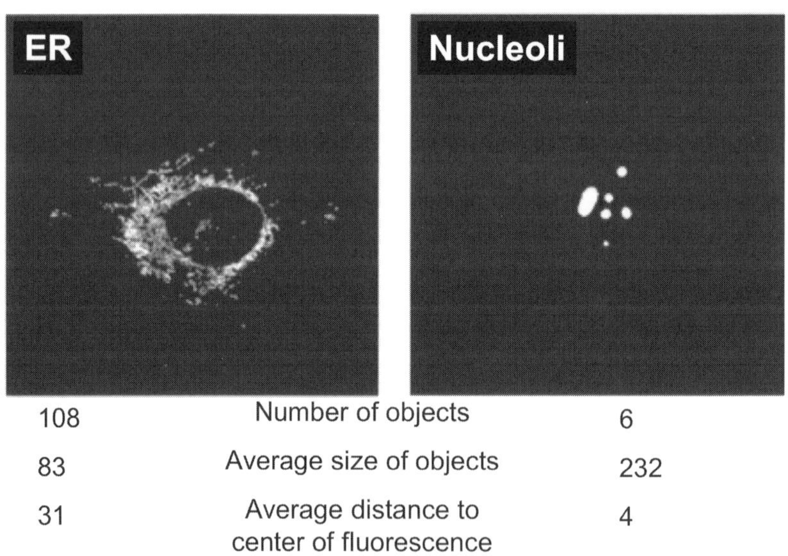

Figure 1. Illustration of feature extraction and classification. Example images of two subcellular patterns are shown, along with the values of three SLF features derived from morphological analysis. Note that any of the three features would be sufficient to distinguish the two patterns. Reprinted with permission from Chen et al. 2006 (39).

of DNA and total protein are available, the seeded watershed algorithm (25) frequently provides good boundaries (26). However, this algorithm usually produces loose contours, which can also include many of the irrelevant background regions (27). In addition, the initial seeding can be error prone, and poor seeding can produce unsatisfactory results. Another approach is to use level set–based algorithms, which numerically optimize the energy in elastic functions (27). This type of segmentation can provide excellent performance, but can be quite computationally expensive.

3.3. Feature Selection

Once each image is segmented into single-cell regions and each cell is reduced to a feature vector, a wide range of machine learning methods can be applied. However, in many cases, the performance of these methods can be weakened if the feature set contains redundant or uninformative features. This is especially true for classification methods that need to find decision boundaries in a potentially large and complex feature space because the presence of extra features increases the complexity of the search. The solution is to reduce the number of features by eliminating features that are uninformative (essentially the same for all classes) or redundant (highly correlated with another feature). Several methods for this purpose have been described, and we have evaluated them for use in the specific problem of subcellular pattern classification (28). We found that the best combination of performance and computational speed was provided by a method known as stepwise discriminant analysis (SDA) (29). Using this method, we have selected a number of different subsets from our 180-feature SLF bank. Each set that we obtained is suited to a slightly different classification problem. To facilitate determination of exactly which features were used for a particular result, we have developed a nomenclature for describing each set and the features within it. A number preceded by the prefix SLF identifies each set, and a particular feature within a set is referred to an index number preceded by the set name and a period (e.g., SLF16.5 is the fifth feature within the set SLF16).

3.4. Classification of Static 2D and 3D Single-Cell Images

Beginning 10 years ago, our group carried out the initial demonstration of the feasibility of automated classification of subcellular location patterns (30,31). For this purpose, we acquired extensive image collections, first for Chinese hamster ovary cells and then for HeLa cells. These collections were obtained for paraformaldehyde-fixed cells using markers (either monoclonal antibodies or fluorescent probes) specific for the major subcellular structures. For HeLa cells, we collected high-resolution 2D images of nine different markers (the 2D HeLa dataset). These included images for proteins whose patterns are similar, such as antibodies against two different Golgi proteins and antibodies against lysosomal and endosomal proteins. In addition, markers for the actin cytoskeleton, the tubulin cytoskeleton, mitochondria, the endoplasmic reticulum (ER), and nucleoli were used. Images of approximately 100 cells were collected for each marker along with a parallel DNA image. The parallel DNA images permitted the calculation of DNA-specific features for each marker, and also were used to define a tenth subcellular pattern in addition to those for the nine markers. Examples of these images are shown in Figure 2. Using the SLF described above, we used these images to

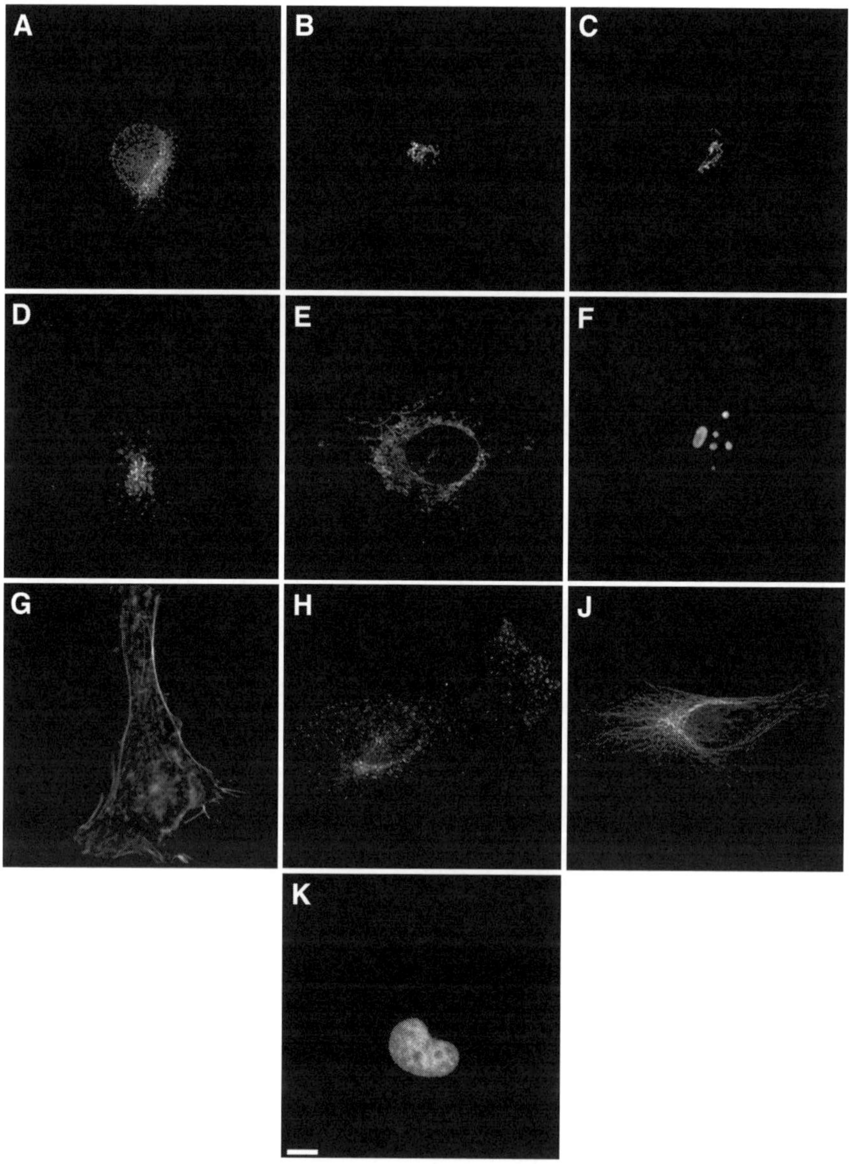

Figure 2. Typical images from the 2D HeLa collection. Images are shown for cells labeled with antibodies against an ER protein (A), the Golgi protein giantin (B), the Golgi protein GPP130 (C), the lysosomal protein LAMP2 (D), a mitochondrial protein (E), the nucleolar protein nucleolin (F), the endosomal protein transferrin receptor (H), and the cytoskeletal protein tubulin (J). Images are also shown for filamentous actin labeled with rhodamine-phalloidin (G) and DNA labeled with DAPI (K). Scale bar = 10 μm. Reprinted with permission from Boland and Murphy, 2001 (9).

produce the first demonstration that all major subcellular location patterns could be automatically recognized with reasonable accuracy. This task is one of *supervised learning* or *classification*, in which each instance (image) is known to belong to one of a set of predefined patterns (classes). We used images of known classes (training data) to design a specific classifier, which can be considered as a function (but often a very complicated function), to predict the class when given an image. The performance of such classifiers is evaluated using images whose class is known, but were not used for training (testing data). A classifier with more predictive power will give higher classification accuracy on the testing data. In our initial work, the 10 subcellular patterns could be recognized with an overall accuracy of 84% using the SLF4 feature set (consisting of 37 features) and a neural network classifier (9). We subsequently improved this performance by adding new features and using different classifiers, with the best performance to date (92% accuracy) having been achieved with a majority-voting ensemble classifier and the SLF16 feature set (consisting of 47 features) (32).

In addition, we showed that the automated system had better distinguishing power than visual inspection by measuring human recognition ability on the same 2D HeLa images (33). The overall accuracy for visual examination was 83%, which is almost 10 percentage points lower than the best performance of a machine classifier. The accuracies for each class by computer and visual analysis are compared in Figure 3.

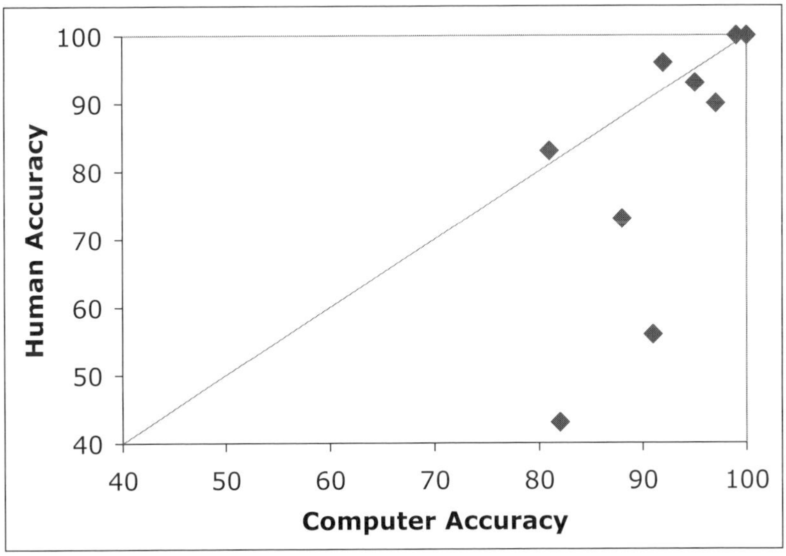

Figure 3. Comparison of accuracy of human and computer classification of subcellular patterns. Each symbol represents the accuracy of classification by computer (19) and human (33) for a different pattern class from the 2D HeLa collection. In increasing order of human classification accuracy, these are gpp130, giantin, LAMP2, TfR, ER, tubulin, mitochondria, nucleolin, and DNA (both at 100% for human and 99% for computer accuracy), and actin (100% for both). Reprinted with permission from Murphy, 2004 (40).

The largest difference in accuracy was found for the two Golgi proteins, which the automated system can distinguish with an average accuracy greater than 86%, whereas visual examination gave an accuracy of approximately 50% (indicating that visual examination could distinguish Golgi proteins from other proteins, but not discriminate between the two Golgi proteins). This led to an important conclusion: automated recognition of subcellular patterns was not only feasible, but it could perform better than human visual inspection.

Because cells are 3D objects, we also examined whether analyzing 3D images would improve the performance of automated recognition. To this end, we collected 3D images using staining methods similar to those for the 2D HeLa dataset, and we designed 3D SLF features to describe them. Some of these features were 3D versions of the 2D features, and some were unique to 3D images. The 3D images were collected by laser scanning confocal microscopy. A parallel DNA image was again collected with each marker, and, in addition, a third parallel image of the total protein distribution was collected after staining with an amine-reactive fluorescent dye (26). The total DNA and total protein images were used to perform automated segmentation of the images into single cell regions using the seeded watershed algorithm (26). The initial system achieved an overall accuracy of 91%, and subsequent additions improved this to 98% (34). In this case, even similar patterns such as the two Golgi proteins could be distinguished with near perfect accuracy. The conclusion from the work to date is that methods for building automated systems to assign proteins (or other macromolecules) to one of the major subcellular structures have been carried out well for 2D and 3D images, and that collection of high-resolution 3D images is desirable to achieve the best results possible.

Subsequent work on applying these methods to images collected by automated microscopy has yielded somewhat lower classification accuracies (35). Whether this is because of possible perturbations of subcellular patterns that might be induced by transfection reagents used to express GFP-tagged markers, the use of undeconvolved wide-field images, the use of lower magnification, or differences in cell line or computational implementation (or some combination of these factors) remains to be determined.

3.5. Classification of Other Types of Images

The features and classifiers described above have been designed for the analysis of single-cell regions of static (single time point) images of live or fixed samples. Subsets of these features that do not require segmentation (and are approximately independent of the number of cells in an image) can also be created. We have shown that these features are able to achieve reasonable classification accuracy on multicell images created from the 2D HeLa dataset.

Some of the SLFs are averages of features that describe the size, shape, and position of individual objects in a cell (in this context, an object is considered to be a contiguous set of pixels that are above an automatically chosen threshold). The unaveraged features can, of course, be used to describe those objects individually, and these can be used to create classifiers that recognize various types of objects (36).

When a time series image (movie) is available (either of a single 2D slice or of an entire 3D stack over time), additional features can be used to characterize the temporal behavior in a model-free manner. We have demonstrated that these features are useful for discriminating between proteins whose static patterns are similar (37).

4. Systematic Comparison and Clustering

Because the work described above shows that the SLFs are effective for representing location patterns, it is reasonable to also use them for other purposes involving interpretation of subcellular patterns. One such purpose that is commonly encountered is the comparison of images or sets of images. To measure the similarity of a pair of images, we can simply calculate the Euclidean distance between their SLF vectors. For comparing sets of images, we can calculate the Mahalanobis distance between their mean SLF vectors; using this distance measure avoids the problems with redundant or uninformative features discussed above in the context of feature reduction. To determine if this interest distance is statistically different from zero, we can use multivariate hypothesis tests such as the Hotelling T^2 test (38). The difference between the two visually indistinguishable patterns giantin and gpp130 (33) can be detected successfully in this way.

The possibility of objective similarity measurement is a critical requirement for the new field of location proteomics, the goal of which is to organize all proteins into a systematic framework based on their location. The task of organization can also be stated as "Given a set of proteins, each with multiple image representations, find a partitioning of the protein set such that images from members in the same partition show a single-location pattern" (10). We have demonstrated how this can be achieved by combing the SLF with clustering methods.

The starting point for clustering proteins by their subcellular location is, of course, a collection of images for many (or all) proteins expressed in a given cell type. Such a collection can be created by any of the methods discussed above for protein tagging. We have extensively analyzed one such collection, which consists of 3D images of CD-tagged cell lines, each expressing a different protein tagged with GFP (17). Single-color GFP images were collected for live cells by spinning disk confocal microscopy (18). The 3D SLFs used for classifying the 3D HeLa images were calculated and SDA was applied to generate an optimal feature set for these data. This resulted in feature set SLF18 consisting of 34 features, including 9 morphological features, 1 edge feature, and 24 texture features.

Just as for comparison of images, the selection of a distance function is also important for clustering. We have evaluated two distance functions, z-scored Euclidean distance, which is calculated after normalizing each feature to have zero mean and unit variance, and Mahalanobis

Table 4. Measurement of agreement between clustering results by various methods and using different distance functions.

Clustering approaches compared	Z-scored Euclidean distance κ	Mahalanobis distance κ
k-means/AIC versus consensus	1	0.5397
k-means/AIC versus ConfMat	0.4171	0.3634
Consensus versus ConfMat	0.4171	0.1977
k-means/AIC versus visual	0.2055	0.1854
Consensus versus visual	0.2055	0.1156

The κ statistic values are shown for comparisons between the four clustering methods: *k*-means/AIC (k-means by AIC optimization), consensus (consensus tree), ConfMat (confusion matrix), visual (visual inspection). The κ statistic is 1 when there is perfect agreement between two methods and 0 when the agreement is not significant relevant to that expected at random. Note that the Euclidean distance shows a better agreement than the Mahalanobis distance.
Source: From Chen and Murphy, 2005 (10).

distance, which scales the contribution of each feature using the covariance matrix.

For clustering the 3D 3T3 dataset, we evaluated a range of clustering methods using both of these distance functions (10). In this comparison, our premise was that a better distance function should produce better agreement among different clustering algorithms. The Cohen κ statistic can be used to measure the agreement between two partitions of data by comparing the observed agreement with the expected agreement (the larger the κ statistic, the higher the agreement). Table 4 shows the κ statistic for comparisons between different clustering methods and distance functions. Four clustering algorithms were evaluated. The *k*-means/Akaike information criteria (AIC) method uses *k*-means clustering for various numbers of clusters followed by calculation of the AIC to select number of clusters. The consensus method uses hierarchical clustering to group many random subsets of the input data and determine which groupings of proteins are stable for many random subsets. A third approach, termed ConfMat, begins by training a classifier for all clones, and then group clones that are hard to distinguish by examining the confusion matrix. The last method is based on visual inspection, and consists of assigning descriptive terms (e.g., "cytoplasmic") to each clone, and then grouping proteins whose terms match completely. From the table, we can conclude that z-scored Euclidean distance is better than Mahalanobis distance, at least from the viewpoint of consistency among automated methods.

Figure 4 shows a consensus subcellular location tree based on the z-scored Euclidean distance. The clustering result is consistent with visual inspection and available information from protein databases. For example, proteins showing a nuclear pattern and those showing a cytoplasmic pattern are well separated into different clusters. It is important to note that no human intervention is needed to create this tree, and that, unlike manually created groupings, the criteria used to construct it (the SLF, the distance function, and the clustering algorithm) are all well-described and can be easily updated or replaced as justified by evaluation of the results. In the limited cases where information on the subcellular location of these 90 proteins is available from protein data-

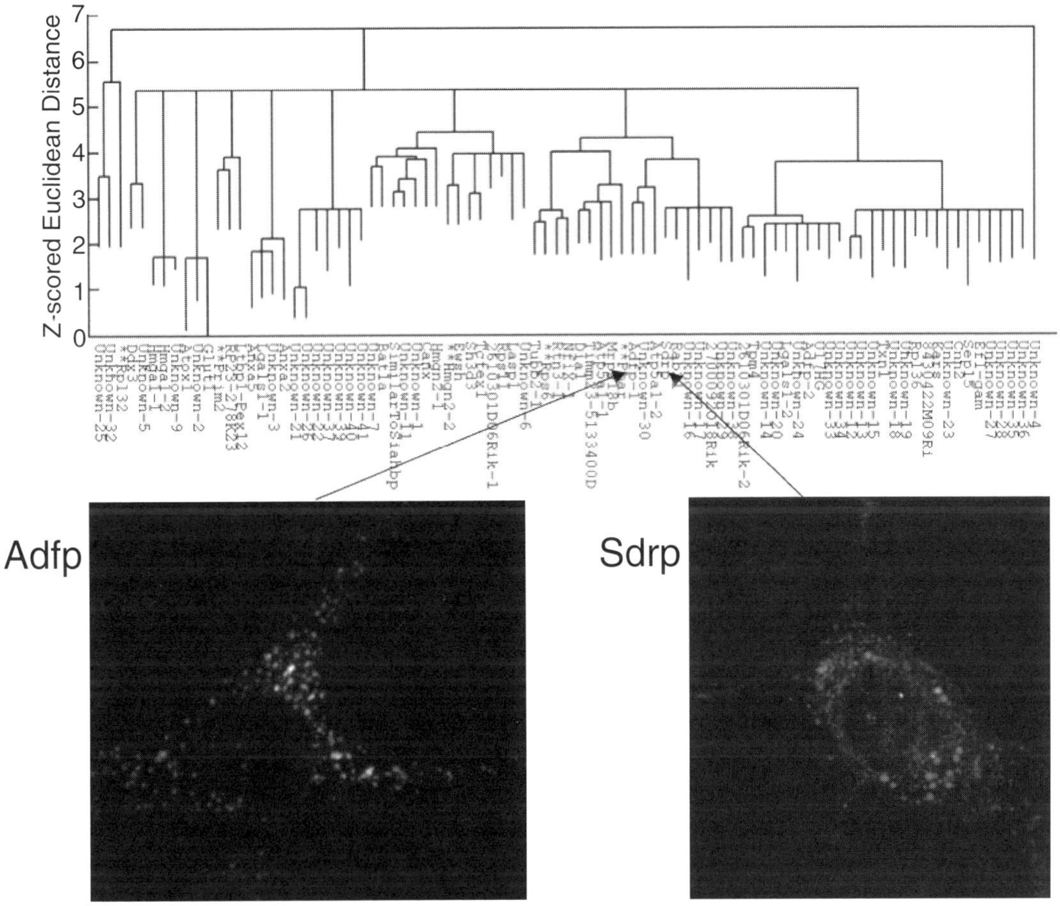

Figure 4. Example of a subcellular location tree. A consensus subcellular location tree is shown for 87 3T3 cell clones expressing different CD-tagged proteins. Numerical features were calculated to describe each image and proteins were grouped into statistically distinguishable groups. An interactive browser (available at http://murphylab.web.cmu.edu/services/PSLID/tree.html) permits viewing of images for particular proteins in order of the degree to which they are representative of the overall pattern. Examples for two proteins within neighboring, but distinct, clusters are shown. Reprinted with permission from Murphy, 2005 (41).

bases, it generally agrees with the automated groupings. The obvious power of automated pattern clustering is that it can be used to cluster thousands of proteins automatically, a task that would challenge even the most dedicated visual curators.

5. Models of Subcellular Patterns

We have shown how location patterns can be described numerically and how they can be interpreted using both unsupervised and supervised learning approaches. The next step is to convert the knowledge we have obtained about protein patterns into a form that can be used for systems biology studies. This step is accomplished through pattern modeling.

5.1. Object-based Modeling

Models are efficient ways to describe a system. The most practical models have a compact form characterized by a small number of parameters. We can distinguish between *descriptive* models, which utilize an irreversible mapping from image to features and cannot easily be used to synthesize new images, and generative models, which capture the essence of a pattern in a manner that can be used to generate new images that are, in principle, indistinguishable from the examples it was built from. A feature matrix is an example of a descriptive model because we cannot (usually) reconstruct patterns from it. Building a generative model may seem daunting given the large amount of data contained in one image. A typical 512 × 512 image with 256 gray levels could contain 250 kilobytes. The task becomes even more challenging when considering the variation between images of the same pattern.

Fortunately, the sufficient representation of features implies that a pattern could be modeled in a much lower dimensional space. Because morphological features describing the properties of objects are powerful features, we considered the possibility that subcellular patterns could be adequately represented by considering them to be composed of distinct combinations of objects of particular types. Our starting point was to use some SLFs that had previously been calculated as averages for all objects to instead describe each individual object in a pattern. These subcellular object features (SOFs) were then used to determine how many statistically distinguishable object types were contained in an image collection using clustering. Once these types were found, a classifier was trained to recognize each type. This permits each cell pattern to be represented as a vector showing how many objects of each type it contains. This yields a two-stage process for modeling any cell image using objects. It consists of assigning a type to each object in the image and then applying a multinomial process to model the object distribution for a given pattern. This process has been applied on the 2D HeLa images to determine how accurately patterns can be recognized using only the objects they contain (36). For this purpose, the SLFs were replaced by a vector containing object frequencies, a vector containing the fraction of fluorescence in each object type, and/or a vector summarizing the SOF of each object type. The accuracy was 81%–82% for the 9 major patterns (giantin and gpp130 were merged into one class) using some or all of these features. This is encouraging because no spatial relation between objects was considered. With this fact, we can model the patterns in a hierarchical way. The first level is the composition of objects with different types, and the next level is the model of spatial relationship between the objects. The models will be complete at the level in which the objects themselves are described.

Such models facilitate further analysis. For example, we have for the first time developed a model to recognize mixture patterns (36). More importantly, the models will provide us with a higher level of understanding of subcellular locations because they could be related to biological entities easily. For example, the object type learning process will put all nuclei into one type and vesicles into another type. The object models can further describe the details of location patterns and provide a high-resolution map for protein location. All of these will lead to generative models that can synthesize data of location patterns for systems biology study, such as cell behavior simulation.

5.2. Generative Models

In the previous section, we discussed building object-based models of location patterns. To develop them into generative models, the descriptions of individual objects and their positions are required. Our preliminary work has shown the possibility of building such generative models for lysosome proteins (Zhao and Murphy, in preparation). We start by building generative models of nuclear and cell shape. The shape of objects containing a specific protein (e.g., LAMP2, a marker for lysosomes) is modeled by a 2D Gaussian distribution, which can then be easily used to synthesize new objects. The position of an object is described by the ratio of its distance to cell membrane and its distance to the nuclear membrane. Combining the object models with the models of nuclear and cell shape, we can synthesize a three-color image with three compartments (cytoplasm, nucleus, and vesicles) where the protein is localized. Figure 5 briefly shows the steps of the procedure. In this procedure, nuclear shape is synthesized before cell shape because the cell shape model is the description of the ratios between the distances of cell membrane to nuclear center and the distances of nuclear membrane to nuclear center.

Generative models of subcellular location patterns will be important for bottom-up modeling in systems biology because they describe the

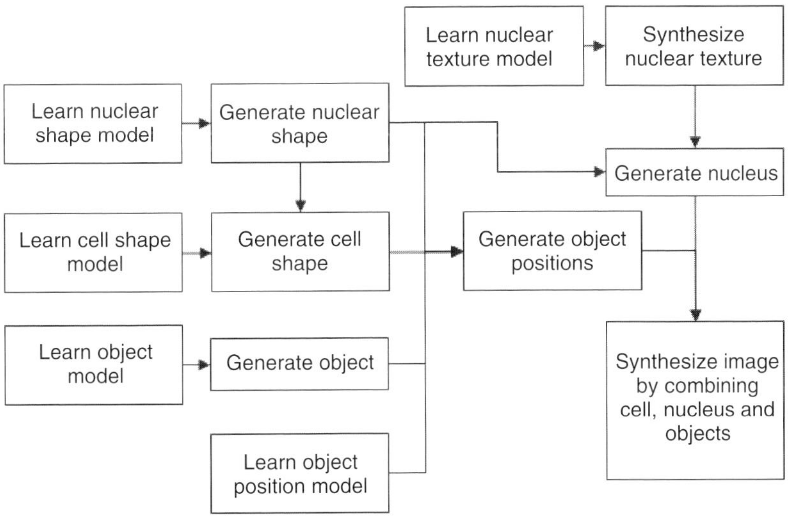

Figure 5. Flowchart for learning object-based models for location patterns and synthesizing an image from the models. Statistical models of the nuclear shape, nuclear texture, cell shape, objects, and object positions are learned from real images. Nucleus, cell membrane and protein objects are then generated and combined based on the statistical models to synthesize an image.

distributions of proteins in a cell with a high resolution. More interestingly, they provide a friendly interface between location proteomics and systems biology by image synthesis. The synthesized images can be used as initial conditions of cell behavior simulation and they could be more reliable than manual settings because the models are learned from real data objectively.

Acknowledgments: Original research from the Murphy group was supported in part by NIH grant R01 GM068845 and NSF grant EF-0331657.

References

1. Karim R, Tse G, Putti T, et al. The significance of the Wnt pathway in the pathology of human cancers. *Pathology* 2004;36(2):120–128.
2. White MA, Anderson RGW. Signaling networks in living cells. *Annu Rev Pharmacol Toxicol* 2005;45:587–603.
3. Harris MA, Clark J, Ireland A, et al. The Gene Ontology (GO) database and informatics resource. *Nucleic Acids Res* 2004;32:D258–D61.
4. Horton P, Nakai K. Better Prediction of Protein Cellular Localization Sites with the *k* Nearest Neighbors Classifier. *Intell Sys Mol Biol* 1997;5:147–152.
5. Nakai K. Protein sorting signals and prediction of subcellular localization. *Adv Protein Chem* 2000;54:277–344.
6. Hua S, Sun Z. Support vector machine approach for protein subcellular localization prediction. *Bioinformatics* 2001;17(8):721–728.
7. Lu Z, Szafron D, Greiner R, et al. Predicting subcellular localization of proteins using machine-learned classifiers. *Bioinformatics* 2004;20(4):547–556.
8. Chou K-C, Elrod DW. Protein subcellular location prediction. *Protein Eng* 1999;12(2):107–118.
9. Boland MV, Murphy RF. A neural network classifier capable of recognizing the patterns of all major subcellular structures in fluorescence microscope images of HeLa cells. *Bioinformatics* 2001;17(12):1213–1223.
10. Chen X, Murphy RF. Objective clustering of proteins based on subcellular location patterns. *J Biomed Biotechnol* 2005;2005(2):87–95.
11. Eisenberg D, Marcotte E, McLachlan AD, et al. Bioformatic challenges for the next decade(s). *Phil Trans R Soc B* 2006;Published online.
12. Dreger M. Proteome analysis at the level of subcellular structures. *Eur J Biochem* 2003;270:589–599.
13. Brunet STP, Gagnon E, Kearney P, et al. Organelle proteomics: looking at less to see more. *Trends Cell Biol* 2003;13(12):629–638.
14. Yates JR 3rd, Gilchrist A, Howell KE, et al. Proteomics of organelles and large cellular structures. *Nat Rev Mol Cell Biol* 2005;6(9):702–714.
15. Huh W-K, Falvo JV, Gerke LC, et al. Global analysis of protein localization in budding yeast. *Nature* 2003;425(6959):686–691.
16. Simpson JC, Wellenreuther R, Poustka A, et al. Systematic subcellular localization of novel proteins identified by large-scale cDNA sequencing. *EMBO Rep* 2000;1(3):287–292.
17. Jarvik JW, Fisher GW, Shi C, et al. In vivo functional proteomics: Mammalian genome annotation using CD-tagging. *Biotechniques* 2002;33(4):852–867.
18. Chen X, Velliste M, Weinstein S, et al. Location proteomics—building subcellular location trees from high resolution 3D fluorescence microscope images of randomly-tagged proteins. *Proc SPIE* 2003;4962:298–306.

19. Chen X, Murphy RF. Location proteomics: determining the optimal groupings of proteins according to their subcellular location patterns as determined from fluorescence microscope images. In: Proceedings of 2004 Asilomar Workshop on Signals, Systems and Computers. 2004: 50–54.
20. Uhlén M, Björling E, Agaton C, et al. A human protein atlas for normal and cancer tissues based on antibody proteomics. *Amer Soc Biochem Mol Biol* 2005;4:1920–1932.
21. Huang K, Murphy RF. From quantitative microscopy to automated image understanding. *J Biomed Optics* 2004;9(5):893–912.
22. Huang K, Murphy RF. Data mining methods for a systematics of protein subcellular location. In: Wang JTL, Zaki MJ, Toivonen HTT, Shasha D, eds. Data Mining in Bioinformatics. London: Springer-Verlag; 2004:143–187.
23. Jones TR, Carpenter AE, Golland P. Voronoi-based segmentation of cells on image manifolds. In: Proceedings of ICCV Workshop on Computer Vision for Biomedical Image Applications. 2005:535–543.
24. De Solorzano CO, Malladi R, Lelievre SA, et al. Segmentation of nuclei and cells using membrane related protein markers. *J Microsci* 2001;201(Pt 3):404–415.
25. Lotufo R, Falcao A. The ordered queue and the optimality of the watershed approaches. In: Goutsias J, Vincent L, Bloomberg DS, eds. Mathematical Morphology and its Application to Image and Signal Processing: Kluwer Academic Publishers; 2000.
26. Velliste M, Murphy RF. Automated determination of protein subcellular locations from 3D fluorescence microscope images. In: Proceedings of 2002 IEEE International Symposium on Biomedical Imaging (ISBI-2002). 2002:867–870.
27. Coulot L, Kirschner H, Chebira A, et al. Topology preserving STACS segmentation of protein subcellular location images. In: Proceedings of 2006 IEEE International Symposium on Biomedical Imaging (ISBI-2006). 2006:566–569.
28. Huang K, Velliste M, Murphy RF. Feature reduction for improved recognition of subcellular location patterns in fluorescence microscope images. *Proc SPIE* 2003;4962:307–318.
29. Jennrich RI. Stepwise discriminant analysis. In: Enslein K, Ralston A, Wilf HS, eds. Statistical Methods for Digital Computers. New York: John Wiley & Sons; 1977:77–95.
30. Boland MV, Markey MK, Murphy RF. Classification of protein localization patterns obtained via fluorescence light microscopy. In: Proceedings of 19th Annual International Conference of the IEEE Engineering in Medicine and Biology Society; 1997:594–597.
31. Boland MV, Markey MK, Murphy RF. Automated recognition of patterns characteristic of subcellular structures in fluorescence microscopy images. *Cytometry* 1998;33(3):366–75.
32. Huang K, Murphy RF. Boosting accuracy of automated classification of fluorescence microscope images for location proteomics. *BMC Bioinformatics* 2004;5:78.
33. Murphy RF, Velliste M, Porreca G. Robust numerical features for description and classification of subcellular location patterns in fluorescence microscope images. *J VLSI Sig Proc* 2003;35(3):311–321.
34. Chen X, Murphy RF. Robust Classification of Subcellular Location Patterns in High Resolution 3D Fluorescence Microscopy Images. In: Proceedings of 26th Annual International Conference of the IEEE Engineering in Medicine and Biology Society. 2004:1632–1635.

35. Conrad C, Erfle H, Warnat P, et al. Automatic identification of subcellular phenotypes on human cell arrays. *Genome Res* 2004;14:1130–1136.
36. Zhao T, Velliste M, Boland MV, et al. Object Type Recognition for Automated Analysis of Protein Subcellular Location. *IEEE Trans on Image Processing* 2005;14(9):1351–1359.
37. Hu Y, Carmona J, Murphy RF. Application of temporal texture features to automated analysis of protein subcellular locations in time series fluorescence microscope images. In: Proceedings of the 2006 IEEE International Symposium on Biomedical Imaging (ISBI 2006); 2006; 2006. pp. 1028–1031.
38. Roques EJS, Murphy RF. Objective evaluation of differences in protein subcellular distribution. *Traffic* 2002;3(1):61–65.
39. Chen X, Velliste M, Murphy RF. Automated interpretation of subcellular patterns in fluorescence microscope images for location proteomics. *Cytometry* 2006;69A:631–640.
40. Murphy RF. Automated interpretation of subcellular location patterns. In: 2004 IEEE International Symposium on Biomedical Imaging (ISBI-2004). 2004. p. 53–56.
41. Murphy RF. Cytomics and location proteomics: automated interpretation of subcellular patterns in fluorescence microscope images. *Cytometry* 2005; 67A:1–3.

Part III

Theoretical and Modeling Techniques

12

Reconstructing Transcriptional Networks Using Gene Expression Profiling and Bayesian State-Space Models

Matthew J. Beal, Juan Li, Zoubin Ghahramani, and David L. Wild

Summary

A major challenge in systems biology is the ability to model complex regulatory interactions. This chapter is concerned with the use of Linear-Gaussian state-space models (SSMs), also known as linear dynamical systems (LDS) or Kalman filter models, to "reverse engineer" regulatory networks from high-throughput data sources, such as microarray gene expression profiling.

LDS models are a subclass of dynamic Bayesian networks used for modeling time series data and have been used extensively in many areas of control and signal processing. We describe results from simulation studies based on synthetic mRNA data generated from a model that contains definite nonlinearities in the dynamics of the hidden factors (arising from the oligomerization of transcription factors). Receiver operating characteristic (ROC) analysis demonstrates an overall accuracy in transcriptional network reconstruction from the mRNA time series measurements alone of approximately a 68% area under the curve (AUC) for 12 time points, and better still for data sampled at a higher rate.

A key ingredient of these models is the inclusion of "hidden factors" that help to explain the correlation structure of the observed measurements. These factors may correspond to unmeasured quantities that were not captured during the experiment and may represent underlying biological processes. Results from the modeling of the synthetic data also indicate that our method is capable of capturing the temporal nature of the data and of explaining it using these hidden processes, some of which may plausibly reflect dynamic aspects of the underlying biological reality.

Key Words: Transcriptional networks; microarrays; state-space models; variational Bayesian; reverse engineering.

1. Introduction

A major challenge in systems biology is the ability to model complex regulatory interactions. Several computational approaches have been developed over recent years to address this challenge. So-called bottom-up approaches begin with a detailed mathematical description of the biophysical processes involved in the regulatory pathway (transcription factor binding, diffusion, mRNA and protein degradation, etc.), usually expressed as differential equations (see Kholodenko et al. (1), for example). To date, these approaches have been applied mainly to small, well-defined biological model systems. In contrast, so-called top-down approaches attempt to "reverse engineer" regulatory networks from high-throughput data sources, such as microarray gene expression profiling. Many of the tools that have been applied in an exploratory way to the problem of reverse engineering genetic regulatory networks from gene expression data have been reviewed by Wessels et al. (2), van Someren et al. (3), de Jong (4), and Friedman (5). These include Boolean networks (6,7,8), time-lagged cross-correlation functions (9), and linear and nonlinear autoregression models (10,11,12). Although these techniques have produced models that appear biologically plausible, based on circumstantial evidence from the biological literature, many have been derived from public domain data with insufficient replication, given that these papers attempt to reconstruct the interactions of large numbers (sometimes thousands) of genes from small data sets, with the consequent likelihood of model overfitting. Because there exists very little gene expression data for which the ground truth regulatory network is known, many authors have turned to *in silico* simulations to test the performance of reverse-engineering methods on data produced from a biologically plausible model system. Care needs to be taken in interpreting the results of such simulations, particularly if the artificially generated data comes from a model that has the assumed model structure of the identification scheme, as the actual structure of the networks generating real data are presumably unknown. In particular, Smith et al. (13), Yeung et al. (14), and Zak et al. (15,16) have attempted to evaluate reverse engineering techniques by the use of simulated gene expression data from *in silico* networks. Zak et al. (15,16) considered linear, log-linear, and nonlinear (squashing function) regressive models and concluded that these methods were unable to identify the generating network from simulated gene expression data alone and constituted little more than curve fitting.

Murphy and Mian (17) were the first to propose the use of a general class of graphical models known as dynamic Bayesian networks (DBNs) to model time series gene expression data. Bayesian networks have a number of features that make them attractive candidates for modeling gene expression data, such as their ability to handle noisy or missing data, to handle hidden variables, such as protein levels, which may have an effect on mRNA steady-state levels, to describe locally interacting processes and the possibility of making causal inferences from the derived models. The application of Bayesian networks to microarray data analysis was first explored experimentally in the pioneering work of Friedman et al. (18). However, this approach ignored the temporal dependence of the gene intensities during trials and went only as far as to infer the causal relationships between the genes within one time step. Their method discretized expression levels and made use of efficient candidate proposals and methods for searching the space of model structures. This approach also assumed that all the possibly interacting variables are observed on the microarray, which precludes the existence of hidden causes or unmeasured genes whose involvement might dramatically simplify the network structure, and therefore ease interpretability of the mechanisms in the underlying biological process. Although microarray technologies have made it possible to measure time series of the expression level of many genes simultaneously, we cannot hope to measure all possible factors contributing to genetic regulatory interactions, and the ability of Bayesian networks to handle such hidden variables would appear to be one of their main advantages as a modeling tool. Husmeier (19) has evaluated the accuracy of gene regulatory network reconstruction from simulated discretized gene expression data using fully observed Bayesian network models and concluded that, at least to a certain extent, local regulatory network structures could be recovered. The fidelity of network reconstruction was found to depend on the prior probabilities used in the Bayesian inference scheme. Following Friedman et al., a number of other authors have described Bayesian network models of gene expression data. Most published work to date has only considered either static Bayesian networks with fully observed data (20) or static Bayesian networks that model discretized data but incorporate some hidden variables (21,22,23,24). Ong et al. (25) have described a dynamic Bayesian network model for *E. coli* which explicitly includes operons as hidden variables, but again uses discretized gene expression measurements. There is, therefore, a clear need for a dynamic modeling approach that can both accommodate gene expression measurements as continuous, rather than discrete, variables and which can model unknown factors and unobserved regulators as hidden variables. This chapter describes one such method. Our focus on modeling unobserved and unknown factors is crucial, as measurements of steady-state mRNA levels are the result of a variety of complex events, including gene transcription and mRNA degradation.

This chapter is concerned with Linear-Gaussian state-space models (SSMs), which are also known as linear dynamical systems (LDS) (26) or Kalman filter models (27). These models are a subclass of dynamic Bayesian networks used for modeling time series data and have been used extensively in many areas of control and signal processing. We will use the terms LDS and SSM interchangeably throughout this chapter, although they emphasize different properties of the model. Linear dynamic system emphasizes that the dynamics are linear; such models can be represented either in state-space form or in input–output form. State-space model emphasizes that the model is represented as a latent-variable model (i.e., the observables are generated via some hidden states). State-space models can be nonlinear in general; here, it should be assumed that we refer to linear models with Gaussian noise unless otherwise stated. State-space models have a number of features that make them attractive for modeling gene expression time-series data. They assume the existence of a set of hidden state variables, from which noisy continuous measurements can be made, and which evolve with Markovian dynamics. In our application, the noisy measurements are the observed gene expression levels at each time point, and a key innovation of our method is that we assume that the hidden variables are modeling effects that cannot be measured in a gene expression–profiling experiment. The effects of genes that have not been included on the microarray, levels of regulatory proteins, the effects of mRNA, and protein degradation are examples of such hidden variables.

We have previously described the application of linear state-space modeling to reverse engineer transcriptional networks from highly replicated expression profiling data obtained from a well-established model of T-cell activation in which we monitored a set of relevant genes across a time series (28,29,30,31). Nachman et al. (32) recently described a related dynamic Bayesian network model that includes hidden states to model unobserved regulator activity levels. Perrin et al. (33) and Wu et al. (34) also described related SSMs for modeling genetic regulatory networks. However, the methods used in these papers to estimate the number of hidden states in these models suffer from several technical drawbacks, not least in that they cannot provide us with posterior distributions over all the parameters of the model, which are needed to quantify our uncertainty. Furthermore, neither the model described by Perrin et al. (33) nor the one described by Wu et al. (34) utilizes inputs to feed back the outputs from the previous time step. Consequently, these models do not allow genes to affect the hidden states and do not allow genes to affect other genes directly. The model described in this chapter does not suffer from these limitations and we have discussed the advantages of our approach in our published work (31).

Some aspects of using the LDS model for this type of problem are not ideal. For example, we make the assumptions that the dynamics and output processes are time invariant, which is unlikely in a real biological system. Furthermore, the times at which the data are taken are not linearly spaced, which might imply that there is some nonlinearity in the rate of the transcriptional process; there may be whole missing time slices that, if they had been included, would have made the dynamics process closer to stationary. There is also the limitation that the noise in the dynamics and output processes may not be Gaussian. Nevertheless, despite the assumptions inherent in linear state-space models, the results described in our published work (28,29,30,31) indicate that they are a very useful tool for investigating gene transcriptional networks. The resulting network models provide excellent examples of the type of testable biological hypotheses that can be generated using reverse engineering approaches. Our models reflect many of the dynamics of an activated T-cell. In particular, they reveal the integrated activation of cytokines, proliferation, and adhesion following activation and place JunB and JunD at the center of the mechanisms that control apoptosis and proliferation.

2. Modeling Time Series with SSMs

2.1. Variables and Topology

In SSMs, a sequence $(\mathbf{y}_1, \ldots, \mathbf{y}_T)$ of p-dimensional real-valued observation vectors, denoted $\mathbf{y}_{1:T}$, is modeled by assuming that at each time step t, $\mathbf{y}_t$ was generated from a k-dimensional real-valued hidden-state variable $\mathbf{x}_t$, and that the sequence of $\mathbf{x}$'s follow a first-order Markov process. The joint probability of a sequence of states and observations is therefore given by:

$$p(\mathbf{x}_{1:T}, \mathbf{y}_{1:T}) = p(\mathbf{x}_1)\, p(\mathbf{y}_1|\mathbf{x}_1) \prod_{t=2}^{T} p(\mathbf{x}_t|\mathbf{x}_{t-1}) p(\mathbf{y}_t|\mathbf{x}_t). \tag{1}$$

The distribution $p(\mathbf{x}_1)$ over the first hidden state is assumed Gaussian. Our approach has focused on models where both the dynamics, $p(\mathbf{x}_t|\mathbf{x}_{t-1})$, and output functions, $p(\mathbf{y}_t|\mathbf{x}_t)$, are linear and time-invariant, and the distributions of the state evolution and observation noise variables are Gaussian, i.e., linear-Gaussian SSMs:

$$\mathbf{x}_t = A\mathbf{x}_{t-1} + \mathbf{w}_t, \qquad \mathbf{w}_t \sim \text{Gaussian}(\mathbf{0}, Q) \tag{2}$$

$$\mathbf{y}_t = C\mathbf{x}_t + \mathbf{v}_t \qquad \mathbf{v}_t \sim \text{Gaussian}(\mathbf{0}, R) \tag{3}$$

where A is the $(k \times k)$ state dynamics matrix, C is the $(p \times k)$ observation matrix, and Q $(k \times k)$ and R $(p \times p)$ are the covariance matrices for the state and output noise variables $\mathbf{w}_t$ and $\mathbf{v}_t$. The parameters A and C are analogous to the transition and emission matrices, that are respectively found in the discrete analogue model, the hidden Markov model (26).

A straightforward and powerful extension of this model is to allow the dynamics and observation models to include a dependence on a series of d-dimensional driving inputs $\mathbf{u}_{1:T}$:

$$\mathbf{x}_t = A\mathbf{x}_{t-1} + B\mathbf{u}_t + \mathbf{w}_t \tag{4}$$

$$\mathbf{y}_t = C\mathbf{x}_t + D\mathbf{u}_t + \mathbf{v}_t. \tag{5}$$

Here, B $(k \times d)$ and D $(p \times d)$ are the input-to-state and input-to-observation matrices, respectively. If these driving inputs $\mathbf{u}_{1:T}$ are augmented with a constant bias, then this model is able to incorporate an arbitrary origin displacement for the hidden state dynamics, and can also induce an arbitrary displacement in the observation space. These displacements can be learned as parameters of the input-to-state (B) and input-to-observation (C) matrices. An input-dependent SSM can be used to model control systems, but another possible way to utilize the input's construction is to *feed back* the data from previous time steps in the sequence into the inputs for the current time step. This means that the hidden state can concentrate on modeling hidden factors, while the Markovian dependencies between successive *outputs* are modeled directly using the output–input feedback construction. We adapt this model for the analysis of gene expression time-series data in Section 2.2.

Without loss of generality we set the hidden state evolution noise covariance, Q, to the identity matrix; this is possible because an arbitrary noise covariance can be incorporated into the state dynamics matrix, A, and the hidden state rescaled and rotated to be made commensurate with this change.

The remaining parameter of a linear-Gaussian SSM is the covariance matrix, R, of the Gaussian output noise $\mathbf{v}_t$. This noise is p-dimensional, and, as applied to the gene expression model, corresponds to the noise variance in the expression level of each of the p genes. We assume R to be diagonal and can learn the scale of these diagonal terms. For notational convenience, we collect the above parameters into a single parameter vector for the model $\boldsymbol{\theta} = (A, B, C, D, R)$.

2.2. A SSM for Gene Expression Time Series

In this chapter, we use the input-dependent SSM, and we *feed back* the gene expression levels from the previous time step into the input for the current time step. In doing this, we attempt to discover gene–gene interactions across time steps, with the hidden state in this model now really representing unobserved variables, such as levels of regulatory proteins or unspotted genes.

A graphical model for this setup is given in Figure 1. When the input is replaced with the previous time step's observed data, $\mathbf{u}_t = \mathbf{y}_{t-1}$, the equations for the SSM can be rewritten from equations (4) and (5) into the form:

$$\mathbf{x}_t = A\mathbf{x}_{t-1} + B\mathbf{y}_{t-1} + \mathbf{w}_t \tag{6}$$

$$\mathbf{y}_t = C\mathbf{x}_t + D\mathbf{y}_{t-1} + \mathbf{v}_t. \tag{7}$$

Here, $\mathbf{y}_t$ denotes the gene expression levels at time step t and $\mathbf{x}_t$, the unobserved hidden factors in the state space. In practice, $\mathbf{y}$ is the suitably normalized and transformed values of the gene expression levels. Table 1 summarizes the roles of the various parameter matrices.

As a function only of the data at the previous time step, $\mathbf{y}_{t-1}$, the data at time t can be written as follows:

$$\mathbf{y}_t = (CB + D)\mathbf{y}_{t-1} + \mathbf{r}_t, \tag{8}$$

where $\mathbf{r}_t = \mathbf{v}_t + C\mathbf{w}_t + CA\mathbf{x}_{t-1}$ includes all contributions from noise and previous states. **Thus, the direct interaction between gene *j* and gene *i***

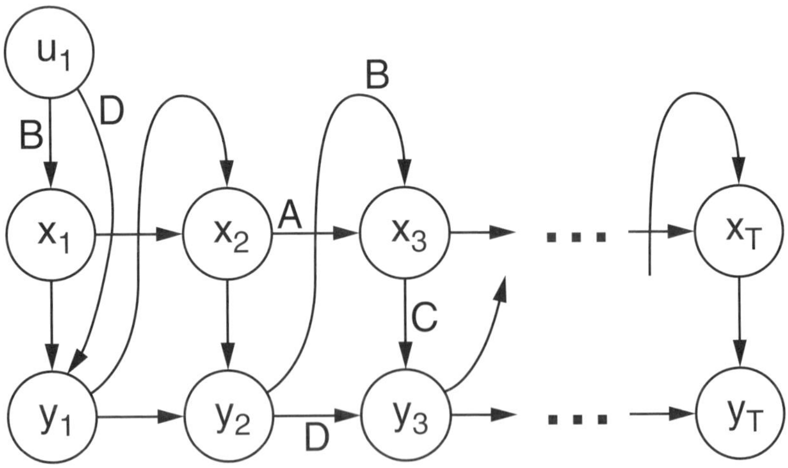

Figure 1. Feedback graphical model with outputs feeding into inputs. Gene expression levels at time t are represented by y_t, while the hidden factors are represented by x_t. Reprinted from Beal et al., (31) by permission of Oxford University Press.

Table 1. Summary of the roles of the various parameter matrices of the SSM with feedback inputs.

param.	models influence of...	on...	and parameter's role is to capture
$[A]_{ij}$	state j at time $t-1$	state i at time t	the state dynamics
$[C]_{ij}$	state j at time t	gene i at time t	effect of the states on gene levels
$[B]_{ij}$	gene j at time $t-1$	state i at time t	effect of gene levels on state
$[D]_{ij}$	gene j at time $t-1$	gene i at time t	causal gene-gene interactions

Here, "state" is used to refer to a particular hidden state dimension.

can be characterized by the matrix element $[CB + D]_{ij}$. Indeed, this matrix need not be symmetric, and the element represents positive or negative regulation from gene j to gene i at the next time step, depending on its sign. This is the matrix we will concentrate our analysis on because it captures all of the information related to gene–gene interaction over one time step. We have also shown that, if the gene expression model is stable, controllable, and observable, then the $[CB + D]$ matrix remains invariant to any coordinate transformations of the state and is, therefore, *identifiable* (30). The identifiability property is important, for without it, it would be possible for different values of the SSM parameters (and hence, different values of $[CB + D]$) to give rise to identically distributed observables, making the statistical problem of estimation ill-posed.

2.3. State Estimation in the SSM

Given a SSM with parameters $\boldsymbol{\theta} = (A, B, C, D, R)$ and a sequence $(\mathbf{y}_1, \ldots, \mathbf{y}_T)$ of observation vectors (the gene expression levels), one may be interested in estimating the hidden state $\mathbf{x}_t$ of the dynamical system. In particular, the hidden state may correspond to known biological processes that might be interesting to estimate from noisy data. There are three scenarios for solving this state estimation problem: filtering, smooothing, and prediction. The *filtering* task attempts to infer the likely values of the hidden variables $\mathbf{x}_t$ that generated the current observation, given a sequence of observations up to and including the current observation $\mathbf{y}_1 \ldots \mathbf{y}_t$. For linear Gaussian models, this problem was solved by the now classic Kalman filter (35). Because all variables are Gaussian and all relationships are linear (which preserves Gaussianity), $p(\mathbf{x}_t|\mathbf{y}_1, \ldots, \mathbf{y}_t)$ is also Gaussian. The Kalman filter, by recursively computing the mean and covariance matrix of $p(\mathbf{x}_t|\mathbf{y}_1, \ldots, \mathbf{y}_t)$, given the analogous quantities at time $t - 1$, therefore, fully represents the distribution of the hidden state variable.

The *smoothing* task infers the likely values of the hidden variables at some point in the past $\mathbf{x}_{t-\tau}$, for $\tau > 0$, given a sequence of observations up to and including the current observation $\mathbf{y}_1 \ldots \mathbf{y}_t$. An extension of the Kalman filer—the Rauch-Tung-Striebel smoother—achieves this recursively for linear Gaussian models (36), again, by computing the mean and covariance matrix of $\mathbf{x}_{t-\tau}$.

Finally, the *prediction* task tries to simulate the unobserved dynamics one or many steps into the future to predict future hidden states or observations. Thus, the quantities of interest are $p(\mathbf{x}_{t+\tau}|\mathbf{y}_1 \ldots, \mathbf{y}_t)$ and

$p(\mathbf{y}_{t+\tau}|\mathbf{y}_1 \ldots, \mathbf{y}_t)$. Simalar recursive algorithms give the means and covariance matrices of these distributions.

The discussion in this section assumes that the parameters of the model are known. For the problems of interest to us in this chapter, the only quantity that is given is the sequence of observed gene expression levels. Therefore, we now turn to the problem of *learning* the parameters of the model from such data. One should keep in mind, however, that to learn the model parameters it is often necessary to solve the state-estimation problems outlined in this section, as a subroutine.

2.4. Parameter Learning In the SSM and Bayesian Learning

A conventional objective function when learning SSMs is the likelihood function, which is the probability of the observed data $\mathbf{y}_{1:T}$ given the parameters $\boldsymbol{\theta}$, written $p(\mathbf{y}_{1:T}|\boldsymbol{\theta})$. The parameters of an SSM can be learned using *maximum likelihood* (ML) methods that involve expectation-maximization (EM) to integrate from the hidden state sequence $\mathbf{x}_{1:T}$ (37). However, in general, the ML approach is prone to overfitting, especially when fitting models with many variables with relatively small amounts of data because a more complex model can always fit the data better than a simpler one. A principled way to avoid this overfitting is to place a prior $p(\boldsymbol{\theta})$ on the parameters of the model that captures our biological intuition for *sparse* network models (often referred to as parameter regularization) and find the *maximum a posteriori* (MAP) parameters by using Bayes' rule. However, as has been well documented in the literature, both ML and MAP learning methods still suffer from overfitting on small datasets, and MAP learning suffers from inconsistencies, such as a dependence on the choice of parameterization (38,39). We have instead turned to a fully Bayesian analysis, which avoids overfitting and provides error bars on all model parameters—in this paradigm, the objective function is simply the probability of the data $p(\mathbf{y})$, which results from integrating the parameters of the model in respect to their prior distribution:

$$p(\mathbf{y}_{1:T}) = \int d\boldsymbol{\theta} p(\mathbf{y}_{1:T}|\boldsymbol{\theta}) p(\boldsymbol{\theta}). \tag{9}$$

Optimizing a model in respect to such an objective function avoids overfitting in the conventional sense. In practice, a Bayesian learning scheme infers *distributions* over all the parameters and makes modeling predictions by taking into account all possible parameter settings. In doing so, we penalize models with too many parameters, embodying an automatic *Occam's Razor* effect.

2.5. Prior Specification for the SSM

This section describes the parameter priors for the SSM, which are important to discuss for three reasons. The first is that no meaningful inferences about the parameters of the model can be made in the absence of some subjective prior specification; the second reason is that the success of the fully Bayesian scheme depends on an advanced prior specification framework known as *automatic relevance determination* (ARD) (38), which we describe briefly in the next paragraph; the third reason is that these priors can serve the important role of articulating expert knowledge about the biological system in question, and so are key to building realistic and powerful mathematical models of the biological processes we wish to study.

In general, the priors on all model parameters are parameterized by *hyperparameters* (*see* Figure 2), which can be optimized so as to adapt to the scale of the data and so as to automatically select relevant and irrelevant variables in the model, i.e., ARD. There are two ways to understand the ARD process. The first is to consider it simply as penalizing or regularizing parameters that are not useful in modeling the observed data. The second way to understand the ARD process is to consider it as a minimum description length principle or the embodiment of an Occam's Razor effect. If we are able to model the data just as well without some of the parameters, then it wastes bits to encode these parameters under the model prior (the ARD prior). During an optimization, the ARD process will endeavor to reduce the description length of the data (the negative log marginal likelihood), and altering the hyperparameters of the prior will dictate that certain groups of parameters tend to values that are ineffectual (this may be zero or a different operating point). This can be understood by realizing that if a parameter setting is excluded by the prior, then the posterior distribution of that parameter, having learned from data, still cannot include that setting. The parameter has effectively been "turned off."

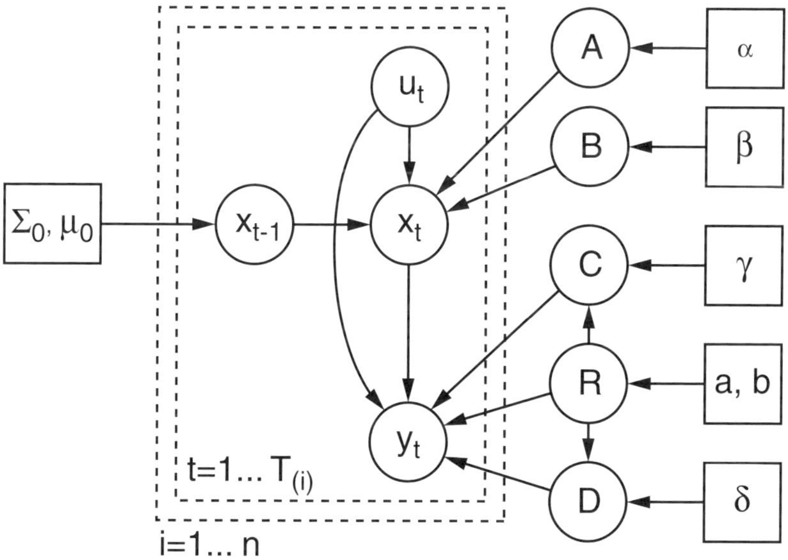

Figure 2. Graphical model representation of a Bayesian SSM. Each sequence $\{\mathbf{y}_1, \ldots, \mathbf{y}_{T_i}\}$ is now represented succinctly as the (inner) plate over T_i pairs of hidden variables, each presenting the cross-time dynamics and output process. The second (outer) plate is over the data set of size n sequences. Here, (α, β, γ, δ, a, b, Σ, μ) represent the *hyperparameters* of the model.

As an example of this framework, consider the collection of parameters in the C matrix (the state-to-output or factor-loading matrix), which is given a separate prior on each of its *k columns*. Letting $\mathbf{c}_{(j)}$ denote the column vector consisting of the p entries in the jth column of the C matrix, the ARD prior on $\mathbf{c}_{(j)}$ is a product of one-dimensional Gaussians, each having mean zero and a precision (inverse-variance) hyperparameter β_j:

$$C = \left(\mathbf{c}_{(1)} \ldots \mathbf{c}_{(k)}\right), \quad p\left(\mathbf{c}_{(j)}\right) = \prod_{q=1}^{p} \text{Gaussian}\left(\mathbf{c}_{jq}; 0, \beta_j\right). \tag{10}$$

Part of our SSM learning algorithm is to optimize the hyperparameters $\{\beta_1, \beta_2, \ldots, \beta_k\}$. Often, we find that for some select j's the values of β_j tend to infinity. What does this correspond to? This implies that the prior for the jth column is now Gaussian with zero mean and infinite precision (or zero variance), which means that the only setting of the parameters $\mathbf{c}_{(j)}$ that would have nonzero probability under $p(\mathbf{c}_{(j)})$ would be the $\mathbf{0}$ vector. In the case of the C matrix, having the jth column all zeros is very interesting: looking at equation (3) this means that the jth hidden dimension is irrelevant to generating any of the p dimensions of the output y. The jth hidden dimension has been *pruned* from the model.

2.6. Determination of State-Space Dimensionality

A first-order Markov chain between observed gene expression levels will likely not capture some of the longer time correlations, and this will be exacerbated if the sampling intervals are long. However, the inclusion of a hidden state in our models allows, in theory, dependencies between the observed gene expression values that are potentially higher than first-order Markov.

An analogy is the use of hidden Markov models (which have an identical graphical model to that shown in Figure 1) for speech recognition. There are obviously high-order correlations between the audio waveforms of speech across multiple time points, but to a good approximation, these can be successfully modeled by a first-order Markov chain in the hidden states, each of which correspond to a different *phoneme*. The more phonemes a language has, the more potential there is for longer-range temporal correlations in the observed speech signals. This leads us to the question of how complex these dependencies should be, which is directly related to the question of what the required dimensionality of the hidden state should be.

The task of deciding upon a suitable dimension for the hidden state-space remains a difficult problem. If too few dimensions are used, then it can be impossible to fully capture hidden dynamics, and, as a result, the model is forced to infer direct gene–gene interactions that are in fact indirect and mediated by an unobserved mechanism (thus, the inferred interaction graph is no longer sparse). If too many dimensions are used, the model will overfit to the noise in the gene expression levels and produce erroneous inferences. In our earlier work (28,29,30), this dimensionality was determined by a cross-validation experiment in which we incremented the number of hidden states and monitored the predictive likelihood using a portion of the data set that had not been used to train the model. However, the holdout set error in a cross-validation experiment is a noisy quantity, and for a reliable measure a very large holdout data set is needed; ideally, we would prefer to utilize all the data for model learning, rather than holding out a portion of it. A variational Bayesian (VB) treatment of these models provides a novel and efficient way of using all the data to learn their structure, i.e., to identify the optimal dimensionality of their state-space. The VB algorithm provides distributions over the model parameters and, as has been shown in a series of experiments on synthetic data, can be used successfully to determine the structure of the true generating model, including inferring the dimensionality of the hidden state-space (39).

3. Results

3.1. Results from Experimental Data Using the Variational Bayesian Model

In this section, we describe how the ARD process described can be applied to gene expression measurements (as outputs of the model), as well as hidden dimensions, just by applying the same sort of flexible Gaussian prior to other parameters of the SSM. By pruning away unneeded interactions using ARD and the Bayesian Occam's Razor effect, we obtain sparsely connected gene regulatory networks that do not overfit the data.

In Beal et al. (31), we examined the gene–gene influences represented by elements of the matrix $[CB + D]$. The VB model provides us with posterior distributions for the parameters C, B, and D. Using the posterior distributions for these parameters, we compute the distribution of *each of the elements* in the combined matrix $[CB + D]$. We consider an element of this matrix as providing evidence for a candidate gene–gene interaction if the element's posterior distribution is positioned *significantly far from the zero point* of no influence. Significance in this scenario corresponds to the zero point being more than n standard deviations from the posterior mean for that entry. Because these distributions are Gaussian (39), and may lie above or below the zero point (corresponding to positive or negative regulation), we can use the standard Z-statistic for normally distributed variables to threshold the connectivity matrix at any desired level of statistical significance. We can consider this as a simple decision problem with two hypotheses:

$$H_0 \quad : \quad [CB + D]_{i,j} = 0 \ (\textit{no connection})$$

$$vs$$

$$H_1 \quad : \quad [CB + D]_{i,j} \neq 0 \ (\textit{connection}),$$

where H_0 is rejected when 0 is not within the confidence interval. Thresholding the $[CB + D]$ matrix using this criterion produces a connectivity matrix or directed graph where the diagonal elements represent self–self interactions at consecutive time steps.

Figure 3 (A and B) shows results from a number of state-space models trained using our VB training algorithm, starting from 10 different

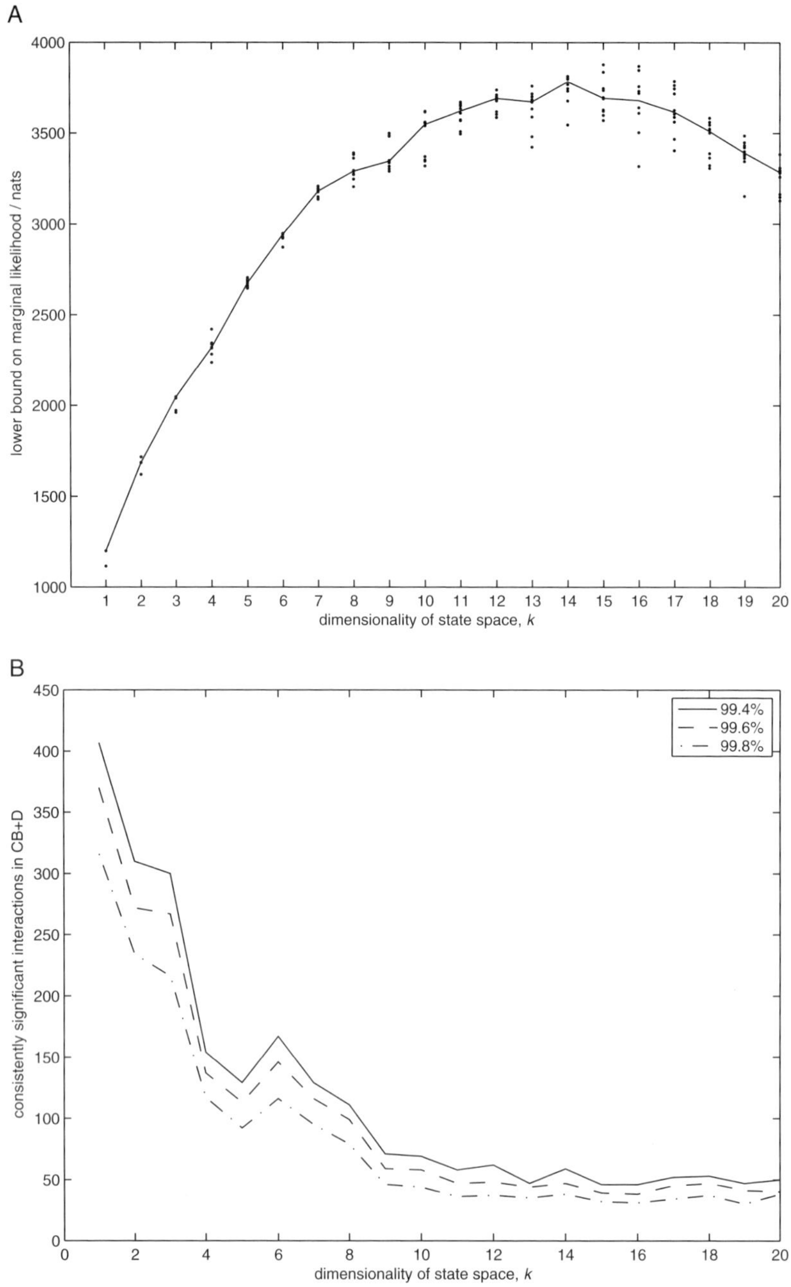

Figure 3. The effect of changing the state-space dimensionality k on (A) the lower bound to the marginal likelihood, and (B) the number of significant gene–gene interactions recovered in the $CB + D$ matrix. (A) Variation of the lower bound on the marginal likelihood, $\mathcal{F}$, with dimensionality of hidden state-space, k, for 10 random initializations of model training. The line represents the median value of $\mathcal{F}$. (B) The number of significant interactions that are repeated in all 10 runs of model training at each value of k. There are 3 plots, each corresponding to a different significance level. Reprinted from Beal et al., (31) by permission of Oxford University Press.

random initializations and with $k = 1, \ldots, 20$ hidden state dimensions. As a byproduct of the VB approach, we can obtain a lower bound on the marginal likelihood, or Bayesian "evidence," which allows us to select an appropriate model containing the optimum number of hidden state dimensions. Figure 3A shows the variation of the median value of the lower bound on the marginal likelihood, which we call $\mathcal{F}$, with hidden state dimension k (also plotted are the individual $\mathcal{F}$ values from each of the 10 random seeds). The median $\mathcal{F}$ value peaks around $k = 14$, indicating the optimal dimensionality of the hidden state-space for this data set. Figure 3B shows the number of significant gene–gene interactions that are consistently repeated in all 10 runs at each value of k; there are three plots, corresponding to three significance levels (chosen to correspond to those used in our previous study [29]). Regardless of the significance level we choose, we can see that the number of significant interactions has leveled off by approximately $k = 14$, which corresponds to the peak in $\mathcal{F}$ graph. Importantly, some interactions appear robustly even in models that incorporate many hidden state dimensions. Note that models with no hidden states, i.e., $k = 0$, equivalent to the linear models of D'Haeseleer et al. (10) and Holter et al. (40), give a much higher estimate of the number of direct interactions, which may result in a very misleading impression of the underlying genetic regulatory networks.

3.2. Results from Synthetic Simulated Data

To evaluate our method to reverse engineer genetic regulatory networks using data collected from gene expression microarray experiments, we simulated a microarray experiment *in silico*, generating a synthetic data set using our gene expression model. Synthetic data was generated from an initial network defined by the connectivity matrix in the matrix $[CB + D]$. In previous work (30), we have demonstrated that we are able to recover the same network using the SSM model and bootstrap procedure described earlier and also estimated the size of the data set needed by the method to reconstruct the underlying network. However, care must be taken in extrapolating these simulation results to experiments with actual gene expression data. Although the simulations were idealized in that the data were actually generated from a linear dynamic system, in real gene expression data one would expect to see nonlinearities, more or less noise, various time scales and delays, and possibly many hidden variables.

3.3. Results from Realistic Simulated Data

Ideally, we would like to test the performance of our reverse-engineering method on data produced from a biologically plausible model system. Because there exists very little gene expression data for which the ground truth regulatory network is known, we have followed a recent approach described by Zak et al. (16), who designed a highly detailed and biologically plausible *in silico* network upon which a formal identifiability analysis can be based. The network is shown in Figure 4, and consists of 10 genes with expression levels derived from realistic interactions between 10 transcription factor proteins (most of which form homodimers), a

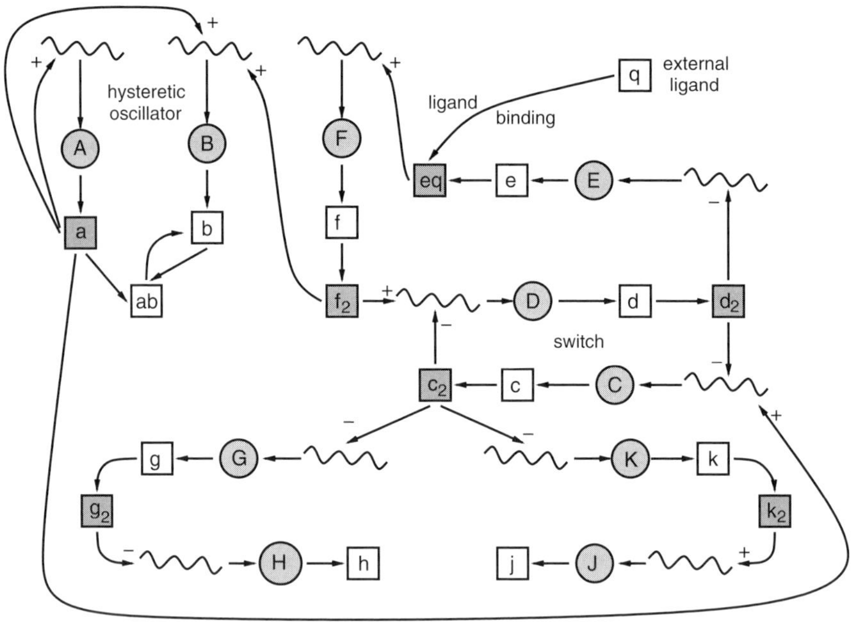

Figure 4. Realistic model gene regulatory network from Zak et al. (16). Wavy lines represent promoters, circles represent mRNAs, and squares represent proteins. Shaded squares represent active transcription factors (mostly dimers). The symbols + and – represent whether a transcription factor acts as an activator or an inhibitor. Adapted from Husmeier (19), by permission of the author and Oxford University Press.

ligand input, and 13 bound promoters. This network was constructed by arranging modules of transcriptional regulation into regulatory motifs drawn from the biological literature, such as a cascade (41), mutual repression (41,42,43), autoactivation, and sequestration (44) and agonist-induced receptor down-regulation (45,46). Model parameters were selected to yield time scales representative of mammalian gene expression (16). *In silico* simulations were carried out using MATLAB code provided by Daniel Zak, integrating the deterministic ordinary differential equations (ODEs), which describe the model based on a spiked ligand input at time 10 h. From these, we constructed time courses of length 58 h and generated incremental sets of replicates of sizes {1, 2, 4, 8, 16, 32} by adding Gaussian noise at each time point to the log-ratio of each of the 10 observed gene expression values. Note that although there are 54 players in this network, we only provide our SSMs with the log-ratios of the mRNA levels. We tested our VBSSMs on the task of reverse-engineering the interactions between the genes, *even though* there are 44 hidden quantities (the previously listed biological players) that our method did not have access to, but were nevertheless required for the ODE simulations.

To investigate the effect of different sampling rates in time, we constructed two complementary data sets from these replicates: the first sampled across 12 equally spaced time points, the second over 120 points. We also investigated the performance of the VBSSM with ARD hyperparameter optimization either enabled or disabled. Note that with ARD hyperparameter optimization enabled, the SSM has a prior that strives for sparse networks; hence, this may drive down the positive arc rate. Figure 5 shows the plots of $\mathcal{F}$ against the dimensionality of the hidden state, k, for varying model conditions (*see* caption for details). We note the following trends. For small numbers of replicates (just 1 or 2), the curve of $\mathcal{F}$ contains a maximum at $k = 1$ (or less), implying that there is not sufficient information in the data to warrant the creation of even a single hidden-state dimension; conversely, in the case of 32 replicates, there is a distinct peak in the curve of $\mathcal{F}$, implying that there is an optimal nonzero state-space dimensionality. Concentrating on 32 replicates, we see that with just $T = 12$ sampled time points, $\mathcal{F}$ peaks at $k = 2$ and $k = 4$ dimensions, for hyperparameter optimization off and on, respectively. However, we also see that for $T = 120$ measured time points there is much more information available and the model supports a larger number of dimensions; with hyperparameter optimization off it supports $k = 5$, and, intriguingly, with hyperparameter optimization turned on it seems as if infinitely large k is supported (this is because the ARD mechanism self-prunes if necessary so that the model with $k = \infty$ is just using of the order $k = 5$ dimensions).

Changing the number of replicates certainly affects the optimal setting of the dimensionality of the state-space, as measured in terms of

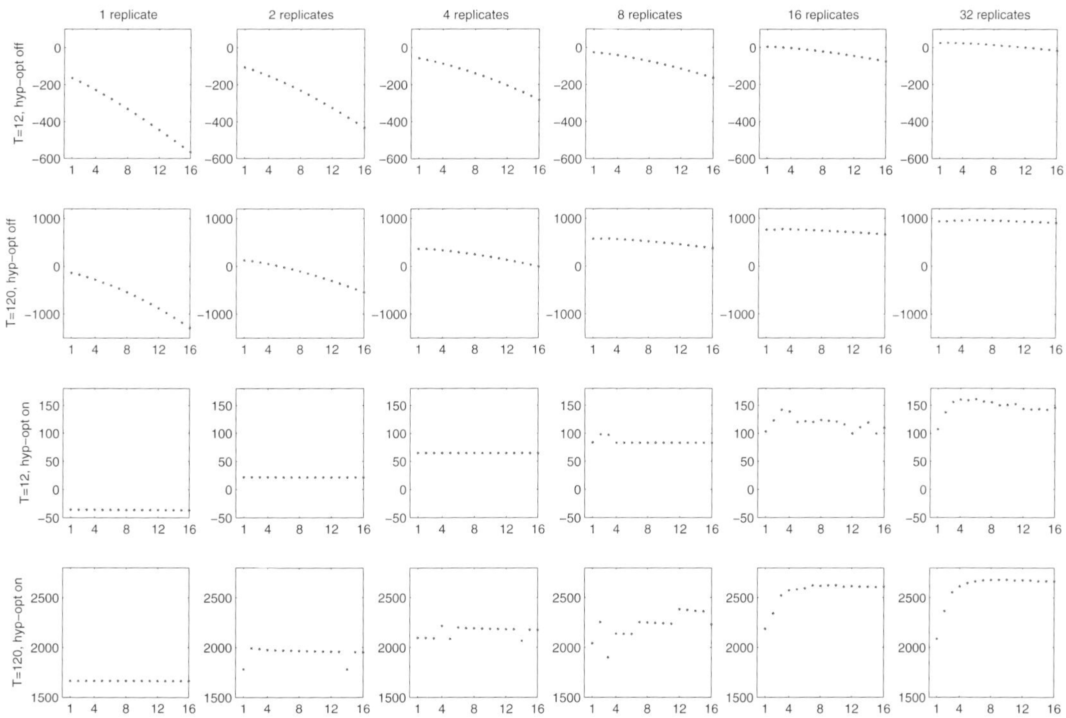

Figure 5. Computations for the values of $\mathcal{F}$, per replicate, for many different VBSSM models. Each row denotes a different combination of number of time samples taken and whether hyperparameter optimization was enabled or not (as shown on y-axis labels). Each column corresponds to a different data set supplied to the model in terms of the numbers of replicates provided.

the highest $\mathcal{F}$ value. However, we would also like to demonstrate that changing the hidden state-space dimensionality does, in fact, have an impact on the model's ability to infer the presence or absence of interaction arcs in the regulatory network. To this end, we have performed a ROC analysis or sensitivity/specificity trade-off, using the known gene regulatory network specified by the model of Zak et al. (16). ROC curves are obtained by computing the hit and false alarm rates for different thresholds (i.e., the confidence level placed on testing individual connections). The hit rate, or *Sensitivity*, is defined as the proportion of recovered true connections, and the *Specificity* is defined as the proportion of correctly identified nonconnections. The false alarm rate, or *Complementary Specificity*, represents one minus the fraction of nonconnections that are correctly identified, as defined below:

$$\text{Sensitivity} = \text{TP}/(\text{TP} + \text{FN})$$

$$\text{Specificity} = \text{TN}/(\text{TN} + \text{FP})$$

$$\text{Complementary Specificity} = 1 - \text{Specificity},$$

where:

TP = number of actual connections that are declared connections
FP = number of nonconnections that are declared connected
FN = number of actual connections that are declared not connected
TN = number of nonconnections that are declared not connected.

Perfectly recovering the network corresponds to a sensitivity of 1 and a complementary specificity of 0. Our definition of a true-positive (TP) is, in fact, more rigorous than in earlier similar studies (19) because, if an interaction is present in the true network, we require it to be recovered *with the correct direction* of influence. In the ROC plots, points in each curve represent the rate values computed based on different thresholds (nominal confidence levels). For each value of k, a trace is plotted consisting of a total of 21 points, each of which corresponds to a confidence level for a Z-statistic from the set $\{0, .5, 1, 1.5, \ldots, 10\}$. On the ROC curve, the vertical "hit" axis represents the fraction of the total actual connections in the network that are correctly identified. The horizontal "false alarm" axis represents one minus the fraction of nonconnections that are correctly identified.

Figure 6 shows the results of our ROC analyses for the data set consisting of 4 replicates, comparing the use of data subsampled at $T = 12$ (left) or $T = 120$ (right) time points for hyperparameter optimization turned off. Each ROC curve is obtained by varying the threshold for the statistical significance of a connection in the network, from infinitely large (top right: no connections considered significant) to very small (bottom left: almost all connections considered significant), and, as such, this analysis *does not require the specification of any particular significance threshold* (e.g., 95.0%, 99.0%, etc.). By comparing subplots in Figure 6, we see that having more time points sampled leads to both higher sensitivity and higher specificity (we move toward the top left of

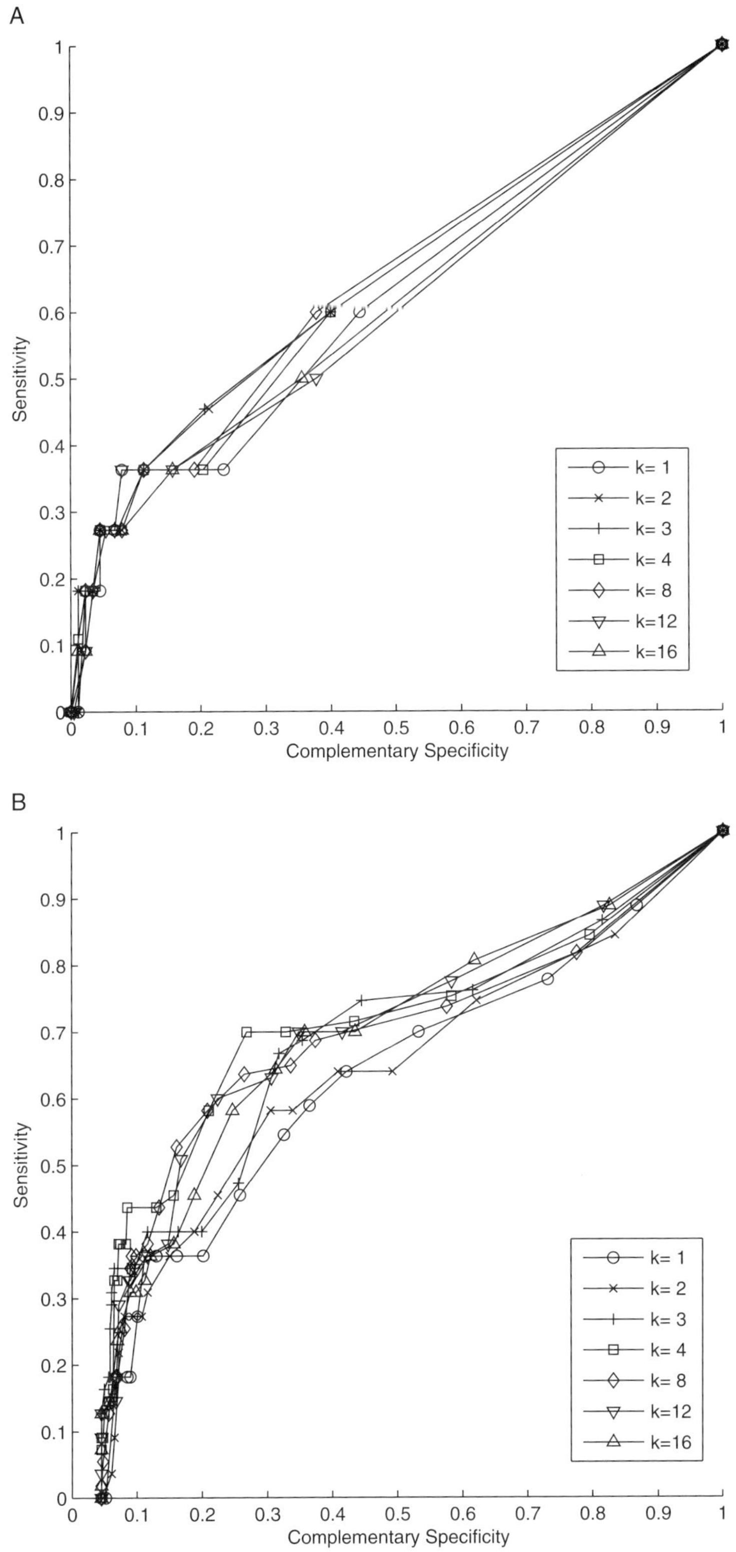

Figure 6. ROC analyses (sensitivity vs. complementary specificity) for (A) $T = 12$ and (B) $T = 120$ sampled time points, for values of the hidden state-space dimensionality, k, ranging from 1, 2, . . . , 16, trained on the data set of size 4 replicates.

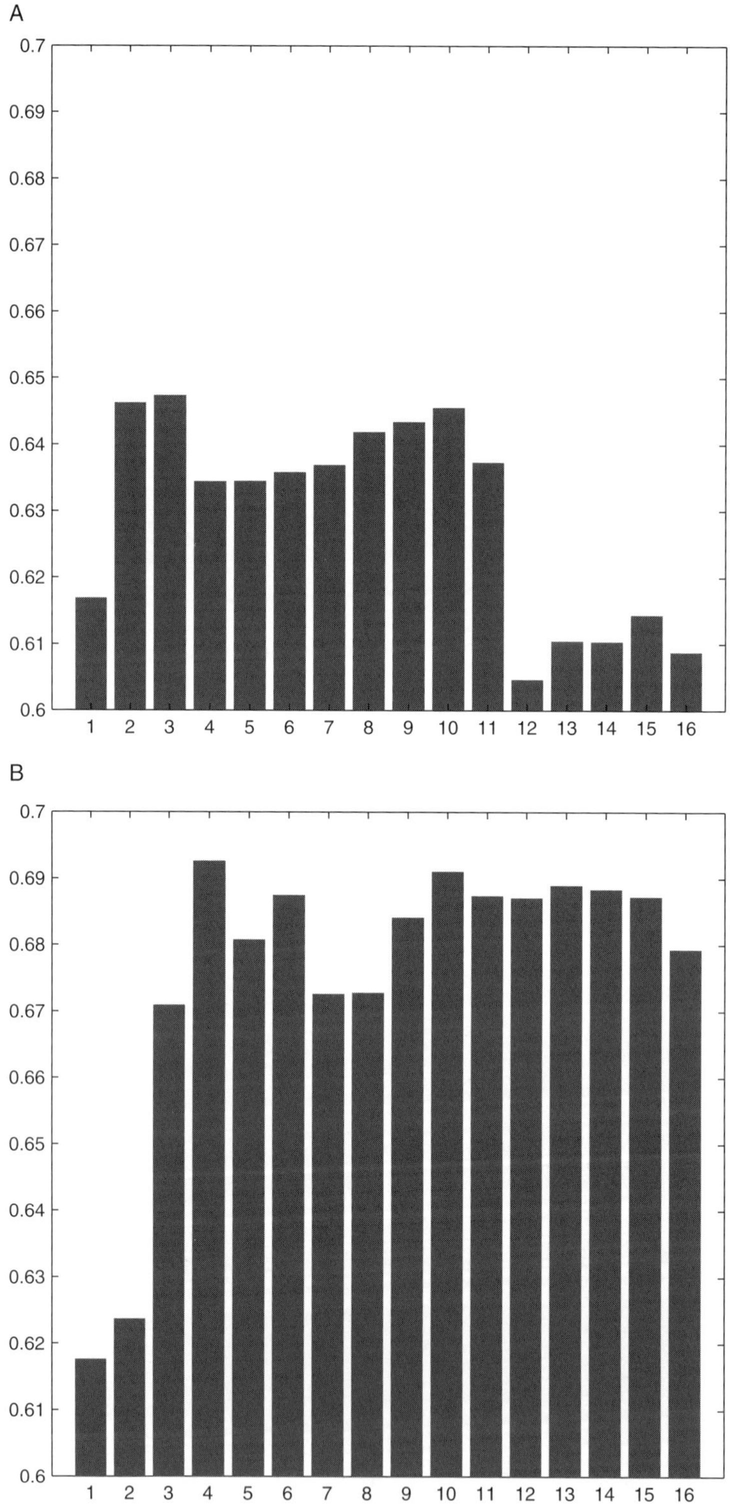

Figure 7. The respective AUC as k is varied, for (A) $T = 12$ and (B) $T = 120$ time points, for the ROC plots shown in Figure 6.

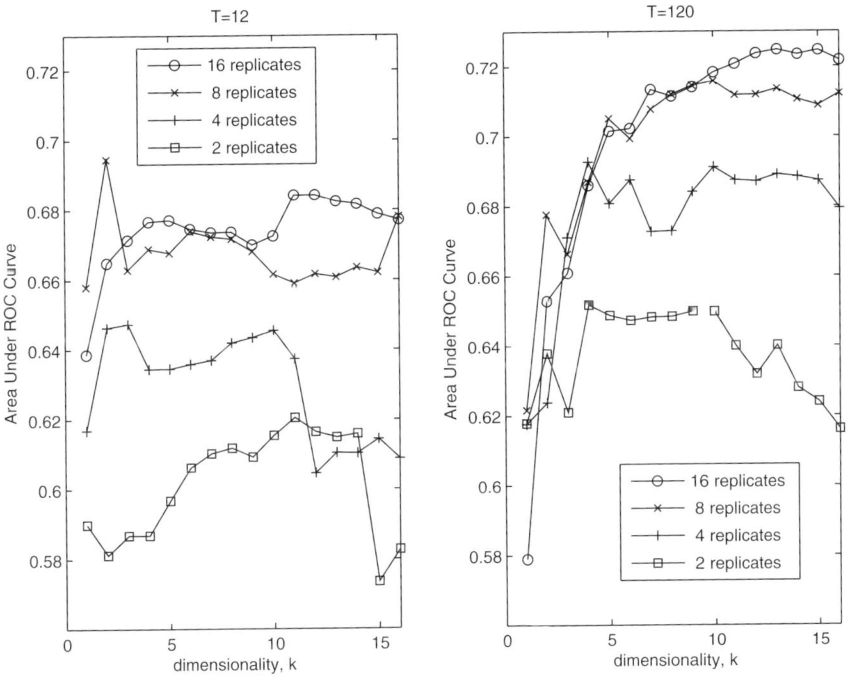

Figure 8. Shows the AUC for (left) $T = 12$ and (right) $T = 120$ as the number of replicates is artificially increased (by adding Gaussian noise to the simulated mRNA levels). Each point on each trace is the AUC according to a VBSSM model for a particular value of k trained on the number of replicates indicated in the legend.

the ROC plot). Moreover, as is seen more clearly in the case of $T = 120$ (right), we improve recovery of the network as we increase the hidden state-space dimension from $k = 1$ to $k = 16$. These observations can be quantitatively substantiated by examining the AUC as k is varied for the two scenarios of T, as shown in Figure 7. In particular, we note that for $T = 120$ the AUC increases by at least 7–8% on increasing the value of k. Note the AUC is the average probability that a particular connection would be correctly inferred as present or not present, with the average taken assuming that any given connection is *equally likely* to be present or not present.[1] Finally, we plot the average performance for $T = 12$ and $T = 120$ for different numbers of replicates, as shown in Figure 8. By noting that the high-k AUC is approximately 68% for both of the scenarios ($T = 12$, 16 replicates) and ($T = 120$, 4 replicates), we can conclude that, at least for this data set and method of generating replicates, a 10-fold increase in temporal sampling allows us to reduce the number of experimental replicates by a factor of 4. We also note that for $T = 120$ and 16 replicates, we can achieve 72% AUC performance.

[1] This assumption of equally likely present and not-present interactions evidently does not hold in real-world gene regulatory networks, which we know to be sparse, having many fewer positives than negatives. As a result, the AUC is an underestimate of performance, because all our models have operating points favoring high specificity at the expense of sensitivity, i.e., TN favored over TP rates.

3.4. Correlations of Hidden States with Unobserved Quantities

An advantage of the time series model that we have proposed is that it explicitly provides a hidden state-space, which has a dynamics that may help capture some nontrivial temporal processes, and in so doing explain the dependencies between the gene expression levels. In particular, using a hidden state allows us to capture non-Markov dependencies between the observed data (mRNA levels), and we may also find that some of the hidden state dimensions correspond to relevant biological processes that are key to the regulatory mechanisms. In the case of this synthetic data set that we have been analyzing, there are players other than the mRNA levels that take part in the regulatory network, such as transcription factors, promoters, etc. It is interesting to analyze the distribution of the inferred hidden state trajectories within our learned models to see if they match, or closely correlate with, some of these other players whose time course profiles are never provided to the VBSSM model. In Figure 9, we show the time courses of the observed mRNA levels and the protein and ligand concentrations, which are not provided to the SSM.

In the case of $T = 120$ and no hyperparameter optimization, we measured the correlation through time between each of the 54 dimensions of the synthetic data and each of the k inferred hidden states of the SSM. The analysis was simplified by examining only the means of the state-space trajectories, which correspond to the most likely hidden state sequence explanation. For each hidden state that had not been pruned by the ARD prior (note that this may and often does occur even with ARD hyperparameter optimization disabled), we obtained the most correlated time course profiles out of the 54 profiles in the complete data set.

We do not find that any of the unpruned hidden state sequences in these models are unambiguously directly correlated with the individual unobserved time series profiles. This may be partly explained by the fact that many of the unobserved players have profiles that almost exactly mimic one or more of the observed mRNA levels, and thus, it would be redundant for the hidden state trajectories to resemble any such player in the model. Instead, we find several smoothly varying hidden state trajectories, and for no model there are more than two such trajectories. Figure 10 depicts a selection of these trajectories. For example, we show that for the models with $k = 7$ and 9, there is one hidden state taking on rather similar forms over time (this is also seen in models of other sizes). However, for the model of size $k = 12$ there are two hidden states, which quite correctly, are different (these happen to be hidden dimensions 5 and 12). Likewise for $k = 15$ there are two quite different remaining hidden state trajectories (hidden dimensions 5 and 8). We can also examine the effect that each of these hidden states is having on the mRNA levels at the current time point by examining the C matrix, plotted in Hinton diagram form at the right of each trace: white and black squares correspond to positive and negative influences, respectively, on the relevant genes, with the size of the square being proportional to the strength of the influence.

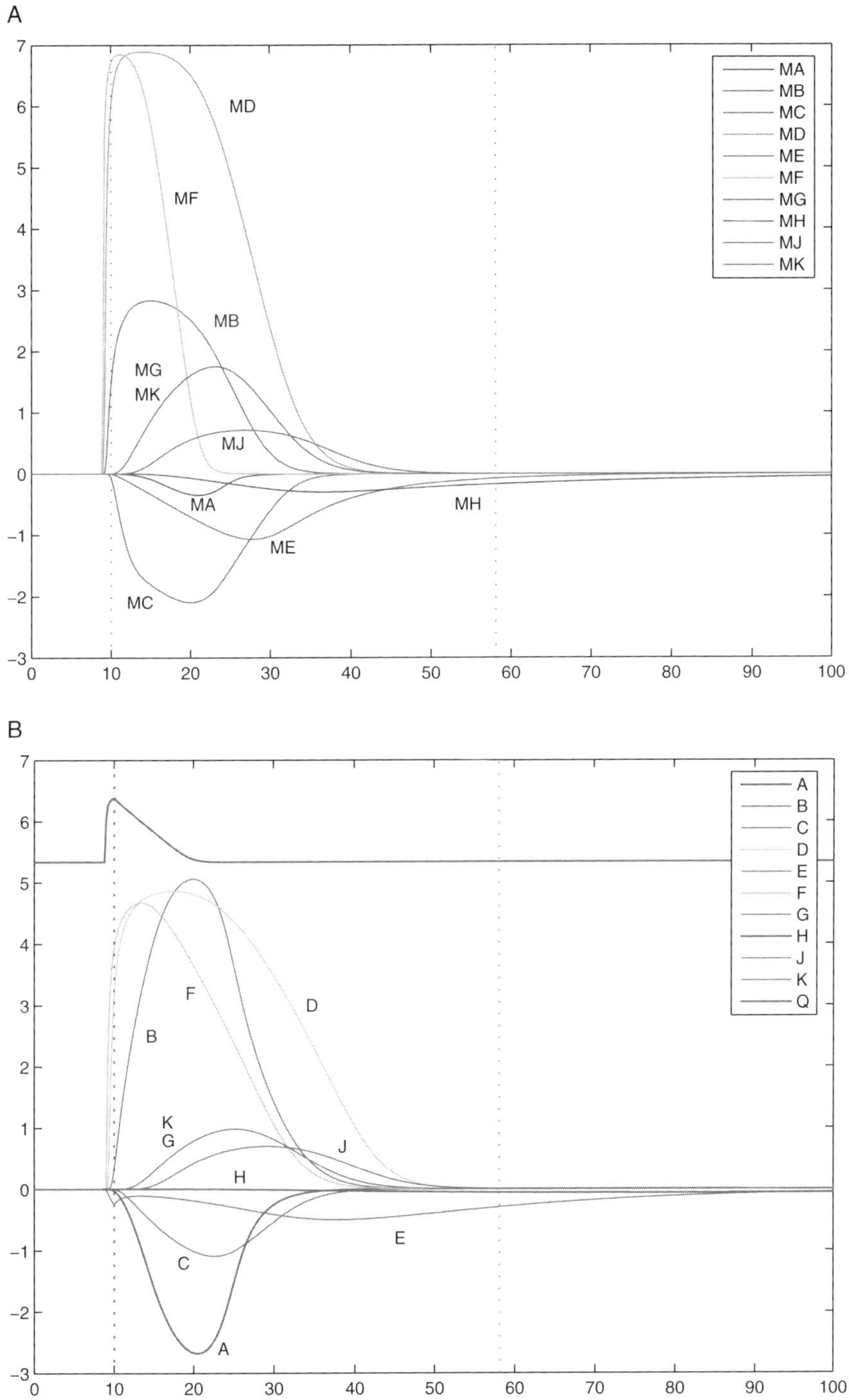

Figure 9. Time course profiles of the synthetic data set generated from the model of Zak et al. (16). We show only the levels of (A) the mRNA concentrations MA, MB, . . . , MK, and (B) the hidden levels of the associated proteins A, B, . . . , K. The vertical axes denote the levels measured in terms of $\log(M/M_0)$, where M is concentration and M_0 is concentration at time zero, and the horizontal axes are time, in hours. The vertical dotted lines in both plots denote the time window of data for which mRNA levels were provided to the VBSSM algorithm. The time window begins at the peak of the ligand injection, which is shown on the lower plot as the bold curve Q (the ligand log concentration is vertically offset for clarity).

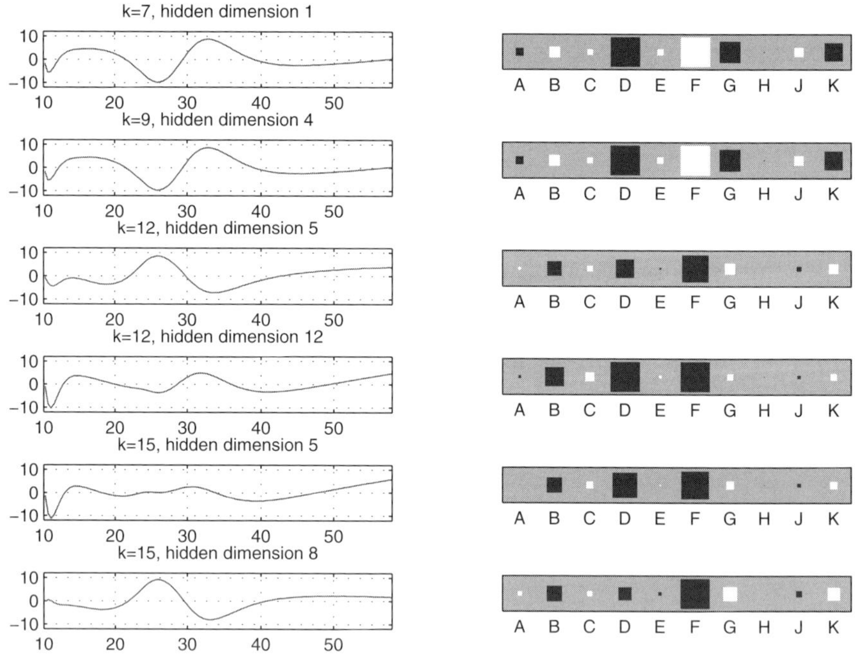

Figure 10. (left) Unpruned hidden state trajectories for a selection of four models of varying state-space size: $k = 7, 9, 12$, or 15; and (right) the influence of each of the hidden states on the 10 observed mRNA levels denoted A through K, expressed as Hinton diagrams. Note that the polarity of the hidden states trajectories for the top two rows are opposite to the others, and this is canceled by the opposite polarity seen in those two Hinton diagrams.

Interestingly, most of these hidden state trajectories show a sharp negative peak just after 10 h, corresponding to injection of the ligand Q and the sharp drop in the concentration of protein E (which models the receptor to which the ligand Q binds). The Hinton diagram indicates that these hidden states have greatest influence on the expression of genes F (which is directly regulated by the bound receptor, EQ), B and D (which are both up-regulated by the transcription factor encoded by gene F). We can thus speculate that one of the hidden states is plausibly capturing some of the dynamics of this hidden process.

4. Conclusions

Despite the assumptions inherent in linear SSMs, the results of the simulation studies described above indicate their usefulness as a tool for "top-down" reverse engineering of gene regulatory networks. Based on synthetic data generated from a model that contains definite nonlinearities in the dynamics of the hidden factors (arising from the oligomerization of transcription factors), we demonstrate an overall accuracy in transcriptional network reconstruction from the mRNA time series measurements alone of approximately 68% AUC for 12 time points, and better still for data sampled at a higher rate. A key ingredient of our models is the inclusion of "hidden factors" that help to explain the correlation structure of the observed measurements. These factors may correspond to unmeasured quantities that were not captured during the experiment and may represent underlying biological processes. Results from the modeling of the synthetic data also indicate that our method is capable of capturing the temporal nature of the data and of explaining it using these hidden processes, some of which may plausibly reflect dynamic aspects of the underlying biological reality.

Acknowledgments: This material is based upon work supported by the National Science Foundation under Grant No. 0524331. Any opinions, findings, and conclusions or recomendations expressed in this material are those of the authors and do not necessarily reflect the views of the National Science Foundation (NSF).

References

1. Kholodenko BN, Kiyatkin A, Bruggeman FJ, et al. Untangling the wires: a strategy to trace functional interactions in signaling and gene networks. *Proc Natl Acad Sci* 2002;99:12841–12846.
2. Wessels LF, van Someren EP, Reinders MJ. A comparison of genetic network models. *Pac Symp Biocomput* 2001;6:508–519.
3. van Someren EP, Wessels LFA, Backer E, Reinders MJT. Genetic network modeling. *Pharmacogenomics* 2002;3:507–525.
4. de Jong H. Modeling and simulation of genetic regulatory systems: A literature review. *J Comp Biol* 2002;9:67–103.
5. Friedman N. Inferring cellular networks using probabilistic graphical models. *Science* 2004;303:799–805.
6. Akutsu T, Miyano S, Kuhara S. Identification of genetic networks from a small number of gene expression patterns under the Boolean network model. *Pac Symp Biocomput* 1999;17–28.
7. Liang S, Fuhrman S, Somogyi R. Identification of genetic networks from a small number of gene expression patterns under the Boolean network model. *Pac Symp Biocomput* 1998;18–29.
8. Thomas R. Boolean formalization of genetic control circuits. *J Theor Biol* 1973;42(3):563–586.
9. Arkin A, Shen P, Ross J. A test case of correlation metric construction of a reaction pathway from measurements. *Science* 1997;277:1275–1279.
10. D'Haeseleer P, Wen X, Fuhrman S, Somogyi R. Linear modeling of mRNA expression levels during CNS development and injury. *Pac Symp Biocomput* 1999;3:41–52.
11. van Someren EP, Wessels LF, Reinders MJ. Linear modeling of genetic networks from experimental data. *Proceedings 9th International Conference on Intelligent Systems for Molecular Biology (ISMB)* 2000;8:355–366.
12. Weaver DC, Workman CT, Stormo GD. Modeling regulatory networks with weight matrices. *Pac Symp Biocomput* 1999;4:112–123.
13. Smith VA, Jarvis ED, Hartemink AJ. Evaluating functional network influence using simulations of complex biological systems. *Bioinformatics* 2002;18(1): S216–S224.
14. Yeung MK, Tegner J, Collins JJ. Reverse engineering gene networks using singular value decomposition and robust regression. *Proc Natl Acad Sci* 2002;99:6163–6168.
15. Zak DE, Doyle FJ, Gonye GE, Schwaber JS. Simulation studies for the identification of genetic networks from cDNA array and regulatory activity

data. In: *Proceedings of the 2nd International Conference on Systems Biology*. Madison, WI: Omipress; 2001:231–238.
16. Zak DE, Gonye GE, Schwaber JS, Doyle FJ, 3rd. Importance of input perturbations and stochastic gene expression in the reverse engineering of genetic regulatory networks: insights from an identifiability analysis of an in silico network. *Genome Res* 2003;13:2396–2405.
17. Murphy K, Mian S. Modelling gene expression data using Dynamic Bayesian Networks. *Proc. Intelligent Systems for Molecular Biology*, August 1999.
18. Friedman N, Linial M, Nachman I, Pe'er D. Using Bayesian networks to analyze expression data. *J Comput Biol* 2000;7:601–620.
19. Husmeier D. Sensitivity and specificity of inferring genetic regulatory interactions from microarray experiments with dynamic Bayesian networks. *Bioinformatics* 2003;19:2271–2282.
20. Pe'er D, Regev A, Elidan G, Friedman N. Inferring subnetworks from perturbed expression profiles. *Proc. 9th International Conference on Intelligent Systems for Molecular Biology (ISMB)*, 2001.
21. Cooper GF, Herskovits E. A Bayesian method for the induction of probabilistic networks from data. *Machine Learning* 1992;9:309–347.
22. Hartemink AJ, Gifford DK, Jaakkola TS, Young RA. Using graphical models and genomic expression data to statistically validate models of genetic regulatory networks. *Pac Symp Biocomput* 2001;422–433.
23. Hartemink AJ, Gifford DK, Jaakkola TS, Young RA. Combining location and expression data for principled discovery of genetic regulatory network models. *Pac Symp Biocomput* 2002;437–439.
24. Yoo C, Thorsson V, Cooper GF. Discovery of causal relationships in a gene-regulation pathway from a mixture of experimental and observational DNA microarray data. *Pac Symp Biocomput* 2002;422–433.
25. Ong IM, Glasner JD, Page D. Modelling regulatory pathways in E. coli from time series expression profiles. *Bioinformatics* 2002;18(1):S241–S248.
26. Roweis ST, Ghahramani Z. A unifying review of linear Gaussian models. *Neural Comput* 1999;11:305–345.
27. Brown RG, Hwang PYC. *Introduction to Random Signals and Applied Kalman Filtering*. New York: John Wiley and Sons; 1997.
28. Rangel C, Wild DL, Falciani F, et al. Modelling biological responses using gene expression profiling and linear dynamical systems. In: *Proceedings of the 2nd International Conference on Systems Biology*. Madison, WI: Omipress; 2001;248–256.
29. Rangel C, Angus J, Ghahramani Z, et al. Modelling T-cell activation using gene expression profiling and state space models. *Bioinformatics* 2004;20:1361–1372.
30. Rangel C, Angus J, Ghahramani Z, Wild DL. Modeling genetic regulatory networks using gene expression profiling and state space models. In: Husmeier D, Roberts S, Dybowski R, ed. *Probabilistic Modelling in Bioinformatics and Medical Informatics*. Springer-Verlag; 2005:269–293.
31. Beal MJ, Falciani F, Ghahramani Z, et al. A Bayesian approach to reconstructing genetic regulatory networks with hidden factors. *Bioinformatics* 2005;21:349–356.
32. Nachman I, Regev A, Friedman N. Inferring quantitative models of regulatory networks from expression data. *Bioinformatics* 2004;20:i248–i256.
33. Perrin BE, Ralaivola L, Mazurie A, et al. Gene networks inference using dynamic Bayesian networks. *Bioinformatics* 2003;19:S138–S148.
34. Wu F, Zhang W, Kusalik A. Modeling gene expression from microarray expression data with state-space equations. *Pacific Symposium for Biocomputing*, 2004;9.

35. Kalman RE. A new approach to linear filtering and prediction problems. *Trans. American Society of Mechanical Engineers, Series D, Journal of Basic Engineering* 1960;82D:35–45.
36. Rauch HE, Tung F, Striebel CT. On the maximum likelihood estimates for linear dynamic systems. Technical Report 6-90-63-62, Lockheed Missiles and Space Co., Palo Alto, California, June 1963.
37. Shumway RH, Stoffer DS. An approach to time series smoothing and forecasting using the EM algorithm. *Journal of Time Series Analysis* 1982;3: 253–264.
38. Neal RM. Assessing relevance determination methods using DELVE. In: Bishop CM, ed. *Neural Networks and Machine Learning*. Springer-Verlag; 1998:97–129.
39. Beal MJ. *Variational Algorithms for Approximate Bayesian Inference* [PhD thesis]. London, UK: University College London; 2003.
40. Holter NS, Maritan A, Cieplak M, et al. Dynamic modeling of gene expression data. *Proc Nat Acad Sci USA* 2001;98:1693–1698.
41. Reinitz J, Sharp D. Mechanism of eve stripe formation. *Mech Dev* 1995;49: 133–158.
42. Alberts B, Bray D, Lewis J, et al. *Molecular Biology of the Cell*. New York: Garland Publishing; 1994.
43. Gardner T, Cantor C, Collins J. Construction of a genetic toggle switch in *Escherichia coli*. *Nature* 2000;403:339–342.
44. Herdegen T, Leah J. Inducible and constitutive transcription factors in the mammalian nervous system: control of gene expression by jun, fos, and krox, and creb/atf proteins. *Brain Res Rev* 1998;28:370–490.
45. Meyer A, Schmidt T. Differential effects of agonist and antagonists on autoregulation of glucocorticoid receptors in a rat colonic adenocarcinoma cell line. *J Steroid Biochem* 1997;62:97–105.
46. Ouali R, Berthelon M, Begeot M, Saez J. Angiotensin ii receptor subtypes at1 and at2 are downregulated by angiotensin ii through at1 receptor by different mechanisms. *Endocrinology* 1997;138:725–733.

13

Modeling Spatiotemporal Dynamics of Multicellular Signaling

Hao Zhu and Pawan K. Dhar

Summary

Molecular interaction in cells is context dependent, rich in semantics, spatiotemporally evolvable, and forms emergent networks. To unveil how molecular-level signaling leads to cell and tissue level phenotype, a computational model must link intra- and intercellular signaling, capture emergent events, and reconstruct network evolution. We have developed a multicellular modeling method under a Linux-based platform that combines features of cellular automata (CA) and object-oriented programming (OOP). In this chapter, we describe the simulation method and supporting tools to capture spatiotemporal dynamics of signaling networks and provide two illustrative examples. We also demonstrate that the order, timing, and networking of signaling within and between cells are fundamental characteristics that must be captured to fully understand signaling networks.

Key Words: Cellular automata; signaling; multicellular; object oriented; notch; planar cell polarity; emergent; somite segmentation; development.

1. Introduction

Signaling networks are both emergent and evolvable (1). "Emergent" indicates that the networking of molecules is context dependent and reprogrammable. A typical example of signaling reprogramming comes from the cell fate transformation in *Drosophila melanogaster* eye development. If the gene *rough* is ectopically expressed in the presumptive R7 cells, the developing ommatidial cells transform their fate from R7 to R1/6 (2). "Evolvable" means networks in cells may undergo significant structural evolution, for example, during cell differentiation in embryogenesis and cell dedifferentiation in carcinogenesis (3). The phenomenon of "evolvability" is typically seen during the differentiation of stem cells. Even in the single-cell organism yeast, the transcriptional regulatory network is significantly altered under different cell conditions (4). Signaling networks in cells of metazoans, which are closely coupled through cell communication and modified during developmental and pathophysiological processes, are more complicated and interesting. Given that different signaling networks impart different functions and identities to cells, an issue of fundamental importance in biology is to reveal when, how, and under what conditions signaling networks change temporally in one cell and spatially in multiple cells (Figure 1).

Studying developmentally regulated networks *in vivo* comes with technical constraints. As molecular interactions within cells cannot be entirely and continually measured due to the prohibitive cost of experiments, tissue level observations provide only limited molecular-level pointers. Moreover, to get data from gene knockout and mutation experiments to infer signaling networks is time consuming; and to describe dynamics of their structural changes under various abnormal conditions is even more challenging. An *in silico* experiment, on the other hand, does not face such technical constraints if key data are available.

Several biological issues must be understood to model signaling processes reliably. First, the existing knowledge of molecular interactions includes both the biochemically confirmed and the genetically inferred. For example, an interaction between molecules A and B may be context dependent and subject to the availability of different unknown *connectors*

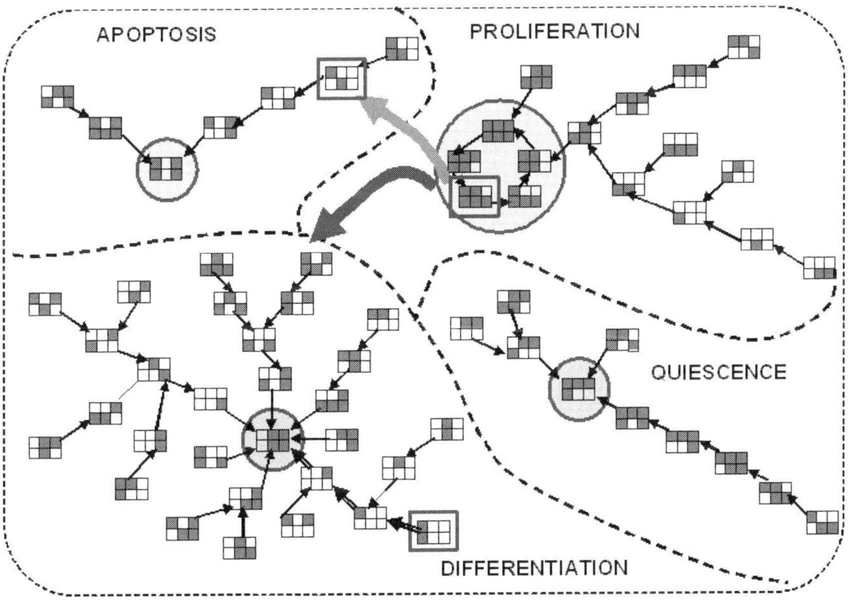

Figure 1. A hypothetic state transition map of a 6-gene network. Four gray circles designate the attractor states, each corresponding to a particular cell fate (differentiation, proliferation, senescence, cell division cycle, and apoptosis). Dotted lines delineate the basins of attraction. The green and orange large arrows denote two attractor state transitions (the cell fate switch from proliferation to differentiation and apoptosis). Simulating how the system evolves from an undifferentiated state (marked in red) into the differentiation attractor may disclose much about gene regulation dynamics. A more interesting case may be from a proliferation state to the basin of apoptosis attractor. (Redrawn from Huang, 2001.)

and *regulators*. A model should therefore be adaptable to new information. Second, during development, signaling processes within cells are tightly coupled to the signaling process among cells, and evolve with differentiation. This demands modeling formalism to be adequately flexible to allow dynamic reconfiguration of links. Third, an interaction known at semantic level, usually symbolized as A→B (activation) or A–| B (inhibition), may be unknown at the biochemical level, i.e., whether it involves phosphorylation, dephosphorylation, binding, cleavage, etc. Fourth, quantitative data is unavailable in a majority of cases. The nature of biological processes therefore calls for an inclusion of both qualitative and quantitative description. Finally, molecular interactions are discrete events.

The aforementioned issues call for special methods and tools. Using a combination of CA and OOP, we developed a new programming language for building *in silico* models of heterogeneous cells (5). Simulation studies demonstrate that the order, timing, and networking of signaling in cells greatly enrich our understanding of the dynamics of pathways and networks.

2. Methods

2.1. Two-Tier Parallelism for the Description of Parallel Networks

Local interactions leading to global complexity is a common feature of both physical and biological systems. Although CA have been widely used to model physical systems of massively simple and identical components (6,7), CA-based biological models, especially the rule-based ones, are not equally successful because of the hierarchy and heterogeneity of multicellular systems. This may change given a renewed interest in the language based CAs in contrast to the rule based ones (8). To describe diverse and multiple molecular species within a cell, we have introduced a new program component *object,* and additional features into the CA language *Cellang.* Our aim was to encapsulate the structure and function of each molecule into an object (5). In addition to the predefined variable *time*, another variable, *msgq* (message queue), was automatically created for each molecule/cell to implement message passing based on molecular interaction and cellular communication. In a message, there can be bound (with a value), unbound (to get a value), and anonymous (unused) variables, each with different roles. Floating point data for quantitative computing, function call and runtime perturbation have also been included. For example, the built-in function *position ()* returns the global address of a cell to allow runtime perturbation to the cell, while *sendmsg* sends messages between two cells. The computational program, shared and executed by all automata cells, consists of a *cell program* (the traditional component) and a new group of *molecule programs* (objects), creating a two-tier mapping between automata and biological cells; and between objects and molecules. These two kinds of cells, and the molecules and objects, are not distinguished from each other in the remaining parts of the chapter. Although all the cells share an identical set of molecule programs, the *if* statement, with a *cell type* field, guides cells to run a particular subset of molecular programs, which run on demand.

The initial distribution of cells in a 2D or 3D space is processed separately with an edit tool. This file is used as the input to a model. Both computational program and cell array file can be freely modified at any time.

2.2. Separating Discrete Molecular Signaling from Continuous Concentration Computation

Molecular interactions, such as ligand-receptor binding, are essentially discrete events. Instead of rigidly wiring molecule A to molecules B and C with an equation

$$dA/dt = f_a(A) + f_b(B) + f_c(C) + A_0 \quad (1)$$

and continuously computing molecular interaction, we adopt a dynamic and discrete binding among them through message passing, which can be formalized as

$$dA/dt = msg(A)\cdot f_a(A) + msg(B)\cdot f_b(B) + msg(C)\cdot f_c(C), \quad (2)$$

where msg(X) means the message from X to A. Therefore, the computation of dA/dt at a particular time step depends on the arrival (or not) of messages from A, B, and C. This allows a molecule to realize different interactions with different counterparts at different times and in different contexts (Figure 2). For example, in molecule B, message *bind* is sent to A in the local cell by

sendmsg(cell.(A, bind, _))

or to A in a neighboring cell located as $[x,y]$ (relative address) by

sendmsg([x,y]·(A, bind, _))

In A, the message is read by

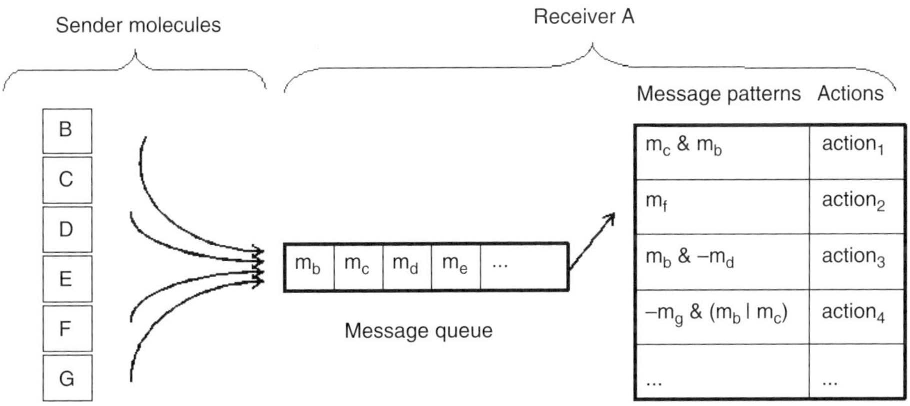

Figure 2. Event-driven computation in an object. A message queue is created by the system in each object to buffer incoming messages. Behaviors of a molecule at any time step are determined jointly by its current state and received messages.

if msgq \$ _.(B, bind, _) then action
end.

The operator "\$" means "contain" and the anonymous variable "_" indicates the address of the sending cell is neglected. Terms f_a, f_b, *and* f_c in equation (2), which are encoded in *action* part and implement quantitative computing without semantic information, are actually not different from those in equation (1). Obviously, cellular and molecular activities in cells are turned on by received messages, instead of by fixed links in formulas (Figure 2). In other words, a system is formalized for computation-on-demand.

2.3. Integrating Signaling Activities into Event-Action Tables

Because a molecule (a protein, a gene, or a binding site) may interact with several molecules (proteins and DNA binding sites), it is necessary to organize all possible interactions in the form of an event-action table, which, akin to a truth table, describes how molecules respond to the incoming signals (Figure 3). We emphasize that, with the rapid increase

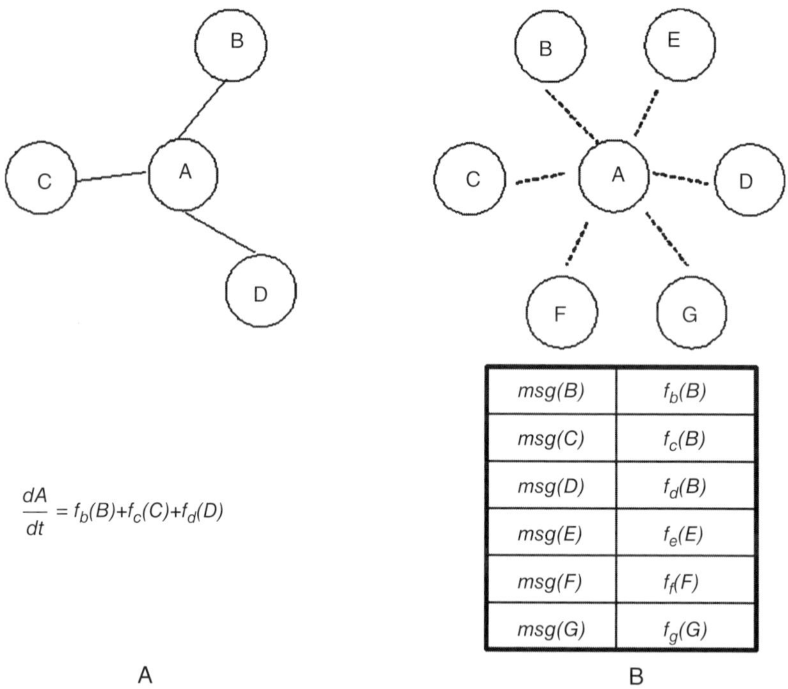

Figure 3. Systematic representation of molecular interactions. Solid lines in (A) indicate interactions happened among a molecule and its three partners under a particular condition, which are hardwired in a differential equation. Dashed lines in (B) indicate all possible interactions of the molecule with all of its partners. To describe interactions under different conditions, an object program is organized, such as an event-action table in which *msg*(X) means the signal from molecule X and f_b(X) gives the equation for computing the contribution from X.

in genomic and proteomic data, the event-action tables can be independently organized, and more importantly, shared by different models. This greatly increases the modeling efficiency and helps build an integrated model of the whole cell.

2.4. Dynamic Capture and Display of Signaling Events

The first step is to automatically capture all message-passing events among molecules continuously. A group of windows, whose number and size are determined by the system based on a particular model, are created at runtime, each corresponding to a message (Figure 4). The chronological signaling events in one cell and the parallel signaling events in all cells are displayed (Figure 4). In a two-dimensional model, the message *XXX* from molecule *A* to *B* in the same cell is captured and assembled as *A_XXX_B_0_0*; *0_0* is used to indicate the relative address of the sender cell at [0,0]; the message from *A* to *B* in the top-left neighboring cell is captured and assembled in the target cell as *A_XXX_B_p1_n*, *p1_n1* indicates the relative address of the sender cell at [+1,–1]. Treating signaling between the same molecule in different cells as different message passing helps effectively reveal the role and property of cell

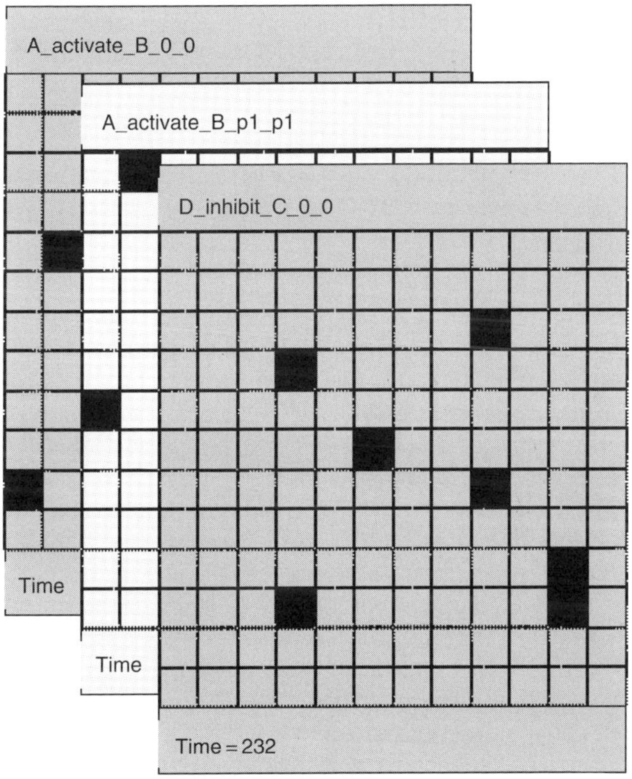

Figure 4. Dynamically captured and displayed signaling events at a particular time step in a two-dimensional model. Each window depicts a message passing event that occurred in the whole-cell space. Cells that received the message are marked in black.

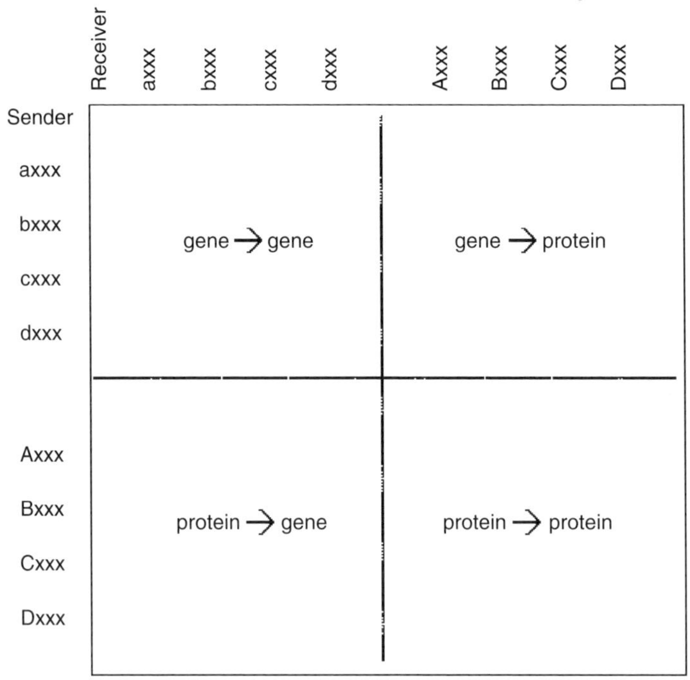

Figure 5. A signaling network in the form of adjacency matrix in a cell at a specific time step.

communication. In addition to message passing, attributes such as concentration of proteins and state of genes are also displayed.

2.5. Dynamic Reconstruction of Evolving Networks

A network is represented by a directed graph. A graph $G = (V, E)$ consists of a set $V = \{v_1, v_2, \ldots, v_n\}$ of vertexes (molecules) and a set $E = \{(v_i, v_j) \mid i, j \in V\}$ of edges (interactions). In the developed system, V is alphabetically ordered by molecular identification to track network evolution. We used the adjacency matrix to represent a directed graph. An adjacency matrix $\mathbf{M} = [m_{ij}]$ of G is an n by n matrix with entries $m_{ij} = 0$ if $(v_i, v_j) \notin E$ and $m_{ij} \geq 1$ if $(v_i, v_j) \in E$. Different values of m_{ij} indicate different interactions. In graph theory, a directed graph G is strongly connected if there is a path in G between every pair of vertices V, and is weakly connected if the underlying undirected graph $\bar{G}$is connected. Apparently, because not all genes are ON at any time and edges in E have not only directions but also semantics, connected graphs, even weak ones, rarely exist. Typically, if genes are named in lowercase (e.g., *notch*) and proteins are named starting with a capital letter (e.g., Notch), the matrix demonstrates four distinct zones reflecting gene–gene, gene–protein, protein–gene, and protein–protein interactions, respectively (Figure 5).

The adjacency matrix provides a structural basis for network reconstruction and analysis. During simulation, an array of matrix, $\mathbf{M}_{[1]}$ to $\mathbf{M}_{[n]}$, n being defined by the user when compiling a model, is created by the system to record network topology in a chosen cell from time step 1 to time step n. The value of m_{ij} in $\mathbf{M}_{[t]}$ is determined by the occurrence or not of the signaling event between molecules i and j at time step t.

3. Applications

3.1. Case 1: One *In Silico* Cell Represents One *In Vivo* Cell—Notch Signaling Propagation

3.1.1. Background

Cells rely on information from neighboring cells, mostly through ligand–receptor interaction, to determine their fate or fulfill their function. A common and important issue in intercellular signaling is that the range of signaling among cells must be precisely controlled. For pathways with secreted ligands, like endothelial growth factor (EGF) signaling, the range of signaling is controlled by different diffusion distance of positive (spitz) and negative (argos) ligands (9). For pathways with membrane-tethered ligands, such as Notch signaling, the mechanism is unclear.

To study Notch signaling propagation in cells, we build a mouse somite segmentation model in a 256 × 256 cell array, with each automata cell corresponding to a biological cell. During embryogenesis, as the tail bud grows caudally, Notch signaling periodically initiates in the tail bud and propagates rostrally in presomitic mesoderm (PSM), each wave of signaling ending with the formation of one somite at the rostral end (Figure 6). Four

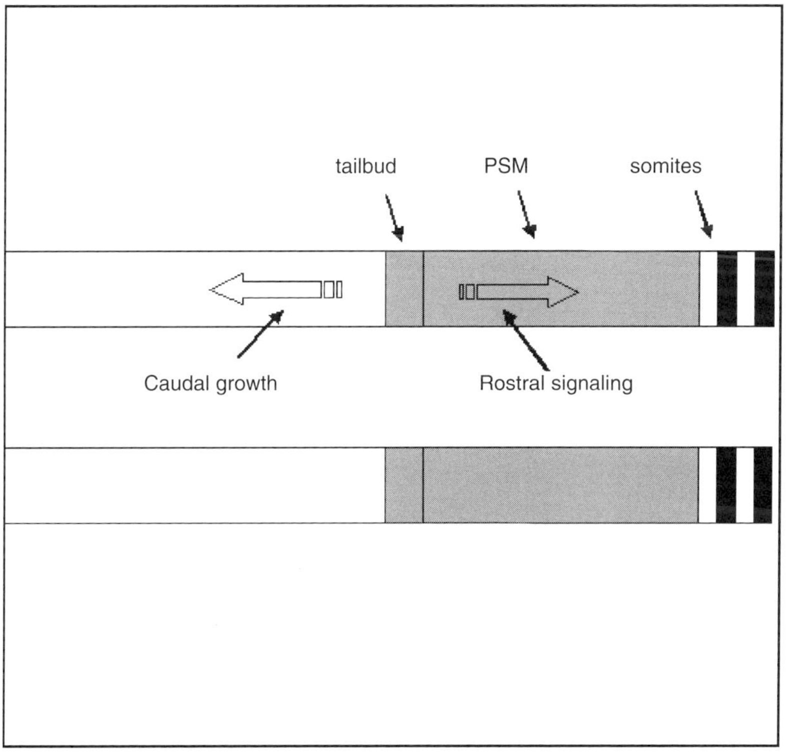

Figure 6. The Notch signaling propagation model in a preexisting cell space.

groups of genes have been identified as key players during the process: i) *fgf8* produces a mRNA and protein gradient in the PSM cells to provide positional information for segmentation (10,11); ii) *Notch* and its ligands *Delta like 1 (Dll1) and Delta like 3 (Dll3)* trigger intercellular communication (12,13); iii) Presenilin1 (Psen1) performs an intracellular Notch cleavage to produce the transcription factor Notch intracellular domain (NICD) that connects inter- and intracellular signaling (14); and iv) NICD-activated, cyclic-expressed genes (*lunatic fringe* [*lfng*] and *hes* in mouse) (15–17) function downstream of, but also feedback onto, *Notch* with roles that are not fully understood. Except for *fgf8*, the other genes belong to the Notch pathway, implementing and controlling Notch signaling in PSM cells. Besides the much-studied periodicity, or segmentation clock, another key feature of Notch signaling in PSM is the unidirectional propagation of signaling from the tail bud to the rostral end. Because both Notch and its Delta ligands are constitutively detectable in PSM cells, theoretically, if there were no control mechanism, ligand–receptor binding would happen liberally in cells, which is not observed under normal conditions.

3.1.2. Method

In the Notch signaling propagation model, 1 time step of automata represents 1 min, and Notch signaling is artificially initiated in the tail bud every 120 min. By default, the state of the gene is ON or OFF, but has multiple expression levels at the ON state. Protein concentration is determined by its synthesis and decay rates. Synthesis is described by first-order growth (18) and decay by first-order decay (19,20). Two versions of the model are built to examine the performance of different computations. In the ordinary differential equation (ODE) version, protein synthesis is described by a variant of Verhulst growth equation

$$\frac{dx}{dt} = k_1 x(1 - x) \tag{3}$$

($\Delta t = 0.1$), and protein decay is described by a variant

$$\frac{dx}{dt} = -\alpha_1 x. \tag{4}$$

In the finite difference equation version, protein synthesis is described by the corresponding finite difference equation

$$x_{t+1} = k_2 x_t (1 - x_t), \tag{5}$$

and protein decay by

$$x_{t+1} = \alpha_2 x_t \tag{6}$$

(21). Concentration of every protein is computed with the same equations, but different parameters. In equation (5), when k_2 has different values, it produces different behaviors, i.e., monotonic approach to a steady state, alternate approach to a steady state, periodic cycles, and aperiodic behavior. We chose the values of k_2 between 1.1 and 2.0 to enable a monotonic approach to the steady state. As expected, we found that both versions produced identical results when run in parallel. α_1, α_2, k_1, k_2 of each protein are from published literatures, but the

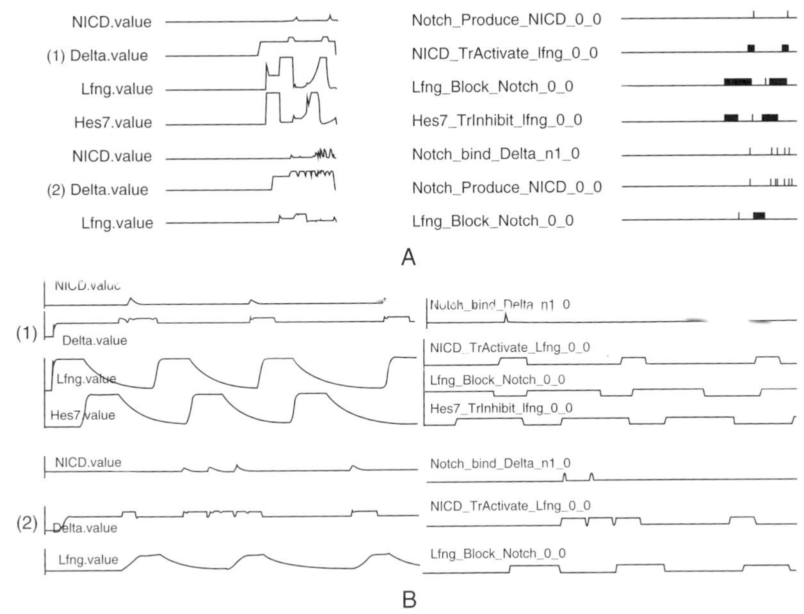

Figure 7. Captured signaling in the Notch signaling model. Caudal is to the left and rostral to the right. (A) Part of instant gene expression (left), protein concentration (left), and signaling profile (right) in a line of cells. (1) The normal case at time step 685. Notch signaling is regular and propagates rostrally. (2) When Lfng synthesis rate is changed from 1.65 to 1.25, inverse Notch signaling occurs. Although Lfng expression profile remains largely unchanged, *Lfng_Block_Notch_0_0* is different from normal. The band of the signal in cells is also narrow and broken. (B) Part of the dynamic process of gene expression (left), protein concentration (left), and signaling profile (right) within a cell from time step 0 to 685. (1) and (2) correspond to (1) and (2) in A, respectively.

concentration at which a protein sends signals to its counterparts is largely unknown. We tested a few assumptions nevertheless. Signaling, as well as protein concentration and gene state, in one cell and in a line of cells (note, in this case signaling happens only between caudorostrally connected cells) are captured and displayed in Figure 7 (some molecules).

3.1.3. Results

The simulation results of Notch signaling are encouraging. First, simulation reveals that the experimentally observed Notch induced Notch block by Lfng suppresses inverse propagation of Notch signaling in PSM cells. Activated Notch, cleaved by Psen1, produces transcriptional factor NICD, which in turn activates the expression of *lfng* and *hes7*. Lfng, probably through Dll3, blocks non–cell-autonomous Notch–Delta interaction. This block is later removed by Hes7. Botch mutations in and overexpression of *lfng* cause atypical signaling within cells, which fails to block Notch binding and leads to reciprocating signaling propagation among cells. The impaired control mechanism leads to changed response of the cells to the positional information established by the Fgf8 gradient, resulting in disordered cellular patterning (Figure 7). Second, signaling exhibits different features at cell and tissue levels. Compared with the chaotic intercellular signaling among cells in *lfng* mutant, signaling within cells does not vary much (Figure 7, A and B). Except for the frequent backward binding of Dll1 to Notch, little anomaly is seen. The interactions among Notch, Dll1, Hes7, and Lfng are accelerated, but still quite regular. Another feature of signaling at tissue level is that the bands of *Lfng_block_Notch* in abnormal cases are narrower and more shattered than those in the normal case, indicating a failed repression of inverse Notch binding in some cells. We note that these tissue level features ostensibly demand multicellular modeling of signaling. As reconstructed signaling at tissue and cellular level illustrates different properties and dynamics, we suggest using *signaling profile* to depict them, which is relevant to but different from *gene expression profile* (Figure 7, A and B). Third, *lfng* mutation demonstrates that, a simple error in cell communication can lead to intricate cellular patterning. Finally, we observed that, as long as the order of signaling does not change, the outcome also does not change, irrespective of changes in molecular concentrations and assumed signaling thresholds. In other words, assumptions, especially those on signaling thresholds, do not tune a model and its corresponding simulation results specific to particular conditions, but rather, help identify general features of molecular signaling. The observed robustness of signaling in cells under various simulated perturbations partly explains the amazing accuracy of molecular systems, even though they are not precisely digitalized as artificial systems.

3.2. Case 2: Multiple *In Silico* Cells Represent One *In Vivo* Cell—Planar Cell Polarity

3.2.1. Background

Epithelial cells show a clear perpendicular polarity in respect to their apical/basal axis. This form of polarity, called planar cell polarity (PCP) (22,23), is important for the function and migration of epithelial cells. In the eye, wing, and abdomen of *D. melanogaster*, this cell polarity is demonstrated by the direction of hair and bristle, and is controlled by the signaling among a group of molecules with a hierarchical relationship (24). The core at the middle layer implementing the conserved function of cell polarization includes *frizzled* (*fz*), *dishevelled* (*dsh*), *prickle-spiny-legs* (*pk*), *flamingo* (*fmi*), *van gogh* (*vang*), and *diego* (*dgo*). Fmi is a membrane protein, and it accumulates at both the distal and proximal edges of cells (25). Vang, which is asymmetrically located at the proximal side of cells, functions in the proximal and distal movement of Pk and Dsh (26). Fz is a transmembrane Wnt receptor with a slightly initial asymmetrical distribution on the cell (27). Dsh is also a Wnt pathway component. Pk is identified to mediate the negative feedback amplification among Pk, Dsh, and Fz, which amplifies the slight initial asymmetry of Fz and the relocation of these proteins in the cell (28). Mutations in these genes lead to domineering nonautonomy. That is, loss-of-function clones of *fz*, *vang*, and *pk* induce changed cell polarity in neighboring wild-type tissue (22,28,29).

In the first phase of PCP during the 0h–8h after prepupa formation (APF), Fmi, Fz, and Dsh uniformly localize around the perimeter of the cell; in the second phase between 18h and 32h APF, these proteins adopt asymmetrical subcellular localization. Fmi is present on both proximal and distal cell boundaries with no detectable asymmetry. During the negative feedback among and mediated by Fz, Pk, and Dsh, Fz and Dsh localize to the distal side and Pk localizes to the proximal side (28). Because Fmi's distribution does not change in the second phase and its contribution to the feedback amplification is unclear, our current PCP model, which focuses on the proximodistal localization of Pk, Dsh, and Fz in the second phase, does not include it. Also, as Vang works with Pk at the proximal side of the cell, shares the same distribution as Pk's, and takes a controversial role (26,30), it is not included. The abnormal signaling among Fz, Dsh, and Pk at the second phase alone can lead to domineering nonautonomy.

3.2.2. *Methods*

Although the feedback amplification hypothesis explains some PCP phenomena, and a partial differential equation (PDE)–based mathematical model improves our understanding of domineering nonautonomy (30), a few issues remain. The key question is whether and how the feedback amplification works under different global directional cues whose properties have not been revealed experimentally. It is feasible to infer the global directional cue based on the existing knowledge of PCP.

A few experimental observations are available for model building: (1) Information in the PCP signaling pathway flows from the receptor Fz to the cytoplasmic protein Dsh (31); (2) the discrete localization of Fz and Dsh appears to result from a cell autonomous feedback amplification mechanism (32,33); (3) Pk functions nonautonomously to generate asymmetry of Fz/Dsh activity (28,34); (4) Proximal localization of Pk depends on intercellular difference in Fz activity (28). Although these observations hint at the existence of feedback amplifications among the three core molecules of PCP signaling, these molecules themselves, along with a few regulators (Fmi, Dgo, and Vang), are insufficient to form a complete network with biochemical details to build a quantitative model with experimentally determined parameters. Therefore, we chose not to use differential equations to bind them together to quantitatively simulate their concentration and physical interaction (30), but instead to use a simple program to simulate the feedback among and the movement of Pk, Dsh, and Fz.

Because a key feature of planar cell polarity is the movement and asymmetric distribution of molecules in a cell, an uncompartmented automata cell may not represent a biological cell accurately. Considering the shape of an epithelial cell, we use six automata cells to represent one biological cell. The two-dimensional model contains 114×114 automata cells, or 2,166 biological cells. These six automata cells share a unique identity number generated at runtime with the random number generator function, and they are linked in a specific manner (Figure 8). However, each automata cell is an independent computational unit. The unique identity number is used to determine whether an intercellular signaling

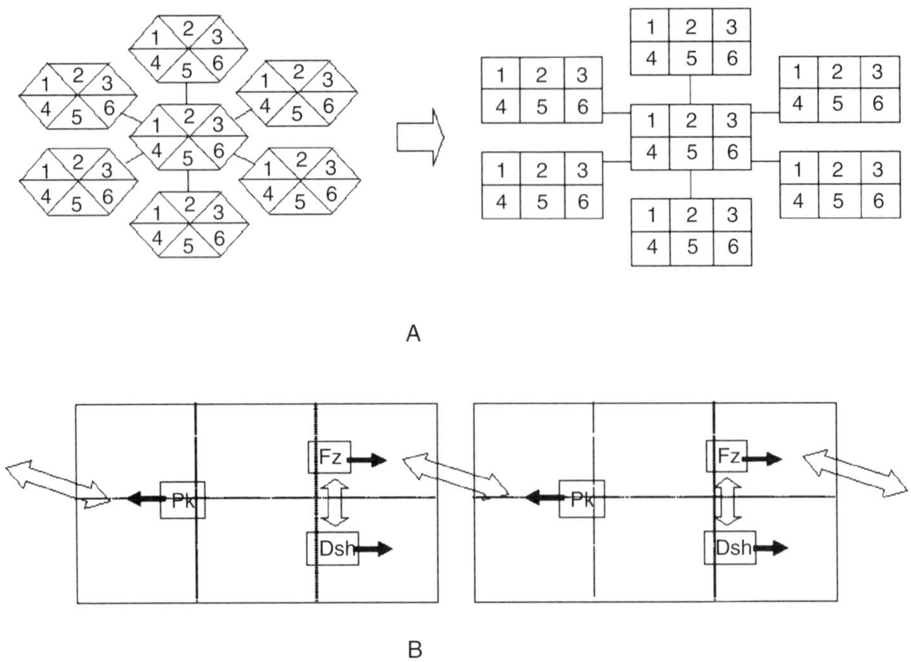

Figure 8. Intercellular communication and intracellular molecular movement during PCP. (A) Each epithelial cell is represented by six automata cells with particular connection. (B) Driven by the mutual interactions among them, Pk, Fz, and Dsh move toward different directions in a biological cell. Double arrows indicate intercellular molecular interaction, with the external Fz cell nonautonomously attracting Pk, and vice versa. Single arrows denote intracellular molecular interaction, with the local Fz cell autonomously attracting Dsh, and vice versa. Proximal is to the left and distal to the right.

between two automata cells is within the same biological cell or between two different ones. The movement of molecules in different automata cells is implemented as intercellular signaling. The message is linked to an attached item to indicate the amount of the moved molecule. For example, in the Fz object in a cell, we use

sendmsg([i,j]·(Fz, move, quantity))
value:= value – quantity

to move a portion of Fz to the automata cell at [i,j]. In the Fz object in the target cell, we use

if msgq $ _.(Fz, move, _amount) then
value:= value + _amount
end

to receive the migrated Fz; the unbound variable *_amount* is used to read the value. Bound and unbound variables in a message are defined freely, except that unbound variables begin with "_." The function *position*(), which returns the global coordinates value of an automata cell, is used to assign cells different initial Fz asymmetry. The initial concentration of molecules and the speed of molecular movement are assumed.

3.2.3. Results

The simple CA-based model shows interesting results. First, simulation produces the PCP phenotypes. The feedback amplification among Fz, Dsh, and Pk, whose regulation mechanisms are unclear, seem to be necessary and sufficient for the generation of polarity repolarization in both wild-type and in mutant cells (29) (Figure 9). Second, we performed simulation with different parameter values, along with a change in molecular concentration, and found that under all conditions, identical results are obtained. This agrees with our hypothesis, for it is the *relationship* among the molecules, not a specific parameter value that determines cell polarity. Parameter independence, along with the various simulated phenotypes, implies that the feedback amplification among Pk, Fz, and Dsh is robust. Third, phenotypes of PCP are slightly different in different tissues because of the tissue-specific regulation and molecular environment (28). Inadequate experimental data exist to explain how different depths of domineering nonautonomy are controlled. Our study suggests that the depth of domineering nonautonomy may be parameter dependent. Finally, we examined cell polarity signaling under different global

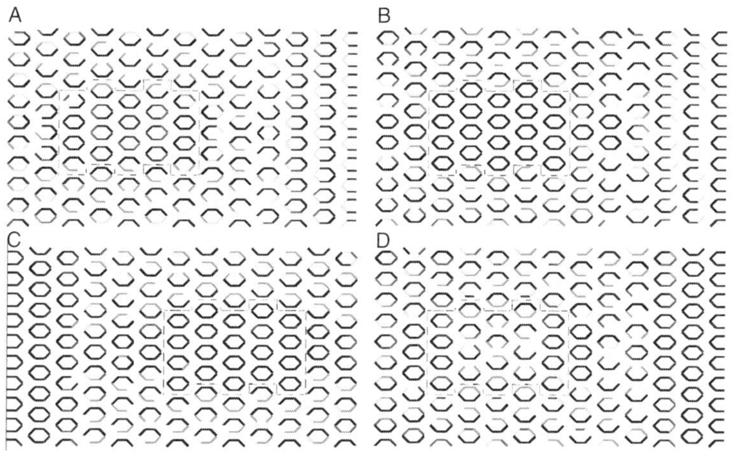

Figure 9. Weak and overexpression of *fz* leads to different domineering nonautonomy. Proximal is to the left and distal to the right, and red color indicates higher protein concentration. In all pictures, molecular slope means the slope in the cell. (A) Pk distribution with clones of fz^- mutation. Pk slope at the outermost layers in the clones points outward to the clones, and Pk slope in wild-type cells distal to the clones is reversed and points outward from the clones. (B) Fz distribution with clones of fz^- mutation. Fz slope at the outermost layers in the clones points inward to the clones, and Fz slope in wild-type cells distal to the clones is reversed and points inward to the clones. Note that Fz slope within the clones (distal side) is also reversed. (C) Fz distribution with *fz* overexpression. Fz slope at the outermost layers in the clones points outward from the clones, and Fz slope in wild-type cells proximal to the clones is reversed and points outward from the clones. (D) Fz distribution with *pk* overexpression. Fz slope at the outermost layers in the clones points inward to the clones, and Fz slope in wild-type cells distal to the clones is reversed and points inward to the clones. Note the similarity between B and D.

directional signals. We observed that the gradient zigzag distribution of Fz produces the best-observed phenotypes, and that polarity propagation can overcome small, local direction errors. Global directional signaling also facilitates PCP and restrains domineering nonautonomy. In the absence of a global directional signal, polarization would propagate ceaselessly in cells.

4. Discussion

Networking and evolution are two important features of cellular signaling. Evolving signaling network in metazoan cells, driven by crosstalk among signaling pathways like Notch, FGF, EGF, Hedghog, TGFβ, and Wnt and perturbed by extracellular signals, is a hot research area. Network analysis has been recognized as an effective method for the understanding of biological systems at the molecular level (35,36). However, on the one hand, assembling signaling network manually from experimental data is difficult; on the other hand, analyzing the complicated "canonical" network, such as that in the study by Kohn (37), is challenging, given the prevalence of temporally and spatially overlapped signaling events in cells. By using well-known biological examples, we describe a computational method for modeling and simulating the dynamic networking of signaling in cells. The reconstructed dynamic evolution of signaling network shows a remarkably high biological accuracy.

Although modeling parallel signaling processes in a multicellular context is still in its infancy, a few methods and tools exist. Besides classic CA and PDE, multicellular modeling with an array of ODE has been reported (38,39). However, it is impractical to incorporate evolving cellular connections and communication in a large, dynamic cell population undergoing growth and patterning, using classic differential equation–based methods. When a model is large (for example, a 100 × 100 cell array contains 10,000 cells), explicit description of multicellularity becomes unrealistic. Although both CA and OOP have been widely used (40,41) the novelty of our method lies in building a hybrid system that combines the most desirable features of each system. Using this hybrid system, it was observed that molecular-level tissue scope modeling can explain complex tissue level phenotypes with captured signaling events and reconstructed signaling networks.

In equation (2), signaling is discrete in time (represented by *msg*(*X*)), but continuous in value (represented by f_a). In case 1, we let it be discrete both in time and value (i.e., occurrence or nonoccurrence). In fact, as shown in case 2, a more flexible description can be realized. A bound variable, named *strength*, which is a floating-point or integer data, can be defined in a message *sendmsg*(*cell.*(*X, transcription, strength*)) to more precisely describe the strength of an interaction. Generally, if signaling is discrete in value, this method involves less independent parameters than differential equation–based biochemical models and provides a workable solution to modeling complex and evolvable signaling in multiple cells. In addition to separating signaling from computation, molecules are computationally programmed in an object-oriented way to facilitate the description of emergent signaling, including those induced by temperature-sensitive alleles and external perturbations. Given that the mathematical traits and biochemical processes of many molecular activities remain unclear, we argue that this method, at least at current stage, is applicable and biologically relevant. The back-end engine, comprising model-independent utilities, captures the occurrence and evolution of signaling during simulation. On one hand, we note that simulating normal morphogenesis process identifies the default signaling thresholds; on the other hand, we find that as long as the order of signaling does not change, the outcome remains unaffected, irrespective of changes in molecular concentrations and signaling thresholds. The robust signaling network explains why natural systems work more efficiently and accurately than artificial systems.

Although network topology hints at the general principles of molecular interaction, connectivity itself doesn't reveal the entire story because of weighted interactions and ontogenetic network evolution. At least two more factors influence signaling processes. The first one is the order of molecular interactions that can decisively influence the course and outcome of signaling. In the first example, we show that it is the wrong order (an undue Notch block and backward Notch/Delta binding), but not the wrong partnership that causes the reverse Notch/Delta signaling and impairs the unidirectional segmentation of presomitic cells. Feedback within and interaction among pathways may generate many different orders of molecular interactions. The second factor is the timing of signaling. A premature signal may drastically change the default-signaling path by emergent formation of key transcription complexes, as happens in many cell fate transformations. The eye-to-antenna transformation in *D. melanogaster* induced by maneuvered *Egfr* and *Notch* signaling is a typical example (42). As signaling at the wrong time and in the wrong order may explain a large number of circuit anomalies (43), the two factors, contributing to *signaling dynamics*, deserve more investigation. The partnership of the molecules, order, and timing interact and impart emergent properties. An *in silico* model of a signaling network must take into consideration all of these factors. In the Notch signaling model, it is difficult to explain how tissue-level phenotype emerges from molecule-level interactions (from protein concentration and gene state data), if we do not consider the order and timing of signaling. Last, we note that for both the captured signaling events and reconstructed signaling networks, a deeper analysis with nontraditional methods can be performed (44,45).

An interesting application of the proposed modeling and simulation method is its potential in studying attractors and basin of attractors in cell fate determination and tissue differentiation (46). Attractors are sets of states that are invariant under system evolution and indicate regions where a dynamic will ultimately end up. A basin of an attractor is a set of states that will evolve toward their corresponding attractor eventually. Although both of them have clear definitions, identifying them in real biological systems, either *in vivo* or *in silico*, remains largely unexplored. Moreover, little is known about the molecular topology of signaling network in respect to the state and basin of attractors and how topological structure drives a cell from basin of attractors to an attractor. With dynamic reconstruction of signaling network during cellular differentiation, there is a strong possibility that these issues can be effectively addressed.

Finally, although our modeling method is a programming language combined with backend (model-independent) supporting utilities, various mathematical formalisms, such as stochastic and deterministic systems, reaction-diffusion systems, differential and difference equations, and Boolean networks (47) can be implemented within the framework of our modeling system. Physical parallelism with built-in OpenMP to run a model on multiprocessor computers will be available. The system runs under Linux and is freely available upon request.

References

1. Bhalla US, Iyengar R. Emergent properties of networks of biological signaling pathways. *Science* 1999;283:381–387.
2. Basler K, Yen D, Tomlinson A, et al. Reprogramming cell fate in the developing Drosophila retina: transformation of R7 cells by ectopic expression of rough. *Genes Dev* 1990;4:728–739.
3. Radisky D, Hagios C, Bissell M. Tumors are unique organs defined by abnormal signaling and context. *Sem Can Biol* 2001;11:87–95.
4. Luscombe NM, Babu MM, Yu H, et al. Genomic analysis of regulatory network dynamics reveals large topological changes. *Nature* 2004;431:308–312.
5. Zhu H, Wu Y, Huang S, et al. Cellular automata with object-oriented features for parallel molecular network modeling. *IEEE Trans NanoBioScience* 2005;4:141–148.
6. Wolfram S. Cellular automata as models of complexity. *Nature* 1984; 311:419–424.
7. Toffoli T, Margolis N. Cellular Automata Machine: A New Environment for Modeling. MIT Press; 1987.
8. Griffeath D, Moore C. New Constructions in Cellular Automata. Oxford University Press; 2003.
9. Freeman M. Cell determination strategies in the *Drosophila* eye. *Development* 1997;124:261–270.
10. Sawada A, Shinya M, Jiang YJ, et al. Fgf/MAPK signalling is a crucial positional cue in somite boundary formation. *Development* 2001;128:4873–4880.
11. Dubrulle J, Pourquie O. fgf8 mRNA decay establishes a gradient that couples axial elongation to patterning in the vertebrate embryo. *Nature* 2004; 427:419–422.
12. Dunwoodie SL, Clements M, Sparrow DB, et al. Axial skeletal defects caused by mutation in the spondylocostal dysplasia/pudgy gene Dll3 are associated with disruption of the segmentation clock within the presomitic mesoderm. *Development* 2002;129:1795–1806.
13. Takahashi Y, Inoue T, Gossler A, et al. Feedback loops comprising Dll1, Dll3 and Mesp2, and differential involvement of Psen1 are essential for rostrocaudal patterning of somites. *Development* 2003;130:4259–4268.
14. Koizumi K, Nakajima M, Yuasa S, et al. The role of presenilin 1 during somite segmentation. *Development* 2001;128:1391–1402.
15. Forsberg H, Crozet F, Brown NA. Waves of mouse Lunatic fringe expression, in four-hour cycles at two-hour intervals, precede somite boundary formation. *Curr Biol* 1998;8:1027–1030.

16. Zhang N, Gridley T. Defects in somite formation in lunatic fringe-deficient mice. *Nature* 1998; 394:374–377.
17. Bessho Y, Sakata R, Komatsu S, et al. Dynamic expression and essential functions of Hes7 in somite segmentation. *Genes Dev* 2001;15:2642–2647.
18. Mehra A, Lee KH, Hatzimanikatis V. Insights into the relation between mRNA and protein expression patterns: I. Theoretical considerations. *Biotechnol Bioeng* 2003;84:822–833.
19. Mosteller RD, Goldstein BE. A mathematical model that applies to protein degradation and post-translational processing of proteins and to analogous processes for other molecules in non-growing and exponentially growing cells. *J Theor Biol* 1984;21:597–621.
20. Wang Y, Liu CL, Storey JD, et al. Precision and functional specificity in mRNA decay. *Proc Natl Acad Sci* 2002; 99:5860–5865.
21. Kaplan D, Glass L. Understanding nonlinear dynamics. London: Springer-Verlag; 1995.
22. Adler PN. The genetic control of tissue polarity in *Drosophila. Bioessays* 1992;14:735–741.
23. Eaton S. Planar polarization of *Drosophila* and vertebrate epithelia. *Curr Opin Cell Biol* 1997;9:860–866.
24. Tree DRP, Ma D, Axelrod JD. A three-tiered mechanism for regulation of planar cell polarity. *Sem Cell Dev Biol* 2002;13:217–224.
25. Usui T, Shima Y, Shimada Y, et al. Flamingo, a seven-pass transmembrane cadherin, regulates planar cell polarity under the control of Frizzled. *Cell* 1999;98:585–595.
26. Bastock R, Strutt H, Strutt D. Strabismus is asymmetrically localised and binds to Prickle and Dishevelled during *Drosophila* planar polarity patterning. *Development* 2003;130:3007–3014.
27. Vinson CR, Conover S, Adler N. A *Drosophila* tissue polarity locus encodes a protein containing seven potential transmembrane domains. *Nature* 1989; 338:263–264.
28. Tree DRP, Shulman JM, Rousset R, et al. Prickle mediates feedback amplification to generate asymmetric planar cell polarity signaling. *Cell* 2002;109:371–381.
29. Lawrence PA, Casal J, Struhl G. Cell interactions and planar polarity in the abdominal epidermis of Drosophila. *Development* 2004;131:4651–4664.
30. Amonlirdviman K, Khare NA, Tree DRP, et al. Mathematical modeling of planar cell polarity to understand domineering nonautonomy. *Science* 2005;307:423–426.
31. Krasnow RE, Wong LL, Adler PN. Dishevelled is a component of the frizzled signaling pathway in *Drosophila. Development* 1995;121:4095–4102.
32. Axelrod JD. Unipolar membrane association of Dishevelled mediates Frizzled planar cell polarity signaling. *Genes Dev* 2001;15:1182–1187.
33. Strutt DI. Asymmetric localization of Frizzled and the establishment of cell polarity in the *Drosophila* wing. *Mol Cell* 2001;7:367–375.
34. Gubb D, Green C, Huen D, et al. The balance between isoforms of the prickle LIM domain protein is critical for planar polarity in *Drosophila* imaginal discs. *Genes Dev* 1999;13:2315–2327.
35. Xiong M, Zhao J, Xiong H. Network-based regulatory pathways analysis. *Bioinformatics* 2004;20:2056–2066.
36. Barabasi AL. Linked: How Everything Is Connected to Everything Else and What It Means. Plume; Reissue edition, 2003.
37. Kohn KW. Molecular interaction map of the mammalian cell cycle control and DNA repair systems. *Mol Biol Cell* 1999;10:2703–2734.

38. Shapiro BE, Levchenko A, Meyerowitz EM, et al. Cellerator: extending a computer algebra system to include biochemical arrows for signal transduction simulations. *Bioinformatics* 2003;19:677–678.
39. Jaeger J, Surkova S, Blagov M, et al. Dynamic control of positional information in the early *Drosophila* embryo. *Nature* 2004;430:368–371.
40. Ermentrout GB, Edelstein-Keshet L. Cellular automata approach to biological modeling. *J Theor Biol* 1993;160:97–133.
41. Johnson CG, Goldman JP, Gullick WJ. Simulating complex intracellular processes using object-oriented computational modeling. *Prog Biophys Mol Biol* 2004;86:379–406.
42. Kumar JP, Moses K. EGF receptor and Notch signaling act upstream of Eyeless/Pax6 to control eye specification. *Cell* 2001;104:687–697.
43. Taipale J, Beachy PA. The Hedgehog and Wnt signaling pathways in cancer. *Nature* 2001;411:349–354.
44. Papin JA, Palsson BO. Topological analysis of mass-balanced signaling networks: a framework to obtain network properties including crosstalk. *J Theor Biol* 2004;227:283–297.
45. Stelling J, Klamt S, Bettenbrock K, et al. Metabolic network structure determines key aspects of functionality and regulation. *Nature* 2002;420:190–193.
46. Huang S. Genomics, complexity and drug discovery: insights from Boolean network models of cellular regulation. *Pharmacogenomics* 2001;2:203–222.
47. De Jong H. Modeling and simulation of genetic regulatory systems: a literature review. *J Comput Biol* 2002;9:67–103.

14

Kinetics of Dimension-Restricted Conditions

Noriko Hiroi and Akira Funahashi

Summary

The intracellular environment is crowded with skeletal proteins, organelle membranes, ribosomes, and so on. When molecular movement is restricted by such environments, some biochemical reaction processes cannot be represented by classical models, which assume that reactions occur in simple Newtonian fluids. Dimension-restricted reaction kinetics (DRRK) modeling is a method that can represent dimension-restricted reactions. We introduce the methods of DRRK in each case of reaction type. DRRK has another advantage in that it can be quantitatively evaluated by biochemical experiments. We also introduce the procedure of applying it for experimental results. This modeling method may provide the basis for *in vivo*–oriented modeling.

Key Words: Fractal kinetics; percolation theory.

1. *In Vivo*–Oriented Modeling

The *in vivo* environment, which is the actual space for biochemical reactions, is very different from ideal conditions, i.e., well-diluted solutions. The organization inside a cell resembles that of a protein crystal with 40% water (1,2) (Figure 1). The assumption of reaction space conditions for classic numerical models, such as the mass-action law and the Michaelis–Menten equation, are ideal conditions (3–5), which differ from the actual conditions of *in vivo* biochemical reactions. Although classic models can be used to compute *in vivo* reactions with sufficient approximation, such approximation is not suitable as a general approach. However, to develop a high-precision model that can be practically used for scientific investigation as well as drug discovery process, a new method that can be applied to biological phenomena occurring in various situations, including nonideal conditions, is required. In fact, there are several biochemical reactions that cannot be represented by classic numerical models, even with optimizing/fitting by experimental data (6–10).

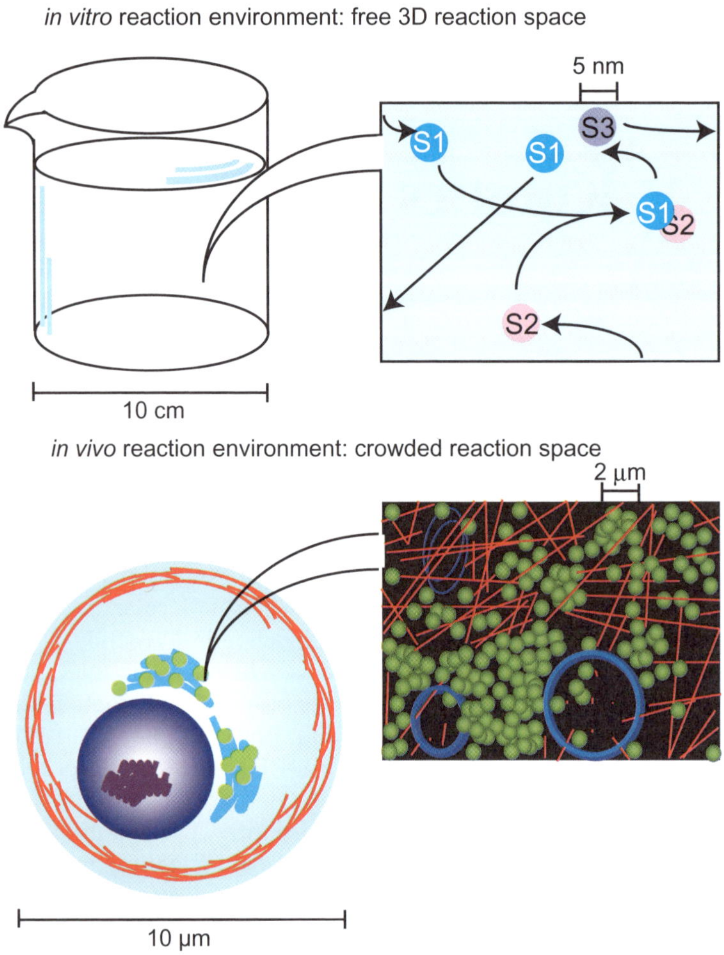

Figure 1. *In vitro* reaction environment versus *in vivo* reaction environment. A typical protein diameter is approximately 10^{22} times smaller than its reaction space. Then reaction species can behave as in Newtonian fluids. Proteins in intracellular environment are packed with only 1.5 times the volume of water. Red, skeletal proteins; blue, organelle membrane; green, ribosomes (1,2).

Dimension-restricted reactions are such a phenomena. Appropriate parameters cannot be determined from the experimental data for a classic, ordinary differential equation (ODE)–based model when the target reaction is a dimension-restricted reaction (6–10). Although stochastic modeling may be able to model probabilistic distributions of molecules in the dimension-restricted space, values for each transition probability are difficult to measure experimentally, and so, such models tend to be qualitative in practice. Partial differential equation (PDE)–based modeling captures spatial distribution, but the space is assumed to be homogeneous, and data to calibrate such models, cannot be measured in practice.

Because of crowding in *in vivo* environments and the restriction of the movements of the molecules in intracellular space, many types of *in vivo* reactions are regarded as dimension restricted. For precise simulation of biological processes, it is critically important to establish a realistic *in vivo* situation method in which parameters can be decided experimentally and simulation results can be verified experimentally.

We have developed a theory of biochemical DRRK that can represent reactions in crowded *in vivo* situations. This chapter reviews and demonstrates some applications, including how this approach can experimentally decide model parameters, and experimentally verifies simulation results.

2. DRRK

2.1. Fundamental Theory and Applications (Figure 2)

Einstein and Smoluchowski formulated the concept and equations of diffusion in a continuous medium. Einstein's equations are based on the assumption that random motion occurs in a homogeneous space (11,12). In 1963, Frish and Hammersley pointed out the drastic effects caused by the inhomogeneity of the medium, and suggested that under such conditions the concept of diffusion should be replaced by that of "percolation" or "random percolation" (13). Random percolation means a random walk on a random lattice, which consists of either "open" or "closed" lattice sites to the motion of the random walker. The lattice sites open or close with a defined probability, so that an inhomogeneity occurs. The difference between diffusive and percolative motion depends on a critical value of a parameter, which represents a critical concentration of open lattice sites. This critical point is defined by the connectivity of the lattice, or defined by the effective range of a migration step (13).

Kopelman continued this theoretical work and conducted a pioneering study on dimension restriction of reaction space. He illustrated how the law of mass action breaks down in heterogeneous environments, and described the kinetics of such reactions as fractal reaction kinetics (14–18).

The first report by Kopelman, in 1980 (14), described simple analytical results for the special case of correlated random hops, and their Monte Carlo calculation results. They considered stochastic and correlated hopping on ordered and random lattices that contained a small fraction

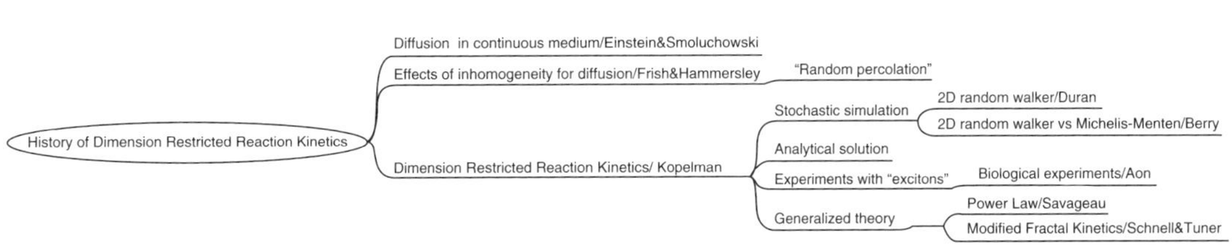

Figure 2. The history of DRRK.

Table 1. Comparison of reactant diameters and volumes.

Molecule	Diameter (nm)	Volume (nm^3)
Mg^{2+}	0.346	0.0217
Naphthalene	$0.2 \times 0.7 \times 1.2$	0.168
*Eco*RV	4.96–6.39	136.56
DNA	$2.0 \times 0.34 \times$ (bp)	4,654.7
E. coli	$10^3 \times 3 \times 10^3$	2.355×10^9

of supertraps and a small number of "hoppers," i.e., excitons (14). Their approach was based on a measurable quantity called "sensor registration probability." This probability is related to time-dependent rate and diffusion constants (14).

Next, they extended the theory to a more general one. They picked up pseudounimolecular and bimolecular reactions of random walkers on large clusters and presented their theory via simple, analytically soluble rate equations (15). The rate coefficient depends on time ($K(t) \propto t^{-h}$), where $1 \geq h \geq 0$. h measures the degree of local heterogeneity; e.g., $h = 0$ means local homogeneity (15). They also reported actual parameter values measured by an experiment with naphthalene crystals (15), as an example of their theory. A more detailed analysis for exciton experiments was given in their subsequent paper (16).

The main works were reviewed by Kopelman (17,18). Because his interest was focused mainly on energy transport through disordered systems, DRRK was applied to the reactions of microscale molecules in Kopelman's work (Table 1) (14–18). Kopelman suggested that DRRK could be applied to the theoretical and experimental analysis of an enzymatic reaction of mesoscopic molecules. We tried to apply the analysis of an enzymatic reaction of mesoscopic molecules, and the results agreed with his suggestion. He showed the power of computer simulations by citing some results of simulations, rather than actual experiments (14–18).

2.2. Development from Basal Theory

Savageau studied fractal kinetics for enzymatic reactions (19–22). He adopted an alternative approach for modeling reaction dynamics in nonhomogeneous environments (19–22), and proposed that, rather than introducing a time dependence to the rate constants k of second- and higher-order reactions, instead the reactant concentrations should be raised to noninteger powers. Thus, his definition could be described as follows: the conventional rate law exhibits a characteristic reduction of the rate constant with time, and this is equivalent to a time-invariant rate law with an increased kinetic order under certain conditions. Savageau argues for the benefit of the power law approach based on its mathematical connection with fractal phenomena. One of the benefits is tractability because of the systematic structure of its formalisms. Another benefit is accuracy because the formalism conforms to actual systems in *Nature*. Savageau's approach has been proven only for homodimeric reactions, because the governing equations quickly become analytically insoluble as the reaction dynamics become more complex. In more complex systems kinetics, orders must be determined experimentally.

Turner pointed out that power law equations produce sigmoid curves for one substrate bonding reaction with each enzyme, especially when the method is applied to the Michaelis–Menten reaction in a quasi–steady state (19). There is no known experimental evidence showing sigmoid kinetics in such a case, so Schnell and Turner suggested that the fractal approach might be better for modeling the actual reaction processes. They presented the modified fractal kinetics (19,23), and proposed a modified form of fractal kinetics, whereby k_1 has the time-dependent form

$$k_1(t) = \alpha\,(t + \beta)^{-h} \qquad (0 < (\alpha,\beta) \in \mathcal{R}), \tag{1}$$

where α and β are constant and to be set by fitting to the experimental data (19).

Vlad et al. showed multiple rate-determining steps for classic, non-classic, and especially fractal kinetics (24). They extended a single rate-determining step in a reaction mechanism to systems with multiple overall reactions for which the elementary reactions obey nonideal, or fractal, kinetics. Their extension requires four assumptions: i) that there exists no constraints which prevent the evolution toward equilibrium; ii) elementary reactions occur in pairs of forward and backward steps; iii) the kinetics of the elementary steps are either nonideal or fractal, and are compatible with equilibrium thermodynamics; and iv) the number of reaction routes is the same as the number of rate-determining steps. Their strategy is limited by these assumptions, so can be applied only to special cases, although their objective is a generalized approach.

The aforementioned works tried to extend the theory of kinetics, whereas some researchers have tried stochastic simulation for fractal-like phenomena.

Duran et al. investigated the fractal *Nature* of the kinetics (25). They simulated aggregated particles visiting on a 2D square grid. This type of investigation has been extended by others researchers, who suggested that some of the physical properties depend on its initial conditions.

In one extension, Berry found that his stochastic simulation analysis of the two-dimensional case showed signs of fractal kinetics (26). He picked up an isolated Michaelis–Menten enzyme reaction case, and represented it by two-dimensional lattices with varying obstacle densities as models of biological membranes. His model indicated that for diffusion on low-dimensional media, the kinetics are of the fractal type. His simulation also indicated that the fractal-like properties are mainly additive. This area requires some new strategies to produce a definite development.

Finally, we introduce a study that tried to evaluate the theory through actual biological experiments.

Aon et al. (27,28) applied the fractal approach with simulations for reactions of small metabolites, for cellular and organ morphology, for the geometry of protein surfaces and organization of macromolecules, and for the spectral analyses of complex signals. Enzymatic kinetics were limited to theory in their study.

This extension is naturally limited to solving one special problem, because the objective of each study is to solve a specific biological problem.

In this chapter, we show the strategy of how to apply DRRK for such studies. We explain a pseudomonomolecular reaction case and a two-reactant bimolecular reaction by extending their theory for a single-reactant bimolecular reaction or a homodimeric reaction.

2.3. DRRK for Bimolecular Reactions

2.3.1. Pseudomonomolecular Reactions in Dimension-Restricted Space

A reaction, which is called a pseudomonomolecular reaction, can be represented as follows:

$$A + C = C. \tag{2}$$

This kind of reaction includes a simple transportation from one compartment to the other, and diffusion phenomena, which are observed as molecules diffusing into a distinguishable area. The differential rate equation for these reactions can be represented as follows:

$$-\frac{d[A]}{dt} = k[A], \tag{3}$$

where k is a time-independent constant.

The formulation of classic mass-action kinetics is universal. The functional form of the rate law does not depend on the dimension of the reaction space or on the mobility of species. Such factors affect only the value of the parameter k. However, the simple functional form of classic mass-action kinetics varies only with its parameter values, which are constants, and is naturally limited to describable phenomena.

By using DRRK in a descriptive manner, the differential rate equation can be written as follows:

$$-\frac{d[A]}{dt} = k(t)[A]. \tag{4}$$

In this case, $k(t)$ is time dependent. The time dependency was indicated by experimental studies on the reaction kinetics of excitons in molecular macroclusters (14–16). Explicitly, this time dependence, $k(t)$ is replaced by

$$k(t) = k_1 t^{-h} \quad \text{with} \quad 0 \le h \le 1 \ (t \ge 1), \tag{5}$$

which is equivalent to

$$\log k = -h \log t + \log k_1. \tag{6}$$

h is a measure of the dimensionality of the systems. h can be calculated from experimental data. First, you should calculate the value of K, which is defined as follows, from experimental data:

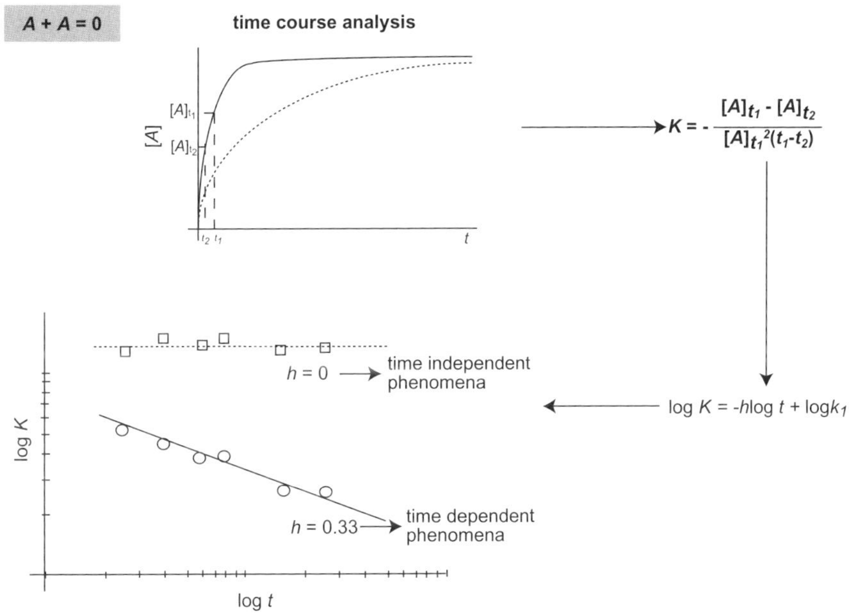

Figure 3. Log–log plot of rate constant k versus time t. Figure shows the case of A + A = 0 (1–reactant bimolecular) reaction. Rate constant K is calculated from time course analysis data. One may plot the log value of K versus log t. The slope of the plot of this value is $-h$. $h = 0$, time-independent reaction process; $h = 0.33$, time-dependent reaction process (14–18).

$$K = -\frac{[A]_{t_1} - [A]_{t_2}}{[A]_{t_1}(t_1 - t_2)}. \tag{7}$$

From this you get the log–log plot of time and K (Figure 3). The slope of the approximated line is $-h$.

The special case of $h = 0$ is the classic case where unconstrained space is assumed. When a reaction environment is made homogeneous by vigorous stirring, h also equals 0. For diffusion-limited reactions that occur in fractal spaces, the theory gives $h > 0$ and time-dependent $k(t)$. For dimensions lower than 2, one finds for elementary bimolecular reactions

$$h = 1 - \frac{d_s}{2}. \tag{8}$$

d_s is a spectral dimension, which appears to be independent of d-dimensional Euclidean space ($d_s = \frac{2d_f}{d_w}$, d_f is a fractal dimension of the aggregate, d_w is the fractal dimension of a random walk on the cluster substrate).

2.3.2. Two-Reactant Bimolecular Reactions in Dimension-Restricted Space

A two-reactant bimolecular reaction can be represented as follows:

$$A + B = P. \tag{9}$$

The differential rate equation in terms of A is written in the classic mass-action kinetics manner as follows:

$$-\frac{d[A]}{dt} = k[A][B]. \tag{10}$$

k is also a time-independent constant in this case.

For a two-reactant bimolecular reaction, the differential rate equation based on DRRK can be written as follows:

$$-\frac{d[A]}{dt} = k(t)[A][B]. \tag{11}$$

The definition of $k(t)$ and the calculation procedure of h from experimental data are the same as in the case of pseudomonomolecular reactions.

On the other hand, the definition of h with d_s for two-reactant bimolecular reactions is as follows, instead of equation (8):

$$h = 1 - \frac{d_s}{4}. \tag{12}$$

When the reaction occurs under steady-state conditions, the *Nature* of dimension-restricted reactions is expressed in the anomalous reaction order X because the "steady state" is time independent by definition. The steady-state reaction rate can be described as follows:

$$\frac{d[A]}{dt} = K[A]^y[B]^z. \tag{13}$$

In the classic case, the overall reaction order X equals:

$$X = y + z = 2, \tag{14}$$

which is described by the sum of partial orders of $A(y)$ and $B(z)$. For the dimension-restricted reaction case,

$$X = 1 + \frac{4}{d_s} = 1 + (1-h)^{-1}. \tag{15}$$

When the A + B reaction occurs in one-dimensional space, it follows that:

$$\left(h = \frac{3}{4}\right), \tag{16}$$

which basically means that when the order is higher, the restriction is stronger.

To apply steady-state DRRK to an enzymatic reaction, we used the steady-state conditions defined by Briggs and Haldane (29) to analyze our experimental results. This approximation technique is the same as quasi-steady-state assumption (QSSA) (19). We can utilize the QSSA described by Segel (5) because, if the concentration of S is high enough, the free enzyme E will immediately combine with another molecule of S. Under these conditions, a steady state is achieved in which the enzyme is always saturated with its substrate. The details will be described later (Appendix A).

The fractal kinetics approach is applicable not just for a reaction in fractal environments, but in many other nonclassic simulations (14–18).

3. Planning the Experiments for the Model

One significant problem is determining what kind of data is required for the precise construction of the model. For application of DRRK, the required data is a time-course fluctuation of reaction species. To apply some *in vitro* experiments for this modeling analysis, we could calculate the required values from the data of well-established procedures (*see* the next section).

Recently, attractive techniques have been developed for observing reactions in living cells (30–33). These kinds of direct observations will promote analysis of molecular behaviors in the *in vivo* environment. One disadvantage of simple observation, however, is that it is limited, especially as a molecule moves in a complex anomalous environment like cytoplasm. It is not possible to determine the extent to which molecular movement is restricted by the environment without taking into account the reaction rate. To apply DRRK to these reactions and transports, and to determine the effect of each actual reaction space, some of the cases that could previously only be guessed at qualitatively can now be described quantitatively and precisely on the basis of experimental data.

3.1. Application for *In Vitro* Experimental Data

This section describes an example of how the DRRK method is applied to real data. The example is the case of *Eco*RV, which is analyzed in an *in vitro* manner, and it shows that it is easy to apply the method and solve this problem, which cannot be solved by a classic approach.

3.1.1. EcoRV

*Eco*RV is a restriction enzyme of *E. coli*, which has a mechanism for protecting bacteria from infection (34,35). The behavior of *Eco*RV is considered anomalous if the enzyme reaction occurs under completely ideal conditions. The behavior of *Eco*RV is anomalous in that the reaction rate with a longer substrate is faster than with a shorter substrate. If the reaction proceeds under ideal conditions, as in a simple Newtonian fluid, which is the premise of classic mass-action models, the reaction rate depends only on the concentration of the reacting species. However, the actual reaction of *Eco*RV does not depend only on the concentration of the species; as a result, under some conditions, the reaction rate is 5 times as fast when the substrate length is 20 times as long, even though the substrate concentrations are the same.

The behavior of *Eco*RV was defined with a stochastic model. Our stochastic model revealed the behavior of this enzyme to be dimension restricted. We then decided to apply DRRK to this reaction as a test case.

We try to represent this phenomenon by DRRK modeling, which cannot be represented by classic modeling methods.

3.1.2. Classic Mass-Action Model Cannot Represent the EcoRV Reaction Process

We constructed an ODE-based model for the *Eco*RV reaction process. We originally constructed a classic mass-action model that could be compared with a model that included DRRK. This model is shown schematically in Figure 4A. The characteristic feature of this model is in the "association step," in which the enzyme associates with its substrate DNA and searches for its target sequence.

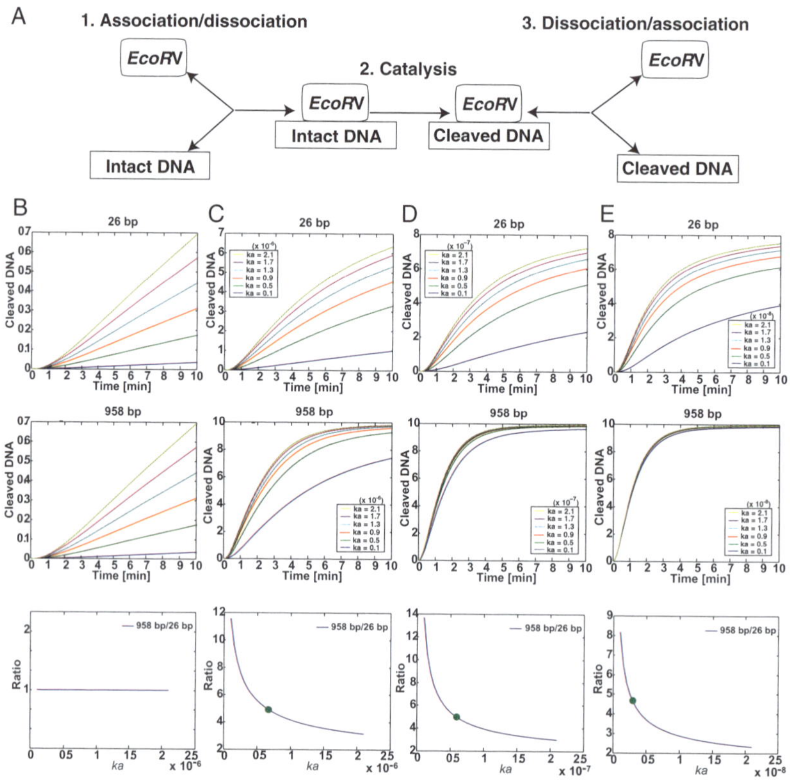

$y + z = 5$ [redacted] }5

Figure 4. Kinetic model for *Eco*RV movement. (A) The kinetic model for the *Eco*RV enzymatic reaction consists of a targeting process, a catalytic process, and a dissociation process. The targeting process includes both of the enzymes that associate with intact DNA to determine their target sequences. This model was described by CellDesigner (42) and analyzed with MATLAB. (B) The case of $y = 1$. Time course simulation of concentrations of cleaved 26- and 958-bp DNA (top and middle panels) and correlation between k_a (x axis) and cleavage reaction rate ratio, 958 bp/26 bp (y axis) (bottom panel). (c–e) The case of $y = 2$(C), $y = 3$(D), and $y = 4$(E). Time course simulation of concentrations of cleaved 26- and 958-bp DNA (top and middle panels) and correlation between the parameter k_a (x axis) and the cleavage reaction rate ratio 958 bp/26 bp (y axis) (bottom panel) Filled circles indicated the appropriate k_a value to reconstruct the reaction rate ratio (43–45). The actual reaction order is not a limited integer. It is a real number.

In the general fractal kinetics, dimension-restricted space for reaction has no limitation other than its dimension, which means that low-dimensional reaction space is infinitely sequential space, like infinitely sequential linear space. On the other hand, substrate DNA, which provides a pseudorestricted space for an enzyme, has the limitation that DNA has its length. In particular, the pseudorestricted space concretely affects the following elements. The ease with which an enzyme finds its target sequence depends on the length of DNA. On the other hand, it is also considered that the ease with which an enzyme can associate its substrates depends not only on the concentration of DNA but also on the length of DNA. Taking into account the effect of DNA length, the flux of the association step can be represented as follows:

$$k_a \times k_1([\text{Intact DNA}] \times \text{length})^y \times [Eco\text{RV}]^z. \tag{17}$$

Parameter k_a represents how likely the enzyme is to associate with DNA, parameter k_1 represents how likely the enzyme is to find its target sequence, "length" is the length of substrate DNA (base pairs), and [Intact DNA] and [*Eco*RV] represent the concentrations of the reaction species.

In the expression for the classic mass-action model, both the substrate order y and the enzyme order z are defined as 1. In this model, all of the steps except the association step are represented by the classic mass-action model to analyze pure DRRK effects for the result. The total model equations are indicated in Table 2, and the parameters are indicated in Table 3.

Table 2. Differential equations for *Eco*RV model

$$\frac{d[\text{Intact DNA}]}{dt} = -k_a \times k_1 \times ([\text{Intact DNA}] \times \text{length})^y \times [Eco\text{RV}]^z + k_{d1} \times [Eco\text{RV\_Intact DNA}]$$

$$\frac{d[Eco\text{RV}]}{dt} = -k_a \times k_1 \times ([\text{Intact DNA}] \times \text{length})^y \times [Eco\text{RV}]^z + k_{d1} \times [Eco\text{RV\_Intact DNA}] + k_{d2} \times [Eco\text{RV\_Cleaved DNA}] - k_a \times [Eco\text{RV}] \times [\text{Cleaved DNA}]$$

$$\frac{d[Eco\text{RV\_Intact DNA}]}{dt} = k_a \times k_1 \times ([\text{Intact DNA}] \times \text{length})^y \times [Eco\text{RV}]^z - k_{d1} \times [Eco\text{RV\_Intact DNA}] - k_c \times [Eco\text{RV\_Intact DNA}]$$

$$\frac{d[Eco\text{RV\_Cleaved DNA}]}{dt} = k_c \times [Eco\text{RV\_Intact DNA}] - k_{d2} \times [Eco\text{RV\_Cleaved DNA}] + k_a \times [Eco\text{RV}] \times [\text{Cleaved DNA}]$$

$$\frac{d[\text{Cleaved DNA}]}{dt} = k_{d2} \times [Eco\text{RV\_Cleaved DNA}] - k_a \times [Eco\text{RV}] \times [\text{Cleaved DNA}]$$

Table 3. Kinetic model rate constants.

k_a	Parameter for association reaction rate
k_l	1/(length of substrate DNA)
Length	Each substrate DNA length
k_{d1}	1.0×10^{-5}
K_{d2}	1.0
k_c	4.16

By this analysis, the classic mass-action model cannot reconstruct the different reaction rate with different-length substrates (Figure 4B), as suggested by other studies (6,8). You may also find that parameter searching could not solve the problem (Figure 4B).

The association step is the dimension-restriction step in this reaction process, so we next applied DRRK to the association step of our model and analyzed the effect on the results.

3.1.3. DRRK Successfully Represents the Reaction Process

To reconstruct the different reaction rates with different-length substrates of *Eco*RV, we applied DRRK to our model. This DRRK-inclusive model successfully represented the differences in reaction rate with different-length substrates (Figure 4C). We applied DRRK in the association step in this case. The other part of this model was similar to the classic mass-action model. In this case, we defined the sum of substrate order y and enzyme order z as equal to 5 (18). The appropriate parameter was searched for k_a to reconstruct the different reaction rates with different length substrates. Simulation results with $y = 3$ and $z = 2$ are shown in Figure 4C. Actually, each value of y and z is not limited in integer. The right panel of Figure 4C shows that we found an appropriate parameter value to reconstruct the different reaction rates with different-length substrates (Figure 4C, right panel).

By using DRRK, we succeeded in reconstructing a phenomenon that could not be reconstructed by a classic mass-action model. We thus accomplished the first task of the numerical model analysis, which is reconstructing the target phenomenon by modeling.

This model can be evaluated experimentally. We estimated the reaction order by means of the experiments discussed in the following sections.

As previously described, the DRRK description expresses the dimension restriction of reaction space by the orders of the species concentration. The orders of species concentration are real numbers; they are not limited to integers. The answers exist in the ranges of both $y > 1$ and $y + z = 5$. It is difficult to define these more specifically by model analysis alone, but a benefit of this model is that the unclear points can be quantitatively defined by experiments. Therefore, as our next step, we introduce experiments to define the reaction order.

3.1.4. Reaction Order Estimation by Experimental Results

To estimate reaction orders, the required data is that the detailed sequential fluctuation of species concentrations analyzed by a time-course experiment (Figure 5). It allows estimating the reaction order directly

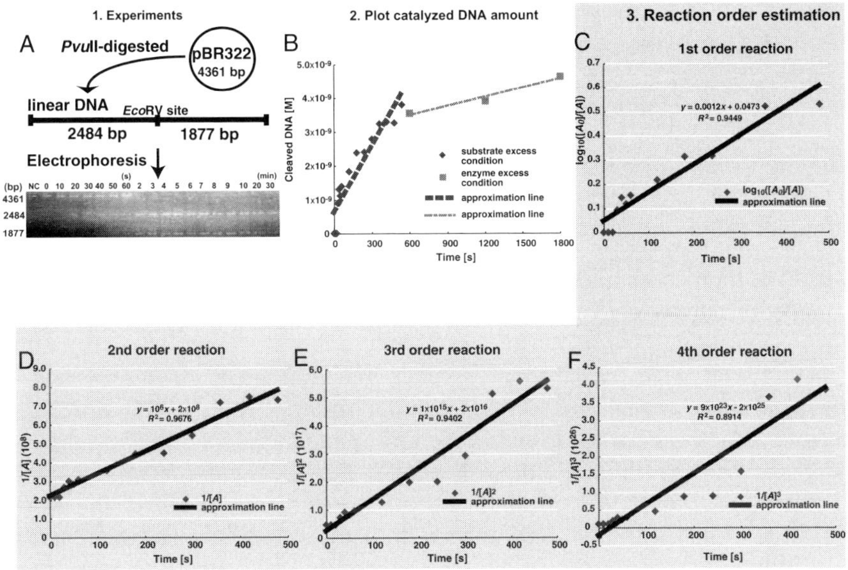

Figure 5. Kinetic order estimation of *Eco*RV reaction. (A) Map of linear pBR322 and photograph of 5% Acryl Amide electrophoretic gel digested with pBR322 for 8 h under 50 mA. The photo was taken after the gel was stained with 100 µg/ml of ethidium bromide in UV illuminator. (B) Time-course result of *Eco*RV–linear pBR322 reaction: *x* axis, time; *y* axis, concentration of catalyzed DNA; blue line, results with excess substrate; red line, results with excess enzyme. Results with excess substrate were used for calculation of reaction orders. Approximations of first-order (C), second-order (D), third-order (E), and fourth-order (F) reactions with time course results (43).

and calculated the rate constant (K) for comparison with the model parameter (k_a) to determine the reaction order in more detail. For the reaction order estimation, we used the data that fulfilled the steady-state conditions (Figure 5B, blue line; Appendix A). In this example, the data also met the condition that the substrate concentration was greater than the enzyme concentration ([S] >> [E]). After time 0, this reaction fulfils the steady-state condition defined by Briggs and Haldane in 1925 (29) (Appendix A) in this range. The reaction order was estimated with these data by plotting the integral value of the reaction rate for intermediate species versus time in each from first to fourth order. If this plot is linear and the intercept value is near the estimated value, the reaction order is suggested as the order. The results of these reaction order estimates are shown (Figure 5C–F). From the simulation results, it is not appropriate to consider this reaction to be a first-order reaction. On the other hand, neither the linearity nor the intercept value offered any evidence to determine whether the reaction is second or third order. For the fourth-order analysis, the intercept value was not appropriate, so this reaction may not be fourth order. These results indicated this reaction is second to third order. We next estimated the reaction order to be nearer to second or third order by the reaction rate constants from the experimental results.

Table 4. Estimated k_a value for each reaction order with experimental conditions.

y	k_a	K
2	3.0×10^{-4}	0.36
3	2.1×10^{-8}	0.36
4	9.0×10^{-13}	0.36

We then estimated each case of k_a in our model with the experimental rate constant K (Table 4). The value of K for each reaction order was calculated with our model (Table 5). Values of K and k_a were compared with each other for each reaction order. Tables 4 and 5 show that, in the case of the reaction orders 2 and 4, the simulated and experimental k_a values differ by a factor of 10^3 to 10^4, and the K values differ by a factor of 1/30 to 2. For the k_a and K values of the third-order reaction, simulation and experiment indicated the same value, or, at the most, a 20% difference. These results showed the reaction to be nearly third order. These experimental results were in agreement with the model analysis and revealed new findings that could not be clarified by either modeling or experiments. We considered whether these estimates were compatible with the premises of the basic theory behind our model, considering the mean of the reaction order in the DRRK model.

3.1.5. Back to the Simulation Results to Check the Compatibility with the Experimental Results

Our model was constructed based on the assumption that the movement of the enzyme is restricted in low-dimension space during the association step. Our experiments indicated that the substrate order y is near 3, which means that the enzyme order z is near 2. However, our simulation had already indicated that the reaction order y could not be 1 to reconstruct the phenomena of different reaction rates with different-length substrates; in other words, $1 < y < 5$ and $0 < z < 4$. These results indicate that the results from experimental results are compatible with the results of our simulation. At the same time, the actual reaction process occurs in one- to two-dimensional restricted reaction spaces, as expected and defined as the premise of our model. Hence, our model is appropriate for representing of this reaction.

Table 5. Estimated K value for each reaction order with the DRRK model.

y	k_a	K
2	7.5×10^{-7}	0.012
3	6.5×10^{-8}	0.469
4	3.0×10^{-9}	0.644

*[1] For aqueous buffers $\eta = \times 10^{-3}$ Pa s.

*[2] k_B = 1.380658 × 10^{-23} J/K: Boltzmann constant, gas constant R (= 8.314510 J/mol·K) divided by Avogadro's number ($N_A = 6.02 \times 10^{23}$).

Thus, we were able to achieve two things that had not been done together before this DRRK study: to represent a reaction under dimension-restricted space by a numerical model and to experimentally evaluate the numerical model.

3.2. Application for *In Vivo* Experimental Data

The conditions for the applying DRRK are a simple process, ATP-independent diffusion, and restricted reaction space. These conditions are applicable for many cases of intracellular diffusion of molecules.

For example, transportation of the Stat1 transcriptional factor from the cell membrane to the nuclei fulfills these conditions (33). As stated above, this transportation process is ATP-independent. At the same time, the diffusion of active protein kinase C (PKC) stacked onto a cell membrane is slower than Stat1 transportation through the cytoplasm. This phenomenon cannot be explained if one assumes that the cytoplasm is a free three-dimensional space, that is, without taking into account the crowding of the cytoplasm. Observed phenomena have suggested that the environment of the cytoplasm restricts molecular movement more than does two-dimensional space, such as the cell membrane. The DRRK method permits the diffusion dimension to be estimated by the rate of movement or reaction. Information about the dimension of diffusion may be useful for the estimation of diffusion constant in a PDE model analysis.

Many other phenomena fulfill the conditions for applying of DRRK (36–38). To develop an *in vivo*-oriented modeling, it will be important to take into account these experimental data. We suggest a candidate to which the DRRK technique could be applied, that is, the data of fluorescence recovery after photobleaching (FRAP) analysis.

To date, many of the studies that analyzed the movement of target molecules by FRAP also calculated the diffusion coefficient of the molecular movement (31,32). However, sometimes the predicted transportation time, which is calculated from the diffusion coefficient, was not compatible with their experimental data. It is conjectured that the incorrectness arises from the assumption of the classic method of calculating diffusion coefficient (39). The original calculation strategy is only valid when the fluorescence recovery arises from 2D diffusion. However, there is no guarantee that the diffusion dimension is 2D, except for the molecules attached to the membrane.

It has been suggested that DRRK can provide the diffusion dimension by calculating d_s from the FRAP data modified for this kind of data analysis. Experimental results suggested that the curve of fluorescence recovery of anomalous (or complex) diffusion has the features of the reaction process following the fractal kinetics (40) (Figure 6).

After deciding the diffusion dimension by experimental data and constructing a model that is compatible with it, we could represent the movement more precisely.

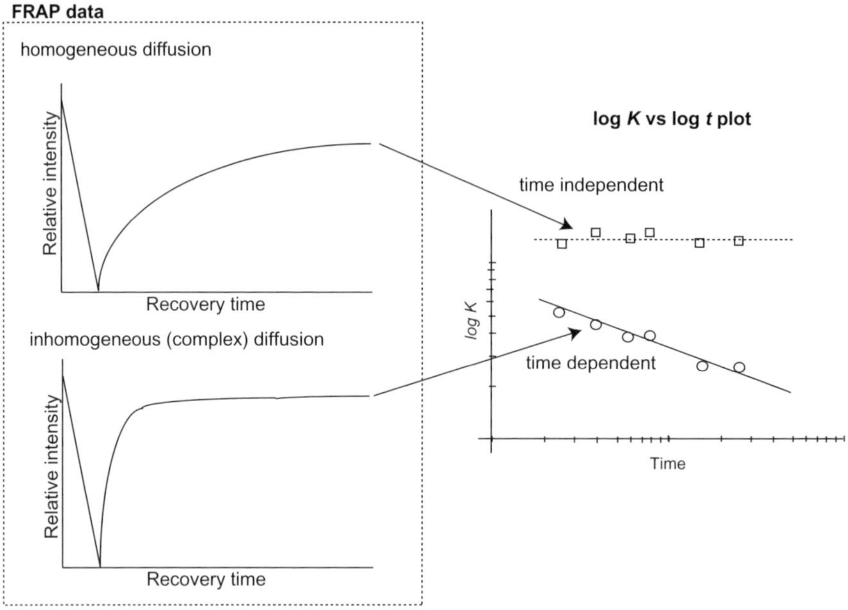

Figure 6. DRRK application for FRAP data. The recovery curve of the diffusion process in free 3D space is suggested as a time-independent process. Some inhomogeneous, complex diffusion processes were observed, and these kinds of diffusion processes are suggested as time-dependent phenomena. We may distinguish these two kinds of processes definitely, and moreover, calculate their concrete reaction order by applying DRRK (14–18,40).

4. Calculation Cost: Another Benefit of the DRRK Model

The calculation costs of *in vivo*–oriented modeling can be huge because of the highly complex network structures. This tendency is shown more clearly when we try to use stochastic and PDE modeling.

The DRRK model is an ODE-based model, and the calculation costs of this kind of model are lower than those of stochastic and PDE models. However, classic ODE-based models cannot represent dimension-restricted reactions, which are suggested to be common in *in vivo* environments. The DRRK model overcomes the problem of the classic model in this regard, and application of the DRRK model to large-scale networks is expected to better reflect the reaction environment in the model.

5. Concluding Remarks

The first step of model analysis is reconstruction of experimentally observed phenomena. The next step is evaluating the model by means of other experiments.

Classic ODE-based models cannot represent dimension-restricted reactions. The probabilities in the stochastic model are difficult to define using experiments. Diffusion constants in the PDE model, which are calculated after researchers have set up the diffusion dimension, do not work when evaluating the model (31,32).

DRRK successfully represents such kinds of phenomena and can be evaluated quantitatively with experimental data.

DRRK could be useful for the precise analysis to know the dynamics of intracellular molecules, not only in the DRRK model itself. The reaction rate constants could be useful for evaluating the model by experiments using models other than DRRK. DRRK will be particularly valuable for PDE modeling after the dimensions of the diffusion have been defined in some other way (41) (Appendix B).

It is difficult to analyze the dynamic character of a network that only has information about the interaction of each node; sometimes the nodes are proteins. It is also difficult to know when and in which order the actual reaction species work in the cell from the information of the relative expression data of messenger RNA. The basic requirements needed to describe the dynamics of a reaction network are the concrete data of "when," "which kind of reaction" occurs, and "how much" of the reactants themselves.

Acknowledgments: We are grateful to Prof. Raoul Kopelman (University of Michigan, USA), Wataru Ohyama (Mie University, Japan), Prof. Stephan Halford (Bristol University, UK), Dr. Tetsuya Kobayashi (Institute of Physical and Chemical Research, Japan), Mineo Morohashi (Human Metabolome, Japan), Dr. Marcus Krantz (Kitano Proj., Japan), Dr. Douglas Murray (Kitano Proj., Japan), and Yukiko Matsuoka (Kitano Proj., Japan) for their kind help and advice during our research. We are very grateful to Prof. Sangdun Choi (Caltech, USA) and Prof. Hiroaki Kitano (Kitano Proj., Japan) for giving us the opportunity to contribute to this book.

This study was supported by Special Coordination Funds for Promoting Science and Technology from the Ministry of Education, Culture, Sports, Science and Technology of Japan, and by Keio University's 21st Century COE Program for Understanding and Control of Life Function via Systems Biology.

Appendix A: Steady-State Conditions and Reaction Order Estimation

For the steady-state conditions for the calculation of the powers, assumption 1 is that enzyme concentration is a constant, and assumption 2 is that k_2 is sufficiently larger than k_1.

In this reaction scheme,

$$[S]+[E]\underset{k_{-1}}{\overset{k_1}{\leftrightarrow}}[SE]\overset{k_2}{\rightarrow}[P], \tag{A1}$$

where $[S]$ is the concentration of substrate, $[E]$ is the concentration of enzyme, $[SE]$ is the concentration of intermediate reactant, and $[P]$ is the concentration of product.

By taking these assumptions into account, the time-dependent fluctuation of the intermediate reactant, *SE*, can be treated the same as the fluctuation of the product. Then, for the case of first-order reaction,

$$\frac{d[SE]}{dt} = \frac{d[P]}{dt} = -\frac{d[S]}{dt} \quad \text{(A2)}$$

This equation is integrated, and the integral value is plotted against time. The integral value is

$$-\frac{\ln[S]}{[S]_0} = -kt, \quad \text{(A3)}$$

where $[S]_0$ is the concentration of *S* at time 0.

When the reaction is a first-order reaction, the plot of this value against time will be linear and its *y* intercept will be 0.

For the case of an *n*th-order reaction,

$$-\frac{d[S]}{dt} = k[S]^n, \quad \text{(A4)}$$

and the integral value is given by

$$\frac{1}{(n-1)} \times \left(\frac{1}{[S]^{n-1}} - \frac{1}{[S]_0^{n-1}} \right) = kt \quad (n \neq 1) \quad \text{(A5)}$$

When the reaction is an *n*th-order reaction, the plot of this value against time will be linear, and the intercept will be

$$\frac{1}{[S]_0^{n-1}}. \quad \text{(A6)}$$

Appendix B: The Relation Between Rate Constant and Diffusion Coefficient

Suppose a protein, which has a diameter d_1, moves through buffer by Brownian motion (41), and that a target molecule of diameter d_2 is present at concentration *c*. When *c* is in units of number of molecules per volume, then the volume per target is

$$V = \frac{1}{c}$$

In this environment, we can expect the two species to be at a distance of roughly

$$r = \frac{1}{c^{1/3}}.$$

After a time of r^2/D, where *D* is a diffusion coefficient, a species will bind a nearby target or it will have diffused away from one target. In the latter case, the binding probability will be pushed up by each r/d_2 time. This gives a total association time of

$$\frac{r}{d_2} \times \frac{r^2}{D} = \frac{r^3}{Dd_2} = \frac{1}{Dd_2c}.$$

The association rate is the inverse of the total association time, Dd_2c. The rate constants per unit concentration is $K = \alpha Dd_2$ (α is a constant for each case). K can be represented based on Smoluchowski's calculation (3),

$$k = 4\pi Dd_2 = \frac{4k_BTd_2}{3\eta d_1}.$$

This rate constant can be estimated as $k \approx 10^8$/M/s. This value indicates that a binary reaction cannot occur at a higher rate than this if the reactants are brought together by unguided 3D diffusion.

The above estimations are for a 3D case. Now let us suppose the case that the diffusion dimension is limited. When a molecule diffuses in a 1D manner and the diffusion distance is called l, then

$$V = l = \frac{1}{c} \text{ and } r \le l.$$

This shows that the total association time can be represented by the one-dimensional case of diffusion coefficient D,

$$\frac{r}{d_2} \times \frac{l}{D} \le \frac{l^2}{Dd_2}.$$

The association rate is larger than the inverse of the total association time, Dd_2c^2. c is a positive real number, and, naturally, the association rate in 1D environment is faster than in a 3D environment.

References

1. Fulton AB. How crowded is the cytoplasm? *Cell* 1982;30:345–347.
2. Medalia O. et al. Macromolecular architecture in eukaryotic cells visualized by cryoelectron tomography. *Science* 2002;298:1209–1213.
3. Smoluchowski MV. Versuch einer mathematischen Theorie der Koagulationskinetik koloider Losungen. *Phys Chem* 1917;92:129–168.
4. Michaelis L, Menten LM. Die Kinetik der Invertinwirkung. *Biochem Z* 1913;49:333–369.
5. Segel IH. Enzyme kinetics. Behavior and analysis of rapid equilibrium and steady-state enzyme systems. New York: John Wiley & Sons, Inc; 1993.
6. Berg OG, Winter RB, von Hippel PH. Diffusion-driven mechanisms of protein translocation on nucleic acids. 1. Models and theory. *Biochemistry* 1981;24:6929–6948.
7. Jelstch A, Pingoud A. Kinetic characterization of linear diffusion of the restriction endonuclease EcoRV on DNA. *Biochemistry* 1998;37:2160–2169.
8. Stanford NP, Szczelkun MD, Marko JF, et al. One- and three-dimensional pathways for proteins to reach specific DNA sites. *EMBO J* 2000;1:6546–6557.
9. Gowers DM, Halford SE. Protein motion from non-specific to specific DNA by three-dimensional routes aided by supercoiling. *EMBO J* 2003;17:1410–1418.
10. Jeltsch A, Urbanke C. Sliding or hopping? How restriction enzymes find their way on DNA. In: Pngoud A, ed. Nucleic Acids and Molecular Biology: Restriction Endonuclease, vol. 14. Heidelberg: Springer-Verlag; 2004:95–110.
11. Einstein A. Investigations on the Theory of the Brownian Movement. New York: Dutton; 1926.
12. von Smoluchowski M. Drei Vorträgeüber Diffusion, Brownische Molekularbewegung und Koagulation von Kolloidteilchen. *Phys Z* 1916;17:557–571, 585–599.

13. Frish HI, Hammersly JM. Percolation process and related topics. *J Soc Indust Appl Math* 1963;11:894–918.
14. Kopelman R, Argyrakis P. Diffusive and percolative lattice migration: Excitons. *J Chem Phys* 1980;72:3053–3060.
15. Klymko PW, Kopelman R. Fractal reaction kinetics: exciton fusion on clusters. *J Phys Chem* 1983;87:4565–4567.
16. Kopelman R, Klymko PW, Newhouse JS, et al. Reaction kinetics on fractals: random-walker simulations and exciton experiments. *Phys Rev B* 1984;29:3747–3748.
17. Kopelman R. Fractal reaction kinetics. *Science* 1988;241:1620–1626.
18. Kopelman R. Exciton microscopy and reaction kinetics in restricted spaces. In: Glass WA, Varma MN, ed. Physical and Chemical Mechanisms in Molecular Radiation Biology. New York: Plenum Press; 1991:475–502.
19. Turner TE. *Stochastic and deterministic approaches to modelling in vivo biochemical kinetics* [masters thesis]. Trinity, England: University of Oxford; 2003.
20. Savageau MA. Influence of fractal kinetics on molecular recognition. *J Mol Recognit* 1993;4:149–157.
21. Savageau MA. Michaelis-Menten mechanism reconsidered: implications of fractal kinetics. *J Theor Biol* 1995;176:115–124.
22. Savageau MA. Development of fractal kinetic theory for enzyme-catalysed reactions and implications for the design of biochemical pathways. *Biosystems* 1998;47:9–36.
23. Schnell S, Turner TE. Reaction kinetics in intracellular environments with macromolecular crowding: simulations and rate laws. *Biophys Mol Biol* 2004;85:235–260.
24. Vlad MO, Popa VT, Segal E, et al. Multiple rate-determining steps for nonideal and fractal kinetics. *J Phys Chem* 2005;109:2455–2460.
25. Duran J, Pelle F, Portella MT. Fractal kinetics of multiparticle diffusion. *J Phys C: Solid State Phys* 1986;19:6185–6194.
26. Berry H. Monte Carlo simulations of enzyme reactions in two dimensions: fractal kinetics and spatial segregation. *Biophys J* 2002;83:1891–1901.
27. Aon MA, Cortassa S. On the fractal nature of cytoplasm. *FEBS Lett* 1994;344:1–4.
28. Aon MA, O'Rourke B, Cortassa S. The fractal architecture of cytoplasmic organization: scaling, kinetics and emergence in metabolic network. *Mol Cell Biochem* 2004;256/257:169–184.
29 Briggs GE, Haldane JBS. A note on the kinetics of enzyme action. *Biochem J* 1925;19:338–339.
30. Kulkarni RP, Wu DD, Davis ME, Fraser SE. Quantitating intracellular transport of polyplexes by spatio-temporal image correlation spectroscopy. *Proc Natl Acad Sci USA* 2005;102:7523–7528.
31. Burack WR, Shaw AS. Live cell imaging of ERK and MEK. *J Biol Chem* 2005;280:3832–3837.
32. Phair RD, Misteli T. High mobility of proteins in the mammalian cell nucleus. *Nature* 2000;404:604–609.
33. Lillemeier BF, Köster M, Kerr IM. STAT1 from the cell membrane to the DNA. *EMBO J* 2001;20:2508–2517.
34. Kabata H, Okada W Washizu M. Single-molecule dynamics of the *Eco*RI enzyme using stretched DNA: its application to in situ sliding assay and optical DNA mapping. *Jpn J Appl Phys* 2000;39:7164–7171.
35. Seidel R, van Noort J, van der Scheer C, et al. Real-time observation of DNA translocation by the type I restriction modification enzyme *Eco*R124I. *Nat Struct Biol* 2004;11:838–843.

36. Solovjeva L, Svetlova M, Stein G, et al. Conformation of replicated segments of chromosome fibers in human S-phase nucleus. *Chromosome Res* 1998;6:595–602.
37. Maly IV, Vorobjev IA. Centrosome-dependent anisotropic random walk of cytoplasmic vesicles. *Cell Biol Int* 2002;26:791–799.
38. Orci L, Ravazzola M, Volchuk A, et al. Anterograde flow of cargo across the Golgi stack potentially mediated via bidirectional "percolating" COPI vesicles. *Proc Natl Acad Sci USA* 2000;97:10400–10405.
39. Axelrod D, Koppel DE, Schlessinger J, et al. Mobility measurement by analysis of fluorescence photobleaching recovery kinetics. *Biophys J* 1976; 16:1055–1069.
40. Verkman AS. Solute and macromolecule diffusion in cellular aqueous compartments. *Trends Biochem Sci* 2002;27:27–33
41. Halford SE, Marco JF. How do site-specific DNA-binding proteins find their targets? *Nucleic Acid Res* 2004;32:3040–3052.
42. Kitano H, Funahashi A, Matsuoka Y, Oda K. The process diagram for graphical representation of biological networks. *Nat Biotechnol* 2005;23:961–966.
43. Hiroi N, Funahashi A, Kitano H. Kinetics for dimension restricted reactions. Submitted;2005.
44. Hiroi N, Funahashi A, Kitano H. Two numerical model analysis for the movement of a restriction enzyme. Foundations of Systems Biology in Engineering (FOSBE 2005). Santa Barbara, CA, USA. August 2005.
45. Hiroi N, Funahashi A, Kitano H. Analysis for dimension restriction kinetics with bacterial endonuclease movement. The 2005 WSEAS International Conference on Cellular and Molecular Biology—Biophysics and Bioengineering. Athens, Greece. July 2005.

15

Mechanisms Generating Ultrasensitivity, Bistability, and Oscillations in Signal Transduction

Nils Blüthgen, Stefan Legewie, Hanspeter Herzel, and Boris Kholodenko

Summary

Stimulus-response curves of signal transduction cascades are often nonlinear; take, for example, sigmoidal curves. Such sigmoidal curves are frequently termed ultrasensitive, as small alterations in the stimulus can elicit large changes in the response. This chapter shall review the importance of ultrasensitivity in signal transduction, with a focus on the activation of the mitogen-activated protein kinase (MAPK) cascade. The major mechanisms that generate ultrasensitivity (Figure 1) are introduced. In particular, zero-order kinetics and multisite phosphorylation are discussed.

Ultrasensitive signaling cascades equipped with positive or negative feedback loops may exhibit complex dynamic behavior. The large body of theory for effects of feedbacks shall be reviewed in this chapter. It is discussed that bistability can emerge from ultrasensitivity in conjunction with positive feedback, whereas adaptation, oscillations, and, surprisingly, highly linear response can arise with negative feedback.

Key Words: Mathematical modeling; control theory; dynamics; bifurcation analysis; signal transduction cascades; zero-order ultrasensitivity; mitogen-activated protein kinase (MAPK) cascade.

1. Introduction

Intracellular signal processing in higher eukaryotes is carried out by signaling networks composed of enzymes that control each other's activities by covalent modification. Signals at the cellular membrane ripple through these signaling networks by covalent modification events to reach various locations in the cell and, ultimately, cause cellular responses. The biochemical building blocks of these networks are so-called covalent modification cycles, where couples of opposing enzymes (e.g., a kinase and a phosphatase) activate and deactivate a target substrate by covalently modifying it at a single or at multiple sites (Figure 2).

The steady-state stimulus-response curves of covalent modification cycles often display strong sigmoidality *in vivo*, as shown, for example,

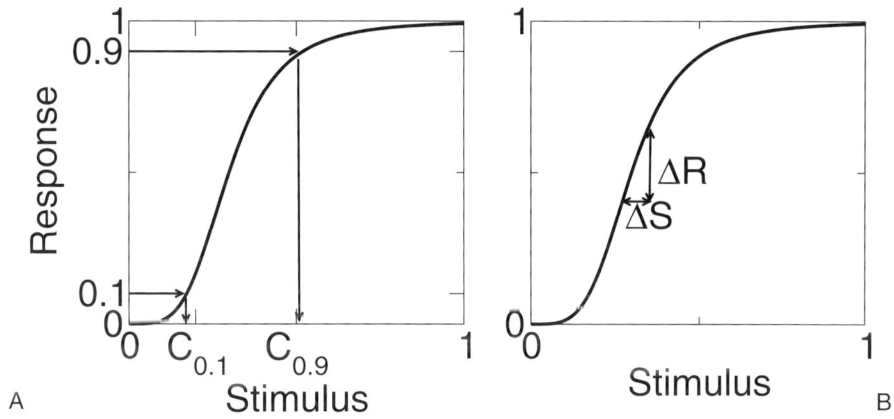

Figure 1. Methods to quantify ultrasensitivity. (A) The Hill coefficient is often estimated by the stimuli needed for getting 10% and 90% of the maximal activation, $C_{0.1}$ and $C_{0.9}$, respectively. (B) In contrast, response coefficients are defined locally, i.e., for a given stimulus, and evaluate the relative change in response upon a relative change in stimulus $\left(R_S^R = \frac{S}{R}\frac{\Delta R}{\Delta S}\right)$.

for the activation of Sic1 (1), a cyclin-dependent kinase inhibitor that controls entry to S phase of the cell cycle. Sigmoidality of the stimulus-response curves has been termed ultrasensitivity, to capture the highly sensitive nature of those systems to changes in signals near a threshold stimulus. Additionally, ultrasensitive behavior was reported in various

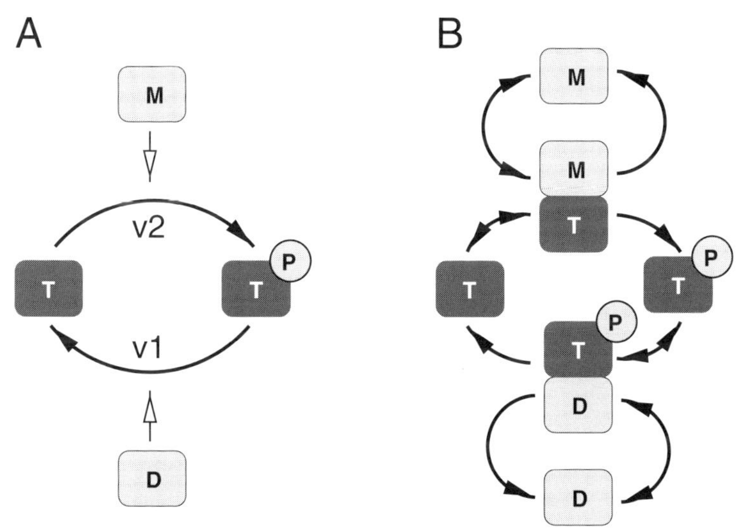

Figure 2. Sketch of a simple covalent modification cycle. (A) The modifier protein M catalyzes the modification of the target T, and the demodifying protein D removes the modification from T^P. The reaction rates are called v_1 and v_2. (B) A mechanistic scheme of the covalent modification cycle, which takes enzyme sequestration into account. The enzymes M and D bind their targets T and T^P reversibly, and release their product irreversibly (irreversible Michaelis–Menten mechanism).

in vitro experiments, e.g., in the phosphorylation of isocitrate dehydrogenase (2), in muscle glycolysis (3), in calmodulin-dependent protein kinase II (CaMKII) activation (4), and in the activation of the MAPK cascade (5). Subthreshold stimuli are damped, whereas superthreshold stimuli are transmitted, which allows for virtually binary decisions (6). Ultrasensitivity can help to filter out noise (7) or can delay responses (8). Mechanisms that lead to ultrasensitive stimulus-response curves include cooperativity, multisite phosphorylation, feed-forward loops, and enzymes operating under saturation. The latter mechanism has been termed zero-order ultrasensitivity because a necessary condition is that the opposing enzymes of a covalent modification cycle display zero-order kinetics. Moreover, oscillations can be observed if an ultrasensitive cascade possesses a negative feedback (9). Bistability (hysteresis) can occur if such an ultrasensitive cascade is equipped with a positive feedback, as has recently been observed in eukaryotic signaling pathways (10). Surprisingly, ultrasensitivity can also lead to highly linear signal transduction in the presence of high load, such as translocation to the nucleus (11).

This chapter shall review the theoretical concepts that have been brought up to understand the appearance of ultrasensitivity, bistability, and oscillations. It introduces methods to quantify ultrasensitivity, explains the means by which ultrasensitivity is generated, and demonstrates how bistability and oscillations arise because of ultrasensitivity and feedback loops.

2. Quantification of Ultrasensitivity

2.1. Hill Coefficient

The first characterization of ultrasensitivity was introduced by Hill as an empirical description of the cooperative binding of oxygen to hemoglobin (12). Hill found that binding was well described by the following relationship, which is now known as the Hill equation:

$$y = \frac{x^h}{K_{0.5}^h + x^h}, \qquad (1)$$

where y is the bound fraction of oxygen, x is the oxygen pressure, $K_{0.5}$ is the oxygen pressure where half of the binding sites are occupied, and h is the Hill coefficient.[1] Enzymes exhibiting positive cooperativity, such as hemoglobin, display ultrasensitivity; that is, their Hill coefficients exceed unity. For example, the Hill coefficient of hemoglobin equals 2.8, and in general, a Hill coefficient of 4 is often thought to be an upper limit for cooperative enzymes (13).

The Hill coefficient is not commonly estimated by fitting the formula to the data, but it is often calculated from the cooperativity index, which is defined as (14):

[1] Originally, Hill used the constant differently: $y = 100\frac{Kx^n}{1+Kx^n}$, with y being in per cent. Thus $K = 1/(K_{0.5})^{1/h}$.

$$R_a = \frac{C_{0.9}}{C_{0.1}}, \tag{2}$$

where $C_{0.9}$ is the stimulus-value generating 90% of the response, and $C_{0.1}$ is the value for 10% of the maximum response. To get the relationship between R_a and the Hill coefficient, h, the following equations need to be solved:

$$C_{0.9} = 9^{1/h} K_{0.5}, C_{0.1} = K_{0.5}/9^{1/h}. \tag{3}$$

Solving for h yields:

$$R_a = \frac{C_{0.9}}{C_{0.1}} = 81^{1/h} \rightarrow h = \frac{\log 81}{\log(C_{0.9}/C_{0.1})}. \tag{4}$$

If the Hill coefficient is 1, an 81-fold increase of the stimulus is needed to increase activation from 10% to 90%. For Hill coefficients higher than 1, the increase of the stimulus needed is smaller and the Hill curve gets sigmoidal.

2.2. Metabolic Control Analysis

Whereas the Hill coefficient quantifies the ultrasensitivity of a stimulus-response curve globally, i.e., over a range of stimuli, there are methods to evaluate ultrasensitivity locally, i.e., for small changes around a certain stimulus. The most commonly used method is metabolic control analysis (MCA) (15–17). Although originally developed to study the control of metabolism, it has been successfully extended to intracellular signal transduction (18–20). MCA is a mathematical framework that connects control properties of the system, e.g., the catalytic activity of an enzyme to a flux or a concentration.

The response of the entire system upon small perturbations in parameters (such as rate constants or total concentrations) is described by so-called response coefficients and defined by the following:

$$R_{p_j}^{S_i} = \frac{p_j}{[S_i]} \frac{\mathrm{d}[S_i]}{\mathrm{d}p_j}. \tag{5}$$

Here, $R_{p_j}^{S_i}$ equals the relative change in the steady-state concentration $[S_i]$ brought about by an infinitesimal relative change in the parameter p_j. Response coefficients higher than 1 correspond to (locally) ultrasensitive systems, which exhibit relative amplification, whereas $R < 1$ implies relative damping and a subsensitive response.

Local kinetic details are captured by elasticities and are defined by the following:

$$\varepsilon_{x_i}^{v_j} = \frac{[x_i]}{v_j} \frac{\partial v_j}{\partial [x_i]}. \tag{6}$$

Elasticities evaluate the relative change in a reaction velocity as a result of an infinitesimal relative change in one of its substrate, product, or effector concentrations (e.g., $[x_i]$). To determine this coefficient, the enzyme is conceptually considered in isolation from the system, and only a single metabolite is perturbed. The elasticities of an enzyme E_i, after irreversible Michaelis–Menten kinetics with the Michaelis–Menten constant K_M, are as follows:

$$\varepsilon_{E_i}^{v_j} = 1, \tag{7}$$

in respect to the enzyme concentration, and

$$\varepsilon_S^{v_j} = \frac{K_M}{[S]+K_M}, \tag{8}$$

for the substrate S.

MCA was designed for the description of steady-state behavior and it is used accordingly in this chapter. However, it was also extended toward transient phenomena and oscillations (19,21,22).

For a Hill function (Eq. 1), the response coefficient reads:

$$R_x^y = h\left(1-\frac{x^h}{K_{0.5}^h+x^h}\right). \tag{9}$$

Thus, the response coefficient R_x^y equals the Hill coefficient h for low stimulation, but decreases to 0 for higher stimuli. Based on this equation, Legewie et al. (23) proposed another method for the quantitative analysis of ultrasensitive systems, which also applies to responses that strongly deviate from the Hill equation.

3. Mechanisms

3.1. Zero-Order Ultrasensitivity

Reversible covalent modification of proteins appears to be a universal regulatory motif in eukaryotic cells, controlling nearly all aspects of cellular life (24). In a series of articles at the beginning of the 1980s, Goldbeter and Koshland have shown that a simple cycle of two opposing enzymes that covalently modify a target protein (Figure 2) can result in highly ultrasensitive responses (14,25). They named this effect zero-order ultrasensitivity because ultrasensitivity in this simple system requires strong saturation of the modifying enzymes, which implies that the reaction velocity is nearly independent of the substrate concentration (zero-order kinetics). The dynamics of the covalently modified form can be described by one ordinary differential equation for the activated target protein (Figure 2):

$$\frac{\mathrm{d}[T^P]}{\mathrm{d}t} = v_1 - v_2, \tag{10}$$

where the reaction rates v_1 and v_2 may follow Michaelis–Menten kinetics:

$$\text{with } v_1 = \frac{V_{\max,1}[M][T]}{[T]+K_{M1}} \text{ and } v_2 = \frac{V_{\max,2}[D][T^P]}{[T^P]+K_{M2}}. \tag{11}$$

Here, $[T]$ and $[T^P]$ are the concentrations of the inactive and active form of the substrate, respectively, $[M]$ and $[D]$ are the concentrations of the activating and deactivating enzymes. If the enzyme–substrate complexes are negligible, the mass conservation reads

$$[T] + [T^P] = [T_{tot}]. \tag{12}$$

Thus, one can express $[T]$ in terms of the total substrate and the active form by $[T] = [T_{tot}] - [T^P]$. In this case, the equation for v_1 can be written as follows:

$$v_1 = \frac{V_{\max,1}[M]([T_{tot}] - [T^P])}{([T_{tot}] - [T^P]) + K_{M1}}. \tag{13}$$

The system is in a steady state if the forward reaction rate v_1 equals the backward reaction rate v_2. This condition can be evaluated graphically, as shown in Figure 3, A and C, where plots of v_1 and v_2 as a function of $[T^P]$ are displayed for two different values of substrate concentrations. In Figure 3A, the substrate concentration is low compared to the K_M values of kinase and phosphatase, whereas the opposite situation is shown in Figure 3C. Consequently, the reaction rates show no saturation in Figure 3A, whereas they are saturated in Figure 3C. The reaction rate v_1 depends linearly on the stimulus, i.e., the modifier concentration $[M]$. These plots illustrate how a slight change in the stimulus $[M]$ can either result in a moderate change in response (Figure 3B) or in a drastic change (Figure 3D).

In terms of a response coefficient, zero-order ultrasensitivity can be expressed by (11):

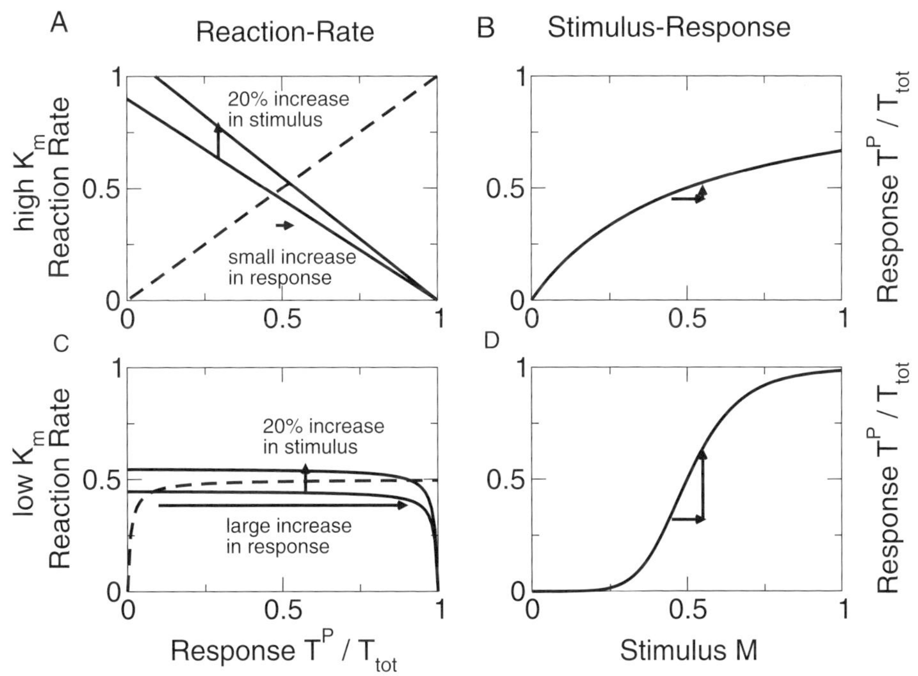

Figure 3. The principle of the Goldbeter–Koshland switch. (A and C) Solid lines represent the modification rate for two different stimuli; dashed lines show the demodification rate. (B and D) Stimulus-response curves. If the enzymes operate under first-order kinetics (A), the stimulus-response curve looks like a Michaelis–Menten curve (B), as opposed to enzymes, which are saturated by the substrate (C), where one can observe ultrasensitivity (D).

$$R_{M_T}^{T^P} = \frac{[T]}{\varepsilon_{T^P}^{V_2}[T] + \varepsilon_T^{V_1}\left[T^P\right]}. \tag{14}$$

One may consider two cases: the enzymes are not saturated with the substrate (i.e., the elasticities $\varepsilon_T^{V_1}$ and $\varepsilon_{T^P}^{V_2}$ equal 1), and the case when they are saturated (i.e., $\varepsilon_T^{V_1}[T^P]$ 1 and $\varepsilon_{T^P}^{V_2}[T]$ 1), compare equation 8. In the first case, the response coefficient equals $R_{M_T}^{T^P} = \frac{[T]}{\varepsilon_{T^P}^{V_2}[T] + \varepsilon_T^{V_1}[T^P]}$, implying that it cannot exceed unity. In contrast, if the enzymes are saturated, the denominator becomes small; therefore, the response coefficient may exceed 1, and the stimulus s-response curve may become ultrasensitive.

This analysis is based on the assumption that the activating and deactivating enzymes are low, and consequently, the enzyme-substrate complex is negligible. In signal transduction cascades, however, the concentrations of enzymes and substrates are often comparable. This may increase the amount of substrate bound to the enzymes (species MT and DT in Figure 2B), which is thereby sequestered. A theoretical analysis of a covalent modification cycle that includes the effects of high enzyme concentrations can be found in (26). It shows that the response coefficient modifies to:

$$R_{M_T}^{T*} = \frac{[T]}{\varepsilon_{T*}^{V_2}[T] + \varepsilon_T^{V_1}[T*] + \varepsilon_{T*}^{V_2}\varepsilon_T^{V_1}([TM] + [T*D])}. \tag{15}$$

Furthermore, the paper shows that zero-order ultrasensitivity disappears if the concentration of the enzymes is comparable to that of the substrate. Thus, there is doubt about the physiological relevance of zero-order ultrasensitivity in signal transduction.

3.2. Multiple Modification Sites

A large fraction of proteins is subject to reversible covalent modification at multiple sites. As early as 1976, where only the phosphorylation of five proteins had been studied in detail, it was realized that three of them possess multiple phosphorylation sites (24). In many cases, multisite phosphorylation is catalyzed by several kinases, each modifying distinct sites, thereby allowing the integration of different information sources (27). In case that substrate activation requires modification at two sites, the substrate may act as a coincidence detector (AND-gate) because activation requires the presence of two kinases. Additionally, modularity can arise if the substrate performs different cellular functions depending on which sites are modified.

In contrast, several proteins, such as the epidermal growth factor receptor (EGFR), and the kinases Erk and Mek have multiple phosphorylation sites that are phosphorylated by the same kinase. It has been widely discussed that the stimulus-response curve of a protein becomes ultrasensitive if activation requires multisite phosphorylation by the same protein (1,28–30). In the following sections, this idea is sketched by a simple model for multisite phosphorylation. To distinguish the effects of multisite phosphorylation and zero-order ultrasensitivity, enzyme saturation and substrate sequestration will be neglected. For simplicity, it is further assumed that the reaction constants for all phosphorylation sites are equal; i.e., the model does not include cooperativity, which is known to enhance ultrasensitivity. Also, it is assumed that the modification sites are modified in a sequential as opposed to a random manner, as, for example, shown for MAPK activation and deactivation by Mek and MKP3, respectively (31,32). A reaction scheme for such a system is depicted in Figure 4. The reaction rate of the reaction $T_{i-1} \rightarrow T_i$ is given by $v_1^{i-1} = k_1[M][T_{i-1}]$ and of the reaction $T_i \rightarrow T_{i-1}$ by $v_2^{i-1} = k_2[D][T_i]$. By applying steady-state conditions, one can express the amount of i-fold modified protein by,

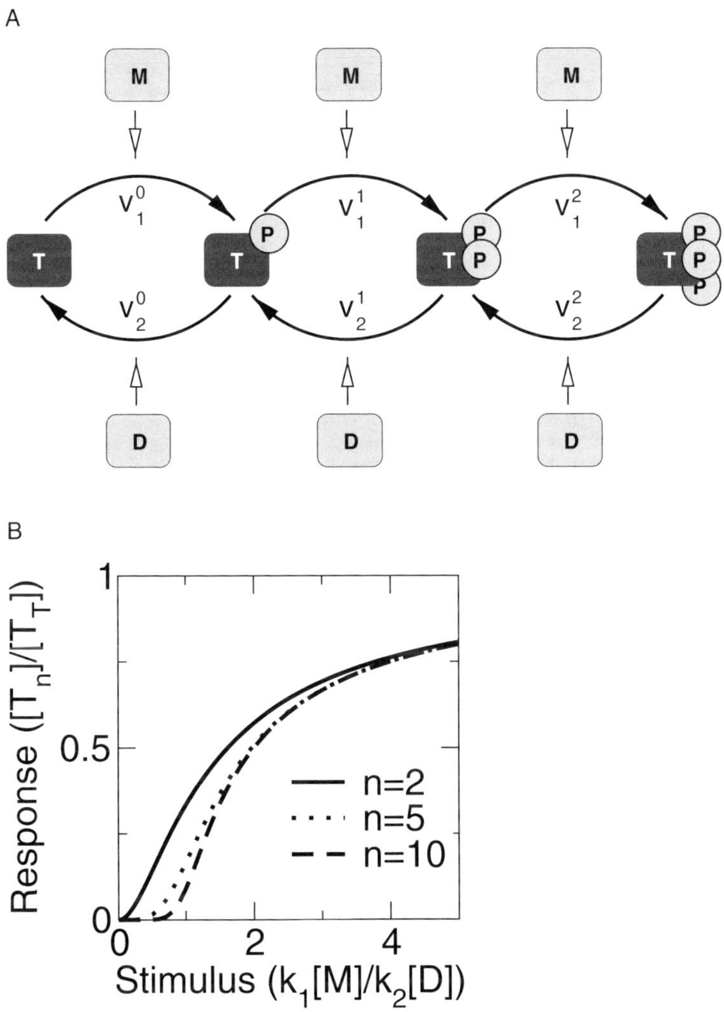

Figure 4. Sketch of a multisite modification cycle, where the sites are modified by the enzyme M and demodified by the enzyme D. The index of the target protein T indicates the number of modified sites. This scheme implies that the modification sites are processed in a sequential manner, e.g., site 1 is the first to be modified, then site 2 is modified, etc. It is assumed in the calculations that all steps have equal rate constants k_1 for modification and k_2 for demodification.

$$[T_i]=\left(\frac{k_1[M]}{k_2[D]}\right)^i[T_0]. \tag{16}$$

Assuming that only the full, n-fold–modified substrate is active, the normalized steady-state activity is given by

$$\frac{[T_n]}{[T_T]}=\left(\frac{k_1[M]}{k_2[D]}\right)^n\frac{\left(\frac{k_1[M]}{k_2[D]}-1\right)}{\left(\frac{k_1[M]}{k_2[D]}\right)^{n+1}-1}, \tag{17}$$

resulting in a response coefficient of

$$R_M^{T_n}=\frac{[M]}{T_n}\frac{\mathrm{d}T_n}{\mathrm{d}[M]}=\frac{1+n}{1-\left(\frac{k_1[M]}{k_2[D]}\right)^{n+1}}-\frac{1}{1-\frac{k_1[M]}{k_2[D]}}. \tag{18}$$

For weak stimulation, the response coefficient equals approximately the number of modification sites, n. As the stimulus-response curve of multisite modification deviates significantly from a Hill curve, the Hill coefficient according to equation 6 is lower than the number of modification sites, e.g., the Hill coefficient of double phosphorylation is 1.38 (33). A more detailed study by Kholodenko (34) on double-phosphorylation also includes saturation effects and yields a response coefficient of

$$R_M^{T_2}=\frac{[T_0]\left(\varepsilon_{T_2}^{v_1^1}+\varepsilon_{T_2}^{v_2^0}\right)+[T_1]\varepsilon_{T_1}^{v_1^0}}{[T_0]\varepsilon_{T_2}^{v_2^1}\varepsilon_{T_1}^{v_2^0}+[T_1]\varepsilon_{T_0}^{v_1^0}\varepsilon_{T_2}^{v_2^2}+[T_2]\varepsilon_{T_0}^{v_2^0}\varepsilon_{T_1}^{v_2^1}}. \tag{19}$$

The response coefficient is approximately 2 for low stimulation and linear kinetics, but may exceed 2 if the enzymes are saturated. However, as the derivation by Kholodenko (9) did not take high enzyme concentrations into account, it applies only as long as the enzyme concentrations are negligible when compared to the substrate concentration, or the K_M values are rather high.

3.3. Other Mechanisms

The earliest discovered mechanism for an ultrasensitive response was cooperative binding, first found in 1904 for the binding of oxygen to hemoglobin (35). Each hemoglobin molecule possesses four binding sites for oxygen. The affinity of a site for oxygen depends on the occupancy of the other binding sites, and increases when these sites are already bound to oxygen. The concentration of oxygen bound as a function of oxygen pressure is well described by a Hill function with coefficient 2.8 (12).

Stoichiometric inhibition and ultrasensitization caused by substrate sequestration are two mechanisms that can give rise to ultrasensitivity by sequestering the target. Both require high affinity binding of the target by the stoichiometric inhibitor (36) or by a phosphatase (37), respectively. The stoichiometric inhibitor or the phosphatase binds the target molecule and prevents it from being active until the target concentration significantly exceeds the phosphatase or the inhibitor.

Another mechanism that generates ultrasensitivity is molecular crowding (38,39). This mechanism is based on the finite size of the molecules and needs very high concentrations of the protagonists. Thus, it is an unlikely effect in signal transduction, where most molecules are present only in low or medium concentrations.

3.4. Sensitivity Amplification by a Cascade

Covalent modification cycles are usually organized in linear signaling cascades. Brown et al. (40) pointed out that the response coefficient of a linear signal-transduction cascade, R_1^n, is simply the product of the individual response coefficients of each kinase with respect to its upstream kinase r_i^{i+1}:

$$R_1^n = \prod_{i=1}^{n-1} r_i^{i+1}. \tag{20}$$

This relation can be derived by applying the chain rule for derivatives. Assuming a three-level cascade, where each level responds like a Hill function

$$[K_i] = \frac{[K_{i,tot}][K_{i-1}]^{h_i}}{[K_{i-1}]^{h_i} + k_{0.5,i}^{h_i}}, \tag{21}$$

the response coefficient of the terminal kinase K_3 upon a stimulus K_0 reads:

$$R_0^3 = \prod_{i=1}^{3} h_i \times \prod_{i=1}^{3}\left(1 - \frac{[K_{i-1}]^{h_i}}{[K_{i-1}]^{h_i} + k_{0.5}^{h_i}}\right) = \prod_{i=1}^{3} h_i(1-f_i), \tag{22}$$

where $f_i = K_i/K_{i,tot}$ is the activated fraction of kinase i. Thus, such cascades can exhibit a sensitivity as high as the product of all Hill coefficients of the kinases if they are weakly activated. In conclusion, a cascade can act as an amplifier of ultrasensitivity.

4. Effect of Feedbacks on Ultrasensitive Cascades

The dynamic behavior of signal transduction cascades is strongly controlled by feedback loops. These feedbacks act on all levels of signal transduction, as illustrated in Figure 5 for the MAPK cascade. Feedbacks arise from autocrine induction of hormones, from the transcriptional regulation of cascade intermediates, from their covalent modification or from receptor internalization. In the following sections, the consequences of such feedback loops on the dynamics are briefly reviewed.

4.1. Bistability Caused by Positive Feedback

Many theoretical and experimental investigations have shown that a positive (or double negative) feedback loop in gene regulatory circuits is a structural condition that allows for bistability (41–45). A bistable system is a system that exhibits two stable steady states, separated by an unstable state. Often the coexistence of the steady states is a function of

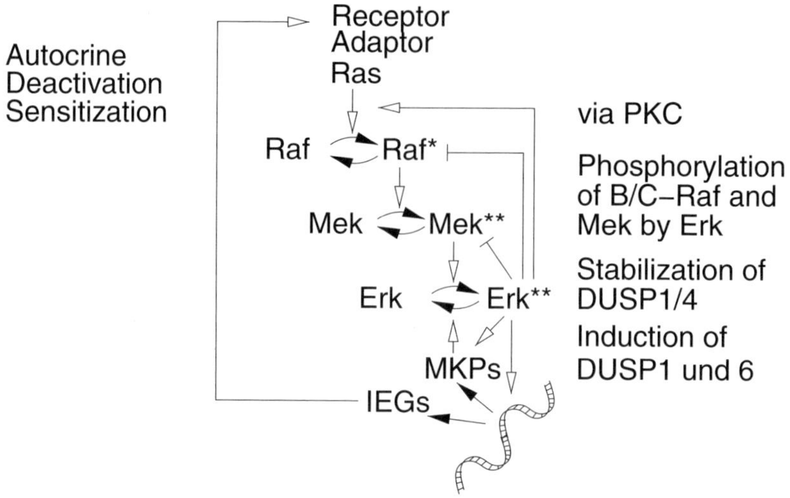

Figure 5. Feedbacks in Ras/Raf/Mek/Erk signal transduction can be found at all levels. They involve direct modification, regulation of the phosphatase stability, autocrine signaling, and changes in expression.

a stimulus; therefore, the system can be switched from one state to another at so-called saddle-node bifurcations (Figure 6). There is now ample evidence that bistability is important in biological signal processing, in the cell cycle, in apoptosis, and in the yeast Gal/Glc network. Also, in the MAPK signal transduction cascade, a positive feedback loop wrapped around a signal transduction cascade can cause bistability (46–48). A prerequisite for observing bistability is that either the cascade or the feedback loop is ultrasensitive (49), as discussed in the following section.

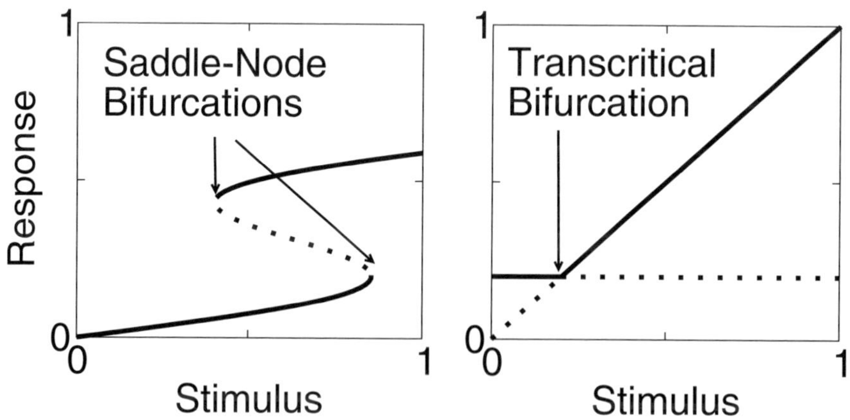

Figure 6. Whereas ultrasensitive cascades equipped with a positive-feedback loop may exhibit two stable steady states (solid line) separated by an uinstable state (dotted line), a cascade lacking ultrasensitivity often displays a transcritical bifurcation, where the stable and unstable steady state exchange their stability at a certain stimulus value.

Bistability provides the means for a biological signaling system to suppress noise, to memorize the signaling history, or to perform all-or-none decisions. It is a mechanism to establish checkpoints, i.e. a threshold, which a stimulus has to exceed before the system is committed into a new state, e.g., cell cycle phase.

An intuitive, graphical way to investigate whether a system is bistable is a combined plot of the stimulus-response curve and of the effect of the feedback loop (open-loop approach) (50,51). This can be done as follows (Figure 7): first, one blocks the feedback loop and records the steady-state output as a function of a steady-state input. Subsequently, one blocks the signal-transduction cascade, perturbs the terminal kinase (the output),

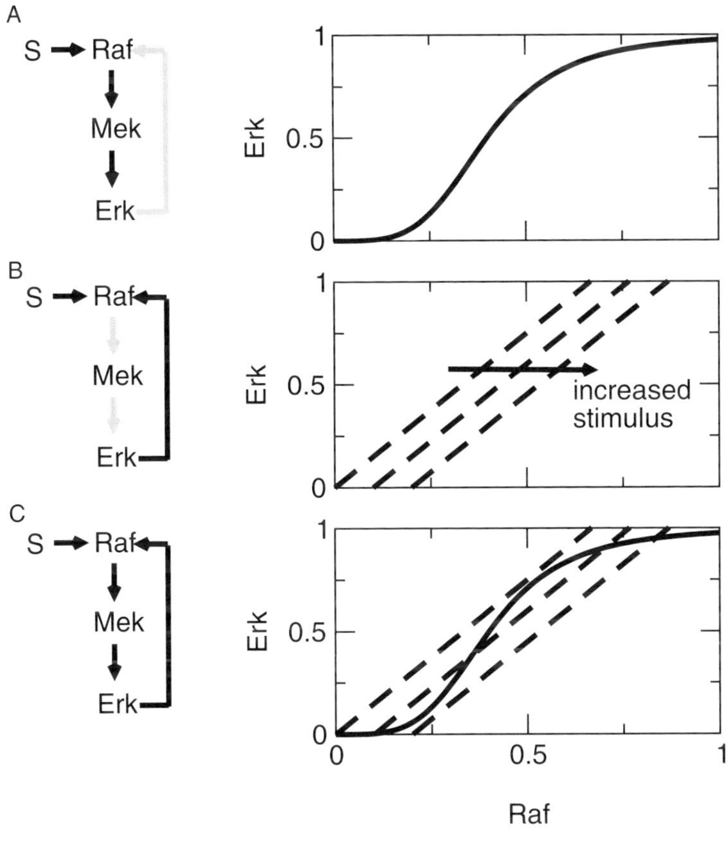

Figure 7. How bistability arises. First, the feedback is inhibited and the stimulus-response curve from Raf to Erk is drawn (A). Second, the cascade (Raf–Erk) is inhibited, and the Erk activity is varied to obtain the steady state of Raf. As Raf is dependent on the stimulus S and Erk activity, the curve shifts toward higher Raf activity for higher stimuli (B). The intersecting points of these curves in the combined plot correspond to the steady states (C). There are three situations: one intersecting point at low Erk-activity (corresponding to one stable off state), three intersecting points (corresponding to a stable off and on state and an unstable state in between), and an intersecting point at high Erk activity (a monostable system with high Erk-activity in steady-state). Depending on the slopes of the curves, all three situations might be reached by varying the stimulus. From the plot it is apparent that at least one curve needs to have a point of inflection to get three intersecting points.

and measures the influence on the first kinase. As the curves represent lines where the cascade or the feedbacks are in steady state, respectively, their intersecting points are the steady state for the entire system.[2] In Figure 7, such curves are displayed for different stimuli. From these curves, it becomes clear that at least one of the curves must have a point of inflection, and thus must be ultrasensitive for the entire system to exhibit three steady states (intersecting points). If both the stimulus-response curve and the feedback loop are not ultrasensitive, the system possesses maximally two steady states, one unstable and one stable, which give rise to a so-called saddle-node bifurcation. For networks containing feedback circuits with nonultrasensitive stimulus-response curves, the authors Binder and Heinrich (52) and Heinrich et al. (53) have investigated the conditions under which the ground state is stable or loses its stability. If the network possesses a stable ground state, the system shows only transient activation. Otherwise the system would exhibit permanent activation, which might contribute to tumorigenesis (53).

4.2. Linear Response, Adaptation, and Oscillation Caused by Negative Feedback

Whereas positive feedback tends to destabilize the stable "off" state and often creates a second stable steady state, the existence of a negative feedback loop usually leads to a stabilization of the steady state (11,43,54). Three emergent dynamic phenomena have been described for ultrasensitive signal-transduction cascades and negative feedback loops: a linearization of the response (11), damped oscillations (interpreted as adaptation [33,55]) and sustained oscillations (9). In the following, these three phenomena will be briefly reviewed.

4.2.1. Linear Response

Sauro and Kholodenko (11) were the first who related the stimulus-response curves of ultrasensitive cascades with those of an operational amplifier, a (negative feedback) device often used in analog electronic circuits to obtain a linear response. They showed that a negative feedback leads to a linear response over a wide range of stimuli. Kholodenko (9) derived the response coefficient of a cascade with a feedback loop:

$$R_1^T = \frac{R_{\text{cascade}}}{1 - r_T^1 R_{\text{cascade}}}. \tag{23}$$

Here, R_{cascade} is the response coefficient of the cascade in isolation, r_T^1 is the local response coefficient of the first kinase in the cascade upon changes in the targets (i.e., the effect of the feedback loop in isolation). In case of a linear negative feedback loop, $r_T^1 = -1$ and thus R_1^T simplifies to:

---

[2] This plot corresponds to null-clines in the phase-plane for a two-dimensional system. The conclusions drawn from these null-clines in higher dimensions than two are only valid in case both stimulus-response curves for the feedback and signal-transduction cascade are monotonic, *see* Angeli et al. (50) for details. Additionally, three intersecting points are not sufficient for bistability if there are other feedbacks within the cascade.

$$R_1^T = \frac{R_{\text{cascade}}}{1+R_{\text{cascade}}}, \tag{24}$$

if the first kinase is not saturated by the stimulus. Surprisingly, the response becomes linear if the response coefficient of the cascade becomes large compared to 1. This result depends strongly on the assumption that $r_T^1 = -1$. Otherwise the response is approximately:

$$R_1^T \approx -r_T^1 \tag{25}$$

Therefore, such a negative feedback loop together with a highly ultrasensitive cascade can be a strategy to "outsource" control of the sensitivity to the feedback. As long as the sensitivity of the cascade is high, only the sensitivity of the feedback determines the sensitivity of the system. Such a strategy might lead to higher robustness if the feedback is rather simple, such as Erk modifying an upstream molecule. Then, only this reaction has to be tightly controlled to yield an amplifier with robust sensitivity.

4.2.2. Adaptation

Often only the information that some concentration has changed is important, whereas information about the absolute value is not important. For example, rising concentrations of growth factors indicate wounding, and neighboring cells need to respond by migration and proliferation. If the signal-transmission is prolonged, however, this behavior might lead to cancer, and thus the response has to be terminated after some time. Another example is the sensing of nutritional gradients in bacterial chemotaxis (56), where the bacteria need to respond to changes only as their size does not permit then to sense a gradient directly. To transduce only the information that something changes, the signal-transduction cascades need to adapt, i.e., they should become less sensitive to higher stimuli when a prolonged stimulus is given, and regain sensitivity if the stimulus drops. Lauffenburger (57) distinguishes two types of adaptation: perfect and partial adaptation. Perfect adapting systems show transient activation but have a steady-state output that is insensitive to the signal. In contrast, partial adapting systems show a peak of activity after stimulation but reach a steady state that is higher than that before stimulation.

A common motif in signal-transduction to gain adaptation is the negative feedback loop (58). A negative feedback can introduce damped oscillations in the cascade. If a stimulation is given to the system the target protein will be activated after some delay and will cause, e.g., feedback desensitization of the receptor, which leads to cascade adaptation. Many Ras-activating receptors and adapter molecules are desensitized this way. It is already apparent from Eq. 23 that a simple negative feedback system, where the terminal kinase desensitizes the receptor cannot perform perfect adaptation. Perfect adaptation requires that the steady state of the terminal kinase is insensitive toward the stimulus, i.e., $R_1^T = 0$. In other words, the cascade itself needs to be insensitive toward the stimulus, i.e., $R_{\text{cascade}} = 0$, but this implies that the cascade itself has to perform adaptation. This is also intuitively clear: if the steady state of the target kinase is insensitive to the stimulus, it cannot provide any information about the stimulus strength that can be used to desensitize the receptor. Therefore, other mechanisms have to be exploited to achieve perfect adaptation. One possibility is to feed the integrated output into the system, e.g., to have a molecule that has a very slow dynamics and integrates the difference between the desired steady-state output and the actual output, as it is realized in bacterial chemotaxis (59). However, perfect adaptation is probably not required in all systems, since, e.g., weak steady-state activation after partial adaptation of Erk may be insufficient to activate downstream targets.

4.2.3. Oscillations

Sustained oscillations play a crucial role to regulate diverse processes including the daily rhythm, also known as the circadian clock. Additionally, the cell-cycle core oscillator ensures that the correct order of events in the cell division cycle is maintained, and thereby prevents carcinogenesis. Finally, oscillations are known to be essential for segmentation during animal development, where spatially alternating structures are generated (60). Sustained oscillations can be observed if the cascade or the feedback loop are strongly ultrasensitive. In Kholodenko's work (9), it was shown that a three-layer cascade can exhibit sustained oscillations, if the overall sensitivity exceeds a threshold value that is determined by the timescales in the cascade. The authors show that oscillations appear if the overall sensitivity (the sensitivity of the cascade and the feedback) exceeds 8, provided that the timescales of all levels of the cascade are equal. Sensitivity of eight in the Raf/Mek/Erk cascade might be reached in conjunction with positive feedback, and sustained oscillations have been reported (11).

5. Discussion

As outlined in the previous sections, ultrasensitivity in signal-transduction gives rise to cellular complex behavior, which allows cells to respond toward external stimuli in an appropriate way. If an ultrasensitive signal-transduction cascade is equipped with one or more positive feedback loops, this cascade can act as a reversible or bistable (irreversible) switch, and thereby establishes checkpoints and cellular memory. In contrast, negative feedback can bring about linear responses and adaptation. If signaling cascades are particularly ultrasensitive, negative feedbacks can even generate sustained oscillations. As such complex phenomena cannot be understood intuitively, and simulation and bifurcation analysis are required to get insights into the behavior of biochemical networks (49).

References

1. Nash P, Tang X, Orlicky S, et al. Multisite phosphorylation of a CDK inhibitor sets a threshold for the onset of DNA replication. *Nature* 2001;414:514–521.
2. LaPorte DC, Koshland DE. Phosphorylation of isocitrate dehydrogenase as a demonstration of enhanced sensitivity in covalent regulation. *Nature* 1983; 305:286–290.

3. Meinke MH, Bishop JS, Edstrom RD. Zero-order ultrasensitivity in the regulation of glycogen phosphorylase. *Proc Natl Acad Sci USA* 1986;83: 2865–2868.
4. Bradshaw JM, Kubota Y, Meyer T, et al. An ultrasensitive Ca2+/calmodulin-dependent protein kinase II-protein phosphatase 1 switch facilitates specificity in postsynaptic calcium signaling. *Proc Natl Acad Sci USA* 2003;100: 10512–10517.
5. Ferrell JE, Machleder EM. The biochemical basis of an all-or-none cell fate switch in *Xenopus* oocytes. *Science* 1998;280:895–898.
6. Ferrell JE. Tripping the switch fantastic: how a protein kinase cascade can convert graded inputs into switch-like outputs. *Trends Biochem Sci* 1996; 21:460–466.
7. Thattai M, van Oudenaarden A. Attenuation of noise in ultrasensitive signaling cascades. *Biophys J* 2002;82:2943–2950.
8. Goldbeter A. A minimal cascade model for the mitotic oscillator involving cyclin and cdc2 kinase. *Proc Natl Acad Sci USA* 1991;88:9107–9111.
9. Kholodenko BN. Negative feedback and ultrasensitivity can bring about oscillations in the mitogen-activated protein kinase cascades. *Eur J Biochem* 2000;267:1583–1588.
10. Bagowski CP, Besser J, Frey CR, Ferrell JE. The JNK cascade as a biochemical switch in mammalian cells: ultrasensitive and all-or-none responses. *Curr Biol* 2003;13:315–320.
11. Sauro HM, Kholodenko BN. Quantitative analysis of signaling networks. *Prog Biophys Mol Biol* 2004;86:5–43.
12. Hill AV. The possible effects of the aggregation of the molecules of haemoglobin on its dissociation curves. *J Physiol* 1910;40:iv–vii.
13. Cardenas ML. Kinetic behaviour of vertebrate hexokinases with emphasis on hexokinase D (IV). *Biochem Soc Trans* 1997;25:131–135.
14. Goldbeter A, Koshland DE. An amplified sensitivity arising from covalent modification in biological systems. *Proc Natl Acad Sci USA* 1981;78: 6840–6844.
15. Kacser H, Burns JA. The control of flux. *Symp Soc Exp Biol* 1973;210: 65–104.
16. Heinrich R, Rapoport TA. A linear steady-state treatment of enzymatic chains. critique of the crossover theorem and a general procedure to identify interaction sites with an effector. *Eur J Biochem* 1974;42:97–105.
17. Fell DA. Metabolic control analysis: a survey of its theoretical and experimental development. *Biochem J* 1992;286 (Pt 2):313–330.
18. Kahn D, Westerhoff HV. Control theory of regulatory cascades. *J Theor Biol* 1991;153:255–285.
19. Hornberg JJ, Bruggeman FJ, Binder B, et al. Principles behind the multifarious control of signal transduction. ERK phosphorylation and kinase/phosphatase control. *FEBS J* 2005;272:244–258.
20. Kholodenko BN, Hoek JB, Westerhoff HV, et al. Quantification of information transfer via cellular signal transduction pathways. *FEBS Letters* 1997; 414:430–434.
21. Hornberg JJ, Binder B, Bruggeman FJ, et al. Control of MAPK signalling: from complexity to what really matters. *Oncogene* 2005;24:5533–5542.
22. Wolf J, Heinrich R. Effect of cellular interaction on glycolytic oscillations in yeast: a theoretical investigation. *Biochem J* 2000;345:321–334.
23. Legewie S, Blüthgen N, Herzel H. Quantitative analysis of ultrasensitive responses. *FEBS J* 2005;272:4071–4079.
24. Cohen P. The regulation of protein function by multisite phosphorylation—a 25 year update. *Trends Biochem Sci* 2000;25:596–601.

25. Koshland DE, Goldbeter A, Stock JB. Amplification and adaptation in regulatory and sensory systems. *Science* 1982;217:220–225.
26. Blüthgen N, Bruggeman FJ, Legewie S, et al. Effects of sequestration on signal transduction cascades. *FEBS J* 2006;273:895–906.
27. Holmberg CI, Tran SE, Eriksson JE, et al. Multisite phosphorylation provides sophisticated regulation of transcription factors. *Trends Biochem Sci* 2002; 27:619–627.
28. Ferrell JE, Bhatt RR. Mechanistic studies of the dual phosphorylation of mitogen-activated protein kinase. *J Biol Chem* 1997;272:19008–19016.
29. Deshaies RJ, Ferrell JE. Multisite phosphorylation and the countdown to S phase. *Cell* 2001;107:819–822.
30. Salazar C, Höfer T. Allosteric regulation of the transcription factor NFAT1 by multiple phosphorylation sites: a mathematical analysis. *J Mol Biol* 2003; 327:31–45.
31. Mansour SJ, Candia JM, Matsuura JE, et al. Interdependent domains controlling the enzymatic activity of mitogen-activated protein kinase kinase 1. *Biochemistry* 1996;35:15529–15536.
32. Zhao Y, Zhang ZY. The mechanism of dephosphorylation of extracellular signal-regulated kinase 2 by mitogen-activated protein kinase phosphatase 3. *J Biol Chem* 2001;276:32382–32391
33. Blüthgen N, Herzel H. MAP-kinase-cascade: Switch, amplifier or feedback controller. In: Gauges R, van Gend C, Kummer U, eds., 2nd Workshop on Computation of Biochemical Pathways and Genetic Networks. 2001:55–62.
34. Kholodenko BN, Hoek JB, Brown GC, et al. Control analysis of cellular signal transduction pathways. In: Larsson C, Pahlman IL, Gustafsson L, eds., Biothermokinetics in the postgenomic era. Goeteborg: Chalmers: 1998: 102–107.
35. Bohr C, Hasselbach KA, Krough A. Über einen in biologischen Beziehungen wichtigen Einfluß, den die Kohlensäurespannung des Blutes auf dessen Sauerstoffbindung übt. *Skand Arch Physiol* 1904;16:402–412.
36. Yeung K, Seitz T, Li S, et al. Suppression of Raf-1 kinase activity and MAP kinase signalling by RKIP. *Nature* 1999;401:173–177.
37. Legewie S, Blüthgen N, Herzel H. Ultrasensitization: switch-like regulation of cellular signalling by transcriptional induction. *PLoS Comput Biol* 2005;1: e54.
38. Gomez Casati DF, Aon MA, Iglesias AA. Ultrasensitive glycogen synthesis in Cyanobacteria. *FEBS Lett* 1999;446:117–121.
39. Aon MA, Gomez-Casati DF, Iglesias AA, et al. Ultrasensitivity in (supra) molecularly organized and crowded environments. *Cell Biol Int* 2001;25: 1091–1099.
40. Brown GC, Hoek JB, Kholodenko BN. Why do protein kinase cascades have more than one level? *Trends Biochem Sci* 1997;22:288.
41. Thomas R, Gathoye AM, Lambert L. A complex control circuit. Regulation of immunity in temperate bacteriophages. *Eur J Biochem* 1976;71:211–227.
42. Arkin A, Ross J, McAdams HH. Stochastic kinetic analysis of developmental pathway bifurcation in phage lambda-infected *Escherichia coli* cells. *Genetics* 1998;149:1633–1648.
43. Hasty J, Pradines J, Dolnik M, et al. Noise-based switches and amplifiers for gene expression. *Proc Natl Acad Sci USA* 2000;97:2075–2080.
44. Gardner TS, Cantor CR, Collins JJ. Construction of a genetic toggle switch in *Escherichia coli*. *Nature* 2000;403:339–342.
45. Becskei A, Seraphin B, Serrano L. Positive feedback in eukaryotic gene networks: cell differentiation by graded to binary response conversion. *EMBO J* 2001;20:2528–2535.

46. Ferrell JE, Xiong W. Bistability in cell signaling: how to make continuous processes discontinuous, and reversible processes irreversible. *Chaos* 2001; 11:227–235.
47. Ferrell JE. Self-perpetuating states in signal transduction: positive feedback, double-negative feedback and bistability. *Curr Opin Cell Biol* 2002;14: 140–148.
48. Shvartsman SY, Muratov CB, Lauffenburger DA. Modeling and computational analysis of EGF receptor-mediated cell communication in *Drosophila* oogenesis. *Development* 2002;129:2577–2589.
49. Tyson JJ, Chen KC, Novak B. Sniffers, buzzers, toggles and blinkers: dynamics of regulatory and signaling pathways in the cell. *Curr Opin Cell Biol* 2003; 15:221–231.
50. Angeli D, Ferrell JE, Sontag ED. Detection of multistability, bifurcations, and hysteresis in a large class of biological positive-feedback systems. *Proc Natl Acad Sci USA* 2004;101:1822–1827.
51. Bhalla US, Iyengar R. Emergent properties of networks of biological signaling pathways. *Science* 1999;283:381–387.
52. Binder B, Heinrich R. Interrelations between dynamical properties and structural characteristics of signal transduction networks. *Genome Informatics* 2004;15:13–23.
53. Heinrich R, Neel BG, Rapoport TA. Mathematical models of protein kinase signal transduction. *Mol Cell* 2002;9:957–970.
54. Becskei A, Serrano L. Engineering stability in gene networks by autoregulation. *Nature* 2000;405:590–593.
55. Asthagiri AR, Lauffenburger DA. A computational study of feedback effects on signal dynamics in a mitogen-activated protein kinase (MAPK) pathway model. *Biotechnol Prog* 2001;17:227–239.
56. Alon U, Surette MG, Barkai N, et al. Robustness in bacterial chemotaxis. *Nature* 1999;397:168–171.
57. Lauffenburger DA. Cell signaling pathways as control modules: complexity for simplicity? *Proc Natl Acad Sci USA* 2000;97:5031–5033.
58. Lauffenburger DA, Linderman JJ. Receptors—models for binding, trafficking, and signaling. Oxford: Oxford University Press; 1993.
59. Yi TM, Huang Y, Simon MI, Doyle J. Robust perfect adaptation in bacterial chemotaxis through integral feedback control. *Proc Natl Acad Sci USA* 2000; 97:4649–4653.
60. Hirata H, Yoshiura S, Ohtsuka T, et al. Oscillatory expression of the bHLH factor Hes1 regulated by a negative feedback loop. *Science* 2002;298: 840–843.

16

Employing Systems Biology to Quantify Receptor Tyrosine Kinase Signaling in Time and Space

Boris N. Kholodenko

Summary

Environmental cues received by plasma membrane receptors are processed and encoded into complex spatiotemporal response patterns of protein phosphorylation networks, which generate signaling specificity. The emerging synergistic, experimental computational approach is presented, which provides insights into the intricate relationships between stimuli and cellular responses. Computational models reveal how positive and negative feedback circuits and other kinetic mechanisms enable signaling networks to amplify signals, reduce noise, and generate complex nonlinear responses, including oscillations, ultrasensitive switches, and discontinuous bistable dynamics; and many of these predictions have been verified experimentally. The analysis of the spatial signaling dynamics highlights an important distinction between electronic and living circuitry and shows how intriguing signaling phenomena are brought about by the heterogeneous cellular architecture and diffusion. Spatial gradients of signaling activities emerge as hallmarks of living cells. These gradients guide pivotal physiological processes, such as cell motility and mitosis, but also impose a need for facilitated signal propagation, which involves trafficking of endosomes and signaling complexes along microtubules and traveling waves of phosphorylated kinases.

Key Words: Epidermal growth factor receptor; computational modeling; feedback; bistable dynamics; MAPK cascades; combinatorial complexity; spatial gradients; reaction-diffusion equations.

1. Introduction

Cells respond to hormones and growth factors using a limited cadre of signaling pathways activated by cell surface receptors, such as G protein–coupled receptors (GPCRs) and receptors with intrinsic tyrosine kinase (RTK) activities. These pathways interact with each other and form the multilevel signaling network that processes and integrates external cues. Several lines of recent evidence indicate that distinct spatiotemporal activation profiles of the same repertoire of signaling proteins result in different gene activation patterns and diverse physiological responses. Therefore, critical cellular decisions ranging from cell survival, growth, and proliferation to differentiation or apoptosis depended on the precise spatiotemporal control of activation response patterns of intracellular signal transducers.

GPCRs control a variety of physiological functions by activation of heterotrimeric G proteins, which gave the name to this largest family of cell surface receptors. After the guanosine diphosphate (GDP)/guanosine triphosphate (GTP) exchange, which is facilitated by an agonist-activated receptor, G proteins dissociate into Gα and Gβγ subunits. These subunits interact with multiple effectors, regulating ion channels and second messenger production, thereby triggering hormonal, sensory, and neurotransmitter signaling pathways (1,2). In addition to this classic paradigm, GPCRs can also stimulate cell proliferation and differentiation, which is traditionally associated with RTK activation. Intriguingly, this overlap of GPCR and RTK signaling pathways can be partially explained by GPCR-mediated "transactivation" of RTKs (3,4).

Signaling by RTKs has long been in the limelight of scientific interest, owing to its central role in the regulation of embryogenesis, cell survival, motility, proliferation, differentiation, glucose metabolism, and apoptosis (active cell death) (5–7). Malfunction of RTK signaling is a leading cause of major human diseases, ranging from developmental defects to cancer, chronic inflammatory syndromes, and diabetes. All RTKs consist of three major domains: extracellular domains of ligand binding and dimerization (collectively called the ectodomain); a membrane-spanning segment; and a cytoplasmic domain, which possesses tyrosine kinase activity and contains phosphorylation sites with tyrosine, serine, and threonine residues. RTKs can be activated by growth factors or transactivated by GPCRs. After ligand binding, RTKs undergo receptor dimerization (e.g., epidermal growth factor receptor (EGFR)) or an allosteric transition (e.g., insulin receptor, IR, and insulin-like growth factor receptor (IGF-1R)) (5,8). These structural transitions result in the activation of intrinsic tyrosine kinase activity and subsequent autophosphorylation. Autophosphorylation of RTKs initiates signal processing through a battery of receptor interactions with adapter and target proteins containing characteristic protein domains, such as Src homology (SH2 and SH3), phosphotyrosine binding (PTB) and pleckstrin homology (PH) domains (reviewed in 5,9,10). These proteins include Src homology and collagen domain protein (Shc); growth factor receptor–binding protein 2 (Grb2); Grb2-associated binder (GAB1/2); GTPase-activating protein (GAP); phosphoinositide-specific phospholipase C-γ (PLCγ); the 85-kDa subunit of phosphatidylinositol 3-kinase (PI3K); cytoplasmic tyrosine kinases, such as c-Src; the protein tyrosine phosphatase (PTP) SHPTP-2; insulin receptor substrates (IRS-1 to IRS-4 for IR or IGF-1R); and others. Subsequent interactions of these proteins with downstream effectors generate complex biochemical circuits, including cascades of protein and lipid phosphorylation and dephosphorylation reactions.

Signal specificity and cellular decisions are outputs of the complex *temporal* and *spatial* dynamics of multiple signaling processes that involve feedback regulation (11). The classic example is distinct functional responses of PC12 cells to EGF and nerve growth factor (NGF) activation, e.g., proliferation versus differentiation, attributed to different temporal patterns of extracellular signal regulated kinase (ERK) activation, i.e., whether ERK activation is transient or sustained (12–14). Likewise, sustained versus transient activation of the mitogen-activated protein kinase (MAPK) cascade was suggested to be a mechanism underlying RTK specificity in EGF- and hepatocyte growth factor–induced keratinocyte migration (15). Interestingly, the activation of Raf-1 has been linked to such opposing responses as the induction of DNA synthesis and growth inhibition (16,17). In NIH 3T3 cells, low Raf-1 activity was shown to induce cell cycle progression, whereas higher Raf activity inhibited proliferation (18). This variety of biological outcomes of MAPK activation is thought to be related to the intricate dynamics of these kinase cascades. MAPK cascades can generate bistable dynamics (where two stable "on" and "off" steady states coexist), abrupt switches, and oscillations (19–28), and MAPK responses depend dramatically on the subcellular localization and recruitment to scaffolds (29,30).

This chapter illustrates the benefits of the application of systems analysis and modeling to the studies of RTK networks. It starts with a discussion of mechanistic modeling of growth factor signaling and challenges that face mechanistic models. A novel "domain-oriented" approach (31–33) that addresses the combinatorial complexity of interacting pathways is presented. Other modeling strategies, such as Bayesian networks or Boolean network modeling, are not discussed here. Next, this chapter surveys the temporal dynamics of information transfer and shows that basic signaling motifs can generate complex nonlinear dynamic phenomena, including bistability and oscillations. Finally, the spatial aspects of intracellular communication are analyzed. The transfer of phosphorylation signals over distances larger than a few micrometers is often hampered by rapid dephosphorylation. We suggest that endosomes and scaffolds, which have bound phosphorylated kinases and are driven by molecular motors, and traveling phosphorylation waves spread signals over large intracellular distances.

2. Challenges of Mechanistic Modeling

2.1. Computational Modeling of the EGFR Network

Experimental data alone are not sufficient to understand and predict signaling dynamics, and faithful computational models are required (34–36). The EGF receptor (EGFR) pathway was one of the first test cases for computational modeling of signal transduction (37). EGFR is a member of the ErbB family of growth factor receptors, which involves ErbB1/EGFR, ErbB2/Neu, ErbB3, and ErbB4 (38). Aberrant signaling by this family often leads to human neoplasia, such as breast, lung, prostate, bladder, and other cancers. The first mechanistic model of the EGFR pathway was published in 1999 and explained the temporal dynamics of signaling events observed in liver cells after the onset of EGF stimulation (39). Despite constant EGF level, cells showed markedly transient phosphorylation of EGFR and selected target proteins including PLCγ, PI3K, and GAB (with peaks reached within the first 15s and rapid decreases to pseudostationary levels by 2–3min), whereas phosphorylation of other signaling proteins, such as Shc and the concentration of the Shc–Grb2–SOS complex, increased almost monotonically (40). Modeling suggests that the transient time-course of EGFR phosphorylation arises from a protection of phosphotyrosine residues against phosphatase activity, while occupied by an adaptor/target protein, whereas the slow dissociation of the complexes formed by the phosphatases and unphosphorylated PLCγ and PI3K may explain transient patterns of tyrosine phosphorylation of these proteins (39,41). In fact, the existence of such complexes between tyrosine phosphatases SHP-1 and SHP-2, as well as PLCγ and PI3K, was reported (42). Surprisingly, phosphorylation of adaptor/target proteins by EGFR was predicted to facilitate their subsequent dissociation from the receptor, e.g., Shc phosphorylation on Tyr 317 was suggested to significantly decrease Shc binding affinity for the EGFR receptor. This was unexpected because Tyr317 is located within the central collagen homology (CH) linker region of Shc at the distances of 53 and 110 residues away from the SH2 and PTB domains that mediate binding to EGFR. However, recent work showed that Tyr317 phosphorylation significantly affects collective motions of Shc domains, increases structural rigidity of the CH linker region, and dramatically decreases the flexibility of the PTB and SH2 domains, thus reducing their capacity to interact productively with EGFR (43).

Several EGFR pathway models that address important aspects of EGF-induced signaling were recently developed, including: (a) nonlinear dependence of the amplitude of MAPK activation on the EGF receptor number (44); (b) complex regulation of transient versus sustained responses of the MAPK cascade "gatekeepers," small GTPases Ras and Rap1, to growth factors (45–47); (c) autocrine positive-feedback loops (48); (d) cross-talk between the MAPK and Akt pathways (49); and (e) integration of EGFR signaling from the plasma membrane and endosomes (50). Hypotheses generated by these models have a certainty and precision, which will further our understanding of signaling dynamics. A variety of software tools can assist in computational modeling (51–57), and several databases of biological models have been developed, offering an interesting environment to generate and test novel hypotheses by using a computer keyboard (58,59).

2.2. Combinatorial Increase in Network Complexity

Major challenges that face mechanistic modeling are the lack of quantitative kinetic data and the *combinatorial* increase in the number of emerging distinct species and states of the protein network being modeled (60,61). The first challenge of experimental uncertainty is beginning to be addressed by nascent quantitative proteomics of posttranslational modification (*see* Chapters 11 and 24). The second challenge arises because RTKs and many adapter proteins display multiple docking sites, serving as *scaffolds* that generate a variety of heterogeneous multi-protein complexes, each involved in multiple parallel reactions. Mechanistic modeling describes the functional states of a multi-domain protein by the function that simultaneously depends on the states of all domains of that protein. Each domain can assume multiple states; for instance, a docking site on a receptor or scaffold can be unphosphorylated and free, phosphorylated and free, phosphorylated and bound to a partner, which in turn can be unphosphorylated and free, or phosphorylated and bound to another protein or lipid, and so on. All of these distinct possibilities multiply and generate hundreds of thousands and millions of "micro-states," which account for potentially formed molecular species, and even for a few initial steps following a ligand binding to a receptor (60). Importantly, these micro-states of a signaling network generate (micro)variables and chemical kinetic equations, one ordinary differential equation (ODE) for every micro-state (species). The problem of the combinatorial complexity for network modeling has been recognized (39,60,61). Several software tools addressing this problem have been proposed, including the "rule-based" ODE simulator BioNetGen, which automatically generates all species and reactions (53,62), and the stochastic simulators StochSim and Moleculizer (54,61) that circumvent the explosion of the number of micro-states by generation of the species and reactions as needed during a stochastic simulation. However, for large networks of many interacting RTKs, scaffolds, and hundreds of other proteins, a microdescription becomes impractical for both deterministic and stochastic simulations.

In contrast, in a domain-oriented framework (31–33), we introduce so-called "macrostates" and a set of "macrovariables," which depend on the control hierarchy of interactions between protein domains (distinct sites). Each macrovariable accounts for the states of a separate site on a protein and the domains that control this site. For instance, ligand binding and dimerization control the state of each docking site on a typical RTK, and the macrovariable associated with that site may depend on the ligand and dimerization state of the receptor. Therefore, instead of a single function that simultaneously depends on the states of all domains of a multidomain protein, a macro-description operates with several separate state functions for each protein domain. Although several macrovariables are associated with a scaffold, a multiplicative (exponential) amplification of the number of microstates is substituted by an additive increase in the number of macrostates (31–33). A necessary prerequisite for the validity of a macrodescription and the reduction of a mechanistic model is the presence of protein domains/sites that do not influence other sites allosterically or through interactions with bound partners. Importantly, the existence of additional sites involved in allosteric interactions does not impede the reduction of combinatorial complexity of multi-component receptor-mediated signal transduction. This domain-oriented framework drastically reduces the number of states and differential equations to be solved and, therefore, the computational cost of both deterministic and stochastic simulations (31,32).

3. Complex Temporal Dynamics of Signaling Networks

3.1. Responses of Signaling Cycles and Cascades

Already simple, basic signaling motifs can exhibit complex dynamic behavior, including multiple steady states and sustained oscillations. A universal motif found in cellular networks is the cycle formed by two or more interconvertible forms of a signaling protein, which is modified by two opposing enzymes. Such enzymes can be a kinase that phosphorylates a target protein on serine/threonine or tyrosine residues (in mammalian cells) and a phosphatase that dephosphorylates these residues. Likewise, a guanine nucleotide exchange factor (GEF), such as SOS, and GAPs, such as RasGAP, catalyze the cyclic conversion for a small G protein, such as Ras (Figure 1). Cascades of such cycles form the backbone of most signaling pathways. A well-known property of these cycles is "ultrasensitivity" to input signals, which occurs when the converting enzymes operate near saturation (63). Depending on the degree of saturation, the response of either interconvertible form ranges from a merely hyperbolic to an extremely steep sigmoidal curve. Sequestration of a signaling protein by converting enzymes significantly decreases sigmoidicity of responses ((28) and Chapter 15 of this book). Likewise, ultrasensitivity can disappear if converting enzymes are inhibited or saturated by their products (64). Importantly, multisite phosphorylation that occurs through a distributive, multicollision mechanism was shown to increase the sensitivity dramatically, resisting the sequestration effect and leading to switch-like responses (25,65–67).

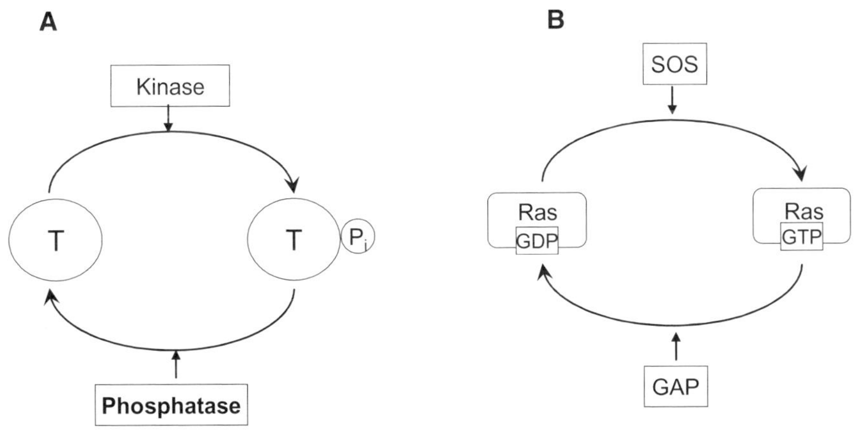

Figure 1. Universal cycle motif in cellular signaling networks. (A) Phosphorylation and dephosphorylation cycle of a target protein (T). The protein T is phosphorylated by a kinase that converts T in its phosphorylated form T_p. An opposing phosphatase dephosphorylates T_p to yield T. (B) A kinetic cycle of the small GTPase Ras. The guanosine nucleotide exchange factor SOS catalyzes the transformation of the inactive GDP-bound form of RasGDP into an active GTP-bound form Ras-GTP). The GTPase–activating protein RasGAP (shown as GAP) is the opposing enzyme that catalyzes the reverse transformation.

3.2. Feedback Loops Determine Input–Output Patterns and Can Bring About Dynamic Instabilities

Several interconnecting cycles acting on each other form a signaling cascade. An increase in the number of cycles or a positive feedback further increases the sensitivity of the target to the input signal (65,68). Positive feedback loops amplify the signal, whereas negative feedback attenuates the response. However, feedback loops not only change steady-state responses, but also favor the occurrence of instabilities. When a steady state becomes unstable, a system can jump to another stable state, start to oscillate, or exhibit chaotic behavior. Negative feedback and ultrasensitivity can bring about oscillations in the concentrations of active kinases in a kinase/phosphatase cascade (21). Positive feedback can cause bistability (24), but, either alone or in combination with negative feedback, it can trigger oscillations; for example, the Ca^{2+} oscillations arising from Ca^{2+}-induced Ca^{2+} release (35) and the cell cycle oscillations (69,70). Such positive-feedback oscillations generally do not have sinusoidal shapes (which is a characteristic for only negative-feedback oscillations) and are referred to as relaxation oscillations. These oscillations operate in a pulsatory manner: a part of a dynamic system is bistable, and there is a slow process that periodically forces the system to jump between "off" and "on" states, generating periodic swings. Importantly, negative feedback not only attenuates the input signal or leads to oscillatory behavior, but also endows signaling pathways with robustness to parameter variations within the feedback loop (71,72). For instance, genetic variability of protein expression or medical drugs that affect processes within a negative feedback loop would have only minor influences on signaling responses, compared with the situation where target processes are outside this loop (71).

4. Spatial Dimension of Cell Signaling

Distinct temporal patterns of downstream response to the activation of GPCRs and RTKs are further regulated by translocation of signaling proteins to diverse cellular locations and their colocalization with various scaffolding and cytoskeletal proteins (73–75). The spatial control of signaling is pivotal for key cellular processes, such as cell division, motility, and migration. During evolution, cells have developed not only the means to control the temporal dynamics of signaling networks, but also mechanisms for precise spatial sensing of the relative localization of signaling proteins. Positional information is critical for signaling from different cellular compartments, including the plasma membrane, the cytoplasm, and intracellular membrane compartments, such as endocytic vesicles and the Golgi complex. The same protein cascades operate in surprisingly dissimilar ways when localized to different cellular compartments (30). We show how basic principles of the control of reaction rates, diffusive movement, and directed transport underlie sophisticated mechanisms to activate signaling by the membrane recruitment of binding partners, to provide spatial cues that guide cellular decisions, and to transmit signals to distant cellular targets.

4.1. Membrane or Scaffold Recruitment of Interacting Partners Switches on Signaling Responses

Growth factor stimulation triggers the mobilization of cytosolic proteins to cellular membranes, scaffolding, and cytoskeletal elements. The classic example is the membrane recruitment of SOS and RasGAP, which activate and inactivate, respectively, the membrane-anchored small G protein Ras, which acts as the "gatekeeper" of the MAPK/ERK cascade (5,76). This recruitment is mediated by RTKs, e.g., EGFR, or scaffolds, e.g., Gab1/2, which bind to the membrane phospholipid PIP_3 (the product of PI3K) (10). Importantly, these interactions do not increase SOS and RasGAP catalytic activities, but only recruit the proteins to the membrane. We are left with the question of why the membrane versus cytoplasm localization facilitates the catalysis. Adam and Delbrück suggested that the reduction in dimensionality might enhance reaction rates between solutes that bind to membranes and membrane-bound species (77); the solutes should not get lost by wandering off into the bulk phase. However, the diffusion-limited rates in the membrane are about two orders of magnitude slower than in the cytosol and, therefore, the membrane recruitment would decrease, and not increase the first-encounter rates, as noticed by Bray (78). In fact, the function of the membrane recruitment has been recently shown to amplify the number of complexes formed between signaling partners (76,79). SOS and RasGAP bound to EGFR or PIP_3–GAB1/2 are confined to a narrow layer below the plasma membrane, approximately 5 nm–10 nm thick, corresponding to the dimension of membrane-anchored proteins. The volume of that layer (V_m) is much smaller than the cytoplasmic volume (V_c). For a spherical cell with a radius of 10 μm and a proximal membrane layer of 10 nm, the ratio of cytosolic volume to proximal membrane volume (V_c/V_m) is between 10^2 and 10^3. This decrease in the reaction volume results in a 10^2–10^3-fold increase in the apparent affinity of SOS and RasGAP for Ras. Simulations demonstrate that this spatial organization of SOS/RasGAP signaling is crucial for effective control of Ras activity (46).

Similar estimates apply for assembly of interacting signaling partners on scaffold proteins and membranes (80). Scaffolds act as templates, bringing together signaling proteins and organizing and coordinating the function of entire signaling cascades (78). Importantly, our results suggest that the number of signaling complexes will increase only if these complexes do not dissociate from a scaffold. Even if the interacting proteins were brought to close vicinity on a scaffold, the dissociation of the protein complex from the scaffold will result in further dissociation of the complex, which will be in thermodynamic equilibrium with its components.

4.2. Emergence of the Spatial Gradients of Signaling Activities Within Cells

In living cells, two opposing enzymes of a universal cycle motif (Figure 1) can be spatially separated. For instance, a kinase can be localized to a scaffold or supramolecular structure, whereas a phosphatase can be homogeneously distributed in the surrounding area of the cytoplasm. In this case, the spatial gradient of a target cytoplasmic protein can occur, with the high level of phosphorylation of this protein in the close vicinity of the scaffold and the low phosphorylation level at distant cytoplasmic areas (81,82). Likewise, the spatial gradient of a GTP-bound form of a small G protein can occur if the GEF for that protein is confined to a supramolecular structure, whereas a GAP freely diffuses in the cytoplasm (83). A variation to the theme of the spatial separation of signaling enzymes is a cycle where one enzyme-modifier is membrane-bound and the opposite enzyme is cytosolic. For a protein phosphorylated by a membrane-bound kinase and dephosphorylated by a cytosolic phosphatase, it was predicted theoretically that there can be a gradient of the phosphorylated form that is high, close to the membrane, and low within the cell (81). Given measured values of protein diffusivity and kinase and phosphatase activities, it was estimated that phosphoprotein gradients can be large within the intracellular space. These theoretical predictions have materialized recently, when fluorescence resonance energy transfer–based biosensors enabled discoveries of intracellular gradients of the active form of the small GTPase Ran (84) and the phosphorylated form of stathmin/oncoprotein 18 (Op18/stathmin) that regulates the microtubule polymerization (85,86).

We will demonstrate how intracellular signaling gradients can arise from chemical transformation and diffusion. Importantly, this analysis explains that the spatial gradients of signaling activities can emerge even in small bacterial cells (87,88). First, we will consider the simplest one-dimensional geometry that corresponds to a cylindrical bacterial cell of the length L. We will assume that a kinase is localized to the membrane at a single pole of this cell (at the spatial coordinate $x = 0$) and a phosphatase is distributed in the cytoplasm. The kinase phosphorylates a target protein with rate v_{kin}^{men} (defined as the surface rate at $x = 0$). The phosphorylated protein diffuses into the cell and gets dephosphorylated by the phosphatase at rate v_p. The spatiotemporal dynamics of the phosphorylated form, c_p, of the interconvertible protein is governed by the reaction–diffusion equation,

$$\frac{\partial c_p}{\partial t} = D\frac{\partial^2 c_p}{\partial x^2} - v_p(c_p), \tag{1}$$

with the following boundary conditions,

$$-D\frac{\partial c_p}{\partial x}\bigg|_{x=0} = v_{kin}^{mem}, \quad \frac{\partial c_p}{\partial x}\bigg|_{x=L} = 0. \tag{2}$$

The boundary conditions stipulate that the diffusive flux equals v_{kin}^{men} at the kinase pole and zero at the opposite pole. When the diffusivities D are equal for the phosphorylated c_p and unphosphorylated c_u forms of the target protein, the total protein concentration is constant across the cell, $c_p + c_u = C_{tot}$ (which is untrue for different diffusivities (82)). The steady-state spatial profile $c_p(x)$ is determined by letting the time-derivative in equation (1) equal zero,

$$D\frac{d^2 c_p}{dx^2} - v_p = 0. \tag{3}$$

When the phosphatase is far from saturation (a reasonable assumption for most cytosolic phosphatases), $v_p = k_p c_p$ ($k_p = V_{max}/K_m$ is the observed first-order rate constant), the analytical solution to equations 2 and 3 reads,

$$c_p(x) = c_p(0)\left(\frac{e^{\alpha x} + e^{2L\alpha}e^{-\alpha x}}{1 + e^{2\alpha L}}\right), \quad \alpha^2 = \frac{k_p}{D}. \tag{4}$$

When $\alpha L << 1$, the phosphoprotein concentration decreases almost linearly, and when $\alpha L \geq 1$, it decreases nearly exponentially $c_p(x)/c_p(0) \approx e^{-\alpha x}$, with distance x from the membrane (the dimensionless parameter αL is recognizable as the square root of the Damkohler number). This provides a straightforward and powerful criterion that demonstrates that large phosphoprotein gradients exist when the dephosphorylation time $1/k_p$ is smaller than the diffusion time L^2/D (82). The kinase activity only influences the concentration $c_p(0)$ near the membrane (81,82). This criterion helps us understand how the cell may control phosphoprotein gradients, an increase in phosphatase expression levels, or activities will make these gradients more precipitous, whereas down-regulation of phosphatase activities or the compartmentalization of phosphatases will decrease the steepness of or even eliminate the gradients.

Spherical symmetry simplifies analysis of signaling in three dimensions. For a cell of the radius L with a kinase located on the cell surface and phosphatase in the cytoplasm, the reaction–diffusion equation that governs the dynamics of the phosphorylated form c_p of the target protein has the following form (cf. equation 1)

$$\frac{\partial c_p}{\partial t} = \left(\frac{D}{L^2}\right)\cdot\frac{1}{x^2}\frac{\partial}{\partial x}\left(x^2\frac{\partial c_p}{\partial x}\right) - v_p, \quad \left.\frac{\partial c_p}{\partial x}\right|_{x=1} = \frac{L}{D}\cdot v_{kin}, \quad \left.\frac{\partial c_p}{\partial x}\right|_{x=0} = 0 \tag{5}$$

Here, x is a dimensionless coordinate equal to the distance from the cell center divided by the cell radius, L (the distance d from the cell membrane is expressed in terms of x as $d = (1 - x) \times L$); v_{kin} and v_p are the rates of the kinase and phosphatase, respectively. As for previous equations, we will assume that the phosphatase is not saturated by a target phosphoprotein, $v_p = k_p c_p$. Next, the steady-state solution to equation 5 can be found readily, and the ratio of the phosphoprotein concentrations at the distance x from the cell center and at the plasma membrane is given by Kholodenko et al. (82,89):

$$\frac{c_p(x)}{c_p(1)} = \frac{e^{\alpha L x} - e^{-\alpha L x}}{x\cdot\left(e^{\alpha L} - e^{-\alpha L}\right)}; \quad \alpha^2 = \frac{k_p}{D}. \tag{6}$$

Therefore, when $\alpha L \geq 1$, the phosphoprotein concentration decreases toward the cell interior approximately exponentially, and the total relative gradient (the ratio $c_p(1)/c_p(0)$) equals $(e^{\alpha L} - e^{-\alpha L})/2\alpha L$ (81).

A similar exponential decrease in the phosphorylation signal $c_p(r)$ may occur when a kinase is bound to a supramolecular structure with the radius s and a phosphatase resides in the surrounding area of the radius L. Assuming spherical symmetry, the steady-state concentration $c_p(r)$ is determined by,

$$\frac{D}{r^2}\frac{d}{dr}\left(r^2\frac{dc_p}{dr}\right) - k_p c_p = 0, \quad -D\left.\frac{dc_p}{dr}\right|_{r=s} = v_{kin}^{mem}, \quad \left.\frac{dc_p}{dr}\right|_{r=L} = 0$$

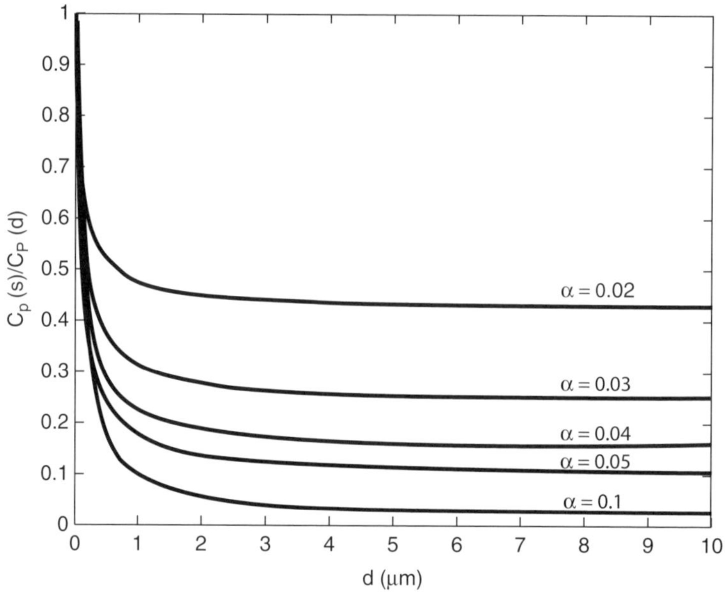

Figure 2. Steady-state activation profile of a phosphoprotein. A target protein in the cytoplasm is phosphorylated by a kinase localized to a supramolecular structure of the radius $s = 0.1\,\mu m$ and dephosphorylated by a homogeneously distributed phosphatase in the cytoplasmic area of the radius $L = 10\,\mu m$. The relative phosphorylated fraction, $c_p(s)/c_p(d)$, declines with the distance d from the kinase that is confined to the spherical structural element (*see* equation 7). The steepness of the gradient (reciprocal of the characteristic length) is determined by the parameter α ($\alpha^2 = k_p/D$ is the ratio of the phosphatase activity k_p and the protein diffusivity D).

$$c_p(r) = c_p(s)\frac{se^{-\alpha r}}{re^{-\alpha s}}\left(\frac{e^{2\alpha r}(\alpha L+1)+e^{2\alpha L}(\alpha L-1)}{e^{2\alpha s}(\alpha L+1)+e^{2\alpha L}(\alpha L-1)}\right), \quad \alpha^2 = \frac{k_p}{D} \qquad (7)$$

Figure 2 illustrates how the concentration $c_p(r)$ decreases for different values of αL. We conclude that signaling gradients cannot be built merely by diffusion, but require the spatial segregation of opposing enzymes.

4.3. Do Phosphoprotein Gradients Exist in MAPK Cascades?

The calculations above suggest that phosphoprotein gradients might exist in kinase/phosphatase cascades where kinases and phosphatases are spatially separated, such as in MAPK cascades. MAPK cascades contain three interconnected cycles of MAPK, MAPK kinase (MAP2K), and MAP2K kinase (MAP3K). Mammalian cells express at least four different MAPK families, including the ERK, the JNK, and p38 MAPK cascades. A kinase cascade that is the best characterized kinetically is the MAPK/ERK cascade that consists of ERK (MAPK), MEK (MAP2K), and Raf (MAP3K). Upon RTK stimulation and Ras activation, the cytosolic Raf is recruited to the cell membrane. At the membrane, Raf undergoes a series of activation steps involving dephosphorylation of the inhibitory phosphorylated Ser 259, interaction with 14-3-3 proteins, and phosphorylation on specific tyrosine residues (90–92). Although the mechanism of activation is not yet completely understood, the association of Raf with membranes appears to be essential for its activation. The Raf kinase phosphorylates the cytosolic kinase MEK (MKK) at the plasma membrane, whereas soluble serine/threonine phosphatases dephosphorylate the activated MEK in the bulk phase. In the cytosol, active MEK kinase phosphorylates ERK on threonine and tyrosine residues, and specific ERK phosphatases are localized to the cytosol and nucleus. Because ppMEK is dephosphorylated predominantly in the cytoplasm, the spatial gradients of ppMEK and therefore ppERK may occur. A typical value for a cell radius is 10 μm, the protein diffusivity D is estimated to be of the order of 10^{-8} cm^2/s, and values for k_p were found to range from roughly 0.1 to 10 s^{-1} (76,82,93). With the k_p values of 0.25 or 1 s^{-1}, the distance from the plasma membrane at which the phosphorylation signal is attenuated by a factor of 10 is equal to 6.8 or 2.6 μm, respectively. Importantly, the exponential character of the decline in the phosphorylation signal with the distance from the cell membrane does not depend on the specific activity and kinetics of the membrane kinase, provided that the phosphatase is far from saturation. More elaborated calculations confirm that MEK and ERK gradients can be precipitous (94, 95), decreasing the strength of the phosphorylation signal to the nucleus. Instructively, the phosphorylation signal reaches further into the cell when the cascade has more levels (81); this may be one reason that cascades exist. The cascades found in eukaryotes tend to be longer than cascades in prokaryotes, which may be related to larger distances of signal propagation in eukaryotes.

5. Facilitated Mechanisms for Intracellular Signal Propagation

5.1. A Novel Role of Endocytosis in Activation of MAPK Signaling

Upon ligand binding and activation, many GPCRs and RTKs internalize via clathrin-coated pits. For instance, in hepatocytes, over 50% of phosphorylated EGFR is transferred to early endosomes during the first 10 min after EGF stimulation (40). Internalization takes receptor–ligand complexes and other signaling proteins from the plasma membrane and brings them inside the cell. Molecules, which were not recycled back to the cell membrane, are degraded in lysosomes. Although internalized GPCRs and RTKs continuously recycle back to the cell surface after dephosphorylation in endosomes, a significant proportion of receptors are located internally (40). Therefore, traditionally, clathrin-mediated endocytosis has been implicated in down-regulation of signaling by plasma membrane receptors. A novel role of endocytosis in "turning on" activation of the ERK cascade by cell surface receptors was first reported for the EGF receptor (96). A conditional defect in endocytosis can be imposed by the regulated expression of a mutant form of dynamin (Dyn1-K44A), a GTPase that is required for clathrin-coated vesicle formation. In HeLa cells, this expression led to a marked decrease in EGF-induced ERK activation, whereas Shc phosphorylation was enhanced in endocytosis-defective cells. Subsequent studies have demonstrated that both GPCR- and EGFR-mediated activation of ERK is sensitive to various distinct inhibitors of clathrin-mediated endocytosis, including monodansylcadaverine, depletion of intracellular K^+ or cholesterol, cytochalasin D, and a mutant dynamin (96–101). Therefore, a possible mechanism of control over signal transduction may engage receptor endocytosis. However, whereas experimental evidence points to an essential role of receptor endocytosis in the activation of MAPK cascades, the reason for the involvement of the endocytic machinery remains poorly understood (98,100). Interestingly, in some cellular systems, endocytosis was not required to activate ERK (102).

The relationship between receptor internalization and ERK activation allows us to suggest that trafficking of signaling intermediates within endocytic vesicles may be an efficient way of propagating the signal (94,103). Endocytic trafficking of active MEK can help to avoid the formation of steep spatial gradients of phosphorylated MEK and ERK because this mechanism overcomes the spatial separation of kinases and phosphatases within the MAPK cascade (94,103–106). Therefore, the endocytosis of phosphorylated MEK (or a protein complex containing activated MEK), rather than of activated receptors, appears to be critical for ERK activation.

5.2. Active Transport of Endosomes and Scaffolds is a Mechanism that Facilitates Signal Propagation

Living cells have developed multiple mechanisms to facilitate the information transfer from the plasma membrane to distant targets. These include trafficking of phosphorylated kinases with endosomes ("signaling endosome") and nonvesicular signaling complexes driven by molecular motors (89,94,103,105–107). Recent evidence indicates that the MAPK cascade components can bind to scaffolding proteins, e.g., MP1 and JIP-1 in mammalian cells (108). Dephosphorylation of kinases assembled on scaffold complexes might be decreased, or even precluded, because of sterical obstructions, as was suggested by Levchenko et al. (109).

Scaffolding also helps to deliver an entire signaling complex containing the MAP kinases to endocytic vesicles. Novel mechanisms have been discovered that link GPCRs to MAPK activation through use of β-arrestin as a scaffold for the ERK and JNK cascades (3,110). Besides its role in GPCR desensitization, β-arrestin has been shown to promote the targeting of the receptor to clathrin-coated pits. As β-arrestins can also recruit and activate Src, it is likely that the entire ERK and JNK cascades can be activated and recruited for clathrin-mediated internalization. Recent data suggest that molecular motors can be involved in transport of signaling complexes. In fact, in nerve cells, scaffolding proteins for the JNK pathway, known as JIPs, is the cargo for the molecular motor kinesin (111). Significantly, it was recently shown that survival signals in neurons are transmitted by a complex of phosphorylated ERK with intermediate filament vimentin and importin, driven by the molecular motor dynein

(107). Motor-mediated movement of the endosomes and kinase complexes along microtubules is remarkably distinct from chaotic diffusive motion and is able to prevent the formation of precipitous reaction–diffusion gradients (89,94,105,106).

6. Outlook

The spatiotemporal organization of mitogenic pathways analyzed here is central for understanding the control over intracellular signal transfer. Quantitative models integrate data on the distinct spatiotemporal dynamics of signaling from different cellular compartments and provide new insight into the connection between external stimuli and the signaling outcome in terms of gene expression responses. A picture is emerging, in which simple diffusion has a limited role in intracellular transport of signaling complexes. Endocytosis, scaffolding, molecular motors, and traveling waves of phosphoproteins appear to be involved in the propagation of signals to different cellular locations. These mechanisms control cellular decisions that determine cell fate.

Acknowledgments: I am grateful to Marc Birtwistle and Anatoly Kiyatkin for discussions and help with illustrative materials.

This work is supported by the National Institutes of Health grant GM59570.

References

1. Hepler JR, Gilman AG. G proteins. *Trends Biochem Sci* 1992;17(10): 383–387.
2. Neves SR, Ram PT, Iyengar R. G protein pathways. *Science* 2002;296(5573): 1636–1639.
3. Pierce KL, Luttrell LM, Lefkowitz RJ. New mechanisms in heptahelical receptor signaling to mitogen activated protein kinase cascades. *Oncogene* 2001;20(13):1532–1539.
4. Schafer B, Gschwind A, Ullrich A. Multiple G-protein-coupled receptor signals converge on the epidermal growth factor receptor to promote migration and invasion. *Oncogene* 2004;23(4):991–999.
5. Schlessinger J. Cell signaling by receptor tyrosine kinases. *Cell* 2000;103(2): 211–225.
6. Gray SG, Stenfeldt Mathiasen I, De Meyts P. The insulin-like growth factors and insulin-signalling systems: an appealing target for breast cancer therapy? *Horm Metab Res* 2003;35(11–12):857–871.
7. Hunter T. Signaling–2000 and beyond. *Cell* 2000;100(1):113–127.
8. De Meyts P, Whittaker J. Structural biology of insulin and IGF1 receptors: implications for drug design. *Nat Rev Drug Discov* 2002;1(10):769–783.
9. Pawson T, Nash P. Assembly of cell regulatory systems through protein interaction domains. *Science* 2003;300(5618):445–452.
10. Schlessinger J. Common and distinct elements in cellular signaling via EGF and FGF receptors. *Science* 2004;306(5701):1506–1507.
11. Shymko RM, De Meyts P, Thomas R. Logical analysis of timing-dependent receptor signalling specificity: application to the insulin receptor

metabolic and mitogenic signalling pathways. *Biochem J* 1997;326(Pt 2): 463–469.
12. Marshall CJ. Specificity of receptor tyrosine kinase signaling: transient versus sustained extracellular signal-regulated kinase activation. *Cell* 1995; 80(2):179–185.
13. Murphy LO, Smith S, Chen RH, et al. Molecular interpretation of ERK signal duration by immediate early gene products. *Nat Cell Biol* 2002;4(8): 556–564.
14. Murphy LO, MacKeigan JP, Blenis J. A network of immediate early gene products propagates subtle differences in mitogen-activated protein kinase signal amplitude and duration. *Mol Cell Biol* 2004;24(1):144–153.
15. McCawley LJ, Li S, Wattenberg EV, et al. Sustained activation of the mitogen-activated protein kinase pathway. A mechanism underlying receptor tyrosine kinase specificity for matrix metalloproteinase-9 induction and cell migration. *J Biol Chem* 1999;274(7):4347–4353.
16. Lloyd AC, Obermuller F, Staddon S, et al. Cooperating oncogenes converge to regulate cyclin/cdk complexes. *Genes Dev* 1997;11(5):663–677.
17. Sewing A, Wiseman B, Lloyd AC, Land H. High-intensity Raf signal causes cell cycle arrest mediated by p21Cip1. *Mol Cell Biol* 1997;17(9): 5588–5597.
18. Woods D, Parry D, Cherwinski H, et al. Raf-induced proliferation or cell cycle arrest is determined by the level of Raf activity with arrest mediated by p21Cip1. *Mol Cell Biol* 1997;17(9):5598–5611.
19. Bhalla US, Iyengar R. Emergent properties of networks of biological signaling pathways. *Science* 1999;283(5400):381–387.
20. Bhalla US, Ram PT, Iyengar R. MAP kinase phosphatase as a locus of flexibility in a mitogen-activated protein kinase signaling network. *Science* 2002;297(5583):1018–1023.
21. Kholodenko BN. Negative feedback and ultrasensitivity can bring about oscillations in the mitogen-activated protein kinase cascades. *Eur J Biochem* 2000;267(6):1583–1588.
22. Bagowski CP, Ferrell JE, Jr. Bistability in the JNK cascade. *Curr Biol* 2001;11(15):1176–1182.
23. Bagowski CP, Besser J, Frey CR, et al. The JNK Cascade as a Biochemical Switch in Mammalian Cells. Ultrasensitive and All-or-None Responses. *Curr Biol* 2003;13(4):315–320.
24. Xiong W, Ferrell JE, Jr. A positive-feedback-based bistable "memory module" that governs a cell fate decision. *Nature* 2003;426(6965):460–465.
25. Markevich NI, Hoek JB, Kholodenko BN. Signaling switches and bistability arising from multisite phosphorylation in protein kinase cascades. *J Cell Biol* 2004;164(3):353–359.
26. Wang X, Hao N, Dohlman H, et al. Computational and experimental analysis of bistability, stochasticity and oscillations in the mitogen activated protein kinase cascade. *Biophys J* 2006;90:1961–1978.
27. Altan-Bonnet G, Germain RN. Modeling T cell antigen discrimination based on feedback control of digital ERK responses. *PLoS Biol* 2005;3(11): e356.
28. Bluthgen N, Bruggeman FJ, Legewie S, et al. Effects of sequestration on signal transduction cascades. *Febs J* 2006;273(5):895–906.
29. Whitehurst A, Cobb MH, White MA. Stimulus-coupled spatial restriction of extracellular signal-regulated kinase 1/2 activity contributes to the specificity of signal-response pathways. *Mol Cell Biol* 2004;24(23):10145–10150.
30. Harding A, Tian T, Westbury E, et al. Subcellular localization determines MAP kinase signal output. *Curr Biol* 2005;15(9):869–873.

31. Borisov NM, Markevich NI, Hoek JB, et al. Signaling through receptors and scaffolds: independent interactions reduce combinatorial complexity. *Biophys J* 2005;89(2):951–966.
32. Borisov NM, Markevich NI, Hoek JB, et al. Trading the micro-world of combinatorial complexity for the macro-world of protein interaction domains. *Biosystems* 2006;83(2–3):152–166.
33. Conzelmann H, Saez-Rodriguez J, Sauter T, et al. A domain-oriented approach to the reduction of combinatorial complexity in signal transduction networks. *BMC Bioinformatics* 2006:7;34.
34. Bray D. Reductionism for biochemists: how to survive the protein jungle. *Trends Biochem Sci* 1997;22(9):325 326.
35. Goldbeter A. Computational approaches to cellular rhythms. *Nature* 2002;420(6912):238–2345.
36. Wolkenhauer O, Sreenath SN, Wellstead P, et al. A systems- and signal-oriented approach to intracellular dynamics. *Biochem Soc Trans* 2005;33(Pt 3):507–515.
37. Wiley HS, Shvartsman SY, Lauffenburger DA. Computational modeling of the EGF-receptor system: a paradigm for systems biology. *Trends Cell Biol* 2003;13(1):43–50.
38. Yarden Y, Sliwkowski MX. Untangling the ErbB signalling network. *Nat Rev Mol Cell Biol* 2001;2(2):127–137.
39. Kholodenko BN, Demin OV, Moehren G, et al. Quantification of short-term signaling by the epidermal growth factor receptor. *J Biol Chem* 1999; 274(42):30169–30181.
40. Di Guglielmo GM, Baass PC, Ou WJ, et al. Compartmentalization of SHC, GRB2 and mSOS, and hyperphosphorylation of Raf-1 by EGF but not insulin in liver parenchyma. *EMBO J* 1994;13(18):4269–4277.
41. Moehren G, Markevich N, Demin O, et al. Temperature dependence of the epidermal growth factor receptor signaling network can be accounted for by a kinetic model. *Biochemistry* 2002;41(1):306–320.
42. Machide M, Kamitori K, Kohsaka S. Hepatocyte growth factor-induced differential activation of phospholipase cgamma 1 and phosphatidylinositol 3-kinase is regulated by tyrosine phosphatase SHP-1 in astrocytes. *J Biol Chem* 2000;275(40):31392–31398.
43. Suenaga A, Kiyatkin AB, Hatakeyama M, et al. Tyr-317 phosphorylation increases shc structural rigidity and reduces coupling of domain motions remote from the phosphorylation site as revealed by molecular dynamics simulations. *J Biol Chem* 2004;279(6):4657–4662.
44. Schoeberl B, Eichler-Jonsson C, Gilles ED, et al. Computational modeling of the dynamics of the MAP kinase cascade activated by surface and internalized EGF receptors. *Nat Biotechnol* 2002;20(4):370–375.
45. Brightman FA, Fell DA. Differential feedback regulation of the MAPK cascade underlies the quantitative differences in EGF and NGF signalling in PC12 cells. *FEBS Lett* 2000;482(3):169–174.
46. Markevich NI, Moehren G, Demin O, et al. Signal processing at the Ras circuit: What shapes Ras activation patterns? *IEE Sys Biol* 2004;1:104–113.
47. Sasagawa S, Ozaki Y, Fujita K, et al. Prediction and validation of the distinct dynamics of transient and sustained ERK activation. *Nat Cell Biol* 2005; 7(4):365–373.
48. Shvartsman SY, Muratov CB, Lauffenburger DA. Modeling and computational analysis of EGF receptor-mediated cell communication in *Drosophila* oogenesis. *Development* 2002;129(11):2577–2589.
49. Hatakeyama M, Kimura S, Naka T, et al. A computational model on the modulation of mitogen-activated protein kinase (MAPK) and Akt

pathways in heregulin-induced ErbB signalling. *Biochem J* 2003;373(Pt 2):451–463.
50. Resat H, Ewald JA, Dixon DA, et al. An integrated model of epidermal growth factor receptor trafficking and signal transduction. *Biophys J* 2003; 85(2):730–743.
51. Goryanin I, Hodgman TC, Selkov E. Mathematical simulation and analysis of cellular metabolism and regulation. *Bioinformatics* 1999;15(9): 749–758.
52. Sauro HM, Hucka M, Finney A, et al. Next generation simulation tools: the Systems Biology Workbench and BioSPICE integration. *Omics* 2003;7(4): 355–372.
53. Blinov ML, Faeder JR, Goldstein B, et al. BioNetGen: software for rule-based modeling of signal transduction based on the interactions of molecular domains. *Bioinformatics* 2004;20(17):3289–3291.
54. Lok L, Brent R. Automatic generation of cellular reaction networks with Moleculizer 1.0. *Nat Biotechnol* 2005;23(1):131–136.
55. Slepchenko BM, Schaff JC, Macara I, et al. Quantitative cell biology with the Virtual Cell. *Trends Cell Biol* 2003;13(11):570–576.
56. Mendes P. Biochemistry by numbers: simulation of biochemical pathways with Gepasi 3. *Trends Biochem Sci* 1997;22(9):361–363.
57. Schaff JC, Slepchenko BM, Loew LM. Physiological modeling with virtual cell framework. *Methods Enzymol* 2000;321:1–23.
58. Sivakumaran S, Hariharaputran S, Mishra J, et al. The Database of Quantitative Cellular Signaling: management and analysis of chemical kinetic models of signaling networks. *Bioinformatics* 2003;19(3):408–415.
59. Olivier BG, Snoep JL. Web-based kinetic modelling using JWS Online. *Bioinformatics* 2004;20(13):2143–2144.
60. Goldstein B, Faeder JR, Hlavacek WS. Mathematical and computational models of immune-receptor signalling. *Nat Rev Immunol* 2004;4(6):445–456.
61. Morton-Firth CJ, Bray D. Predicting temporal fluctuations in an intracellular signalling pathway. *J Theor Biol* 1998;192(1):117–128.
62. Faeder JR, Blinov ML, Goldstein B, et al. Rule-based modeling of biochemical networks. *Complexity* 2005;10:22–41.
63. Goldbeter A, Koshland DE, Jr. An amplified sensitivity arising from covalent modification in biological systems. *Proc Natl Acad Sci USA* 1981; 78(11):6840–6844.
64. Ortega F, Acerenza L, Westerhoff HV, et al. Product dependence and bifunctionality compromise the ultrasensitivity of signal transduction cascades. *Proc Natl Acad Sci USA* 2002;99(3):1170–1175.
65. Ferrell JE, Jr. How responses get more switch-like as you move down a protein kinase cascade [letter; comment]. *Trends Biochem Sci* 1997;22(8): 288–289.
66. Kholodenko BN, Hoek JB, Brown GC, Westerhoff HV. Control analysis of cellular signal transduction pathways. In: Larsson C,. Pahlman I-L, Gustaffson L, eds. BioThermoKinetics in the Post Genomic Era. Guterborg; 1998:102–107.
67. Salazar C, Hofer T. Allosteric regulation of the transcription factor NFAT1 by multiple phosphorylation sites: a mathematical analysis. *J Mol Biol* 2003; 327(1):31–45.
68. Kholodenko BN, Hoek JB, Westerhoff HV, et al. Quantification of information transfer via cellular signal transduction pathways [published erratum appears in FEBS Lett 1997 Dec 8;419(1):150]. *FEBS Lett* 1997;414(2): 430–434.

69. Pomerening JR, Sontag ED, Ferrell JE. Building a cell cycle oscillator: hysteresis and bistability in the activation of Cdc2. *Nat Cell Biol* 2003;5(4): 346–351.
70. Sha W, Moore J, Chen K, et al. Hysteresis drives cell-cycle transitions in *Xenopus laevis* egg extracts. *Proc Natl Acad Sci USA* 2003;100(3):975–980.
71. Sauro HM, Kholodenko BN. Quantitative analysis of signaling networks. *Prog Biophys Mol Biol* 2004;86(1):5–43.
72. Stelling J, Sauer U, Szallasi Z, et al. Robustness of cellular functions. *Cell* 2004;118(6):675–685.
73. Carraway KL, Carraway CA. Signaling, mitogenesis and the cytoskeleton: where the action is. *Bioessays* 1995;17(2):171–175.
74. Bauman AL, Scott JD. Kinase- and phosphatase-anchoring proteins: harnessing the dynamic duo. *Nat Cell Biol* 2002;4(8):E203–206.
75. Sorkin A, Von Zastrow M. Signal transduction and endocytosis: close encounters of many kinds. Nat Rev *Mol Cell Biol* 2002;3(8):600–614.
76. Haugh JM, Lauffenburger DA. Physical modulation of intracellular signaling processes by locational regulation. *Biophys J* 1997;72(5):2014–2031.
77. Adam G, Delbruck M. Reduction of dimensionality in biological diffusion processes. In: Rich A, Davidson N, eds. Structural Chemistry and Molecular Biology. San Francisco: W. H. Freeman and Co.; 1968:198–215.
78. Bray D. Signaling complexes: biophysical constraints on intracellular communication. *Annu Rev Biophys Biomol Struct* 1998;27:59–75.
79. Kholodenko BN, Hoek JB, Westerhoff HV. Why cytoplasmic signalling proteins should be recruited to cell membranes. *Trends Cell Biol* 2000;10: 173–178.
80. Winters MJ, Lamson RE, Nakanishi H, Neiman AM, Pryciak PM. A membrane binding domain in the ste5 scaffold synergizes with gbetagamma binding to control localization and signaling in pheromone response. *Mol Cell* 2005;20(1):21–32.
81. Brown GC, Kholodenko BN. Spatial gradients of cellular phospho-proteins. *FEBS Lett* 1999;457(3):452–454.
82. Kholodenko BN, Brown GC, Hoek JB. Diffusion control of protein phosphorylation in signal transduction pathways. *Biochem J* 2000;350 Pt 3:901–907.
83. Carazo-Salas RE, Guarguaglini G, Gruss OJ, et al. Generation of GTP-bound Ran by RCC1 is required for chromatin-induced mitotic spindle formation. *Nature* 1999;400(6740):178–181.
84. Kalab P, Weis K, Heald R. Visualization of a Ran-GTP gradient in interphase and mitotic *Xenopus* egg extracts. *Science* 2002;295(5564):2452–2456.
85. Niethammer P, Bastiaens P, Karsenti E. Stathmin-tubulin interaction gradients in motile and mitotic cells. *Science* 2004;303(5665):1862–1866.
86. Caudron M, Bunt G, Bastiaens P, et al. Spatial coordination of spindle assembly by chromosome-mediated signaling gradients. *Science* 2005; 309(5739):1373–1376.
87. Lipkow K, Andrews SS, Bray D. Simulated diffusion of phosphorylated CheY through the cytoplasm of Escherichia coli. *J Bacteriol* 2005;187(1): 45–53.
88. Rao CV, Kirby JR, Arkin AP. Phosphatase localization in bacterial chemotaxis: divergent mechanism, convergent principles. *Phys Biol* 2005;2:148–158.
89. Kholodenko BN. Four-dimensional organization of protein kinase signaling cascades: the roles of diffusion, endocytosis and molecular motors. *J Exp Biol* 2003;206(Pt 12):2073–2082.

90. Kubicek M, Pacher M, Abraham D, et al. Dephosphorylation of Ser-259 regulates Raf-1 membrane association. *J Biol Chem* 2002;277(10):7913–7919.
91. Dhillon AS, Meikle S, Yazici , et al. Regulation of Raf-1 activation and signalling by dephosphorylation. *EMBO J* 2002;21(1–2):64–71.
92. Dhillon AS, Kolch W. Untying the regulation of the Raf-1 kinase. *Arch Biochem Biophys* 2002;404(1):3–9.
93. Zhao Y, Zhang ZY. The mechanism of dephosphorylation of extracellular signal-regulated kinase 2 by mitogen-activated protein kinase phosphatase 3. *J Biol Chem* 2001;276(34):32382–32391.
94. Kholodenko BN. MAP kinase cascade signaling and endocytic trafficking: a marriage of convenience? *Trends Cell Biol* 2002;12(4):173–177.
95. Maly IV, Wiley HS, Lauffenburger DA. Self-organization of polarized cell signaling via autocrine circuits: computational model analysis. *Biophys J* 2004;86(1 Pt 1):10–22.
96. Vieira AV, Lamaze C, Schmid SL. Control of EGF receptor signaling by clathrin-mediated endocytosis. *Science* 1996;274(5295):2086–2089.
97. Ceresa BP, Kao AW, Santeler SR, et al. Inhibition of clathrin-mediated endocytosis selectively attenuates specific insulin receptor signal transduction pathways. *Mol Cell Biol* 1998;18(7):3862–3870.
98. Ceresa BP, Schmid SL. Regulation of signal transduction by endocytosis. *Curr Opin Cell Biol* 2000;12(2):204–210.
99. Daaka Y, Luttrell LM, Ahn S, et al. Essential role for G protein-coupled receptor endocytosis in the activation of mitogen-activated protein kinase. *J Biol Chem* 1998;273(2):685–688.
100. Kranenburg O, Verlaan I, Moolenaar WH. Dynamin is required for the activation of mitogen-activated protein (MAP) kinase by MAP kinase kinase. *J Biol Chem* 1999;274(50):35301–35304.
101. Rizzo MA, Kraft CA, Watkins SC, et al. Agonist-dependent traffic of raft-associated Ras and Raf-1 is required for activation of the mitogen-activated protein kinase cascade. *J Biol Chem* 2001;276(37):34928–34933.
102. Leof EB. Growth factor receptor signalling: location, location, location. *Trends Cell Biol* 2000;10(8):343–348.
103. Miaczynska M, Pelkmans L, Zerial M. Not just a sink: endosomes in control of signal transduction. *Curr Opin Cell Biol* 2004;16(4):400–406.
104. Howe CL. Modeling the signaling endosome hypothesis: why a drive to the nucleus is better than a (random) walk. *Theor Biol Med Model* 2005;2:43.
105. Howe CL, Mobley WC. Signaling endosome hypothesis: A cellular mechanism for long distance communication. *J Neurobiol* 2004;58(2):207–216.
106. Howe CL, Mobley WC. Long-distance retrograde neurotrophic signaling. *Curr Opin Neurobiol* 2005;15(1):40–48.
107. Perlson E, Hanz S, Ben-Yaakov K, et al. Vimentin-dependent spatial translocation of an activated MAP kinase in injured nerve. *Neuron* 2005;45(5): 715–726.
108. Garrington TP, Johnson GL. Organization and regulation of mitogen-activated protein kinase signaling pathways. *Curr Opin Cell Biol* 1999;11(2): 211–218.
109. Levchenko A, Bruck J, Sternberg PW. Scaffold proteins may biphasically affect the levels of mitogen-activated protein kinase signaling and reduce its threshold properties. *Proc Natl Acad Sci USA* 2000;97(11):5818–5823.
110. Miller WE, Lefkowitz RJ. Expanding roles for beta-arrestins as scaffolds and adapters in GPCR signaling and trafficking. *Curr Opin Cell Biol* 2001;13(2):139–145.
111. Verhey KJ, Rapoport TA. Kinesin carries the signal. *Trends Biochem Sci* 2001;26(9):545–550.

17

Dynamic Instabilities Within Living Neutrophils

Howard R. Petty, Roberto Romero, Lars F. Olsen, and Ursula Kummer

Summary

The oscillatory metabolism of human neutrophils is used as a prototype biochemical subsystem to illustrate the ability of computational biology to both explain data and to predict biochemical mechanisms. Our work focuses upon the events surrounding neutrophil adherence and activation, which are features of many diseases. Cell activation is associated with increases in either or both the metabolic oscillatory frequency and amplitude. Our experimental studies and computational simulations have provided evidence that the frequency increase is linked to hexose monophosphate shunt (HMS) activation. Surprisingly, the increase in frequency is accounted for by a reduction in glycolytic activity. Increases in metabolic amplitude may also be observed during neutrophil activation and have been linked with the peroxidase cycle. Cell activation is independently regulated by these two pathways. The clinical relevance of this work is illustrated by frequency changes associated with febrile temperatures and diabetic levels of glucose. It is also demonstrated by neutrophil regulation during pregnancy, wherein high frequency oscillations are not observed and high amplitude oscillations are observed in the absence of cell activation stimuli. In this case, translocation of HMS enzymes to the centrosome accounts for the reduction in its activity, whereas translocation of myeloperoxidase (MPO) to the cell surface accounts for heightened peroxidase cycle activity during pregnancy. Hence, systems biology can be used to understand cell properties in complex clinical settings.

Key Words: Neutrophils; computational models; metabolism; disease mechanisms.

1. Introduction

Self-organization is a key concept in modern biology. On a macromolecular level, self-organization is seen in protein folding, the pairing of nucleic acid strands, and lipid bilayer formation, which are driven largely by entropy. At a larger distance scale, thermodynamics may also play a role in the formation of chemical patterns. Living cells are complex nonlinear systems maintained far from equilibrium by the constant flux of matter and energy. Consequently, intracellular chemistry may lose stability to form temporal oscillations and propagating concentration waves. We propose that these dynamic emergent structures are more than a physical consequence of cellular chemistry; they represent a means of information processing and distribution. These dynamic instabilities bear a striking resemblance to model chemical systems, such as the Belousov–Zhabotinskii reaction.

1.1. The Belousov–Zhabotinskii Reaction

The Belousov–Zhabotinskii reaction is the most thoroughly studied oscillatory chemical system. In this reaction, a transition metal ion catalyzes the oxidation and bromination of an organic dicarboxylic acid by bromate ions in an acidic environment. The reaction is held far from equilibrium for an extended period of time by including excess substrate. Due to chemical nonlinear kinetics and chemical feedback (i.e., the products of later steps being substrates of earlier steps), this system displays chemical oscillations and traveling waves (1). This reaction is sometimes considered as a model for glycolysis in cells.

1.2. Chemical Oscillations in Cells

Both prokaryotic and eukaryotic cells exhibit chemical oscillations (2). These oscillations take many forms, including glycolysis, cAMP, calcium, cytoskeletal assembly, mitosis, and other biorhythms. Human peripheral blood neutrophils are one useful model system for exploring the nature of cell oscillations and their potential clinical relevance. Morphologically polarized human neutrophils have been shown to exhibit oscillations in many parameters, such as membrane potential, calcium, metabolism, and receptor proximity (3). Moreover, oscillations in cell functions, including cell velocity, shape change, and pericellular proteolysis, as well as the production of superoxide anions, hydrogen peroxide and NO, have been reported (3). Interestingly, the periods of many of the chemical oscillations match the periods of the functional oscillations, suggesting a relationship between the two. Indeed, in some cases, such as nicotinamide adenine dinucleotide (phosphate) (NAD[P]H) oscillations and periodic superoxide release, both the period and phase of the oscillators match.

1.3. Neutrophils as a Model System

Neutrophils represent a particularly good model system for the study of cell oscillations. One important advantage of neutrophils is that they are semiautonomous cells; they are able to respond to information, migrate through tissues, and carry out their physiologic responsibilities without requiring other cell types. This is in sharp contrast to most cells, which are spatially fixed and require interactions with other cells for their maintenance or function. Changes in cell phenotype often require changes in gene expression. However, neutrophils can differentiate to the activated state within seconds. Importantly, enucleated neutrophils, which are called cytoplasts, are able to perform many neutrophil functions, such as phagocytosis, oxidant production, and migration (4), suggesting that many functions are hardwired in the cytoplasmic compartment. Finally, in contrast to most cell types, neutrophils rely primarily upon glycolysis for energy production (5).

One oscillator, the production of NAD(P)H, is closely associated with oxidant production. Superoxide production begins with the uptake of extracellular glucose, which is required for its synthesis (6,7). Indeed, cell activation is associated with an increase in the affinity of the glucose transporter for glucose (8). Superoxide is produced by the NADPH oxidase according to:

$$\frac{1}{2}\text{NADPH} + O_2 \rightarrow \frac{1}{2}\text{NADP}^+ + \frac{1}{2}\text{H}^+ + O_2^- \qquad (1)$$

NO production also begins with the donation of electrons from NADPH:

$$\text{L-arginine} + \text{NADPH} + \text{H}^+ + O_2 \rightarrow$$
$$\text{N}^G\text{-hydroxy-L-arginine} + \text{NADP}^+ + H_2O$$

and

$$2\text{N}^G\text{-hydroxy-L-arginine} + \text{NADPH} + \text{H}^+ + O_2 \rightarrow$$
$$2\text{L-citrulline} + \text{NADP}^+ + 2H_2O + 2\text{NO} \qquad (2)$$

Superoxide and NO may then yield additional downstream reactive oxygen metabolites and reactive nitrogen intermediates. Oscillations in NADPH, superoxide, and NO production are correlated with one another and vary in their frequency and amplitude, which can now be understood at a fundamental level based on the results of systems biology experiments.

In this chapter, we will combine recent work in experimental biophysics and computational biology to understand the dynamic behavior of metabolic systems in living neutrophils. The clinical ramifications of this work will be demonstrated through the analysis of *in vitro* model systems and the study of clinical samples. This interdisciplinary synthesis illustrates how systems biology may contribute to understanding complex biological problems.

2. Computational Biology of Neutrophil Oscillators

As illustrated in Figure 1, neutrophils exhibit metabolic oscillations, which have been detected using the autofluorescence of NAD(P)H and of mitochondrial flavoproteins (3,9). Cells that are not morphologically polarized exhibit very low amplitude oscillations with a period of roughly three minutes. Morphologically polarized neutrophils exhibit oscillations that vary in period (roughly 10 s or 20 s) and amplitude. These two parameters of polarized cells may yield four types of metabolic oscillations: low frequency and low amplitude, low frequency and high amplitude, high frequency and low amplitude, and high frequency and high amplitude (Figure 1A). We will begin by exploring computational simulations of these oscillators.

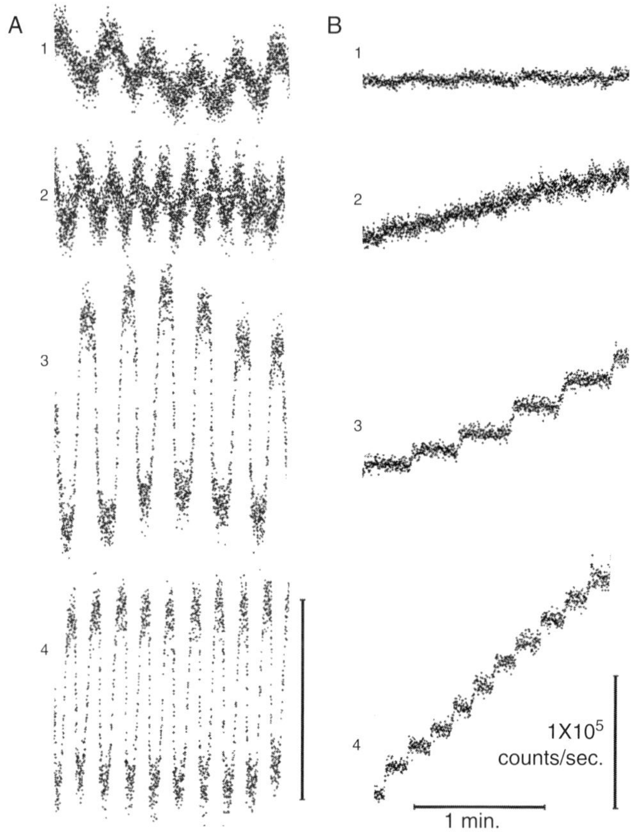

Figure 1. Representative kinetic traces of autofluorescence NAD(P)H oscillations (column A) and oxidation of H_2-TMRose by individual morphologically polarized human neutrophils. To ascertain local oxidant release, H_2-TMRose was contained in a gelatin matrix surrounding the neutrophils attached to a microscope slide. Short time courses are shown for illustrative purposes. Polarized neutrophils exhibit oscillations of approximately 20 s, and very low levels of oxidant release (trace 1 in A and B, respectively). Neutrophil activation with FMLP increases the frequency of NAD(P)H oscillations and oxidant release (trace 2). Exposure to IFN-γ led to oscillations of higher amplitude (traces 3). When cells were treated with both IFN-γ and FMLP, high-frequency/-amplitude NAD(P)H oscillations and high rates of oxidant production were observed. Vertical bars, 10^5 counts/s in columns A and B. The horizontal bar represents 1 min.

2.1. Role of MPO

2.1.1. Theory

To better understand the dynamic aspects of neutrophil biology, we set up model equations (ODEs) to describe neutrophil glycolysis (10) and its HMS (11), as well as its peroxidase cycle (12), which is coupled via the NADPH oxidase to the rest of the network. These equations and their parameters are described in detail in the cited references.

The nonlinear properties of MPO kinetics (13) and its abundance in neutrophils (14) suggest that it participates in the system's oscillatory

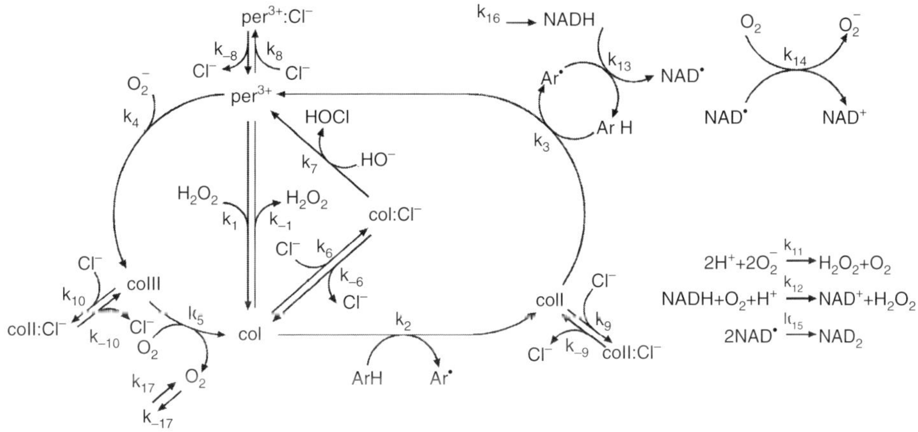

Figure 2. The MPO cycle is illustrated.

behavior. Therefore, we initially investigated its role during the activation of neutrophils (11). Like many peroxidases, MPO is able to catalyze the oxidation of NAD(P)H by molecular oxygen. This reaction involves a highly complex mechanism (Figure 2) that exhibits oscillations (12). In neutrophils, MPO is present either in membrane-enclosed vesicles (acidic granules), other vesicles formed by the fusion of several different types of vesicles (14), and the phagosome after phagocytosis, or it is associated with the outer plasma membrane surface or extracellular milieu after degranulation. NAD(P)H, on the other hand, is not present at these sites. Therefore, we studied if separation of the substrate and the enzyme allows the nonlinear enzymatic cycle to proceed within a membrane-enclosed compartment when electrons are shuttled by organic compounds like melatonin. We showed that this is indeed the case. Moreover, the concentration of this organic compound has a profound influence on the amplitude of the oscillations, which increased with concentration.

2.1.2. Experimental Verification

The computational prediction that melatonin enhances the amplitude of the oscillations in neutrophils has been experimentally verified, as shown in Figures 3 and 4. However, MPO is not the sole origin of NAD(P)H oscillations in neutrophils. This has been shown using MPO inhibitors such as salicylhydroxamic acid and MPO knockout mice (unpublished data). Although these conditions prevent large-amplitude oscillations, they do not block the oscillations per se. Hence, the role of neutrophil MPO in metabolic dynamics seems to be the generation of large amplitudes.

2.2. Role of Glycolysis

2.2.1. Theory

As glycolysis exhibits oscillations in some cell types, we set up a model for glycolysis in neutrophils (10). We were able to observe oscillations with a frequency fitting the experimental observations. Surprisingly, this

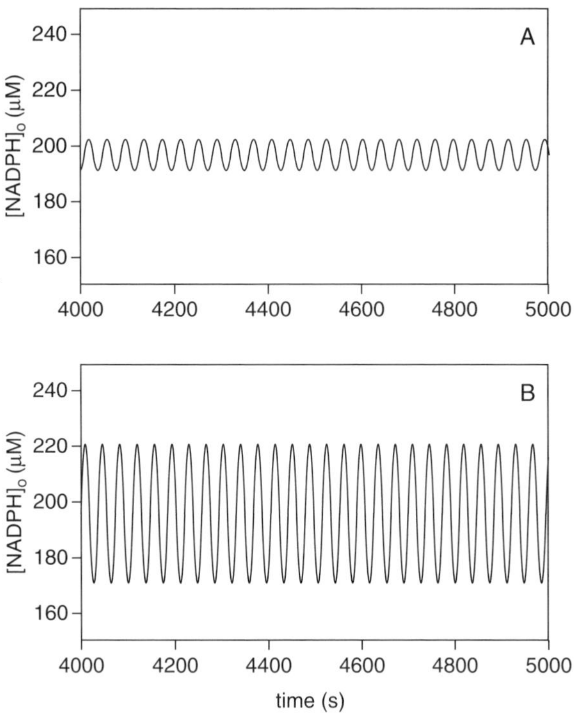

Figure 3. Simulating the effect of increasing concentrations of melatonin. Time series of NADPH in the presence of an initial concentration of melatonin of (a) 300 mM and (b) 350 mM. From Olsen et al., 2003 (11).

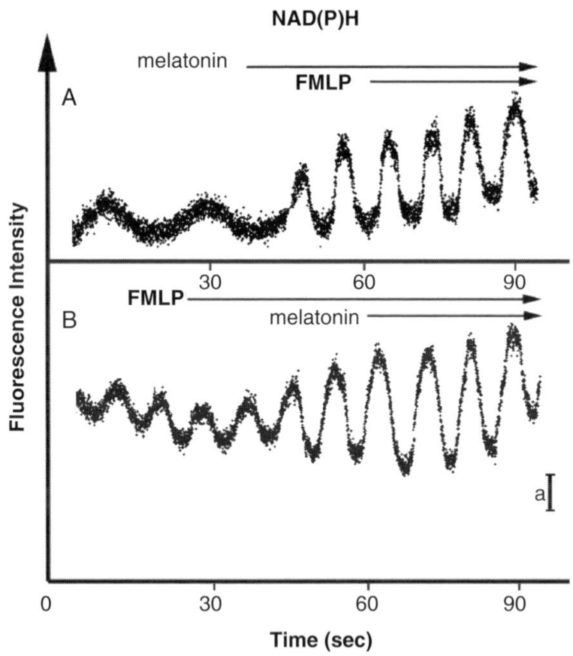

Figure 4. Experimental measurements of the effect of increasing concentrations of melatonin on NAD(P)H oscillations. (A) Although melatonin alone has no effect on NAD(P)H oscillations, when FMLP is added to the cells both the frequency and amplitude of the oscillations are increased. (B) When these same reagents are added in the opposite order, FMLP first increases the frequency, followed by a melatonin-dependent increase in amplitude. From Olsen et al., 2003 (11).

frequency has a strong tendency to increase when the flux through the system is lowered; for example, by enhancing the consumption of glucose-6-phosphate by the HMS (10). Because we and others provided experimental evidence that HMS activity is drastically increased by certain stimuli, this likely explains the change of frequency in NAD(P)H concentration oscillations observed upon activation.

2.2.2. *Experimental Verification*

Our computational findings, again, have been verified experimentally. First, experimental studies have shown that the HMS inhibitors 6-aminonicotinamide and dexamethasone block the formation of higher frequency NAD(P)H oscillations (11). To demonstrate the perturbation in glycolysis, we measured the autofluorescence of flavoproteins in neutrophils before and after exposure to high (12 mM) glucose concentrations, which activate the neutrophil's HMS (10). As the amount of flavoprotein autofluorescence is inversely related to electron transport rate (9), it can be used as an indicator of metabolic activity. Glucose addition increases the steady-state autofluorescence level, which reflects a decrease in mitochondrial activity corresponding to the aforementioned computational predictions. These findings were further supported by addition of a HMS inhibitor 6-aminonicotinamide (15), which prevented the increase in autofluorescence (10).

The combination of computational and experimental investigations allowed us to show a crucial role for MPO for increased amplitudes, whereas glycolysis is largely responsible for a shift in frequencies. We will now discuss the biological and medical implications of these findings.

3. Biomechanisms

In the preceding discussion, we have provided experimental evidence to support the existence of dynamic oscillations in human neutrophils and the underlying mechanisms using computational methods. The following paragraphs will explore some of the biological ramifications of these changes.

3.1. Effects of Exogenous Factors on Metabolic Oscillations

Human leukocytes respond to a wide variety of exogenous factors by undergoing activation. These factors may be of biological or synthetic origin. As leukocytes provide host defense against infectious agents, bacterial components often influence the activation status of cells. For example, lipopolysaccharide (LPS), which is a component of the outer wall of gram-negative bacteria, activates neutrophils (16). LPS promotes high-frequency NAD(P)H oscillations in neutrophils (16). Similarly, the peptide N-formyl-met-leu-phe (FMLP), which is similar or identical to certain bacterial peptides, also promotes neutrophil activation and high frequency NAD(P)H oscillations (Figure 1A, trace 2). Hence, bacterial substances known to activate neutrophils and the HMS alter the frequency of NAD(P)H oscillations. In addition to lipids and peptides, bacteria may possess unique nonmethylated CpG DNA sequences, which are recognized by cell receptors. Model CpG DNA oligonucleotides have the interesting property of being unable to stimulate neutrophil activation unless cells have been previously exposed to other exogenous or endogenous factors that promote high-amplitude NAD(P)H oscillations (17). Another well-known exogenous activator of leukocytes is the synthetic tumor promoter phorbolmyristate acetate (PMA). PMA is believed to activate cells via protein kinase C. In contrast to LPS and FMLP, PMA increases the amplitude of NAD(P)H oscillations, rather than the frequency of these oscillations. A wide variety of exogenous substances that activate neutrophils influence their NAD(P)H oscillations.

3.2. Effects of Endogenous Factors on Metabolic Oscillations

In response to infectious agents, the body produces a variety of molecules that mediate communication among immune cells. These molecules carry out a variety of tasks, such as recruiting leukocytes to a particular location and promoting their activation. Examples include the interleukin (IL) and interferon (IFN) families of molecules. A variety of relevant endogenous molecules were tested and found to influence the NAD(P)H frequency or amplitude. IL-8 and tumor necrosis factor-α (TNF-α) induce high-frequency NAD(P)H oscillations in neutrophils, just as FMLP and LPS do. These agents apparently act on biological signaling pathways that activate the HMS. In contrast, the regulatory molecules IL-12 and IFN-γ increase the oscillatory amplitude without affecting the frequency. Interestingly, two endogenous molecules, IL-2 and IL-6, have been found to be amplitude-dependent frequency modulators (18). These molecules have no effect on neutrophil oscillations or oxidant release unless a factor that enhances amplitude, such as IFN-γ or PMA, is present; in this case, the cell expresses high frequency and amplitude changes. Hence, many endogenous regulators of neutrophil function affect the HMS or peroxidase cycles.

Although melatonin is best known as a pineal hormone, it is also a biosynthetic product of leukocytes (19). It promotes priming, but does not directly activate neutrophils (20); that is, melatonin increases the production of oxidants, but only in the presence of another molecule that activates oxidant production. When melatonin is added to neutrophils, there are no effects on NAD(P)H oscillations or oxidant release. However, if FMLP is added to melatonin-treated cells, both the frequency and amplitude of NADPH oscillations change, as if FMLP has been combined with IFN-γ. Computational simulations predicted these effects of melatonin and the role of the peroxidase cycle in these processes (11).

3.3. Temperature-Dependent Changes in Frequency: Fever

Fevers are often associated with illness, especially infectious diseases. Fever is a complex physiological response involving both biochemical and temperature changes within the host. A variety of endogenous signals and exogenous substances, such as bacterial components, are capable of inducing a rise in body temperature. Although the biochemical changes have been well described, the physiological relevance of the thermal component of fever has been poorly understood. The temperatures found during human fever do not affect bacterial growth, although host defense is enhanced (21). However, it is unclear just how temperature regulates immunity. Recently, we have found that the frequency of NAD(P)H oscillations is a function of temperature (22): as the temperature increases, the oscillatory frequency increases. This is especially pronounced near 37°C. As NAD(P)H oscillations are coupled to the periodic production of superoxide anions and NO, the rates of the production of these reactive metabolites increase with temperature. As these molecules promote the destruction of bacteria, we propose that the thermal component of fever is a systemic signal acting to nonspecifically increase host defense capability throughout the host.

3.3. Glucose-dependent Changes in Frequency: A Possible Model of Diabetic Tissue Damage

As described above, increases in NAD(P)H oscillatory frequency are associated with cell activation, including activation of the HMS. One key feature of neutrophil activation is an increase in glucose transport, which is required for the activation of a neutrophil's HMS (6–8). In eukaryotes, glucose uptake is mediated by facilitated diffusion; hence, heightened external glucose concentrations may affect glucose transport and metabolism. In normal, healthy individuals, fasting serum glucose levels are roughly 1 mM to 2 mM, which supports the generation of oxidative molecules as described above. However, much higher glucose concentrations (>12 mM) may be found in poorly controlled diabetic patients. Therefore, we compared the effects of normal and diabetic glucose levels on NAD(P)H oscillations and cell function. Just as molecules such as FMLP and IL-8 increase the NAD(P)H oscillation frequency and oxidant production by neutrophils, heightened extracellular glucose levels increased NAD(P)H oscillation frequency, HMS activity, and oxidant production (10). In other words, diabetic levels of glucose cause the nonspecific activation of neutrophils. Computational studies (*see* section 2.2) support the conclusion that HMS activation is accompanied by an increase in the NAD(P)H oscillation frequency. Importantly, neutrophils produce as many oxidants at 12 mM glucose as they do in response to many immunologic stimuli. Indeed, heightened glucose influx caused by mass action may simply recapitulate the increase in glucose flux normally associated with neutrophil activation; this suggests that glucose flux is both a necessary and sufficient condition for cell activation. This is substantiated by the fact that the drug LY-83583, which enhances glucose uptake, activates neutrophils.

These computational and experimental findings are consistent with clinical observations. Two clinical manifestations of diabetes are nonspecific tissue damage, which is thought to be mediated by reactive oxygen metabolites, and increased susceptibility to infectious disease. When neutrophils were kinetically examined, 12 mM glucose activated metabolic changes and oxidant release, as described in the previous paragraph. However, after about 30 min, cells returned to a 20-s period oscillation and were refractory to further stimulation. Hence, nonspecific tissue damage may be caused by glucose-mediated cell activation. The cells become exhausted, perhaps by self-inflicted damage, and are unable to respond appropriately to infectious agents.

3.4. Immunomodulation in Pregnancy

It is well known that a pregnant woman's immune system undergoes changes to protect the fetal semiallograft, yet maintain significant resistance to infectious disease. For example, T helper type 2 cells appear to be enhanced during pregnancy, whereas the inflammatory activity of neutrophils is reduced (23). Specifically, the production of superoxide anions is reduced during pregnancy (24). These cellular changes are believed to account for the remission of autoimmune diseases, such as multiple sclerosis, rheumatoid arthritis, uveitis, etc., during pregnancy. To better understand the reduced neutrophil effector functions during pregnancy, we have studied the production of superoxide and NAD(P)H, which is the substrate-driving oxidant formation.

3.4.1. Biochemical Mechanisms Underlying Systems Behavior During Physiologic Regulation

We have found that neutrophils from pregnant women produce an intermediate level of ROMs: fewer ROMs than fully activated nonpregnant cells, but more than resting cells from nonpregnant women. Dynamic studies characterized the NAD(P)H oscillations of nonpregnancy cells as: resting (low amplitude, low frequency), primed (low frequency, high amplitude), activated (high frequency, low amplitude), and fully activated (high frequency, high amplitude). Under normal conditions, pregnancy neutrophils primarily express low-frequency, high-amplitude oscillations despite conditions that would fully activate cells from nonpregnant women (16). Several lines of evidence indicate that the high frequency oscillations are caused by activation of the HMS. We demonstrated that the HMS was depressed in neutrophils from pregnant women, thus accounting for the reduction in superoxide produced by pregnancy neutrophils in comparison to cells from nonpregnant women, which is consistent with the computer simulations described above. Although this explains why ROM production is reduced, it does not explain how this is achieved. As receptor signaling pathways likely remain intact during pregnancy, we hypothesized that metabolism is a key element controlling ROM production. When the intracellular locations of a large panel of metabolic enzymes was analyzed, enzymes of the HMS, including G-6-PDase, 6-PGDase, and transaldolase, were found at the periphery of cells from nonpregnant women, but at the centrosome (or cell center) of cells from pregnant women (16,25,26) (Figure 5). In contrast, the intracellular distribution of glycolytic enzymes was not affected by pregnancy. Furthermore, resonance energy transfer and other studies suggested that the HMS enzymes form a multienzyme complex that is transported by dynein along microtubules (unpublished data). Thus, the HMS could be disengaged by translocating the HMS enzyme complex away from the source of G-6-P, hexokinase, which is found at the periphery of the cell (16). In this way, G-6-P remains available to glycolysis, and the metabolic

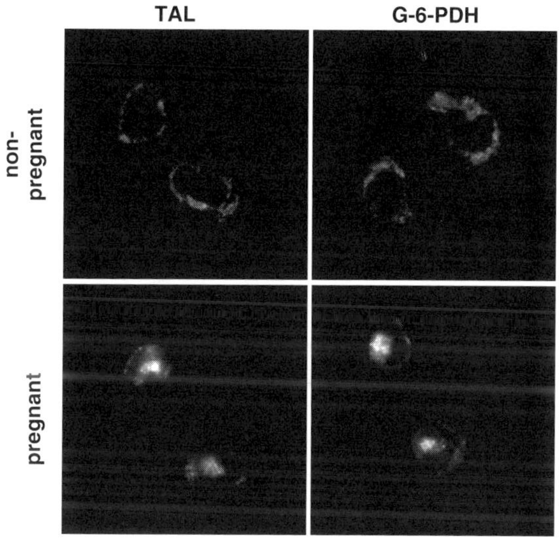

Figure 5. The trafficking of HMS enzymes in neutrophils from pregnant women. G-6-PDase (G-6-PDH) and transaldolase (TAL), key enzymes of the oxidative and nonoxidative arms of the HMS, were localized within neutrophils obtained from nonpregnant and pregnant women. These enzymes were found predominantly at the periphery of cells from nonpregnant women, but near the centrosome in cells from pregnant women.

frequency is not affected. This enzyme translocation mechanism provides an important example of metabolic microcompartmentalization.

We next sought to explain the increase in amplitude found for pregnancy cells (16). Computational and experimental studies (11) indicated that the amplitude changes could be accounted for by activation of the peroxidase cycle during pregnancy. To test this hypothesis, a panel of MPO inhibitors was added to pregnancy neutrophils. These inhibitors blocked the high amplitude oscillations. As the peroxidase cycle in human leukocytes is driven by MPO, we examined NAD(P)H oscillations in pregnant normal and MPO knockout mice. We found high-amplitude oscillations in neutrophils from normal pregnant mice, but low-amplitude oscillations in MPO knockout mice (27). Furthermore, we could switch the oscillatory phenotype of nonpregnancy cells to pregnancy cells by adding exogenous MPO. Conversely, we could reverse the pregnancy phenotype to nonpregnancy by removing MPO from the cell surface or by adding MPO inhibitors. Hence, the increased oscillatory amplitudes observed during pregnancy can be accounted for by the translocation/activation of MPO in pregnancy neutrophils (Figure 6). This seems reasonable from a clinical point of view, as activation of the peroxidase cycle may help offset the reduction in HMS activity to provide some level of oxidative defense against infectious agents during pregnancy.

Diabetes is a potentially severe complication of pregnancy that can lead to birth defects and other health concerns. In section 3.3, we described

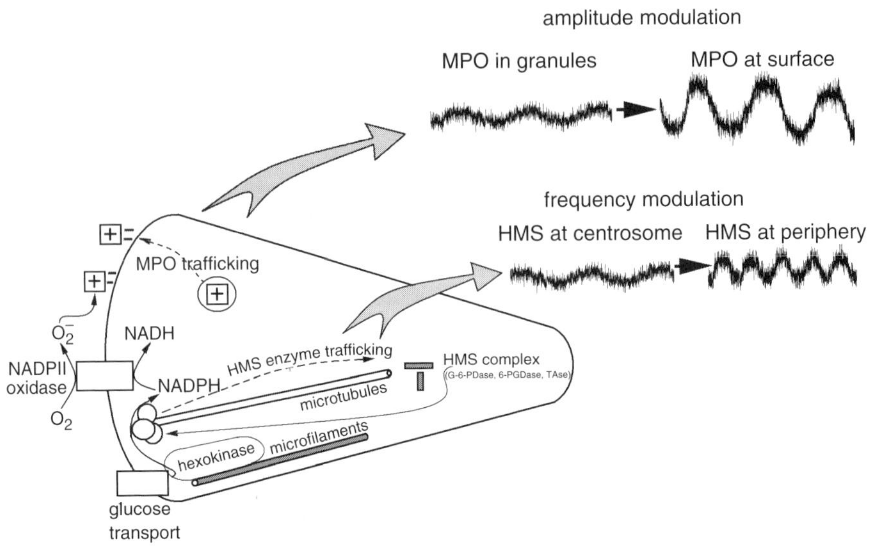

Figure 6. Relationship between NADPH oscillations and enzyme trafficking. High frequency oscillations are found when the HMS is near the cell periphery in cells exposed to agents such as FMLP, but not when it is located at the centrosome in similarly treated cells. Intracellular MPO does not affect oscillations, but when it is translocated to the cell surface, the peroxidase cycle is engaged, thus increasing NAD(P)H amplitudes. Enzyme trafficking accounts for the phenotypic changes in pregnancy cells.

how diabetic levels of glucose could promote dynamical changes and enhance ROM production in normal neutrophils. As the peroxidase cycle is independently regulated by MPO trafficking, heightened glucose levels during pregnancy may allow G-6-P to escape the glycolytic apparatus to reach the HMS enzymes sequestered at the centrosome, thereby supplying NADPH to support ROM production. As the peroxidase cycle is already engaged during pregnancy, glucose-mediated activation of the HMS could lead to anomalously high levels of oxidants (28), thereby accounting for many aspects of this disorder.

3.4.2. Cell Biological Mechanisms of Immunoregulation During Pregnancy

Although the changes in NAD(P)H dynamics, oxidant production, and enzyme trafficking in cells from pregnant women have been described, the biological pathways controlling these dynamic changes have not been identified. One important difference between a pregnant and nonpregnant woman is the presence of the placenta. Within the placenta, trophoblasts, which form a barrier between the mother and fetus, are thought to be important regulators of immune cell activity. We have recently found that the NAD(P)H oscillations and oxidant production of activated neutrophils rapidly revert to a normal phenotype upon contact with trophoblast cell lines, cytotrophoblasts of patients, and placental villi (29). Trophoblasts mediate these dramatic changes in neutrophil activation via at least two metabolic pathways affecting how cells

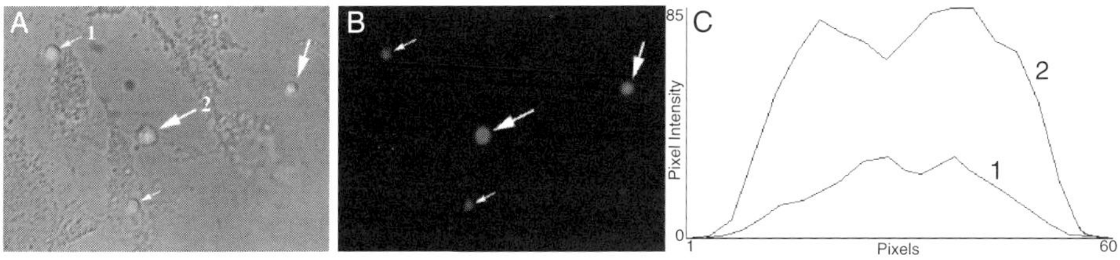

Figure 7. Trophoblast membranes affect the transport of NBD-glucose into neutrophils. Subconfluent trophoblast cultures on glass coverslips were ruptured with distilled water, and then washed. Neutrophils and FMLP were then added. Neutrophils in the proximity of trophoblast fragments (small arrows) internalized less NBD-glucose than those cells unattached to trophoblasts (large arrows). This difference is shown quantitatively in C, which is a line profile analysis of the cells labeled 1 and 2 in A. This suggests that trophoblast contact can regulate glucose transport in neutrophils.

handle glucose (29). In the first mechanism, contact with trophoblast membranes reduces the neutrophil's glucose transport, as illustrated by the reduced uptake of NBD-glucose in Figure 7. This strategy allows the placenta to rapidly protect both the fetus and itself from activated neutrophils. The second strategy, described above for maternal neutrophils, is the translocation of HMS enzymes from the cell periphery to the centrosome, which also affects the nature of intracellular glucose metabolism. Protection from harmful oxidative radicals is apparently important enough to merit two lines of trophoblast protection. The trophoblast molecules mediating these changes are currently under study.

4. Conclusions

Although molecular biology has provided a wealth of information concerning biological structures, it has been less successful in providing an understanding of the behaviors of biological systems, especially their dynamic behaviors. One chief concern of systems biology is to integrate multiple levels of biological research into a coherent vision of biological behavior. For example, consciousness does not arise from a single gene; this behavior emerges from a large set of neurons each containing a large number of participating molecules; the system expresses characteristic dynamical properties at all levels. In this chapter, we have described recent work combining computational and experimental studies designed to better understand the nature of neutrophil differentiation from the resting to activated phenotypes. This amenable model system is relevant to numerous human clinical conditions. In addition to its widely known roles in host resistance to infectious disease, neutrophil activation also participates in host resistance to cancer, tissue damage during ischemia–reperfusion injury, such as heart attacks and transplantation, autoimmune disease, etc. We found that the integration of computational and experimental biology offers unprecedented insight into understanding the dynamics of neutrophil metabolism, which, in turn, provides novel routes in drug discovery. Surprisingly, we have found that very simple epigenetic mechanisms are crucial in understanding metabolic systems dynamics; glucose transport, enzyme trafficking, and temperature are key variables in understanding neutrophil activation.

Previous studies have identified comparatively slow (τ = 3 min) and fast (τ = 10 s) oscillations of NAD(P)H autofluorescence in cells and tissues (2,30). Although the relatively slow oscillations of NAD(P)H have been attributed to the action of phosphofructokinase, the origins of other oscillatory properties, such as the amplitude and higher frequency oscillations, have been unknown. Computational biology has not only explained certain oscillatory properties but it has also predicted cell biological properties that were subsequently confirmed. Using a model based on the MPO–NADPH oxidase reaction (11), computational simulations predicted that neutrophil activation is accompanied by increased NAD(P)H frequencies and amplitudes. The roles of the NADPH oxidase and MPO were confirmed using inhibitors. Inhibitors and MPO knockout mice support the role of the peroxidase cycle in affecting metabolic amplitudes. The role of the HMS in promoting higher frequency oscillations is supported by inhibitors of glucose-6-phosphate dehydrogenase, including 6-aminonicotinamide and dexamethasone. To understand the mechanism accounting for the change in period from τ = 20 s to τ = 10 s, a model of the upper part of glycolysis, including terms for the HMS, was created. Using computational parameters mimicking those of activated neutrophils, such as HMS and glucose transport activation, the frequency was found to increase, as observed experimentally (10). These simulations predicted that the increase in frequency was caused by competition between the HMS and glycolysis for glucose-6-phosphate. This prediction is consistent with the subsequent experimental findings that: i) mitochondrial flavoprotein autofluorescence is increased at τ = 10 s (suggesting a reduction in electron transport, which is downstream from glycolysis) and ii) at least for a brief period of time, inhibition of glycolysis promotes oscillations of τ = 10 s. Although these ideas were developed in the context of neutrophil activation, they may be more generally applicable in cell biology.

Neutrophil activation represents a spectrum of biological states and chemical conditions. The metabolic dynamics of resting cells is dominated by phosphofructokinase. Adherent cells have a distinct phenotype, and can generate large amounts of oxidants (31). Unstimulated adherent cells display very low levels of oxidant release, and τ = 20 s. NAD(P)H autofluorescence oscillations. Neutrophils may demonstrate an activated HMS, which is accompanied by τ = 10 s oscillations. On the other hand, if cells exhibit an "activated" MPO activity (e.g., MPO translocation to the cell surface), higher amplitude oscillations are observed. Finally, both HMS enzymes and MPO can be activated to yield very high levels of oxidant production. Importantly, these pathways can be independently regulated by exogenous (e.g., LPS) and endogenous (e.g., IFN-γ) compounds. This provides a new rational framework encompassing both computational and experimental biology that provides a more quantitative understanding of neutrophil activation and priming.

Our findings suggest that many complicated diseases, all involving neutrophil activation at some level, may be understood at a more basic level using the tools of systems biology. The evolutionary advantage of febrile temperatures is that NAD(P)H oscillations and oxidant release increase in frequency, thereby delivering more oxidants to pathogens or other targets. It is not yet known if this change can be accounted for by some simple change, such as an increase in glucose transport, or some other factor affecting the system's behavior. Enhanced glucose transport is required for receptor-mediated neutrophil activation (6–8). Importantly, when glucose transport is nonspecifically enhanced by raising the extracellular glucose concentration, neutrophil activation is observed. This may explain the increase in nonspecific tissue damage during diabetes, as well as the depression in neutrophil function.

The concepts developed in these basic computational and experimental studies can be further applied to clinical samples. The immunological changes during pregnancy are poorly understood. For example, evidence has indicated that human neutrophils are inhibited during pregnancy, whereas other data indicate that they are activated. Our findings suggest that both of these statements are true. In normal pregnant women, the HMS is inhibited by translocation of HMS enzymes to the centrosome. In contrast, the peroxidase cycle is activated by translocation of MPO to the cell surface. At least some of these changes appear to be influenced by trophoblasts, as they contribute to HMS enzyme translocation and glucose transport. Thus, it is now possible to understand neutrophil activation in complex physiological states quantitatively, via computational simulations, and qualitatively, via cell biological experiments.

References

1. Epstein IR, Pojman JA. An Introduction to Nonlinear Chemical Dynamics: Oscillations, Waves, Patterns, and Chaos. New York: Oxford University Press; 1998.
2. Goldbeter A. Biochemical Oscillations and Cellular Rhythms. Cambridge, UK: Cambridge University Press; 1996.
3. Petty HR. Neutrophil oscillations: temporal and spatiotemporal aspects of cell behavior. *Immunologic Res* 2001;23:125–134.
4. Malawista SE, Van Blaricom G. Cytoplasts made from human blood polymorphonuclear leukocytes with or without heat: preservation of both motile function and respiratory burst oxidase activity. *Proc Natl Acad Sci USA* 1987;84:454–458.
5. Roos D, Balm AJM. The oxidative metabolism of monocytes. In: Sbarra AJ, Strauss RR, eds. The Reticuloendothelial System: A Comprehensive Treatise. New York: Plenum Press; 1980:189–229.
6. Kiyotaki C, Peisach J, Bloom BR. Oxygen metabolism in cloned macrophage cell lines: Glucose dependence of superoxide production, metabolic and spectral analysis. *J Immunol* 1984;132:857–866.
7. Naftalin RJ, Rist RJ. The relationship between sugar metabolism, transport and superoxide radical production in rat peritoneal macrophages. *Biochem Biophys Acta* 1993;1148:39–50.
8. Tan AS, Ahmed N, Berridge MV. Acute regulation of glucose transport after activation of human peripheral blood neutrophils by phorbol myristate acetate, fMLP, and granulocyte-macrophage colony-stimulation factor. *Blood* 1998;91:649–655.

9. Kindzelskii AL, Petty HR. Fluorescence spectroscopic detection of mitochondrial flavoprotein redox oscillations and transient reduction of the NADPH oxidase-associated flavoprotein in leukocytes. *Eur Biophys J* 2004;33:291–299.
10. Kummer U, Zobeley J, Naxerova K, Brasen JC, Fahmy R, Kindzelskii AL, Petty AR, Petty HR. Elevated glucose concentrations promote receptor-independent activation and exhaustion of adherent human neutrophils. *Biophys J* 2007;92:2597–2607.
11. Olsen LF, Kummer U, Kindzelskii AL, et al. A model of the oscillatory metabolism of activated neutrophils. *Biophys J* 2003;84:69–81.
12. Brasen JC, Lunding A, Olsen LF. Human myeloperoxidase catalyzes an oscillating peroxidase-oxidase reaction. *Arch Biochem Biophys* 2004;431: 55–62.
13. Sorensen O, Borregaard N. Methods for quantitation of human neutrophil proteins, a survey. *J Immunol Meth* 1999;232:179–190.
14. Karlsson A, Dahlgren C. Assembly and activation of the neutrophil NADPH oxidase in granule membranes. *Antioxid Redox Signal* 2002;4:49–60.
15. Bender JG, Van Epps, DE. 1985. Inhibition of human neutrophil function by 6-aminonicotinamide: the role of the hexose monophosphate shunt in cell activation. *Immunopharmacology* 1985;10:191–199.
16. Kindzelskii AL, Huang JB, Chaiworapongsa T, et al. Pregnancy alters glucose-6-phosphate dehydrogenase trafficking, cell metabolism and oxidant release of maternal neutrophils. *J Clin Invest* 2002;110:1801–1811.
17. Adachi Y, Kindzelskii AL, Petty AR, et al. IFN-γ primes RAW264 macrophages and human monocytes for enhanced oxidant production in response to CpG DNA via metabolic signaling: roles of TLR9 and myeloperoxidase trafficking. *J Immunol* 2006;176:5033–5040.
18. Adachi Y, Kindzelskii AL, Ohno N, et al. Amplitude and frequency modulation of metabolic signals in leukocytes: Synergistic role in interferon-γ and interleukin-6-mediated cell activation. *J Immunol* 1999;163:4367–4374.
19. Finocchiaro LM, Nahmod VE, Launay JM. Melatonin biosynthesis and metabolism in peripheral blood mononuclear leucocytes. *Biochem J* 1991;280: 727–731.
20. Morrey KM, McLachlan JA, Serkin CD, et al. Activation of human monocytes by the pineal hormone melatonin. *J Immunol* 1994;153:2671–2680.
21. Jiang Q, Cross AS, Singh IS, et al. Febrile core temperature is essential for optimal host defense in bacterial peritonitis. *Infect Immun* 2000;68:1265.
22. Rosenspire AJ, Kindzelskii AL, Petty HR. Febrile temperatures dramatically enhance local oxidant release by adherent neutrophils in response to lipopolysaccharide. *J Immunol* 2002;169:5396–5400.
23. Crocker IP, Baker PN, Fletcher J. Neutrophil function in pregnancy and rheumatoid arthritis. *Ann Rheum Dis* 2000;59:555–564.
24. Cotton DJ, Seligmann B, O'Brian B, et al. Selective defect in human neutrophil superoxide anion generation elicited by the chemoattractant N-formylmethionylleucylphenylalanine in pregnancy. *J Infect Dis* 1983;148: 194–199.
25. Kindzelskii AL, Ueki T, Michibata H, et al. 6-Phosphogluconate dehydrogenase and glucose-6-phosphate dehydrogenase form complexes in human neutrophils and traffic to the centrosome in pregnant, but not non-pregnant, women. *J Immunol* 2004;172:6373–6381.
26. Huang J-B, Espinoza J, Romero R, et al. Human neutrophil transaldolase undergoes retrograde trafficking during pregnancy, but anterograde trafficking in cells from non-pregnant women. *Metabolism* 2005;54:1027–1033.

27. Kindzelskii A, Clark AJ, Espinoza J, et al. Myeloperoxidase accumulates at the neutrophil surface and enhances cell metabolism and oxidant release during pregnancy. *Eur J Immunol* 2005:36;1619–1628.
28. Petty HR, Kindzelskii AL, Chaiworapongsa T, et al. Oxidant release is dramatically increased by elevated D-glucose concentrations in neutrophils from pregnant women: Apparent role of dynamic metabolic oscillations. *J Maternal Fetal Med* 2005:55;279–281.
29. Petty HR, Kindzelskii A, Espinoza J, et al. Trophoblast contact de-activates human neutrophils. *J Immunol* 2005:176;3205–3214.
30. Vern BA, Schuette WH, Leheta B, et al. Low-frequency oscillations of cortical oxidative metabolism in waking and sleep. *J Cereb Blood Flow Metab* 1988;8:215–226.
31. Nathan CF. Respiratory burst in adherent human neutrophils: triggering by colony-stimulating factors CSF-GM and CSF-G. *Blood* 1989;73:301–306.

18

Efficiency, Robustness, and Stochasticity of Gene Regulatory Networks in Systems Biology: λ Switch as a Working Example

Xiaomei Zhu, Lan Yin, Leroy Hood, David Galas, and Ping Ao

Summary

Phage λ is one of the most studied biological models in modern molecular biology. Over the past 50 years, quantitative experimental knowledge on this biological model has been accumulated at all levels: physics, chemistry, genomics, proteomics, functions, and more. All of its components are known in great detail. The theoretical task has been to integrate its components to make the organism work quantitatively and in a harmonic manner. This tests our biological understanding, and would lay a solid foundation for further explorations and applications, which is an obvious goal of systems biology. One of the outstanding challenges in doing this has been the so-called stability puzzle of the λ switch; the biologically observed robustness and the difficulty in mathematical reconstruction based on known experimental values. In this chapter, we review the recent theoretical and experimental efforts on tackling this problem. An emphasis is put on the minimum quantitative modeling, where a successful numerical agreement between experiments and modeling has been achieved. A novel method, tentatively named stochastic dynamical structure analysis, emerged from such study, and it is also discussed within a broad modeling perspective.

Key Words: Phage λ; genetic switch; robustness; efficiency; cooperation; stochastic processes; dynamical landscape; systems biology.

1. Introduction

The completion of the Human Genome Project prompts biological and medical research into a new phase, one that has never been experienced in biology. It is evident that a vast uncharted territory lies ahead, with tremendous promise in store (1). Great questions with important conceptual and practical implications have been asked and discussed (2–4). Speculations on the general principles underlying those great questions, and general methodologies to solve them, have been extensively debated since the beginning of this century (5,6). One of the authors of this chapter has been steadily promoting such exposition and contributing to this trend (7). Such efforts are needed not only as "a call to arms," they also help to define the various emerging fields. Nevertheless, in practical research, a full range of endeavors has to be explored. New tools will be invented to solve new problems and to take on old problems. In this chapter, we therefore turn our attention to the other side of consideration, not as an indication to underestimate the value of grand themes, but as an example to balance the grandeur. Instead of asking general questions and receiving limited answers, we wish to ask limited questions on a limited system and to find as complete answers as possible, along with a few general answers. We have been attempting this for the past few years. Such a methodology has been very effective since the dawn of modern science, and was first exemplified by Galileo. Specifically, we will focus our attention to the robustness and stability of a genetic switch (8,9) in phage λ, arguably the biological model that jump-started modern molecular biology (10).

In the modern information age, switch-like structures are building blocks in all architectures. It is the realization of the binary digit, the unit of information, and the "atom" today. As biology has been increasingly viewed as an information science (11–13), it would be desirable to have a thorough understanding of this building block. Indeed, detailed analyses have demonstrated that the response of a complexity network is often dealt with by various switches (14) and that genetic networks were shown to have the computational ability (15). By drawing a close analogy to the integration circuitry in an electronic wiring board, this methodology has been successfully employed in the modeling of genetic regulation during the earlier developmental stages in sea urchins (16). Currently, the study of switching in biology has been ranging from responses to environmental changes (17,18), developmental biology (19–21), neural networks (22,23), physiological response (24,25), genetic regulation (26–29), signal transductions (30), memory effect (31,32), olfactory perception (33), synthetic biology (34), biotechnological applications (35–38), to photosynthesis (39) and many other areas (40–43). Even in cell cycle processes, if viewing such a process not as driving by a cycling engine, but as what is controlled by a traffic light, the switch-like structure is likely to play a dominant role (44–46). Switch has indeed established itself as one of the fundamental elements in biological processes and as a paradigm for both experimental and theoretical studies in biology.

Why then has so much effort been expended on studying a particular virus genetic switch, the λ switch? To paraphrase Ptashne (8), this is a fair question desiring a clarification at the beginning. After all, every case in biology is at least partly accidental and specific, the workings of every organism having been determined by its evolutionary history, and the precise description we give of a process in one organism will probably not apply in detail to another. Thus, both robustness and stochasticity in biostructure must be included and carefully studied. This has been well illustrated in the context of the fundamental biological processes, such as those mentioned in the previous paragraph. As already indicated above, at various stages of development, depending in part on environmental signals, cells choose to use one or another set of genes, and thereby to proceed along one or another developmental pathway. It would be of great value to know what molecular mechanisms determine these choices. Hence, the λ life cycle is indeed a prototype for this problem, with the structure of feedback loops and the effect of stochasticity. In addition, we have a nearly complete understanding of all its parts; its genome was in fact known (47) long before the completion of the Human Genome Project, and the corresponding quantitative knowledge has been accumulated at all levels: physics, chemistry, DNA, protein, and functions (8,48,49). Despite such a long history of quantitative studies, the stability and robustness of the λ switch remained as one of the outstanding puzzles for computational biology at least until 2004 (9,50,51). The theoretical challenge has been to put all its components together as a harmonic working organism, one of the major tasks of systems biology.

In addition, one might wonder at the value of using quantitative and detailed modeling. Biological theories are generally known for their descriptive nature. For example, when Darwin presented his evolutionary theory, no single equation had been used. It was rather remarkable that though one of Darwin's main predictions, the age of Earth, was in direct conflict with known physics at his time, it was physics, not Darwin's theory that later went through a fundamental transformation to resolve this glaring contradiction, to the good of both physics and biology. Nevertheless, it would be wrong to conclude that a quantitative method would be of no use in biology. In fact, some subfields in biology, such as physiology and population genetics, are among the most mathematical in the natural sciences (3). As biology is becoming an information science (11–13), more subfields would be likely to do so in the future. The important question is: What would be the right framework of mathematical description (9,52)? It is true that an excessive use of mathematical language, which might be attractive to a modeler, generally does not enhance the understanding of a specific biological phenomenon (53). For example, with excessive parameters, any phenomenon can be described by a set of equations. Such a situation is not acceptable under Ockham's razor. The other extreme is to look for an effective description, with the hope of capturing the biological essence. The latter description is necessarily gross and qualitative, although extremely popular and particularly successful in biology. However, many features are obviously left behind by such an approach. It would be desirable to have a detailed quantitative study that can bridge those two approaches. The phage λ genetic switch provides precisely one of the excellent opportunities in biology to do so (54); one side is an on/off Boolean-type description for the genetic switch, and the other side is the detailed physical and chemical equations.

The rest of the review is organized as follows. Salient biological experimental studies on phage λ switch are summarized in section 2. Its key biochemical modeling elements are summarized in section 3. The stochastic dynamical structure analysis method is discussed in section 4, within the minimal quantitative model of phage λ. Calculated results and the comparison to biological data are discussed in section 5. In section 6, we summarize what has been done and place the minimum quantitative modeling methodology in a broader context. In section 7, the research effort on λ switch is put into an optimistic outlook.

2. Phage λ Genetic Switch

2.1. Phage λ Life Cycle

Bacteriophage λ is a virus that grows on a bacterium (8,55,56). It is one of the simplest living organisms. Almost all its parts have been known for the past 50 years. The genome of phage λ consists of a single DNA molecule wrapped in a protein coat. Upon infection of the host *Escherichia coli* cell, the phage λ injects its genome inside the bacterium and leaves the protein coat outside. Inside the bacterium, it chooses one of two modes of growth (Figure 1). Phage λ uses molecular-genetic apparatus of the cell for running and executing its own ontogenetic

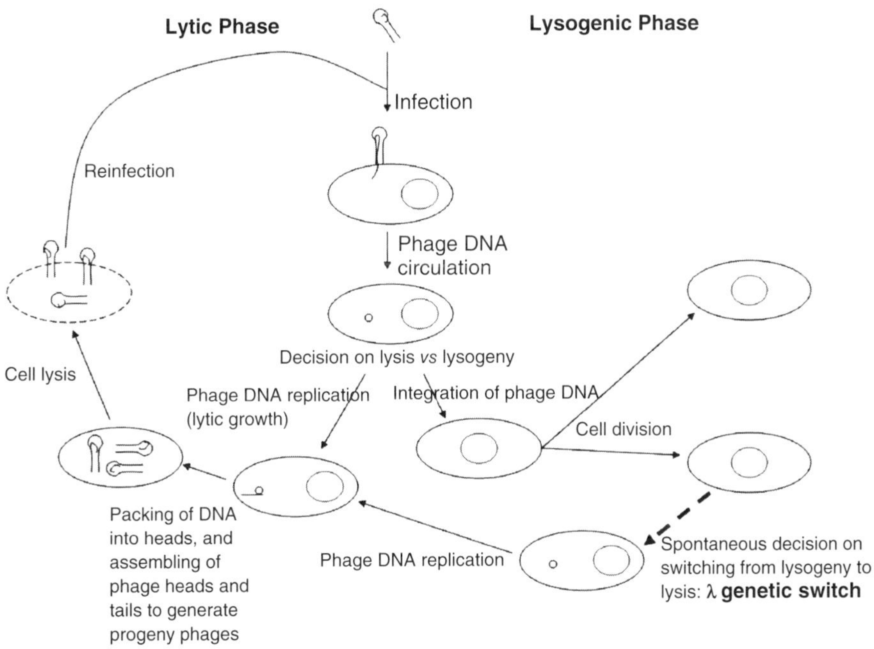

Figure 1. The schematic diagram of phage λ life cycle. The stochastic dynamics of a rare spontaneous decision on switching from the lysogenic phase to the lytic phase is the focus of this chapter. The induction by UV light and other SOS processes are not considered here.

subprograms to produce new λ phage particles, resulting in the lysis of the cell. Or, it can establish dormant residency in the lysogenic state, integrating its genome into the DNA of its host and replicate as a part of the host genome. In these two different life cycles, different sets of phage genes are expressed as a result of molecular interactions. Realistic modeling of the robustness and stability of such a process has remained one of the most important challenges in biocomputation and bioinformatics.

Because the analogous molecular interactions to phage λ are likely to underlie many developmental (8) and epigenetic (56) processes, one wishes to acquire deep understanding of the regulation of major biologic functions on the molecular level through the study of the genetic switch of phage λ. One of these functions is the programming of the epigenetic states; the ways the phage decides if it is going to follow a lysogenic or lytic growing state. Over the past five decades, extensive biological investigations have provided a fairly good qualitative picture in this respect. There exists a plausible scenario to guide the understanding of the experimental observations (8).

The maintenance and operation of the genetic switch is another function performed by the gene regulatory network (8,55,56). The phage growing in the lysogenic state remains latent, unless it is provoked. For example, switching to the lytic state happens when a signal is sent to activate RecA proteins, which cleave CI monomer, sending the phage into lytic growth. It is observed that the phage λ genetic switch is both highly stable and highly efficient. When the phage grows in a lysogenic state, it remains latent for many generations. Spontaneous induction happens less than once in 1 million cell divisions (Figure 1). Once the phage is exposed to an appropriate signal, it changes to the lytic state at a rate of almost 100%. Such a coexistence of stability and efficiency of the genetic switch in phage λ has been considered a mystery from the theoretical and mathematical modeling viewpoint.

2.2. Modeling Effort

There have been continuous mathematical and numerical activities on modeling phage λ. The rationale is rather straightforward; the biological functions should emerge as the systems properties from the model based on the molecular mechanism of phage regulatory elements and their independently measured parameters. The elegant physical–chemical model formulated by Shea and Ackers (48) for gene regulation of phage λ has become the foundation for later studies. However, soon afterward, Reinitz and Vaisnys (57) pointed out that the inconsistency between the theoretical results and experimental data may suggest additional cooperativity. Arkin et al. (58) performed stochastic simulation on phage λ development for the decision of lysogeny in the very early stage, demonstrating that this process is stochastic. Recently, Aurell and Sneppen (59) analyzed the robustness of phage λ genetic switch, using a method based on the Onsager–Machlup functional (58), and concluded that their theoretical analysis could not reproduce the robustness of the phage λ genetic switch. Their further study confirmed earlier results (51). Similar modeling was found mathematically from a different perspective (50).

The coexistence of the switch stability and switching efficiency is an apparent inconsistency for the following reasons. The lysogenic state is exceptionally stable. The fluctuations in the growth environment, the so-called extrinsic fluctuations, and the intrinsic fluctuation in the genetic switch, which are caused by the discrete nature of chemical reactions, do not easily and accidentally flip the switch. Then when the phage is "threatened," how can the switching process become so complete with so little outside intervention? The question about internal inconsistencies in these models naturally arises; whether the easily operable induction, or highly efficient switching, in Shea and Ackers' work (48) is a result of sacrificing the robustness of the genetic switch. Phrased differently, if a model were so constructed that it faithfully reproduces the observed robustness of the genetic switch, would it lose the efficiency of the switch? Undoubtedly, a credible model of phage λ should reproduce the properties of robustness, stability, and efficiency of the genetic switch simultaneously. From such a model we should also be able to calculate the observed quantities of phage development, such as the protein numbers and lysogenization frequencies. We hope to show that a foundation for such a mathematical framework against the experimental data is there, thanks to recent theoretical efforts on phage λ (9,48,57–59).

2.3. Modeling Strategy

Our procedure is to first summarize a minimal quantitative model for the phage λ genetic switch, a model that is motivated by both first principles and biological observations. We then ask the question of whether or not this minimal modeling can be successfully used to quantitatively reproduce various experimental results, and whether it is qualitatively correct in biology. If successful, the necessary modifications of molecular parameters in the modeling may be viewed as the *in vivo* and *in vitro* differences. Additional or different molecular processes inside a cell should be responsible for such differences. Some of them may be identifiable by current experimental techniques. If the answer to the above question would be negative, we would conclude that the minimum quantitative modeling would not be enough. More biological causes should be looked for instead. We will show that the answer so far is positive. By combining a newly developed powerful nonlinear dynamics analysis method, which takes the stochastic force into account (61–63) and classifies the stochastic dynamical structure into four different elements, with the previously established physical–chemical model (48), a novel mathematical framework was formulated to calculate the following quantitative characteristics of epigenetic states and developmental paths (9,64): the protein numbers in one bacterium, the protein number distributions, the lifetime of each state, and the lysogenization frequencies of mutants using the wild type as reference. We should emphasize that our review is focused on a specific biological system, though we have made an effort to put such work in perspective.

3. Towards Quantitative Modeling

3.1. Binding Configurations

The genetic switch controlling and maintaining the function of phage λ consists of two regulatory genes, *cI* and *cro*, and the regulatory regions O_R and O_L on the λ DNA. Established lysogeny is maintained by the protein CI, which blocks operators O_R and O_L, preventing transcription of all lytic genes, including *cro* (8,55,56). In lysogeny, the CI number functions as an indicator of the state of the bacterium; if DNA is damaged, e.g., by UV light, the protease activity of RecA is activated, leading to degradation of CI. A small CI number allows for transcription of the lytic genes, starting with *cro*, the product of which is the protein Cro.

The decision making, or the switching, is centered on operator O_R, and consists of three binding sites, O_{R1}, O_{R2}, and O_{R3}, each of which can be occupied by either a Cro dimer or a CI dimer (55,56). As illustrated in Figure 2, these three binding sites control the activity of two promoters P_{RM} and P_R for *cI* and *cro* transcriptions, respectively. The transcription of *cro* starts at P_R, which partially overlaps O_{R1} and O_{R2}. The transcription of *cI* starts at P_{RM}, which overlaps O_{R3}. The affinity of RNA polymerase for the two promoters, and subsequent production of the two proteins, depends on how Cro and CI bound to the three operator sites, and thereby establishes lysogeny with approximately 500 CI molecules per bacterium. If, however, the CI number becomes sufficiently small, the increased production of Cro flips the switch to lysis.

There have been numerous quantitative experimental studies on the stability in the switching of bacteriophage λ. Recently, the frequency of spontaneous induction in strains deleted for the *recA* gene has been

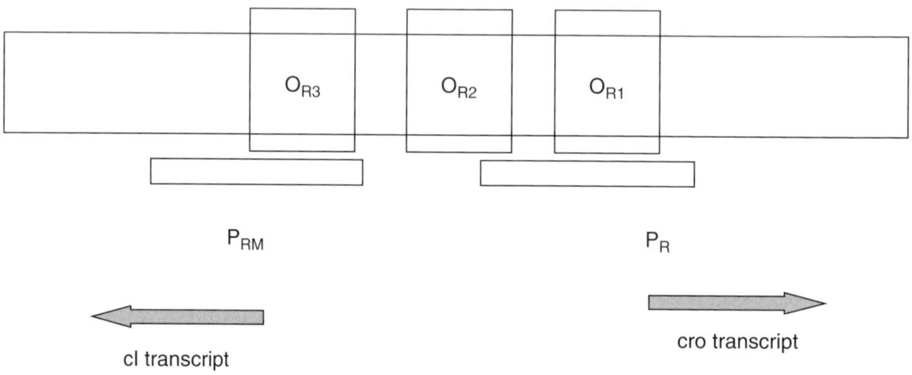

Figure 2. The O_R controlling region of the phage λ genetic switch. The mathematical studies (9,64) indicate that the cooperative binding of CI dimers at the O_{R1} and O_{R2} sites is a key to the robustness of the genetic switch. Such a cooperative binding enhances the positive CI feedback. When the CI positive feedback is turned on by the existing CI dimers, CI proteins are synthesized. The phage evolves to the lysogenic state. Otherwise, Cro proteins are synthesized and the phage evolves to the lytic state.

reported independently by three groups (65–67), which was reviewed by Aurell et al. (67). They all confirmed two earlier important observations: that there is a switching behavior and that the switch is stable. In addition, they all obtained consistent numerical values for the switching frequency, in spite of the use of different strain backgrounds done on different continents and at different times. However, computational and mathematical attempts to quantitatively understand this behavior have not been successful, even permitting the possibility that the wild type may be more stable (59,67).

More recent data (9, 64) suggest that the wild type may be two orders of magnitude more stable than previously observed (66); the switching rate to the lytic state may be less than 4×10^{-9} per minute. In addition to the call for more experimental studies, this puts the theoretical modeling in a more challenging position. This wild-type data was used as the main input to further fix the model in the works of Zhu et al. (9). The previous data were also discussed to illustrate a pronounced exponential sensitivity in such a modeling, which is summarized in the following sections.

The CI and Cro protein molecules in the cell are assumed to be in homeostatic equilibrium. There are not always the same numbers of CI and Cro dimers bound to the operators at any particular time. These numbers are fluctuating, and the equilibrium assumption should give the size of these fluctuations. The key inputs are CI and Cro dimerization constants, and the Gibbs free energies for their bindings to the three operator sites O_R1, O_R2, and O_R3 (68–74) (*see* the legends of Tables 1 and 2 for a more detailed description).

Following Ackers et al. (75) and Aurell et al. (67), we encode a state s of CI and/or Cro bound to O_R by three numbers (i,j,k) referring to O_{R3}, O_{R2}, and O_{R1}, respectively. The coding for s is 0 if the corresponding site is free, 1 if the site is occupied by a CI dimer, and 2 if the site is occupied by a Cro dimer. The probability of a state s with i(s) CI dimers and j(s) Cro dimers bound to O_R is in the grand canonical approach of Shea and Ackers (48)

$$p_R(s) = Z^{-1}\ [CI]^{i(s)}\ [Cro]^{j(s)}\ [RNAp]^{k(s)}\ \exp(-\Delta G(s)/RT). \qquad (1)$$

For example, if CI occupies O_{R1} and Cro occupies O_{R2n} and O_{R3}, we have $i(s) = 1$, $j(s) = 2$, $k(s) = 0$, and $p_R(s) = p_R(221)$. RNA polymerase (RNAp) can occupy either O_{R1} and O_{R2}, or O_{R2} and O_{R3}, not other configurations. There are a total 40 states represented by *s* (Table 1). The normalization constant Z is determined by summing over *s*: $Z = \Sigma_s[CI]^{i(s)}\ [Cro]^{j(s)}\ [RNAp]^{k(s)}\ esp(-\Delta G(s)/RT)$. Here, [] denotes the corresponding protein dimer concentration in the bacterium, $\Delta G(s)$ the binding energy for binding configuration s, R the gas constant, and T the temperature.

3.2. Deterministic Model

We further simplify the expression of $p_R(s)$ by noticing that CI and Cro control the operator (8,55,56). If O_{R1} and O_{R2} are unoccupied by either CI or Cro, RNAp binds to them with a probability determined by RNAp binding energy. The idea that RNAp first binds to O_{R1} and O_{R2}, followed

Table 1. The 40 configurations corresponding to right operator.

State	P_{RM}	O_R3	O_R2	O_R1	P_R	i(s)	j(s)	k(s)
1						0	0	0
2				R_2		1	0	0
3			R_2			1	0	0
4		R_2				1	0	0
5			R_2	R_2		2	0	0
6		R_2	R_2			2	0	0
7		R_2		R_2		2	0	0
8		R_2	R_2	R_2		3	0	0
9				C_2		0	1	0
10			C_2			0	1	0
11		C_2				0	1	0
12			C_2	C_2		0	2	0
13		C_2	C_2			0	2	0
14		C_2		C_2		0	2	0
15		C_2	C_2	C_2		0	3	0
16			C_2	R_2		1	1	0
17		C_2		R_2		1	1	0
18		C_2	C_2	R_2		1	2	0
19		C_2	R_2			1	1	0
20			R	C_2		1	1	0
21		C_2	R_2	C_2		1	2	0
22		R_2	C_2			1	1	0
23		R_2		C_2		1	1	0
24		R_2	C_2	C_2		1	2	0
25		C_2	R_2	R_2		2	1	0
26		R_2	C_2	R_2		2	1	0
27		R_2	R_2	C_2		2	1	0
28					RNA_P	1	0	1
29		R_2			RNA_P	1	0	1
30		C_2			RNA_P	0	1	1
31	RNA_P		R_2			1	0	1
32	RNA_P		R_2	R_2		2	0	1
33	RNA_P		R_2	C_2		1	1	1
34	RNA_P					0	0	1
35	RNA_P			R_2		1	0	1
36	RNA_P			C_2		0	1	1
37	RNA_P		C_2			0	1	1
38	RNA_P		C_2	R_2		1	1	1
39	RNA_P		C_2	C_2		0	2	1
40	RNA_P				RNA_P	0	0	2

R_2 stands for CI (λ repressor) dimer and C_2 for Cro dimer.
(*Sources*: Darling et al. (74), Ackers et al. (75), Capp et al. (76))

by blocking CI and Cro binding, is excluded based on the assumption that only CI and Cro controls the regulatory behavior. In addition to experimental observation, this assumption is justifiable if the time scale associated with CI and Cro binding is shorter than the RNAp binding. Except for an overall constant, which we include into the rate of transcription, the RNAp binding is no longer relevant. We therefore take it out of the expression $p_R(s)$. The total number of states is reduced to 27. This simplification was first used by Aurell and Sneppen (59). We will drop the subscript R for binding probability p_R. We should point out

Table 2. Parameters used in the modeling.

RT	0.617 kcal/mol
Effective bacterial volume	0.7×10^{-15} *l*
E_{cro}	20
E_{cI}	1
T_{RM}	0.115/s
$T_{RM}{}^{u}$	0.0105/s
T_R	0.30/s
τ_{CI}	2.9×10^3 s
τ_{Cro}	5.2×10^3 s
Converting factor between protein number and concentration	1.5×10^{-11}
in vitro free energy differences for wild-type λ	
ΔG (001)	–12.5 kcal/mol
ΔG (010)	–10.5 kcal/mol
ΔG (100)	–9.5 kcal/mol
ΔG (011)	–25.7 kcal/mol
ΔG (110)	–22.0 kcal/mol
ΔG (111)	–35.4 kcal/mol
ΔG (002)	–14.4 kcal/mol
ΔG (020)	–13.1 kcal/mol
ΔG (200)	–15.5 kcal/mol
ΔG(cooperative)	–2.7 kcal/mol
dimerization energy	
ΔG_{CI2}	
ΔG_{Cro2}	–11.1 kcal/mol
	–7.0 kcal/mol
in vitro free energy differences for O_R3' binding	
ΔG (100)	–10.5 kcal/mol
ΔG (200)	–13.7 kcal/mol
in vivo free energy differences—*in vitro* free energy differences	
ΔG_{CI}	–2.5 kcal/mol
ΔG_{Cro}	–4.0 kcal/mol
ΔG (cooperative)	–3.7 kcal/mol

CI dimer affinities to O_{R1}, O_{R2}, and O_{R3} are from Darling et al. (73,74). Cro dimer affinities to O_{R1} and O_{R2} are from Takeda et al. (69,79), Jana et al. (71), Kim et al. (70), and Aurell et al. (67). The CI dimerization energy is taken from Koblan and Ackers (80), Cro dimerization energy from Jana et al. (71,72). The bacterial volume is taken from Bremmer and Dennis (99). E_{CI} and E_{Cro} are taken from Shean and Gottesman (100), Ringquist et al. (101), and Kennell and Riezman (102). The *in vitro* parameters have been summarized by Aurell et al. (67), which we largely follow. However, we here point out two differences: (a) our effective bacterial volume, estimating from the typical size of the bacterium, assuming a tube of about 0.7 μm in diameter and 2 μm in length, is approximately a factor 3 smaller; and (b) the normalization factor for the concentrations, calculated against the number of water molecules, is approximately a factor of 60 smaller. The difference between *in vivo* and *in vitro* values is consistent with qualitative experimental observation (8). The wild-type data (9) is used to determine the *in vivo* and *in vitro* difference.

that previous experimental and theoretical results had been concisely reviewed by Aurell et al. (67), whose convention we shall follow.

The dimer and monomer concentrations are determined by the formation and de-association of dimers, which gives the relation of dimer concentration to the total concentration of proteins as follows:

$$[CI] = [N_{CI}]/2 + \exp(\Delta G_{CI}/RT)/8 - ([N_{CI}]\exp(\Delta G_{CI}/RT)/8 + \exp(2\Delta G_{CI}/RT)/64)^{1/2} \quad (2)$$

Here, $\Delta G_{CI} = -11.1$ kcal/mol is the dimer association free energy for CI.

A similar expression for [Cro] is as follows:

$$[Cro] = [N_{Cro}]/2 + \exp(\Delta G_{cro}/RT)/8 - ([N_{Cro}]\exp(\Delta G_{Cro}T)/8 + \exp(2\Delta G_{Cro}T)/64)^{1/2}. \quad (3)$$

Here, $\Delta G_{Cro} = -7$ kcal/mol is the dimer association free energy for Cro. $[N_{CI}]$ and $[N_{Cro}]$ are the monomer concentrations of CI and Cro, respectively.

CI and Cro are produced from mRNA transcripts of CI and Cro, which are initiated from promotor sites P_{RM} and P_R. The rate of transcription initiation from P_{RM} when stimulated by CI bound to O_{R2} is denoted T_{RM}, and when not stimulated it is denoted T^u_{RM}. The number of CI molecules produced per transcript is E_{cI}. The overall expected rate of CI production is as follows:

$$f_{CI}(N_{CI}, N_{Cro}) = T_{RM}E_{cI}[p(010) + p(011) + p(012)] + T^u_{RM}E_{cI}[p(000) + p(001) + p(002) + p(020) + p(021) + p(022)]. \quad (4)$$

Here, N_{CI} and N_{Cro} are the protein numbers for CI and Cro inside the bacterium respectively. The converting factor between the protein concentration and the corresponding protein inside the bacterium is listed in Table 2. Similarly, the overall expected rate of Cro production is

$$f_{Cro}(N_{CI}, N_{Cro}) = T_R E_{cro}[p(000) + p(100) + p(200)]. \quad (5)$$

We use T_{RM}, E_{cI}, E_{cro}, and T^u_{RM} from Aurell and Sneppen (59), which were deduced from the resulting protein numbers in lysogenic and lytic states.

The free energies ΔG(s) are determined from *in vitro* studies, that is, they are obtained outside of the living bacterium. The *in vivo* conditions, inside a living bacterium, could be different. The measured protein–DNA affinities could sensitively depend on the ions present in the buffer solutions, as well as other factors. This observation will be important in our comparison between theoretical results and experimental data. On the other hand, the *in vivo* effects of such changes should be compensated for, e.g., changed KCl concentrations are attributable to putrescine (76), other ions, and crowding effects (77). We note that Record et al. (77) already observed that there may exist a significant difference between *in vivo* and *in vitro* molecular parameters. The data quoted in Darling et al., (73,74) was obtained at KCl concentration of 200 mM, which resembles *in vivo* conditions. Therefore, though we expect a difference between the *in vivo* and *in vitro* data, the difference may not be large, typically within 20%–30% of the *in vitro* values.

The mathematical model that describes the genetic regulation in Figure 2 is a set of coupled equations for the time rate of change of numbers of CI and Cro in a cell (57):

$$dN_{CI}(t)/dt = F_{CI}(N_{CI}(t), N_{Cro}(t))$$
$$dN_{Cro}(t)/dt = F_{Cro}(N_{CI}(t), N_{Cro}(t)), \quad (6)$$

where the net production rates are

$$F_{CI} = f_{CI}(N_{CI}, N_{Cro}) - N_{CI}/\tau_{CI}$$
$$F_{Cro} = f_{Cro}(N_{CI}, N_{Cro}) - N_{Cro}/\tau_{Cro}. \quad (7)$$

Equations (6) and (7) represent the minimum deterministic model. Here, dN/dt is the rate N changes. The production terms f_{CI} and f_{Cro} are functions of CI and Cro numbers in the bacterium. With no Cro in the system, the curve of f_{CI} versus CI number has been experimentally measured (78). As reviewed in Aurell et al. (67) these measurements are consistent with the best available data on protein–DNA affinities (69,70,79), dimerization constants (80), initiation rates of transcriptions of the genes, and the efficiency of translation of the mRNA transcripts into protein molecules. The decay constant τ_{CI} is an effective lifetime, proportional to the bacterial lifetime, as CI molecules are not actively degraded in lysogeny, whereas τ_{Cro} is approximately 30% smaller (81). We comment that there is considerably more experimental uncertainty in the binding of Cro, both to other Cro and to DNA, than the binding of CI; e.g., the work of Darling et al. (73,74). As a minimal mathematical model of the switch, we take τ_{CI} and τ_{Cro} from data and deduce f_{CI} and f_{Cro} at a nonzero number of both CI and Cro with a standard set of assumed values of all binding constants, which are summarized by Aurell et al. (67) and are adopted here (Table 2, with differences in cell volume and converting factor, as well as the *in vivo* and *in vitro* differences).

3.3. Positive and Negative Feedbacks: *In Vivo* versus *In Vitro*

Both positive and negative feedbacks are employed in this genetic switch. For CI (Cro), it has a positive feedback effect on itself and a negative feedback effect on the production of Cro (CI) (8). Evidently, these feedbacks are systems effects; breaking them down into disintegrated parts would cause the feedback effects to disappear. They emerge only when a proper integration is done. Such a systems effect is well known in engineering (82).

In establishing the minimum deterministic model, another major implicit assumption is on the time scales. We have assumed that the dimerization process is a fast process on the scale of Cro and CI production, and hence can be treated as algebraic constraints. The dimerization has been subjected to continuous experimental (83) and theoretical (84) studies. It was concluded by *in vitro* experiments (83) that the Cro dimerization is slower than that of CI. Nevertheless, the Cro dimerization time is on the order of fractions of a minute (83), which is much smaller than the typical time, on the order of 20 min, used in our modeling (Table 2). We may be able to apply the useful quasi–steady-state approximation (85). Thus, the algebraic constraints appear to be a reasonable assumption for such a minimum modeling as that used in previous works (9,59,64,67). Other cellular processes have also been implicitly assumed to be fast. All their residual effects will be treated as extrinsic stochastic effect contribution, and will be incorporated into the minimum modeling in the name of intrinsic versus extrinsic noises, which will be discussed in the following sections.

We further specify the meaning of minimum deterministic modeling. First, such a model should be viewed as what the system might be, not as what it must be. Many features are not explicitly contained in it, such as the nonspecific binding (86) and the looping (87,88). The nonspecific binding was already demonstrated not to be crucial, but the looping may well be, which we will come back to later. Nevertheless, we point out that by assuming the minimum deterministic modeling, we tentatively and tactically assume it has captured all the essential features of the λ switch by aggregating molecular processes around it. In doing so, it suggests another understanding of the difference between *in vivo* and *in vitro*; all the parameters we adopted from *in vitro* measurement would indeed take a different value *in vivo*, because there exist numerous other biological processes inside the cell that contribute to this difference. If the minimum model is essentially correct, it should be able to account for experimental data in a quantitative manner, along with predictions that need to be further tested. We will show that we have indeed achieved this goal after several decades of theoretical efforts.

4. Stochastic Dynamical Modeling

4.1. Minimum Quantitative Model

Stochasticity is ubiquitous in biology. For this modeling, it is particularly easy to motivate it. If the numbers of CI and Cro were macroscopically large, then equation (1) would be an entirely accurate description of the dynamics, because the fluctuation in numbers is an order of $N^{1/2}$ and the correction is an order of $1/N^{1/2}$, which would be negligibly small when N was very large. However, the numbers are only in the range of hundreds. Hence, the fluctuation is not negligible. The actual protein production process is influenced by many chance events, such as the time it takes for a CI or a Cro in solution to find a free operator site, or the time it takes an RNA polymerase molecule to find and attach itself to an available promoter, suggesting more stochastic sources. As a minimal model of the network with finite N noise, we therefore consider the following system of two coupled stochastic differential equations, with two independent standard Gaussian and white noise sources:

$$\begin{aligned} dN_{CI}/dt &= F_{CI} + \zeta_{CI}(t) \\ dN_{Cro}/dt &= F_{Cro} + \zeta_{Cro}(t) \end{aligned} \qquad (8)$$

We further assume that the means of the noise terms are zero, i.e., $\langle \zeta_{CI}(t) \rangle = \langle \zeta_{Cro}(t) \rangle = 0$, with the variance

$$\begin{aligned} \langle \zeta_{CI}(t)\zeta_{CI}(t') \rangle &= 2D_{CI}\delta(t-t') \\ \langle \zeta_{Cro}(t)\zeta_{Cro}(t') \rangle &= 2D_{Cro}\delta(t-t') \\ \langle \zeta_{CI}(t)\zeta_{Cro}(t') \rangle &= 0 \end{aligned} \qquad (9)$$

Equations (8) and (9) consist of the present minimum quantitative model. Here, the symbol < . . . > denotes the average over noise. Equation (9) defines a 2×2 diffusion matrix D. The noise strength may contain contributions from the production and decay rates, assuming each is dominated by one single independent reaction, as used by Aurell and Sneppen (59). Such a noise may be called the "intrinsic" noise. Other noise sources, "extrinsic" noises, also exist (89–92). We treat the noise to incorporate both intrinsic and extrinsic sources: All are assumed to be Gaussian and white. The consistency of this assumption should be tested experimentally, as we will do in the following paragraphs. Certain probability events, however, may not behave as Gaussian and white in the present context of modeling, which can be determined by separate biological experiments, such as the $p_{RM}240$ mutation (9) to be discussed later in the text.

It has been demonstrated (61–63) that there exists a unique decomposition, such that the stochastic differential equation, equation (8), can be transformed into the following form, the four dynamical element structure:

$$[\Lambda(\mathbf{N}) + \Omega(\mathbf{N})]d\mathbf{N}/dt = -\nabla U(\mathbf{N}) + \xi(t), \qquad (10)$$

with the semipositive definite symmetric 2×2 matrix Λ defining the dissipation (degradation), the antisymmetric 2×2 matrix Ω defining the transverse force, the single-valued function U defining the potential landscape, and the noise vector $\xi(t)$, and the two-dimensional vectors:

$$\mathbf{N}^{\tau} = (N_{CI}, N_{Cro});$$
$$\nabla = (\partial/\partial N_{CI}, \partial/\partial N_{Cro});$$
$$\xi^{\tau} = (\xi_{CI}, \xi_{Cro}), \qquad (11)$$

Here, τ means the transpose of the vector. The connection between the noise ξ and the matrix Λ is similar to that of ζ and D of equation (9):

$$\langle \xi(t) \rangle = 0$$
$$\langle \xi_{CI}(t)\xi_{CI}(t') \rangle = 2\Lambda_{CI}\delta(t - t')$$
$$\langle \xi_{Cro}(t)\xi_{Cro}(t') \rangle = 2\Lambda_{Cro}\delta(t - t')$$
$$\langle \xi_{CI}(t)\xi_{Cro}(t') \rangle = 0 \qquad (12)$$

The decomposition from equations (8) and (9) to equation (10) and (12) is determined by the following set of equations:

$$\nabla \times [(\Lambda + \Omega)\mathbf{F}] = 0 \qquad (13)$$

$$(\Lambda + \Omega)D(\Lambda - \Omega) = \Lambda. \qquad (14)$$

One may solve for Λ, Ω in terms of $\mathbf{F} = (F_{CI}, F_{Cro})^{\tau}$ and D from equations (13) and (14). Indeed, this can be formally done. Once Λ and Ω are known, the requirement that equation (10) can be reduced to equation (8) gives $(\Lambda + \Omega)\,\mathbf{F} = -\nabla U(\mathbf{N})$, which is used to obtain U. In general, this decomposition is an involved mathematical and numerical endeavor. Further simplification follows from the simplification of friction matrix. Typically, the diffusion matrix D is unknown biologically. There are not enough measurements to fix the noise explicitly. Therefore, we may treat the semipositive, definite symmetric matrix Λ as parameters to be determined experimentally. In our calculation, we assume that D is a diagonal matrix. Following from equation (14), Λ is a diagonal matrix for two-dimensional case. The experimentally measured fraction of *recA*$^{-1}$ lysogens that have switched to lytic state is used to determine the elements of Λ.

Here, we would like to give an intuitive interpretation of the mathematical procedure. Equation (8) corresponds to the dynamics of a fictitious massless particle moving in two-dimensional space formed by the two protein numbers N_{CI} and N_{Cro}, with both deterministic and random forces. It is easy to check that, in general, $\nabla \times \mathbf{F}(\mathbf{r}) \neq 0$ and $\nabla \cdot \mathbf{F}(\mathbf{r}) \neq 0$. Therefore, $\mathbf{F}(\mathbf{r})$ cannot be simply represented by the gradient of a scalar potential, due to both the force transverse to the direction of motion and force of friction. The simplest case in two-dimensional motion when both transverse force and friction exist is an electrically charged particle moving in the presence of both magnetic and electric fields, which is precisely in the form of equation (10).

Proceeding from equation (10), we note that we may interpret the semi-positive definite symmetric Λ matrix as the friction matrix, and the antisymmetric matrix Ω as the result of a "magnetic" field. The friction matrix represents the dissipation in physics. It is analogous to the degradation in biology. The scalar function U takes the role of a potential function, which would determine the final steady distribution of the phage. The global equilibrium will be reached when the final distribution function is given by

$$\rho(N_{CI}, N_{Cro}) = \exp(-U(\mathbf{N})) \Big/ \int dN_{CI} \int dN_{Cro} \exp(-U). \qquad (15)$$

The potential U, the landscape of the system, is depicted in Figure 3 (*cf* Figure 5).

The phage sees two minima and one saddle point in the potential landscape. Those two minima correspond to the lytic and lysogenic states. Once the phage is at one of the minimum, the probability rate for it to move into another minimum is given by the Kramers rate formula in the form (93,94):

$$P = \omega_0 \exp(-\Delta U_b) \qquad (16)$$

with the potential barrier height $\Delta U_b = (U_{saddle} - U_{initial\ minimum})$, the difference in potential between the saddle point and the initial minimum, and the time scale, the attempt frequency ν_0, determined by the friction, the curvatures of potential and values of transverse force around the saddle and the local minimum. We remark here that the attempt frequency is, in general, a complicated function of dynamical quantities in equation (10). Its form will be determined empirically in this chapter. We refer readers to Hanggi et al. (94) for the general mathematical discussions.

4.2. Stochastic Dynamical Structure Analysis

Equation (10) gives the dynamical structure of the gene regulatory network in terms of its four dynamical components: the friction, the

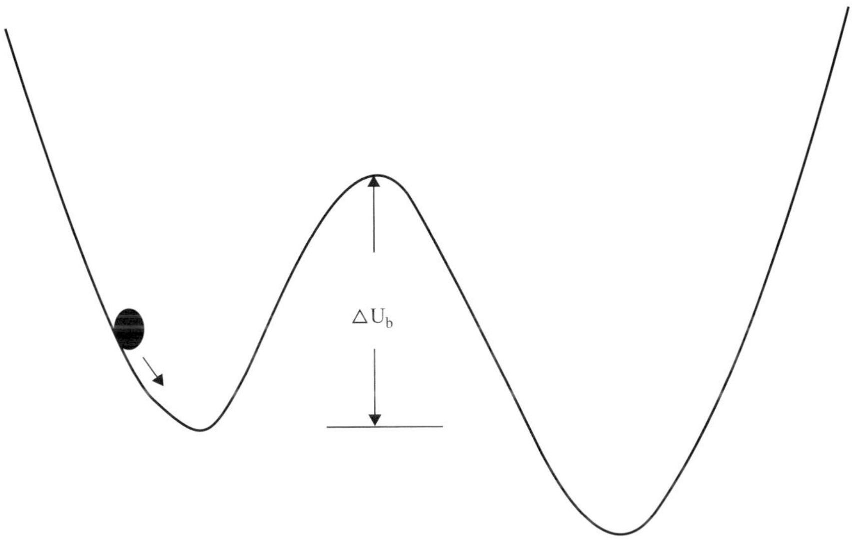

Figure 3. Illustration of the dynamical structure of a genetic switch. The dynamic state of the network is represented by a particle whose position is given by instantaneous protein numbers. The potential function maps a landscape in the protein number space. For a genetic switch, there are two potential minima corresponding to two epigenetic states. The area around each of the minima forms the attractive basin. The state of the network always tends to relax to one of the minima. The fluctuation may bring the network from one minimum to another with a rate given by Kramers rate formula (93,94).

potential gradient of the driving force, the transverse force, and the stochastic force. Such a dynamical structural classification serves two main purposes. It provides a concise description for the main features of the genetic switch by itself and it provides a quantitative measure to compare different gene regulatory networks, for instance, between the wild phage and its mutant. Such an analysis of experimental data based on equation (10) will be tentatively named the dynamical structure analysis.

The potential may be interpreted as the landscape map of the phage development. Each of the epigenetic states is represented by a potential minimum and its surrounding area forms an attractive basin (Figure 3). The dissipation represented by the friction gives rise to the adaptivity of the phage in the landscape defined by the potential. The phage always has the tendency to approach the bottom of the nearby attractive basin. The potential change near the minimum, together with the friction, gives the time scale of relaxation: The time it takes to reach equilibrium after the epigenetic state is perturbed. Once we know the friction and the potential around the minimum, we have a good grasp of the relaxation time, $\tau = \eta/U''$; here η is the strength of friction and U'' is the second derivative of potential, both in one-dimensional approximation along a relevant axis. The relaxation time is independent of the amplitude of the perturbation near the potential minimum when U'' is a constant.

Two remarks are in order here. First, the meaning of friction matrix is the same as in mechanics; if there is no external driving force, the system tends to stop at its nearby minimal position. The closest corresponding concept in biology is “degradation.” There is always a natural protein state under given conditions. Second, it turns out that the transverse force is not a dominant factor in the present switch-like behavior. However, its existence is the necessary condition for oscillatory biological behaviors, which will not be discussed further in this chapter.

Another time scale provided by the potential is the lifetime of the epigenetic state, which is given by the Kramers rate formula, equation (16), through the potential barrier height. Such a scale measures the stability of the epigenetic state in the presence of a fluctuating environment. In the case of phage λ, the lifetime for the lysogenic state is very long, unless the phage is mutated at its operator sites. When the phage is provoked, the height of the potential barrier separating lysogenic and lytic states is reduced. The lifetime of the lysogenic state is drastically reduced because of its exponential dependence on the barrier height and switching takes place. Looking at it from a different angle, the stochastic force gives the phage ability to search around the potential landscape by passing through saddle points, and it drives the switching event. The Kramers rate formula is a quantitative measure of this optimization ability.

5. Quantitative Comparison Between Theory and Experiment

5.1. Determining *In Vivo* Parameters

First, we need to decide the free energies to be used in the theoretical model. Without exception, all the binding energies measured so far for phage λ are determined from *in vitro* studies. The difference between the *in vivo* condition and the *in vitro* condition could include the ion concentration in the buffer solutions and the spatial configuration of the genomic DNA, for instance, looping (95–97). The relative large change of the cooperative energy from *in vitro* to *in vivo* in Table 1 may be partly due to the looping effect, though there is no direct consideration of looping in the present model. We note that in the *in vivo* conditions, all the operators are in the same kind of environment, including the ion condition and the DNA configuration. The reason for the latter is that the operators are located close to each other in the genome. If there is a bending of the genomic DNA that increases or decreases DNA–protein bindings, these closely located and short operator sites will most likely experience the same amount of change. Therefore, we assume that in addition to the *in vitro* DNA–protein binding energy, overall binding energy differences are added to all the CI and Cro protein respectively:

$$\textit{in vivo}\text{ binding energy for CI (Cro)} = \textit{in vivo}\text{ binding energy for CI} + \Delta G_{CI}(\Delta G_{Cro}).$$

To determine $\Delta G_{CI}(\Delta G_{Cro})$, we need more experimental input than the *in vitro* measurement. To avoid unnecessary uncertainty in the model, we try to include a minimal number of parameters. The cooperative binding between two CI dimers is included. The cooperative bindings between two Cro dimers, between CI and Cro dimers, the unspecific CI and Cro bindings are not included. Our later calculation verifies that CI cooperative binding is essential to the genetic switch properties, while the bindings we ignore do not have significant influence on the calculated results. There are three parameters we need to adjust: the difference between *in vivo* and *in vivo* binding energy for CI (ΔG_{CI}), for Cro (ΔG_{Cro}) and for the cooperativity of CI dimers (ΔG(cooperative)). We first use the CI numbers of both wild-type and mutant $\lambda O_R 121$ to determine ΔG_{CI}, then we determine ΔG(cooperative) and ΔG_{Cro} by requiring that both the lytic and lysogenic states of wild type are equally stable, calculated from Kramers rate formula. The adjusted *in vivo* binding energies and other parameters we use for the modeling are given in Table 1. Using these adjusted parameters, the robustness of the phage's genetic switch is reproduced (shown in Figure 4).

The mutant $\lambda O_R 3'23'$ studied by Little et al. (66) was characterized by Hochschild et al. (98) for binding to $O_R 3'$. To produce the desired protein level, we found that the binding energy between $O_R 3'$ and Cro protein is 1.8 kcal/mol smaller than that of the $O_R 3$ and Cro protein, which is consistent with the result of Hochschild et al. The CI binding energy from $O_R 3$ to $O_R 3'$ is slightly increased, 1 kcal/mol, which is also consistent with the measurement.

We assume that friction matrix λ is a diagonal constant matrix. Similar to Aurell and Sneppen (59), we assume the stochastic fluctuations in equation (2) scale with the square root of protein number divided by relaxation time: $D_{CI} = \text{Const} \times \tau_{CI}/N_{CI,lysogen}$, and $D_{Cro} = \text{Const} \times \tau_{Cro}/N_{Cro,lysis}$, where $N_{CI,lysogen}$ is the CI number at the lysogenic state and $N_{Cro,lysis}$ is the Cro number at the lytic state. The constant is to be determined by experiments. In equation (7), we note that if the antisymmetric matrix Ω is small, that is, $|\det(\Omega)| << \det(\Lambda)$, then Λ is the inverse of D. We calculate Ω assuming $\Lambda = D^{-1}$ and find that, indeed, in the regions of concern, i.e., the potential valley connecting two potential minima through the saddle points, Ω is negligible. The final parameters we have used are

$$\Lambda_{11} = 0.056 \times \tau_{CI}/N_{CI,lysogen}$$
$$\Lambda_{22} = 0.040 \times \tau_{Cro}/N_{Cro,lysis}. \quad (17)$$

5.2. Stochastic Dynamical Structure Analysis of λ Switch

The original problem, described by equation (8), may be interpreted as a set of two-dimensional differential equations describing a particle motion, if we view the protein number N_{CI} and N_{Cro} as the coordinates and the particle position to be ($N_{CI}(t)$ $N_{Cro}(t)$) at time t. There is a deterministic force $\mathbf{F}^{\tau} = (F_{CI}, F_{Cro})$ and a stochastic force acting on such a particle. The deterministic force has the characteristics of a friction, a potential force, and a transverse force at the same time. The decomposition we have discussed earlier, equation (10) allows us to separate these components. We discuss them here.

The wild-type phage λ and some of its mutants sees two minima and one saddle point in the potential energy landscape (Figure 5). Those two minima correspond to the lytic and lysogenic states (*cf* Figure 3). The

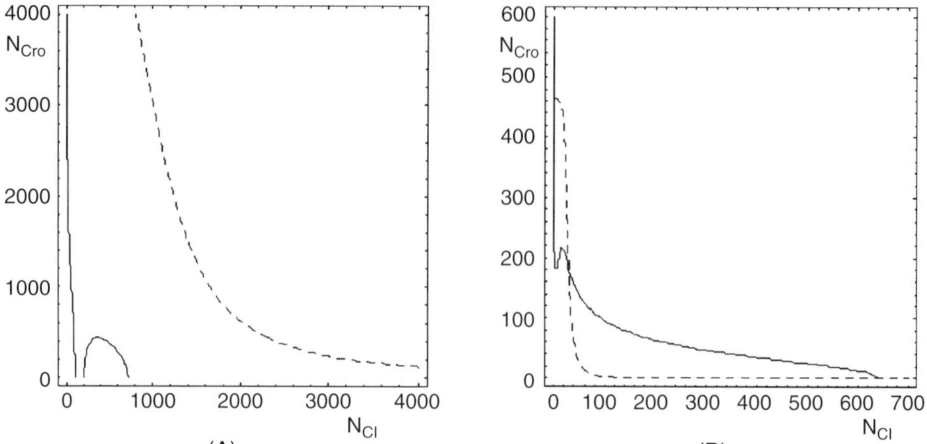

(A) (B)

Figure 4. Lines of $d\langle_{N_{CI}}\rangle/dt = 0$ (solid) and $d\langle_{N_{Cro}}\rangle/dt = 0$ (dashed). Here, $\langle\rangle$ is the average to stochastic force, for (a) the wild-type phage, λO_R321, with parameters taken directly from *in vitro* measurement, and (b) the wild-type phage with parameters adjusted allowing *in vivo* and *in vitro* differences. For mutants λO_R121 λO_R323, *see* Zhu et al. (9). For b, these two lines have three intersections. These three fixed point in equation (2) coincide with the potential extrema, minima, and saddle point in equation (10).

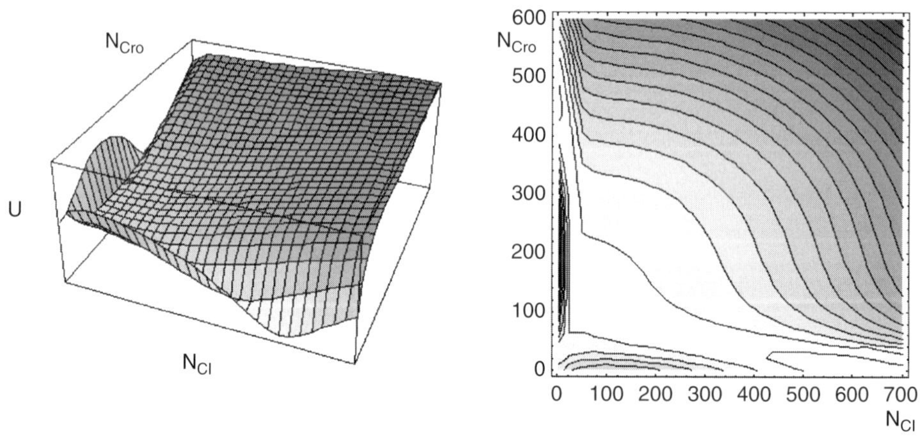

(A) (B)

Figure 5. The potential U of wild-type phage plotted on a logarithmic scale (a) and as a contour map (b). There are two potential minima corresponding to the lysogenic and lytic states. Connecting these two states is a narrow potential valley. The highest point along this valley is the saddle point. The most probable state of the phage is at either of the potential minima. The fluctuation may bring the phage from the original potential minimum, moving along the valley and across the saddle point to reach another potential minimum. The rate for such a switching event is given by the Kramers rate formula.

positions of the potential minima give the average protein number for the lytic and lysogenic states. There is a relatively narrow valley connecting these two minima. The highest point along this valley is the saddle point. Because the areas with large potential are not easily accessible and the low-lying potential region forms a valley, we may visualize the potential along the valley and illustrate it in a one-dimensional graph, as shown in Figures 3 and 6.

The antisymmetric matrix Ω may be represented by a single scalar, B, along the z direction $\Omega\mathbf{F} = Bz \times \mathbf{F}$, assuming $x = N_{CI}$, $y = N_{Cro}$. The transverse field B for the wild type is obtained by numerically solving equation (13). This field is small except at the region along two axes. Along the two axes, the transverse B field has no effect because the motion is guided by the steep potential to a valley. Once the phage evolves away from origin, when both CI and Cro number are small, in the later development the transverse force may be taken out of the equation (10) without changing the dynamics of the phage. In both the calculation of the relaxation time and the lifetime of lysogenic state, we may ignore the transverse force for the above reason.

The protein number distributions of Cro and CI have also been calculated by Zhu et al. (9). We refer readers there for details, as well as for the analysis of other quantities, such as the robustness and stability.

5.3. Switch Efficiency

Efficiency is an important feature that, so far, has received relatively less attention in literature. We present the discussion in some detail here.

The analysis of robustness of phage λ genetic switch demonstrates that its epigenetic states are stable against the variations in parameters and robust against major changes in terms of mutations. Then how does the switching take place? From the theoretical point of view, there are two channels that the phage can be induced from lysogenic growth to lytic growth. In reality, phage seems to use both of these strategies. For clarity, we begin by discussing these two channels separately.

The first channel of induction is to increase the noise level of CI protein number, while keeping all the other conditions intact. Mathematically, it means to increase ζ_{CI} in equation (8) and D_{CI} in equation (9), while keeping all the other terms in equation (8) and equation (9) unchanged. The friction matrix Λ is changed through the decomposition procedure. As a result, the potential energy U is also changed. Therefore, for a different noise level, the phage moves in a different potential landscape. Such a change of noise level has a drastic effect. It changes the minima of the potential well of lysogens by making it shallower. As a good approximation, the barrier height of the lysogen potential well scales inversely with the noise strength. Doubling the noise level reduces the potential barrier by half. As a result, the increased noise level drastically decreases the lifetime of the lysogenic state, as shown in Figure 6. The lifetime of the lytic state, on the other hand, remains unchanged. The combination of these two changes in the potential landscape brings the phage to lytic growth efficiently.

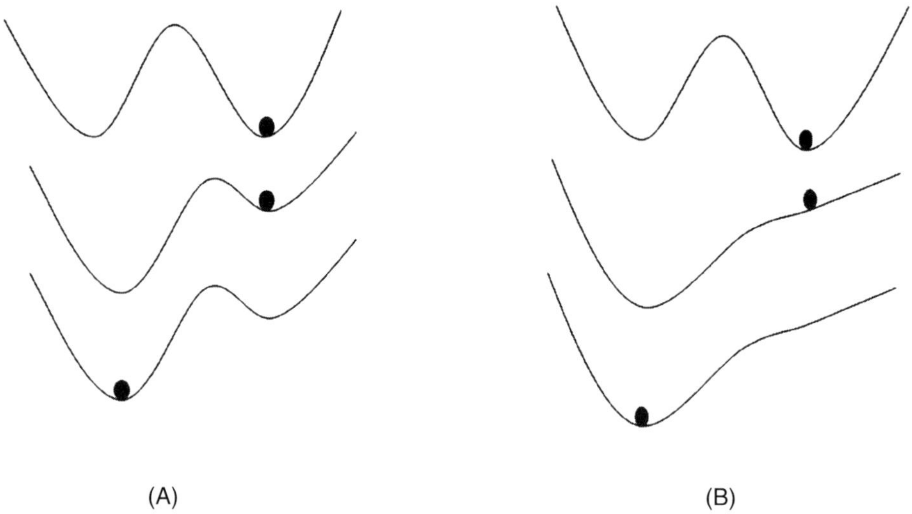

Figure 6. (A) Illustration of the switching mechanism from the current work. Before switching, the phage grows in lysogenic state. The potential barrier separating the lysogenic state and the lytic state is high. When *recA* is activated, this barrier is lowered. The lifetime of lysogenic state reduces drastically, and the phage switches to lytic state. (B) Switching mechanism of Shea and Ackers (48). In their work, fluctuation was not included. Switching was possible only when the lysogenic state is no longer a potential minimum. When stochastic effect is included, the switching happens when the lysogenic potential minimum becomes too shallow to confine fluctuation.

The second channel is through the deterministic terms in equation (8). For the deterministic terms, e.g., introducing CI monomer cleavage is equivalent to substitute N_{CI} in equation (8) with αN_{CI}, α is a factor that represents cleavage strength. If α is smaller than 0.02, we find that lysogenic state is no longer stable, i.e., no longer a potential minima. The interpretation of such a small α is that almost every CI monomer is cleaved. If α is small, let's say 0.1, meaning 90% CI monomers are cleaved, the lysogenic state is still stable, with a lifetime almost unchanged. Apparently, a uniform CI cleavage alone without introducing extra noise to CI levels is not an efficient way for induction.

Phage may have used both of these two channels. The second channel is obviously used because RecA cleaves CI monomers. The strong indication that the first channel is also used comes from the observations that without external stimulus, the *recA*$^+$ phage shows a much shorter lifetime for the lysogenic state compared to *recA*$^-$ phage. Such a significant reduction of lysogen lifetime without activating RecA proteins on an observable scale can be explained by doubling of the CI noise level. Figure 6 gives schematic explanations of the switching process.

In the early work by Shea and Ackers (48), stochastic effect was not included. In their model, even a shallow lysogenic potential minimum would confine the phage to continue growing in lysogenic state. Switching happens only when the lysogenic potential minimum disappears completely. For the parameters they used, they found that when 20% of

CI monomers were cleaved, such a switch would happen. As pointed out by Aurell et al. (67), in these early works, the genetic switch modeling results do not show the observed robustness. After we require that the genetic switch should demonstrate the observed robustness, the disappearance of the lysogenic potential minimum is pushed down to 2%. However, the actual switching happens before the disappearance of lysogen potential minimum, when the lysogen potential minimum is too shallow to confine the fluctuations. If 10% of CI monomer is cleaved, at least we expect a 10-fold increase in D_{CI} due to the reduced CI monomer numbers. The potential barrier for the lysogenic state reduces to less than 1, and therefore becomes too shallow to allow continual lysogenic growth.

5.4. Quantitative Comparison with the Experiment of Little et al.

We summarize the calculation results related to the measurements of Little et al. (66) in Table 3 because their data are most up-to-date and systematic. In their experiment, they measured the free phage per lysogenic cell for both *recA*$^+$ and *recA*$^-$ phage, but did not convert the *recA*$^+$ into fraction of lysogens that switched to the lytic state. If we assume the burst size for both the *recA*$^+$ and the *recA*$^-$ phage are similar, our calculation for the RecA$^+$ protein agrees with their measurements quantitatively.

In Table 3, the bistability of the gene switch in phage λ is assumed, and the protein levels in the lytic state are calculated. This is, of course, not the case for the wild type, hence, it posits a question to test the calculated Cro level experimentally. One way to realize the bistability may be by suppressing the lyses, achieving the so-called antiimmune phenotype (103,104).

As discussed in section 4 of the formulation of the present stochastic model, we have made the simplified assumption of treating all chance or probability events as Gaussian white noise. This assumption affects two testable biological quantities: the lifetime of lysogenic state (equation

Table 3. Comparison between the calculation and the experimental data (in parentheses) of Little et al. (66).

Phage genotype	Relative CI level in lysogen Theoretical (experimental)	Relative Cro level in lysis Theoretical	Switching frequency to lytic state (*recA*$^-$) per minute Theoretical (experimental*⁾)	Switching frequency to lytic state (*recA*$^+$) per minute Theoretical
λ^+	100% (100%)	100%	1×10^{-9} (2×10^{-9})	1×10^{-5}
λO_R121	20% (25–30%)	100%	3×10^{-6} (3×10^{-6})	3×10^{-5}
λO_R323	70% (60–75%)	70%	7×10^{-5} (2×10^{-5})	1×10^{-4}
$\lambda O_R3'23'$	50% (50–60%)	130%	1×10^{-7} (5×10^{-7})	2×10^{-5}

Here, *⁾ indicates that the estimated wild-type data (9) is used. The wild-type biological data were used to find out the difference between *in vivo* and *in vitro* molecular parameters, as listed in Table 1. The relative CI level and switch rate of λO_R121 were used to fine-tune parameters. Rest of theoretical entries are then calculated directly from our model.

[16]), and the shape of CI number distribution in lysogenic state (equation [15] and Figure 7). Simultaneous measuring of both can be used as a consistent check to the Gaussian white noise assumption. For example, we have treated the effect of *recA*$^+$ to switching dynamics as that of a Gaussian white noise to simplify our calculation, in the same reasoning of minimal modeling approach in this chapter. In fact, we have assumed that with *recA*$^+$, the total noise strength doubles. Using this assumption, we calculated the lytic switching rates, represented by the last column of Table 3. The CI distribution with *recA*$^+$ should be twice as broad as in the case with *recA*$^-$. Both results are subject to further experimental testing.

There may be some chance events that cannot be treated as Gaussian white noise in the present formulation. One example has been already suggested in biological experiments (9), the $p_{RM}240$ mutation, which greatly weakens the promoter, and therefore the ability to produce CI as well. This mutation makes the lysogens barely stable, and is estimated to be responsible for at least 99% of observed lytic switching in the wild type. We have used this input for both Tables 2 and 3. We have recalculated the switching rates to lytic state of all strands, assuming the same minimal model, with the same forms of functions for the switching rate, but with the previous experimental data (66). The switching rates obtained in this way are: wild type(λ^+), 2×10^{-7}; λO_R121, $\lambda O_R2 \times 10^{-6}$; λO_R323, 7×10^{-5}; and $\lambda O_R3'23'$, 5×10^{-7}. Indeed, the stability of the wild type decreases by more than 2 orders of magnitude. The overall noise strength is increased by 60% for the wild type, resulting in a broader CI distribution in the lysogenic state. There is no appreciable change in other quantities, such as the protein level. The only noticeable overall change in molecular parameters is the *in vivo* cooperative energy, from −6.4 kcal/mol to −6.7 kcal/mol. A good overall quantitative agreement exists between modeling and experiment.

It is a fact that any mathematical modeling in natural science should have empirical input to completely fix its mathematical structure. For the modeling of phage λ, there is an already large body of molecular data, which enables us to nearly pin down our model. The additional freedom in our parameters is fixed by data from wild type, such as the switching frequency. Above the less-than-expected sensitivity of our mathematical structure to this frequency that a few percentage of change in molecular parameters can result in 2 orders of magnitude of change in frequency is a remarkable demonstration of the internal consistency of our modeling. It demonstrates that the switching is exponentially sensitive to some molecular parameters. In addition to more theoretical effort to go beyond our present minimal modeling, it is clear that more experiments are needed in this direction to test the present model: The precise *in vivo* molecular parameters and the distributions and time-correlation of protein numbers in our model should be viewed as predictions.

5.5. Experimental Determination of Dynamical Elements

We have introduced four dynamical quantities for a gene regulatory network: friction, potential, the transverse force, and the stochastic force.

The friction and the strength of the stochastic force are related. For the genetic switch, the transverse force is irrelevant to the dynamic properties. Therefore, the two crucial quantities for a genetic switch are the friction and the potential. Those quantities can be calculated from the more microscopic modeling with molecular parameters. The present quantitative success lies in the allowance of the *in vivo* and *in vitro* differences, and of various noise contributions. However, those four quantities may be directly determined biologically.

There are three different types of experimental data to determine the dynamical elements of the local potential function for a genetic switch, the degradation (the friction in physical sciences), and the barrier height (Figure 7). The first type is the protein distribution around each of the epigenetic states. It is given mathematically by

$$\rho(\mathbf{N}) = \rho_0 \exp(-U(\mathbf{N})), \tag{18}$$

where ρ_0 is a normalization constant. The protein distribution is explicitly measurable. Once $\rho(\mathbf{N})$ is measured experimentally, the local potential $U(\mathbf{N})$ near the potential minima can be determined as: $U(\mathbf{N}) = -\ln(\rho(\mathbf{N})) + \ln(\rho_0)$.

The second type of experimental data is the relaxation time, a measure of how long it would take the system to return to its local equilibrium after a small perturbation. It is determined by both the potential near the minimum and the degradation,

$$\tau = \eta / U''. \tag{19}$$

Here, η is the strength of friction, which gives the friction matrix Λ along the path of relaxation. U'' is the second derivative of potential. Because potential can be obtained from $\rho(\mathbf{N})$, relaxation time can be used to obtain friction: $\eta = \tau U''$.

The third type of experimental data is the lifetime of epigenetic states, the measure of the switching rate from one state to another. The probability of phage evolving from one epigenetic state of growth to another is given by the Kramers rate formula, our equation (16), $P = \omega_0 \exp(-\Delta U_b)$, where ΔU_b is barrier height and ω_0 is the attempt frequency. ω_0 is given by the friction and the curvature of the potential barrier. The curvature of the potential is related to the height of the potential barrier and the shape of the potential near its minimum. Therefore, ΔU_b can be determined from the lifetime of its epigenetic state: $\Delta U_b = \ln(\omega_0) - \ln(P)$.

The genetic switch for phage λ is a complex dynamical system. It took decades of ingenious experimental research and laborious work to collect the parameters needed for this mathematical modeling. For a more complicated system, resources and time may limit the ability to study each molecular element in detail. A method that is less demanding on the details, yet still can capture the main features, is of great interest. Dynamical structure analysis provides guidance to build such a phenomenological model, as illustrated in Figure 7.

We emphasize that the quantities introduced in dynamical structure theory, the friction, the potential gradient, the transverse force, and the stochastic force associating with the friction, are all measurable quantities at the given description level. These are quantities similar to temperature,

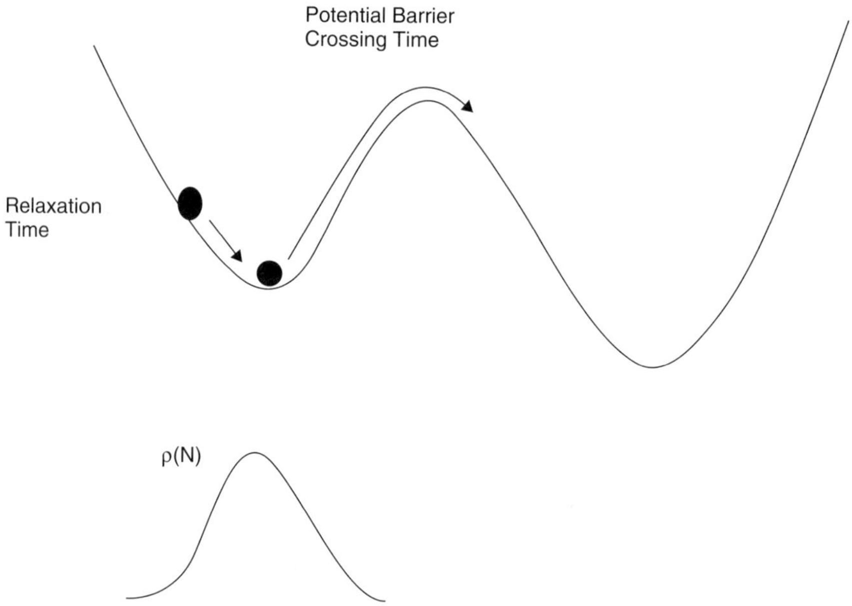

Figure 7. Three types of experiments directly probe the dynamical structure of a genetic switch and determine the dynamic components. The measurement of protein number distribution determines the potential function at each of the epigenetic states. The additional information on the relaxation time determines the strength of friction. The lifetime of each epigenetic state determines the height and shape of the potential barrier.

pressure, and free energy in thermodynamics, which can be determined by microscopic details, but can also be measured independent of those details. Once the relationships between these quantities are established, as shown in equation (10), we are ready to write down an effective equation of motion for the network without resorting to details.

5.6. Stochasticity, Robustness, Cooperation, and Efficiency

Starting from realization that noise is important in the initiation of transcription in phage λ (58), stochasticity has been increasingly viewed as one of the most important elements in the dynamical modeling of biological processes (105–107). Both the intrinsic and extrinsic noises are shown to exist in biological processes and are analyzed theoretically (89–92). The parameterization in the form of equation (17) is a way to account for both noise contributions.

Robustness has been viewed as one of the central features in biological processes (108–110). Numerous recent studies have established its importance (111–116). Combined with stochasticity, the work reviewed here (9,64) established a quantitative criterion, equation (16), for the robustness. The potential landscape function, U(**N**) in equation (10), which emerged from the stochastic dynamics, provides a graphic representation of the robustness. Those results, again, confirm the importance of noise.

Finally, we wish to point out that a complete rigid system, that is, an absolute robust system with no flexibility, is not viable from the evolutionary point of view (117,118). Such a structure would not survive the stringent evolutionary process. This may be illustrated in the switch efficiency discussed above. It is also implied by the ability of the phage λ to switch from the lysogenic to lytic state when provoked (8), a feature that can be captured based on the present minimum model, though no explicit and detailed mathematical analysis has been published yet along this direction. Thus, the stochasticity seems to provide the critical link to understanding both robustness and flexibility. We believe such a feature can, indeed, be understood from the evolutionary point of view (118).

6. Perspective on Mathematical Modeling

6.1. Major Prediction of the Minimum Quantitative Modeling

We have shown that, thanks to continuous theoretical and experimental efforts, the minimum quantitative modeling has achieved the status of quantitative agreement with experimental biological data. New predictions, such as protein distributions, are made and discussed in Zhu et al. (9). There is one prediction on cooperative energy that stands out as an excellent indicator for success of both recent theoretical and experimental efforts.

Since the 1980s, it has been found that it is rather difficult to model the stability of the λ switch with known parameter constraints (48,57), even allowing the possibility of up to 30% difference between *in vivo* and *in vitro* parameter values (51). It has been found that the cooperative energy would play an important role (119,120). Thus, it has been hypothesized that additional effects beyond the minimum model would be needed. One of the most promising ones is a stronger cooperative effect. Indeed, independent of theoretical need, an additional effect, the looping, has been found experimentally (88,96).

A brief account of this effort may be relevant. Four years ago, four of the present authors began a mathematical study on phage λ. Although we tried all methods known to us at that time, we could not solve the stability puzzle. Effectively, we were in the same situation as that reported by Reinitz and Vaisnys and Aurell et al. (57,67). One of the major problems for us was that even allowing the possibility to vary the parameter values drastically in the name of *in vivo* and *in vitro* difference, the parameter space appears too big for an effective research. This difficulty was partially verified in retrospect in Bakk et al. (51), where the change of parameter values appeared too small to explain the "experimental observed robustness" within the minimum quantitative model. Out of this frustration, it was realized 3 years ago that one must have an effective quantification criterion for stability. It turned out that the landscape idea, rooted deeply in both physics and biology, appears to be such a candidate. Driven by this need for quantification, a mathematically consistent construction method for such a landscape function was quickly discovered. With this new method, it was relatively easy to explore bigger parameter space. One critical parameter, the cooperative energy, was then found to double its value to have the desired stability, the value in bold face for ΔG(cooperative) of −3.7 kcal/mol in Table 2. We note this value is about twice of that tried in Bakk et al. (51).

Interestingly, such a value was indeed observed in an independent biological experiment (88). In writing this review, we further noticed that such a big value was suggested to be possible in an independent theoretical investigation based on thermodynamic consideration (87). Given all of the independent efforts, theoretical (9,87) and experimental (88,96), successful (64) and failed (51), the authors believe that such an agreement between the theoretical prediction (64) and experimental value (88) on the cooperative energy may not be accidental. It indicates that the minimum quantitative model may have, indeed, captured the essential biological features of this genetic switch, with its first nontrivial and verified prediction.

6.2. Relation to Other Modeling Methodologies

There has been a tremendous amount of literature on biomodeling and biocomputation. It is impossible to give an adequate survey of various methodologies in this chapter. Nevertheless, we would like to present the following two classification schemes, according to mathematical and scientific structures, to place our method in a broader context (Figure 8).

From a mathematical point of view, a modeling may be classified according to whether it is discrete or continuous and whether it is deterministic or stochastic. The classic modeling of deterministic and discrete is the Boolean logic circuit (121). The works of Shea and Ackers (48), as well as others (57,122,123), are the finest examples of deterministic and continuous modeling. Examples of stochastic and continuous modeling are Arkin et al. (58) and Aurell and Sneppen (59). Naturally, there are methods of combining various features. One such example is the hybrid of continuous and discrete modeling of Tchuraev and Galimzyanov (124). Of those modeling methods, according to this classification, the simplest one is that based on the Boolean logic circuit. It is clearly an approximation, but in many cases it serves specific biological purposes well. It is, in fact, currently the dominant modeling and presentation methodology in biology. The most difficult, but most detailed, modeling is the continuous and stochastic formulation. Many of its predictions are necessarily probabilistic in nature, corresponding nicely to biological phenomena. Our present method belongs to this last category. Nevertheless, we should point out that no method would be definitely better than the rest. The choice of modeling method must be appropriate to the biological questions being addressed.

From the scientific structure point of view, the modeling may be classified into first principle modeling and completely empirical modeling. It is believed that chemical reactions and other physical processes lie behind various biological processes. Hence, it should be possible to predict biological functions based on the physical-chemical principles. This first principle modeling methodology has been followed by Shea and Ackers (48), by Reinitz and Vaisnys (57), and by many others (58,59,122). Ours is also of this type. The advantages of first principle modeling are that it shows the unity of the sciences and that additional

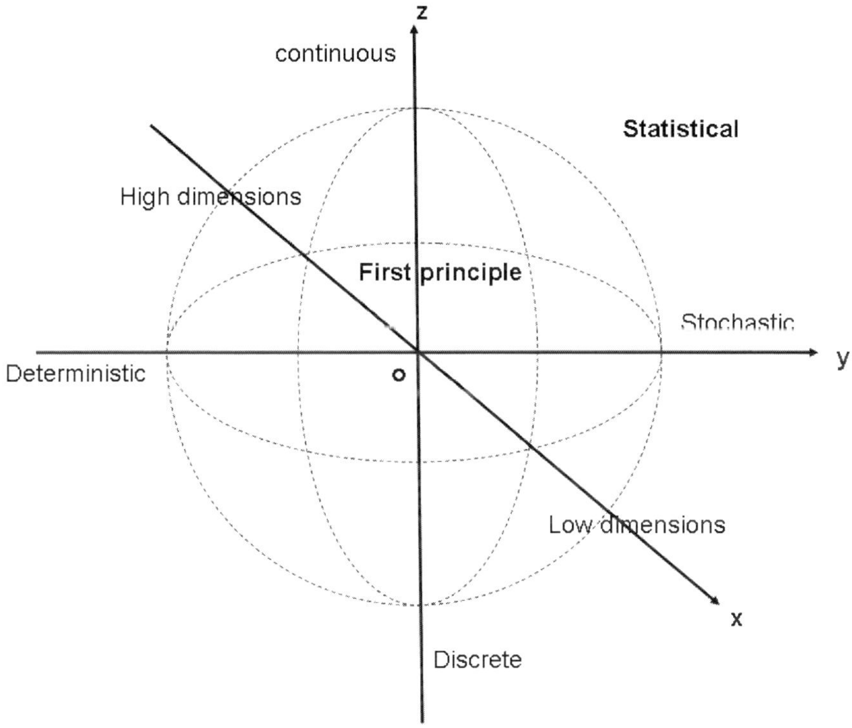

Figure 8. The schematic diagram of modeling methodologies. The coordinates are classified according to mathematical descriptions, which define a three-dimensional space: many degrees of freedom (high dimensions) to few degrees of freedom (low dimensions) (x direction); from deterministic to stochastic (y direction); and from discrete to continuous (z direction). The broken lines, which define a sphere, are classified according to scientific understandings: from first principles (inside) to statistical methods (outside). Both scientific methodologies encompass all mathematical descriptions.

insight and information can be obtained from the lower level scientific descriptions. The evident disadvantage of this modeling, in addition to the difficulty of specifying all needed microparameters, is that higher level processes often show emerging phenomena. It is difficult to make predictions based on the properties of the system's components. The outstanding stability puzzle of phage λ genetic switch (51,59), one of the simplest possible living genetic switches, and the enormous effort (9,48,51,57–59,67) to quantitatively model its behavior, clearly illustrate this situation.

The other extreme, compared with first principle modeling, is to treat the system in question autonomously, inferring its properties completely from empirical studies. Statistical analysis methods play a dominant role in this approach. The advantage of this modeling is that it establishes the independent role of the investigating scientific layer. It is consistent with the view that at each scientific level autonomous laws can be uncovered. Equipped with biological insights, it has been employed successfully in numerous biological studies. The proposals of Waddington (125) on developmental landscape and of Monod and Jacob (126) on gene regulation mechanism are fine examples. In reality, particularly in molecular biology, what is typically encountered is in between those two extremes, as demonstrated by the Boolean logic circuit modeling (121) and by others (122–124,127). Interestingly, even in the empirical modeling setting, we have briefly discussed the direct and transparent connection of our method to empirical data in subsection 5.5, though it is rooted in first principle modeling. This suggests that our method can be used in this extreme example, too. Further investigation in this direction should be carried out.

The statistical analysis also suggests an important issue in mathematical modeling: the number of variables and the associated problem of the "curse of dimension." Regarding the matching of data to appropriate modeling methodologies, there is another issue of "parsimony of experimental data," which is particularly acute in real-time modeling at present. We will not discuss those issues here.

To summarize the unique characteristics of the present novel method, the stochastic dynamical structure analysis, mathematically, is a continuous and stochastic formulation with four dynamical elements. In terms of scientific structure, it is equivalent to first principle modeling. Nevertheless, it can be directly related to empirical data to establish its autonomy.

6.3. Literature Sampling

First, the book by Ptashne (8) on phage λ is a must read; an excellent review on experimental work up to 2004 can be found. Because both the λ repressor and the *lac* operon are instrumental in the current molecular and synthetic biological studies, the book by Mulller-Hill (128) is another must read. A good summary of earlier study on phage λ can be found in Hendrix et al. (129). A broader and recent general review can be found in Birge (130).

For a recent phage λ study reviewed from the switch point of view, Wegrzyn and Wegrzyn (131) is a good start. Five switches were identified there for the developmental process. The stability puzzle was put into sharp focus in Little et al. (66). Looping study has been theoretically studied in Vilar and Leibler (87), and experimentally in Doff et al. (88) and Revet et al. (96). The phage was studied from an evolutionary point of view in Svenningsen et al. (133). More studies on the role of Cro can be found in Jia et al. (83) and Bundschuh et al. (84). More interesting dynamical behaviors were reported experimentally in Svenningsen et al. (133) and Kobiler et al. (134). The effect of degradation time on stability was considered theoretically in Buchler et al. (135). Various other features on genetic switch have been recently explored (136–141).

7. Third Age of Phage

Because of its enormous biomass in the biosphere on Earth, the importance of phage has already been recognized in the current ecological study (47,142,143), and the title of this section is borrowed from Mann's study (143). The presentation of theoretical effort in this chapter has shown that the phage has been playing an important role in the study of fundamental biology in the post-Human Genomic era, too, along with the experimental effort (145).

The quantitative and detailed modeling of gene regulatory networks is evidently at its beginning. There are numerical possibilities to go beyond the minimal quantitative modeling reviewed here. For example, even for the phage λ genetic switch, it is a simplification. It would be desirable to have quantitative demonstration on how some of the *in vivo* and *in vitro* differences arise by incorporating more degrees of freedom, such as the left operon and what would be the quantitative differences. A further extension would be to model all five switches of the λ developmental processes (131), to obtain a comprehensive quantitative understanding of the whole process. Deep biological questions, such as why the phage chooses such a structure or what evolution principles guided this choice (66,127), have not yet been discussed adequately by any measure. Nevertheless, we do wish to point out that the study of phage λ genetic switch has revealed a novel mathematical structure (9) and has already put one of the deep-rooted concepts in biology, the landscape (125,145–147), back on a firm mathematical and biological ground. Numerous recent quantitative phage studies (9,49–51,58,59,66,67, 73–75,131–141) have pushed our understanding of systems biology onto another level, complementary to high-throughput and large-scale analyses. It is optimistic that its study will generate more new biological understandings and will have an influence beyond biology, as biology has already inspired the theories of general systems (148) and cybernetics (149). In view of its past successes (e.g., one may count how many Nobel Prize winners in physiology, medicine and chemistry have appeared at http://www.asm.org/division/m/blurbs/secrets.html#top.), we are confident that more discoveries are waiting ahead.

Acknowledgments: We thank J.J. Collins, J.W. Little, B. Muller-Hill, S.M. Stoylar, and G. Wegrzyn for critical comments and M. Mossing for an update on literature. This work was supported in part by the Institute for Systems Biology (L. Hood and D. Galas) and by a National Institutes of Health grant HG002894 (P. Ao).

References

1. Collins FS, Green ED, Guttmacher AE, Guyer MS. A vision for the future of genomics research: a blueprint for the genomic era. *Nature* 2003; 422:835–847.
2. Kitano H. Systems biology: a brief review. *Science* 2002;295:1662–1664.
3. Cohen JE. Mathematics is biology's next microscope, only better; biology is mathematics' next physics, only better. *PloS Biology* 2004;2:2017–2023.
4. Mesarovic MD, Sreenath SN, Keene JD. Search for organizing principles: understanding in systems biology. *Syst Biol* 2004;1:19–27.
5. Kirschner MW. The meaning of systems biology. *Cell* 2005;121:503–504.
6. Williamson MP. Systems biology: will it work? *Biochem Soc Transact* 2005; 33:503–506.
7. Hood L. Systems biology: integrating technology, biology, and computation. *Mech Aging Dev* 2003;124:9–16.
8. Ptashne M. A Genetic Switch: Phage λ revisited. 3rd edition. 2004. Cold Spring Harbor; Cold Spring Harbor Laboratory Press.

9. Zhu XM, Yin L, Hood L, Ao P. Robustness, stability and efficiency of phage lambda genetic switch: dynamical structure analysis. *J Bioinform Comput Biol* 2004;2:785–817.
10. Cairns J, Stent GS, Watson JD. Phage and the Origins of Molecular Biology. expanded edition. Cold Spring Harbor: Cold Spring Harbor Laboratory Press; 1992.
11. Olson MV. The human genome project: a player's perspective. *J Mol Biol* 2002;319:931–942.
12. Lander E. Biology as information. In: Research in Computational Molecular Biology, Lecture Notes in Bioinformatics 3,500. Miyano S, Mesirov J, Kasif S, Istrail S, Pevzner P, Waterman M, eds. Berlin: Springer; 2005.
13. Hood L, Galas D, Dewey G, et al. Biological Information and the Emergence of Systems Biology. Roberts and Co; 2006.
14. Bar-Yam Y, Epstein IR. Response of complex networks to stimuli. *Proc Natl Acad Sci USA* 2004;101:4341–4345.
15. Ben-Hur A, Sigelmann HT. Computation in gene networks. *Chaos* 2004; 14:145–151.
16. Levine M, Davidson EH. Gene regulatory networks for development. *Proc Natl Acad Sci USA* 2005;102:4936–4942.
17. Schroder A, Persson L, de Roos AM. Direct experimental evidence for alternative stable states: a review. *OIKOS* 2005;110:3–19.
18. Balaban NQ, Merrin J, Chait R, et al. Bacterial persistence as a phenotypic switch. *Science* 2004;305:1622–1625.
19. Laurent M, Kellershohn N. Multistability: a major means of differentiation and evolution in biological systems. *TIBS* 1999;24:418–422.
20. Freeman M, Gurdon JB. Regulatory principles of developmental signaling. *Annu Rev Cell Dev Biol* 2002;18:515–539.
21. Kurakin A. Self-organization vs watchmaker: stochastic gene expression and cell differentiation. *Dev Genes Evol* 2005;215:46–52.
22. Hayer A, Bhalla US. Molecular switches at the synapse emerge from receptor and kinase traffic. *PloS Comp Biol* 2005;1:137–154.
23. O'Conner DH, Wittenberg GM, Wang SS-H. Graded bidirectional synaptic plasticity is composed of switch-like unitary events. *Proc Natl Acad Sci USA* 2005;102:9679–9684.
24. Markevich NI, Hoek JB, Kholodenko BN. Signaling switches and bistability arising from multisite phosphorylation in protein kinase cascades. *J Cell Biol* 2004;164:353–359.
25. Qian H, Reluga TC. Nonequilibrium thermodynamics and nonlinear kinetics in a cellular signaling switch. *Phys Rev Lett* 2005;94:028101.
26. Miller P, Zhabotinsky AM, Lisman JE, Wang X-J. The stability of a stochastic CaMKII switch: dependence on the number of enzyme molecules and protein turnover. *PloS Biol* 2005;3:705–717.
27. Hernday AD, Braaten BA, Low DA. The mechanism by which DNA Adenine methylase and PapI activate the Pap epigenetic switch. *Mol Cell* 2003;12:947–957.
28. Travers A. Transcriptional switches: the role of mass action. *Phys Life Rev* 2004;1:57–69.
29. Loayza D, de Lange T. Telomerase regulation at the telomere: a binary switch. *Cell* 2004;117:279–280.
30. Biggar SR, Crabtree GR. Cell signaling can direct either binary or graded transcriptional response. *EMBO J* 2001;20:3167–3176.
31. Casadesus J, D'Ari R. Memory in bacteria and phage. *Bioessays* 2002; 24:512–518.

32. Acar M, Becskei A, van Oudenaarden A. Enhancement of cellular memory by reducing stochastic transitions. *Nature* 2005;435:228–232.
33. Shykind BM, Rohani SC, O'Donnell S, et al. Gene switching and the stability of odorant receptor gene choice. *Cell* 2004;117:801–815.
34. Kaern M, Elston TC, Blake WJ, Collins JJ. Stochasticity in gene expression: from theories to phenotypes. *Nature Rev Genet* 2005;6:451–464.
35. Lidstrom ME, Meldrum DR. Life on a chip. *Nature Rev Microbiol* 2003;1: 158–164.
36. Kramer BP, Viretta AU, Baba MD-E, et al. An engineered epigenetic transgene switch in mammalian cells. *Nature Biotech* 2004;22:867–870.
37. Ozbudak EM, Thattai M, Lim HN, et al. Multistability in the lactose utilization network of *Escherichia coli*. *Nature* 2004;427:737–740.
38. Sauer M. Reversible molecular photoswitches: a key technology for nanoscience and fluorescence imaging. *Proc Natl Acad Sci USA* 2005;102: 9433–9434.
39. Habuchi S, Ando R, Dedecker P, et al. Reversible single-molecule photoswitching in the GFP-like fluorescent protein Dronpa. *Proc Natl Acad Sci USA* 2005;102:9511–9516.
40. Cherry JL, Adler FR. How to make a biological switch. *J Theor Biol* 2000; 203:117–133.
41. Slepchenko BM, Terasaki M. Bio-switches: what makes them robust? *Curr Opin Genet Dev* 2004;14:428–434.
42. Goutsian J, Kim S. A nonlinear discrete dynamical model for transcriptional regulation: construction and properties. *Biophys J* 2004;86:1922–1945.
43. Warren PB, ten Wolde PR. Chemical models of genetic toggle switches. *J Phys Chem B* 2005;109:6812–6823.
44. Ferrell JE, Machleder EM. The biochemical basis of an all-or-none cell fate switch in *Xenopus* oocytes. *Science* 1998;280:895–898.
45. Angeli D, Ferrell JE, Sontag ED. Detection of multistability, bifurcations, and hysteresis in a large class of biological positive-feed back systems. *Proc Natl Acad Sci USA* 2004;101:1822–1827.
46. Chen KC, Calzone L, Csikasz-Nagy A, et al. Integrative analysis of cell cycle control in budding yeast. *Mol Biol Cell* 2004;15:3841–3862.
47. Hendrix RW. Bacteriophage genomics. *Curr Opin Microbiol* 2003;6: 506–511.
48. Shea MA, Ackers GK. The OR control system of bacteriophage lambda—A physical-chemical model for gene regulation. *J Mol Biol* 2005;181:211–230.
49. Dodd IB, Shearwin KE, Egan JB. Revisted gene regulation in bacteriophage lambda. *Curr Opin Genet Dev* 2005;15:145–152.
50. Santillan M, Mackey MC. Why the lysogenic state of phage lambda is so stable: a mathematical modeling approach. *Biophys J* 2004;86:75–86.
51. Bakk A, Metzler R, Sneppen K. Sensitivity of O_R in phage lambda. *Biophys J* 2004;86:58–66.
52. Vilar JMG, Guet CC, Leibler S. Modeling network dynamics: the *lac* operon, a case study. *J Cell Biol* 2003;161:471–476.
53. May RM. Uses and abuses of mathematics in biology. *Science* 2004;303: 790–793.
54. Hwa T. A genetic switch. *Science* 2004;305:345.
55. Ptashne M, Jeffrey A, Johnson AD, Mauer R, Meyer BJ, Pabo CO, Roberts TM, Sauer RT. How the λ repressor and Cro work. *Cell* 1980;19:1–11.
56. Riggs AD, Porter TN. Overview of epigenetic mechanisms. in Russo VEA, Martienssen RA, and Riggs AD (ed), *Epigenetic Mechanisms of Gene Regulation*. Cold Spring Harbor, NY: Cold Spring Harbor Laboratory Press; 1996:29–45.

57. Reinitz J, Vaisnys JR. Theoretical and experimental analysis of the phage lambda genetic switch implies missing levels of cooperativity. *J Theor Biol* 1990;145:295–318.
58. Arkin A, Ross J, McAdams HH. Stochastic kinetic analysis of developmental pathway bifurcation in phage lambda-infected *Escherichia coli* cells. *Genetics* 1998;149:1633–1648.
59. Aurell E, Sneppen K. Epigenetics as a first exit problem. *Phys Rev Lett* 2002; 88:048101.
60. Freidlin MI, Wentzell AD. Random perturbations of Dynamical Systems, 2nd edition. Berlin: Springer; 1998.
61. Ao P. Stochastic force defined evolution in dynamical systems and complex networks. In: Zhao XG, Jiang S, Yu XJ, eds. Computational Physics, Proceedings of the Joint Conference of ICCP6 and CCP2003. Paramus, NJ: Rinton Press; 2005:12–18). (eprint–physics/0302081: http://it.arxiv.org/PS_cache/physics/pdf/0302/0302081.pdf).
62. Ao P. Potential in stochastic differential equations: novel construction. *J Phys A* 2004;37:L25–L30.
63. Kwon C, Ao P, Thouless DJ. Structure of stochastic dynamics near fixed points. *Proc Natl Acad Sci USA* 2005;102:13029–13033.
64. Zhu XM, Yin L, Hood L, Ao P. Calculating biological behaviors of epigenetic states in phage λ life cycle. *Funct Integr Genomics* 2004;4:188–195.
65. Rozanov DV, D'Ari R, Sineoky SP. RecA-independent pathways of lambdoid prophage induction in *Escherichia coli*. *J Bacteriol* 1998;180: 6306–6315.
66. Little JW, Shepley DP, Wert DW. Robustness of a gene regulatory circuit. *EMBO J* 1999;18:4299–4307.
67. Aurell E, Brown S, Johanson J, Sneppen K. Stability puzzle in phage λ. *Phys Rev E* 2002;65:051914–1–9.
68. Koblan KS, Ackers GK. Site-Specific Enthalpic Regulation of DNA Transcription at Bacteriophage λ O_R. *Biochemistry* 1992;31:57–65.
69. Takeda Y, Sarai A, Rivera VM. Analysis of the sequence-specific interactions between Cro repressor and operator DNA by systematic base substitution experiments. *Proc Natl Acad Sci USA* 1989;86:439–443.
70. Kim JG, Takeda Y, Matthews BW, Anderson WF. Kinetic Studies on Cro Repressor-Operator DNA Interaction. *J Mol Biol* 1987;196:149–158.
71. Jana R, Hazbun TR, Mollah AKMM, Mossing MC. A folded monomeric intermediate in the formation of Lambda Cro dimer-DNA complexes. *J Mol Biol* 1997;273:402–416.
72. Jana R, Hazbun TR, Fields JD, Mossing MC. Single-chain lambda Cro repressors confirm high intrinsic dimer-DNA affinity. *Biochemistry* 1998;37: 6446–6455.
73. Darling PJ, Holt JM, Ackers GK. Coupled energetics of λ cro repressor self-assembly and site-specific DNA operator binding I: Analysis of cro dimerization from nanomolar to micromolar concentrations. *Biochemistry* 2000;39:11500–11507.
74. Darling PJ, Holt JM, Ackers GK. Coupled energetics of λ cro repressor self-assembly and site-specific DNA operator binding II: cooperative interactions of Cro dimers. *J Mol Biol* 2000;302:625–638.
75. Ackers GK, Johnson AD, Shea MA. Quantitative model for gene regulation by λ phage repressor. *Proc Natl Acad Sci USA* 1982;79:1129–1133.
76. Capp MW, Cayley DS, Zhang W, et al. Compensating Effects of Opposing Changes in Putrescine (2+) and K^+ Concentrations on *lac* Repressor- *lac* Operator Binding: *in vitro* Thermodynamic Analysis and *in vivo* Relevance. *J Mol Biol* 1996;258:25–36.

77. Record Jr. MT, Courtenay ES, Cayley S, Guttman HJ. Biophysical compensation mechanisms buffering *E. coli* protein-nucleic acid interactions against changing environments. *Trends Biochem Sci* 1998;23:190–194.
78. Hawley DK, McClure WR. Mechanism of activation of transcription initiation from the lambda P_{RM} promoter. *J Mol Biol* 1982;157:493–525.
79. Takeda Y, Ross PD, Mudd CP. Thermodynamics of Cro protein-DNA interactions. *Proc Natl Acad Sci USA* 1992;89:8180–8184.
80. Koblan KS, Ackers GK. Energetics of Subunit Dimerization in bacteriophage lambda cI Repressor: Linkage to Protons, Temperature and KCL. *Biochemistry* 1991;30:7817–7821.
81. Pakula AA, Young VB, Sauer RT. Bacteriophage λ cro mutations: Effects on activity and intracellular degradation. *Proc Natl Acad Sci USA* 1986;83:8829–8833.
82. Bechhoefer J. Feedback for physicists: A tutorial essay on control. *Rev Mod Phys* 2005;77:783–836.
83. Jia H, Satumba WJ, Bidwell III GL, Mossing MC. Slow assembly and disassembly of lambda Cro repressor dimers. *J Mol Biol* 2005;350:919–929.
84. Bundschuh R, Hayot F, Jayaprakash C. The role of dimerization in noise reduction of simple genetic networks. *J Theor Biol* 2003;220:261–269.
85. Briggs GE, Haldane JBS. A note on the kinetic of enzyme action. *Biochem J* 1925;19:338–339.
86. Bakk A, Metzler R. Nonspecific binding of the O_R Repressors CI and Cro of bacteriophage lambda. *J Theor Biol* 2004;231:525–533.
87. Vilar JMG, Leibler S. DNA looping and physical constraints on transcription regulation. *J Mol Biol* 2003;331:981–989.
88. Dodd IB, Shearwin KE, Perkins AJ, Burr T, Hochschild A, Egan JB. Cooperativity in long-range gene regulation by the lambda CI repressor. *Gene Dev* 2004;18:344–354.
89. van Kampen NG. Stochastic processes in physics and chemistry. Amsterdam: Elsevier; 1992.
90. Elowitz MB, Levine AJ, Siggia ED, Swain PS. Stochastic gene expression in a single cell. *Science* 2002;297:1183–1186.
91. Ao P, Yin L. Towards the understanding of stability puzzles in phage lambda. (2003; eprint: cond-mat/0307747: http://arxiv.org/PS_cache/cond-mat/pdf/0307/0307747.pdf)
92. Raser JM, O'Shea EK. Control of stochasticity in eukaryotic gene expression. *Science* 2004;304:1811–1814.
93. Kramers HA. Brownian motion in a field of force and the diffusion model of chemical reactions. *Physica* 1940;7:284–304.
94. Hanggi P, Talkner P, Borkevic M. Reaction-rate theory: Fifty years after Kramers. *Rev Mod Phys* 1990;62:251–341.
95. Dodd IB, Perkins AJ, Tsemitsidis DT, Egan JB. Octamerization of λ CI repressor is needed for effective repression of P_{RM} and efficient switching from lysogeny. *Genes Dev* 2001;15:3013–3022.
96. Revet B, von Wilcken-Bergmann B, Bessert H, et al. Four dimers of λ repressor bound to two suitably spaced pairs of λ operators form octamers and DNA loops over large distances. *Curr Biol* 1999;9:151–154.
97. Pray TR, Burz DS, Ackers GK. Cooperative non-specific DAN binding by octamerizing λ CI repressors: A site-specific thermodynamic analysis. *J Mol Biol* 1998;282:947–958.
98. Hochschild A, Douhan III J, Ptashne M. How λ repressor and λ Cro distinguish between O_{R1} and O_{R3}. *Cell* 1986;47:807–816.

99. Bremmer H, Dennis PP. Modulation of chemical composition and other parameters of the cell by growth rate. In: Neidhardt FC, ed. *Escherichia coli* and *Salmonella*. ASM Press; 1996:1553–1569.
100. Shean CS, Gottesman ME. Translation of the prophage lambda cI transcript. *Cell* 1992;70:513–522.
101. Ringquist S, Shinedling S, Barrick D, et al. Translation initiation in Escherichia coli: sequences within the ribosome-binding site. *Mol Microbiol* 1992;6:1219–1229.
102. Kennell D, Riezman H. Transcription and translation initiation frequencies of the Escherichia coli lac operon. *J Mol Biol* 1977;114:1–21.
103. Eisen H, Brachet P, Pereira da Silva L, Jacob F. Regulation of repressor expression in λ. *Proc Natl Acad Sci USA* 1970;66:855–862.
104. Calef E, Avitabile LDG, Marchelli C, et al. The genetics of the anti-immune phenotype of defective lambda lysogens. In: Hershey AD, ed. The Bacteriophage Lambda. Cold Spring Harbor, NY: Cold Spring Harbor Laboratory Press; 1971:609–620.
105. Bialek W. Stability and noise in biochemical switches. In: Advances in Neural Information Processing 13. Leen TK, Dietterich TG, Tresp V, eds. Cambridge: MIT Press; 2001:103.
106. Paulsson J. Summing up the noise in gene networks. *Nature* 2004;427: 415–418.
107. Casci T. Systems biology—Noise is golden. *Nature Rev Genet* 2005;6: 346–346.
108. Hartwell LH, Hopfiedl JJ, Leibler S, Murray AW. From molecular to modular cell biology. *Nature* 1999;402:C47–C52.
109. Nijhout HF. The nature of robustness in development. *Bioessays* 2002;24: 553–563.
110. Kitano H. Biological robustness. *Nature Rev Genet* 2004;5:826–837.
111. Ma L, Iglesias PA. Quantifying robustness of biochemical network models. *BMC Bioinformatics* 2002;3:38.
112. Bluthgen N, Herzel H. How robust are switches in intracellular signaling cascades? *J Theor Biol* 2003;225:293–300.
113. Kerszberg M. Noise, delays, robustness, canalization and all that. *Curr Opin Genet Dev* 2004;14:440–445.
114. Goulian M. Robust control in bacterial regulatory circuits. *Curr Opin Microbiol* 2004;7:198–202.
115. Stelling J, Sauer U, Szallasi Z, et al. Robustness of cellular functions. *Cell* 2004;118:675–685.
116. Li FT, Long T, Lu Y, et al. The yeast cell-cycle network is robustly designed. *Proc Natl Acad Sci USA* 2004;101:4781–4786.
117. Flintoft L. Gene networks—The flexible network. *Nature Rev Genet* 2005;6:252–252.
118. Ao P. Laws in Darwinian evolutionary theory. *Phys Life Rev* 2005;2: 117–156.
119. Hill TL. Cooperativity Theory in Biochemistry: Steady State and Equilibrium Systems. New York: Springer; 1985.
120. Ben-Naim A. Cooperativity in binding of proteins to DNA. *J Chem Phys* 1997;107:10242–10252.
121. Glass L, Kauffman SA. The logical analysis of continuous, non-linear biochemical control networks. *J Theor Biol* 1973;39:103–129.
122. von Dassow G, Meir E, Munro EM, Odell GM. The segment polarity network is a robust developmental module. *Nature* 2000;406:188–192.
123. Vohradsky J. Neural network model of gene expression. *FASEB J* 2001;15: 846–854.

124. Tchuraev RN, Galimzyanov AV. Parametric stability evaluation in computer experiments on the mathematical model of *Drosophila* control gene subnetwork *In Silico*. *Biology* 2003;3:0100.
125. Waddington CH. Organisers and Genes. Cambridge: Cambridge University Press; 1940.
126. Monod J, Jacob F. General conclusions: teleonomic mechanisms in cellular metabolism, growth and differentiation. *Cold Spring Harbor Symp Quant Biol* 1961;26:389–401.
127. Savageau MA. Design principles for elementary gene circuits: Elements, methods, and examples. *Chaos* 2001;11:142–159.
128. Muller Hill B. The *lac* Operon: a short history of a genetic paradigm. Berlin: Walter de Gruyter; 1996.
129. Hendrix RW, Roberts JW, Stahl FW, Weisberg RA. Lambda II. Cold Spring Harbor, New York: Cold Spring Harbor Laboratory; 1983.
130. Birge EA. Bacterial and Bacteriophage Genetics. 4th edition. Berlin: Springer; 2000.
131. Wegrzyn G, Wegrzyn A. Genetic switches during bacteriophage lambda development. *Prog Nucl Acid Res Mol Biol* 2005;79:1–48.
132. Atsumi S, Little JW. Regulatory circuit design and evolution using phage lambda. *Gene Dev* 2004;18:2086–2094.
133. Svenningsen SL, Costanitno N, Court DL, Adhya S. On the role of Cro in lambda prophage induction. *Proc Natl Acad Sci USA* 2005;102:4465–4469.
134. Kobiler O, Rokney A, Friedman N, et al. Quantitative kinetic analysis of the bacteriophage lambda genetic network. *Proc Natl Acad Sci USA* 2005;102: 4470–4475.
135. Buchler NE, Gerland U, Hwa T. Nonlinear protein degradation and the function of genetic circuits. *Proc Natl Acad Sci USA* 2005;102:9559–9564.
136. Roma DM, O'Flanagan RA, Ruckenstein AE, Sengupta AM. Optimal path to epigenetic switching. *Phys Rev E* 2005;71:011902.
137. Baak K, Svenningsen S, Eisen H, et al. Single-cell analysis of lambda immunity regulation. *J Mol Biol* 2003;334:363–372.
138. Bintu L, Buchler NE, Garcia HG, et al. Transcriptional regulation by the numbers: application. *Curr Opin Genet Dev* 2005;15:125–135.
139. Tian T, Burrage K. Bistability and switching in the lysis/lysogeny genetic regulatory network of bacteriophage lambda. *J Theor Biol* 2004;227:229–237.
140. Warren PB, ten Wolde PR. Enhancement of the stability of genetic switches by overlapping upstream regulatory domains. *Phys Rev Lett* 2004;92:128101.
141. Walczak AM, Sasai M, Wolynes PG. Self-consistent proteomic field theory of stochastic gene switches. *Biophys J* 2005;88:828–850.
142. Campbell A. The future of bacteriophage biology. *Nature Rev Genet* 2003;4: 471–477.
143. Mann NH. The third age of phage. *PloS Biol* 2005;3:753–755.
144. Friedman DI, Court DL. Bacteriophage lambda: alive and well and still doing its thing. *Curr Opin Microbiol* 2001;4:201–207.
145. Wright S. The roles of mutation, inbreeding, crossbreeding and selection in evolution. Proceedings of the 6th International Congress of Genetics. 1932;1:356–366.
146. Delbruck M. Discussion in Unités biologiquées de continuité génétique (International Symposium CNRS *8*), Paris; 1949:33–35.
147. Thom R. Mathematical Models of Morphogenesis. New York: Wiley; 1983.
148. von Bertalanffy L. General System Theory: Foundations, Development, Applications. New York: G. Braziller; 1968.
149. Wiener N. Cybernetics: or, Control and Communication In the Animal and the Machine, 2nd edition. Cambridge: M.I.T. Press; 1961.

19

Applications, Representation, and Management of Signaling Pathway Information: Introduction to the SigPath Project

Eliza Chan and Fabien Campagne

Summary

This chapter reviews current approaches for managing signaling pathway information. After a brief introduction to signaling pathways and their computational uses in support of biomedical research, the chapter covers the data representation paradigms currently used to store and compute information about signaling pathways. File formats, ontologies, and databases are considered and compared. The chapter includes a description of the SigPath project, which is an information management system for signaling pathway information.

Key Words: Information management; signaling pathways; ontology; database; information management system.

1. Introduction

A signaling pathway refers to the cellular apparatus that a cell uses to receive, transmit, integrate, and act upon signals. Various types of cells respond to various types of signals. Some signals are endogenous chemicals; for instance, hormones carried in the blood flow from one tissue to another (e.g., insulin, prostaglandin, etc.). Other signals are exogenous chemicals (for instance, saccharose, which is sensed by cells of our taste buds through specific receptors (1)). In other cases, a signal can be sensed without being encoded by a molecule. This is the case in the outer rod cells of the retina, where light is the signal, or in nerves of vertebrate endotherms and ectotherms that sense temperature and make thermal regulation possible (2). Although these examples are drawn from higher organisms, signaling pathways are also essential to single-cell organisms (e.g., bacteria and yeast) and the conservation of pathways across organisms is a field of study in itself. These examples illustrate the variety of signals that cells can sense and respond to and the importance of the study of signaling pathways in most branches of biology. These examples also hint at the types of questions that biologists tend to ask when studying signaling pathways, including the following:

- What signal does a given signaling pathway sense?
- What is the end response (the phenotype) of the pathway upon receiving the signal?
- How can a signaling pathway be modulated by a drug-like molecule?
- How does the cell integrate multiple signals to determine what is the appropriate response to generate?
- Do signaling pathways obey recurrent organizational principles?
- What are these principles and what is their advantage to the organism?

It is to answer these questions, and many others that scientists use information about signaling pathways.

This chapter briefly reviews how signaling pathways are represented computationally and what types of questions have been asked to leverage this information. We put an emphasis on the advantages and drawbacks of the various ways to represent pathways computationally for most types of questions that scientists have asked of these data. We hope that this presentation will illustrate the intimate interplay between data representation and data analysis. We then proceed to describe SigPath, an information management system aimed at providing an electronic way to store, edit, and manage information about signaling pathways. Using SigPath as an example, we illustrate the requirements and challenges in designing and constructing an information management system for cell signaling pathways.

2. Signaling Pathways

Figure 1 presents a diagram of a signaling pathway as it is often rendered in textbooks or scientific articles. This diagram depicts molecules (sometimes called components or molecular species of the pathway) and their interactions. Molecules are represented as named shapes on the diagram. When a signaling pathway is represented by a cartoon, different shapes may represent different types of molecules (e.g., small molecules, proteins, or different protein families, such as receptors and enzymes). Some shapes may also represent biological concepts (such as a gene) that do not have a one-to-one mapping to a molecular species. Other shapes may represent molecular complexes with defined or undefined stoichiometry. The spatial arrangements of the molecules on the diagram (i.e., proximity), and sometimes lines or arrows, represent interactions between the molecules (for instance, the line between Raf and MEK indicates that Raf phosphorylates and activates MEK (3)). Interactions are chemical and biochemical reactions or biochemical processes (e.g., transport across a membrane) in which components of a pathway participate. Figure 1 also indicates which cellular processes are activated when certain components of a pathway interact. In this chapter, activation of such processes (e.g., apoptosis and differentiation) is called a phenotype to stress

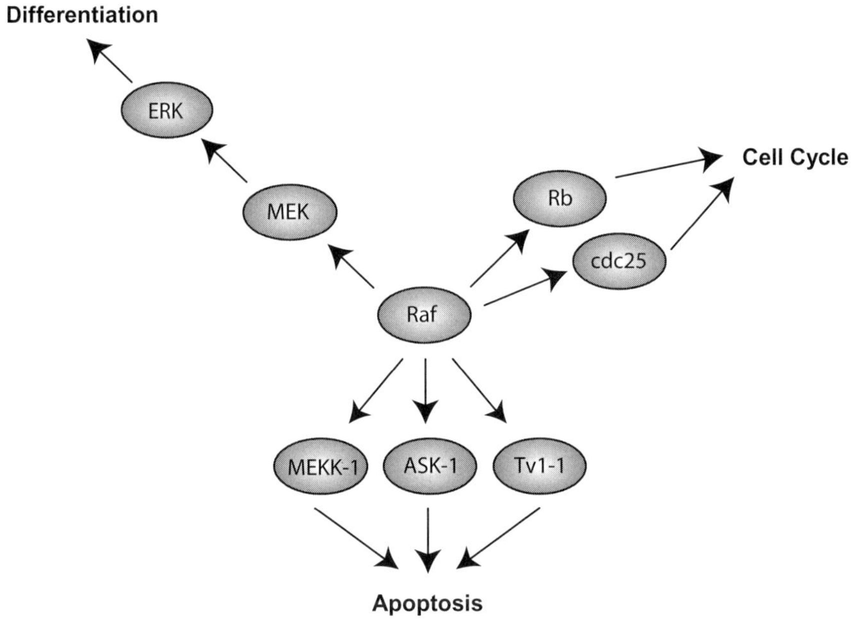

Figure 1. Cartoon representation of a signaling pathway. This diagram shows activation of various molecular species with respect to cellular phenotypes (differentiation, cell cycle, apoptosis). Oval-shaped symbols represent molecular species. An arrow from one species to another indicates that the first species activates the second one. Arrows can also indicate that the activation of a species leads to a given phenotype.

the fact that activation of a process is under the control of genetic and environmental factors.

3. Knowledge Representation and Modeling

It should be clear from the previous section that a signaling pathway is a part of a biological system. Yet, to reason about signaling pathways, it is often convenient to represent them in such a way that "what if" scenarios can be explored, hypotheses formulated, and current knowledge summarized. Because it is often not possible to alter the biological system directly, it is convenient to experiment with a surrogate to the real thing. This surrogate is a model of the signaling pathway. Figure 2 illustrates the interplay between biological system, model, and knowledge. When the surrogate consists of data about the signaling pathway and is used to reason about the pathway, the term knowledge representation may be used in place of model (4). Constructing a knowledge representation or a model is similar, and shares the same trade-offs. First of all, to be useful, the model must be a simplification of reality. If the model was not simplified, it would be as complicated to probe the model as to probe the real signaling pathway. Because the model is a simplification of reality, one would not expect that it would behave exactly as the real system would. The art of model building is in deciding which attributes of reality can

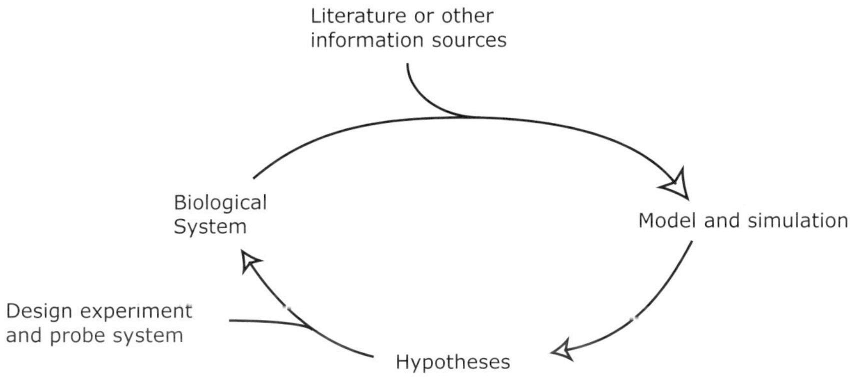

Figure 2. An illustration of the steps involved in the construction of a model based on existing knowledge of a biological system.

be ignored *for a certain application*. Similarly, the art of model testing is in devising tests that check that ignoring certain attributes do not impact the conclusions derived when probing the model.

A direct consequence of the previous considerations is that a model is tailored to a specific application. Indeed, a model that does well for one type of application may fail to reproduce the behavior of the real system in other applications.

We now review the various research applications of signaling pathway information, so that knowledge of applications will be readily available when we discuss the types of signaling pathway models that are in common use. The brief descriptions that follow are meant to be illustrative, rather than complete, of the variety of research uses for signaling pathway information. Research in these areas has been very active in recent years and can only be glanced over in this introductory text. Yet, each section references review articles when appropriate to encourage readers to learn more about each topic.

4. Applications of Signaling Pathway Information

This section reviews the most common applications of signaling pathway data and models.

4.1. Browsing and Looking Up Facts

In this application, a scientist who is not familiar with a molecular species looks up information about it. When the species is retrieved, information about the species and about the interactions the species is engaged in is displayed. Because only one species is looked up at a time, the display can be tailored to human users, using natural language or illustrations as appropriate. Looking up a species in a database can often be a challenge because authors do not name species consistently (very different names can represent the same molecule), and because one name may refer to different molecules (orthologs are sometimes named the same way in different organisms). For these reasons, bioinformaticians have found that assigning accession codes to each biological species helps with many tasks, including fact lookup. Accession codes can be recorded in a notebook and used at a later time to retrieve information without ambiguity. They are preferable to names when precision is required.

4.2. Putting a List of Genes into a Biological Context

Experimental and computational high-throughput methods can identify lists of hundreds of genes. For instance, the analysis of a microarray experiment may reveal that 300 transcripts are differentially expressed between two different experimental conditions. Such a gene list can be potentially very useful in understanding how the cells in the conditions of interest differ, but it is not practical to look up facts for 300 individual transcripts. Therefore, in these cases, scientists prefer methods that can analyze a set of species and create synthetic reports about the species in the set. Several types of analyses are of interest to scientists, e.g., determining what the species have in common, or putting them in the context of current biological knowledge. We have used microarray experiments as an example to illustrate this application, but many high-throughput proteomics or bioinformatics methods can produce gene lists and require techniques that can process them automatically. Methods of this type differ in the type of information that they associate with each species, and in the way the information is collected.

4.2.1. With Gene Ontology

One type of approach leverages information created and maintained by the Gene Ontology (GO) Consortium (http://www.geneontology.org/). Members of the GO consortium organize biological concepts into a concept graph and map genes, transcripts, and proteins to these concepts. For example, the human protein Rhodopsin was mapped to the concepts listed in Table 1. Because many proteins can be associated to the same GO concept, it is possible to identify concepts that appear in a gene list at a greater frequency than would be expected if the gene list was built by a random sampling of all the genes of an organism. Examples of such tools include GoMiner (5), David (6), and EASE (7). Application of these tools can reveal that certain cellular functions and processes are active in a certain condition (*see* Ford et al. (8) for an application of EASE to define cellular processes involved in transient focal stroke). However, analyses based on GO concepts generally do not yield the

Table 1. Gene Ontology concepts associated with protein Rhodopsin (as of January 2006).

GO:0005887 Cellular component: integral to plasma membrane.
GO:0004930 Molecular function: G-protein coupled receptor activity
GO:0007186 Biological process: G-protein coupled receptor protein signaling pathway.
GO:0007603 Biological process: phototransduction, visible light.
GO:0016056 Biological process: rhodopsin mediated signaling.

detailed mechanistic hypotheses that are needed to plan experimental work. GO analyses are, thus, often followed up with detailed literature searches about the genes associated with the function that the analysis has highlighted.

4.2.2. With a Database of Interactions and Appropriate Analysis Tools

The Ingenuity Pathway database is a commercial system that claims to provide access to approximately 1.3 million pairwise interactions (as of 2005). Interactions were extracted from the literature by a scientific staff and are stored in a format that supports automated gene list annotations. When provided with a gene list, the system searches for dense networks of interactions that involve the species in the gene list. Results are summarized as interactive diagrams that present the interactions in the networks and let users access detailed information about the species and the interactions they are involved in.

4.3. Statistical Analyses to Discover Pathway and Network Properties

Analyses of interaction networks do not need to be limited to a given gene list, and several groups have studied what can be learned when information about a large number of interactions is integrated (for instance all the signaling interactions known for a given organism). These studies aim to discover whether signaling pathways are organized according to general principles, and in the affirmative, to describe these principles.

4.3.1. Structural Properties

An example of such a study includes the discovery of motifs in the transcriptional regulation network of *Escherichia coli* (9). A transcriptional regulation network is a network of interactions that includes transcription factor–promoter interactions. In this case, general principles were found in the structural properties (motifs) of the transcriptional network. Figure 3 depicts a few network motifs. With analogy to motifs in protein or DNA sequences, network motifs are sets of interactions sharing common components that are found frequently in a biological network. Usually, how many motif occurrences are needed is determined with statistical methods, comparing the frequency of the motif in a biological

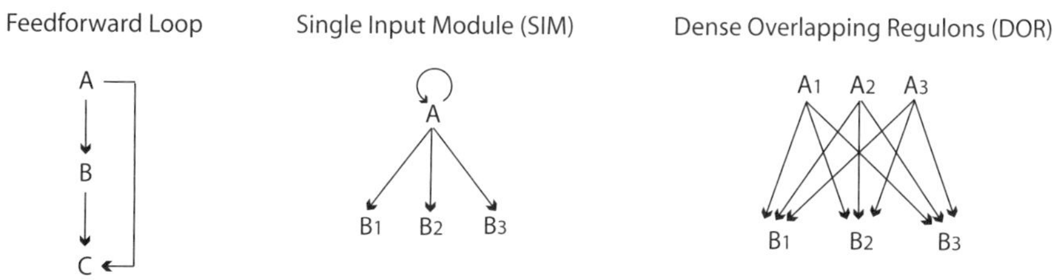

Figure 3. An illustration of network motifs found in transcriptional regulation networks. (Adapted from Shen-Orr et al. (9))

network to the frequency of the same motif in a so-called null-hypothesis network. A null-hypothesis network is a network similar to the biological network under test, usually obtained by randomizing the biological network in some way, so that evolutionary pressure is not expected to influence the presence of motifs in the null-hypothesis network. The significance assigned to discovered motifs is dependent on how the null-hypothesis network is constructed, and constructing unbiased null-hypothesis networks can be challenging (10).

4.3.2. General Network Properties

Large networks can also be analyzed for general properties. Such properties can include statistical measures about the properties of components, interactions, and paths through the network (e.g., distribution of the number of interactions per component). Because computer science has popularized the concept of graphs and developed several algorithms for these data structures, many studies that calculate statistics about large biological networks rely on graph theory at some level. A review of these types of study and their application to improving our knowledge of cell biology is given in a study by Albert (11). Ma'ayan et al. conducted a recent study of the network of known signaling interactions in the CA1 hippocampal neuron (12).

4.4. Detailed Biochemical Modeling and Time-Course Simulations

In the applications that we have already discussed, a qualitative description of signaling pathways and networks was sufficient (how components were connected). The applications that we discuss in this section require information about the dynamics of the interactions that form pathways. This means that detailed kinetic information (knowledge of how fast reactions proceed), or an approximation of this information, is needed for each interaction in a model. Given rate parameters and initial conditions (estimates of the concentration of components at the start of the simulation), dynamical models can simulate how the concentrations of the components in the model evolve with time. The result of the simulation is similar to the result of a time-course experiment, but has distinct advantages over an experiment. First, simulations can track the concentration of each molecule in the model, for each simulated time point. Second, the experimental design can be changed at will by manipulating the inputs or components of the model before or during the simulation. For instance, to simulate how a pathway is activated by a ligand, the concentration of the ligand can be increased for a short period of time, or knockouts can be simulated by forcing the concentration of a component to zero during the entire simulation. These differences with an experimental system make models of signaling pathways a useful tool for scientific hypothesis testing when used as a complement to experiments (13). It should be noted that detailed biochemical models have been used both to make specific predictions about a signaling pathway (13), and to study general properties of pathways (e.g., robustness [14,15], or the property of certain pathways to function similarly under a wide range of specific rate parameters).

5. Representing Signaling Pathways to Support Biomedical Research

In this section, we review the major types of models that have been used to represent and reason with signaling pathways. At one extreme of the spectrum are traditional, natural representations. Textual (interactions are described in sentences) and cartoon representations (interactions are described graphically) were used long before electronic data representation was available. We call these representations natural representations. Electronic, structured representations are aimed at supporting applications where data about many interactions, components, or phenotypes need to be integrated. This is because structured representations make it possible to implement algorithms to process the data, so that programs can perform the repetitive tasks necessary for data integration analysis or modeling.

5.1. Natural Representations

5.1.1. Text Representations

"Spitz and Gurken have been genetically confirmed to activate the EGF receptor, but Keren is uncharacterized" (16). The previous sentence illustrates that natural language is commonly used to represent signaling pathway interactions. This information can be accessed by querying with text search tools such as PubMed (PubMed gives researchers access to approximately 16 million abstracts of biomedical articles) or Twease (see http://www.twease.org). Text representations are best for fact lookup and browsing. However, many interesting applications of signaling pathway information require structured data in an electronic form (*see* section 5.2.). For this reason, various research groups have explored automated methods to extract structured data directly from the literature.

For instance, PubGene mines Medline abstracts to identify species that co-occur in abstracts more than as expected by chance (17). When provided with a gene list, PubGene produces graphs of connected species to indicate that two species frequently appear in the same abstract. Although useful, because co-occurrence in abstracts correlates with biologically meaningful relationships (17), graphs produced by this approach must be carefully inspected to remove spurious connections (connections between species that appear frequently in the same abstracts, but are not functionally related). Other methods attempt to match sentences to predefined language domain grammars to parse natural language and extract interactions, or use more sophisticated information extraction techniques. GENIE is an example of this type of system and has been used to extract interactions from a large number of full-text articles (18).

The evaluation of automatic extraction methods is often challenging because of the lack of a freely available truth standard. In the GENIE study, the gold standard was taken to be a single review article published in Cell, and annotated by a human expert. Performance of text-mining software can vary widely from article to article (as illustrated by the results obtained in our protein name extraction study (19), where precision evaluated on a set of 14 articles from the same journal varied between 51% and 93%). For this reason, it appears preferable to compare several methods on the same dataset, rather than relying on absolute performance measures (19).

Because it is unclear how well fully automatic information extraction systems work, and because of the limited dissemination of these tools (most of them being proprietary and not readily available to the research community), several research groups have preferred to extract information about interactions manually from the literature. An example of this choice is the Human Protein Reference Database (HPRD) (20,21). HPRD aims to offer a comprehensive resource about human protein interactions. A staff of approxiamtely 50 scientists reads articles and populates the HPRD database. The 2005 edition of the database contained approximately 35,000 interactions. HPRD is freely available for research use and licensed for commercial use. HPRD supports browsing and viewing of the information that it offers, but does not provide gene list–oriented tools.

5.1.2. Cartoon Representations

Diagrams are another natural way to present pathway information. In the literature, diagrams of signaling pathways come in various styles and generally do not follow specific conventions. Such diagrams are often called models, in which case they depict the interpretation of the data presented in an article that the authors (and reviewers) feel is the most likely interpretation of the data at the time the article is written. As such, signaling pathway cartoons are very useful summaries of hypotheses and facts and are used extensively throughout the biomedical literature. Informal cartoons of signaling pathways are sometimes accompanied with a legend that explains the meaning of each arrow and shape and other elements shown on the cartoon. However, it is quite common for such legends to be missing, and in these cases, cartoons are ambiguous and difficult to interpret for scientists who are not already familiar with most of the material presented in the cartoon.

In an effort to address the ambiguity problem, several groups have designed graphical languages for metabolic or signaling pathways, hoping to develop a lingua franca for signaling pathways. In 1999, Kohn presented a set of graphical conventions and applied them to diagrams of the cell cycle (22). Kitano et al. have recently proposed the use of process diagrams for the graphical representation of signaling pathways (23). It is worth noting that a vast majority of biologists are not using these conventions and continue to create informal cartoons. Similarly, scientists who can read formal diagrams are a minority in the scientific community, but learning how to read formal diagrams is much easier than learning how to create them.

These considerations strongly suggest that graphical conventions would be most useful if, in addition to being nonambiguous, an algorithm existed to automate the rendering of signaling pathway information according to these conventions.

5.2. Structured Representations

Many types of electronic representations are structured in the sense that they are organized to facilitate computational access to the data.

Structuring data often removes much of the flexibility that natural representation offer. This can sometimes be felt as an impediment to expressing the nuances of scientific data. However, the loss in flexibility is compensated by gains obtained in data consistency (structured data dramatically reduces ambiguity). In turn, data consistency makes large-scale analyses possible. It appears as if, and this may be counterintuitive at first, too much flexibility in how data are represented can hinder system biology studies that require integration and analysis of large amounts of data. In this section, we present structured representations of signaling pathway information.

5.2.1. File Formats

File formats are a broad type of structured information. A file format is a structured representation if it can be described using context-free grammar. This restriction guarantees that the content of a file expressed in the language described by the format can be parsed without ambiguity (24). Various file formats have been developed to represent signaling pathway information. A file format is described in a specification, which is a document that formally describes what rules and conventions a document must follow to adhere to the format. Specifications include both syntactic and semantic rules.

5.2.1.1. HUPO PSI: A file format for protein interaction data, HUPO PSI standardizes how protein–protein interaction data can be represented (25). HUPO documents link protein entries to entries in protein databases, offering users ways to obtain additional data about the proteins observed in the proteomics experiment.

5.2.1.2. CellML: The Cell Markup Language (CellML) aims to provide an encoding of models from the intracellular level to the tissue and organ levels (26). The goal of CellML is to provide for the smooth integration and composition of representations at the various space and time scales necessary to model various aspects of a tissue or organ.

5.2.1.3. SBML: A popular format is the Systems Biology Markup Language (SBML) (27). SBML files can present qualitative and quantitative biochemical models. SBML is developed as part of a community effort (through a mailing list, regular forums, and meetings).

5.2.1.4. Standards and Format Evolutions: Both SBML and CellML are evolving formats. For instance, SBML is being developed by a community of users with various research interests. As research needs evolve, so must the standard. To allow SBML to evolve without breaking existing implementations (tools that can read and write SBML documents), SBML is released in different levels and versions. Levels are major revisions to the standard, and versions group smaller changes within a level. At the time this chapter was written, SBML levels 1 and 2 were available, and Level 3 was being planned. With file formats, the responsibility of evolving data from one version of the file format to the next lies with the provider of the data. File repositories usually complement file formats to address evolution problems. The team responsible for the repository converts data and information from an older version of the file format to the most current version.

5.2.2. Ontologies

The term "ontology" has gained popularity in the biomedical sciences. Perhaps because of this success, this term has been defined in many different ways, some contradictory. In this chapter, we follow Gruber (28) and define an ontology as a set of classes (or concepts) and the relations that exist among them. Those classes and relations often form a hierarchical structure (e.g., the GO (29)), but more complex and semantically rich structures are possible (e.g., EcoCyc [30] and, more recently, the Sequence Ontology (31)). A good introduction to the construction of ontologies is given by Noy and McGuinness (32) (http://www.ksl.stanford.edu/people/dlm/papers/ontology-tutorial-noy-mcguinness.pdf). The key idea is that an ontology formalizes the choices that data modelers have made to represent an aspect of reality. Building an ontology is an efficient and commonly accepted method of creating an explicit and formal specification of a data representation. Ontologies can be used to support semantically rich queries (for example, the TissueInfo ontology is used to expand tissue queries (33)), to structure information during data collection (34), or as formal documentation for data represented in file formats (i.e., linking an item of data in a structured file to a concept of an ontology explicitly defines what the data are about with respect to the other concepts in the ontology). Some tools, such as Protégé, support loading ontologies in one of the formal ontology languages (i.e., RDFS (35) and OWL (36)) and generating data entry forms (37). These tools are very useful to model an ontology and start entering data. However, for reasons of performance, scalability, or stability, most bioinformatics projects are still implemented with traditional database backend systems.

5.2.2.1. BioPax: A file format and associated ontology to represent signaling, metabolic, and genetic interactions. The format aims to be a standard for exchange of these data between databases and tools (reviewed in Stromback and Lambrix (38)).

5.2.3. Databases

Biological databases are a collection of Web-based tools used to present and visualize certain types of biological data and information. Databases generally offer flexible search facilities and may offer services such as downloading of data in one or multiple different file formats (for more information about biological data management, see Srdanovic et al. (39)).

5.2.3.1. Virtual Cell: The Virtual Cell is primarily a modeling tool, but also offers a database of models where users can save different versions of the models they work with, or publish models for sharing with other users (40).

5.2.3.2. BioModels.net: This database is a repository of published, curated models available in SBML format (41).

5.2.3.3. SigPath: The SigPath information management system lets users manage their own signaling pathway data. SigPath is a central, publicly available database, but is also a set of distributed databases and associated tools to support the exchange of data between these databases and with other tools (43). The latest developments to the SigPath project are described in the next section.

6. The SigPath Project

SigPath (http://www.sigpath.org) is a project to create a management system to organize and store biochemical information for signaling pathways (34). SigPath offers a Web interface to a database to facilitate storing and browsing of biochemical data (Figure 4). The project also offers the SigPath Navigator as a standalone user interface to help navigate information in SigPath, transfer data across different SigPath databases, and help with a variety of curation tasks. This section provides an introduction to SigPath and illustrates how SigPath Navigator facilitates the interaction and exchange of data between multiple distributed SigPath systems.

6.1. System

6.1.1. System Architecture

Figure 5 presents the architecture of a Web-based SigPath application. SigPath leverages open standards such as XML, XML schemas, or Java Data Objects (JDO) (39). The source code of the software artifacts

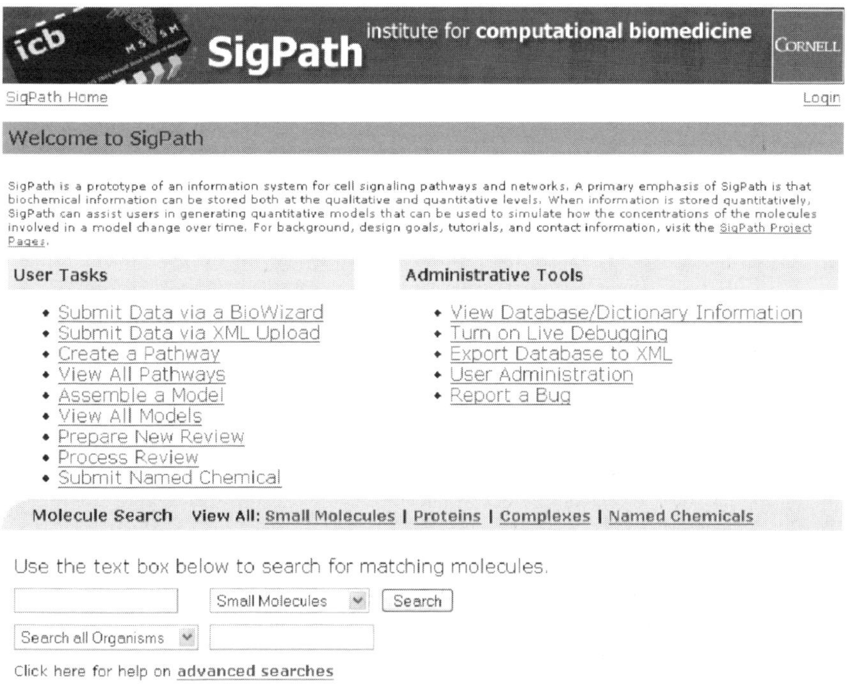

Figure 4. Web-based interface of the SigPath information management system.

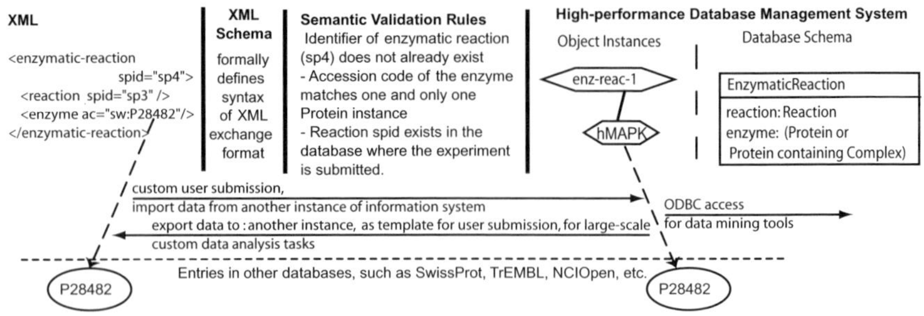

Figure 5. Overview of the information management approach and integration with external databases. The SigPath XML schema, semantic validation rules, and database schema together fully specify the SigPath ontology.

developed in the project is distributed under the Gnu General Public License (GPL).

6.1.2. File Format

Data submitted to SigPath can be downloaded in the SigPath XML exchange format. This format is fully documented on the project Web site. The SigPath XML exchange format can also be used to submit or import data into SigPath. The architecture presented in Figure 5 facilitates maintaining data in SigPath through evolution of the SigPath exchange format, thereby limiting the amount of manual conversions. The SigPath XML exchange format is used as a medium for data transfer between SigPath databases. This format can be changed rapidly to support requirements introduced when developing new versions of SigPath. As such, the SigPath XML exchange format is unlike file formats developed via community discussions which evolve into standards and must remain relatively stable. Indeed the format is tightly linked to the state of development of SigPath and has greatly evolved since its initial version in May 2002 (features are only added to the SigPath XML format when the corresponding feature has been implemented in SigPath).

6.1.3. Ontology

Data in SigPath is organized according to the SigPath ontology (34). Figure 6 describes the main concepts of this ontology that are mentioned in this section. The SigPath ontology is implemented in a database schema and a set of semantic validation rules (Figure 5).

6.1.4. Data

We typically deploy different instances of the SigPath system. SigPath production (see link on the top right of the SigPath project Web page) is an instance of the SigPath system open to data submission from the public, which offers a public dataset at any given time. Data in the production version of SigPath are not actively curated, so that data submitters are ultimately responsible for the quality of their submissions.

SigPath beta (sp-beta) is an instance of the system that we deploy when new developments are incorporated into SigPath. This system is used mostly for testing the various components of SigPath with real data. SigPath beta offers a copy of the data available in SigPath production at

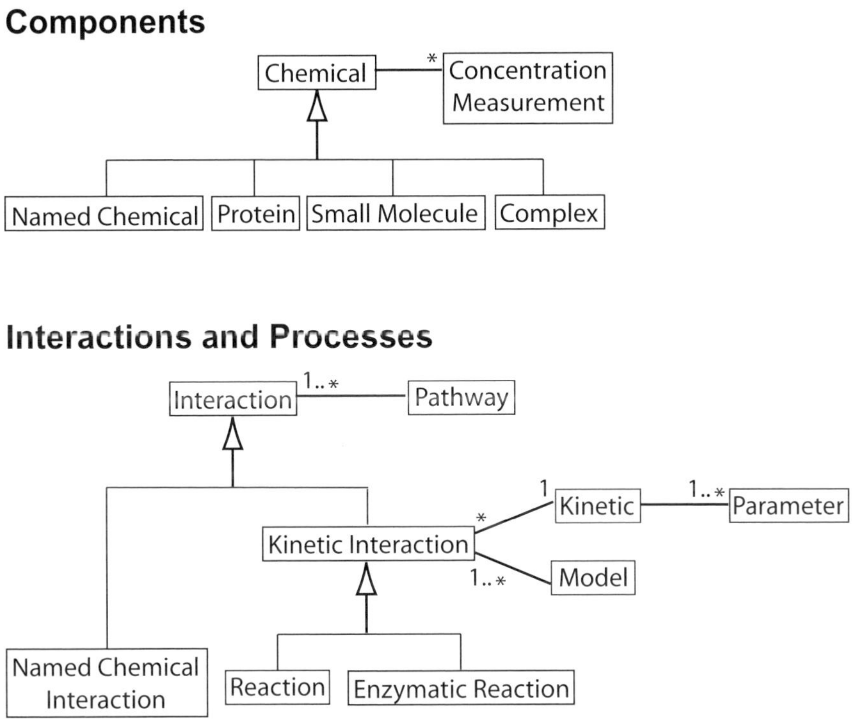

Figure 6. Fragment of the SigPath ontology. Concepts of this ontology can represent components, interactions and processes. The figure follows the Unified Modeling Language conventions. Arrows with an open end indicate that the concept pointed to is more general than the concept where the arrow originates. Lines without arrows indicate relations between concepts. Signs 1, *, 1..* indicate cardinality of these relations according to UML conventions. The line between Pathway and Interaction, for instance, indicates that the concept Pathway makes sense only in relation to one or more (1..*) interactions.

the time sp-beta was set up. Data submitted to sp-beta is discarded when testing of the new software is complete.

Because SigPath is distributed under the GPL, other groups can download the SigPath application, to compile and configure it locally. At any given time, there can therefore be several SigPath instances available, each containing partially overlapping or disjointed datasets.

6.2. Levels of Data Representation

Figure 6 presents what type of information SigPath can currently represent and manage. Figure 7 sorts these concepts differently to relate each concept to a type of application discussed in this chapter. Applications of signaling pathway information are organized from the less specific to the more specific. The following sections describe how each level is supported in the SigPath ontology.

6.2.1. Literature Level

The lower levels of information stored in SigPath allow for some ambiguity. For instance, Named Chemicals are concepts of the SigPath ontology

More Specific

Applications		Components	Interactions	Processes
Biochemical modeling and time course simulations		Concentration Measurement	Kinetic Interaction, Kinetic	Model
Browsing and fact lookup	Discovery applications (e.g., gene lists)	Small Molecule, Protein	Interaction (Reaction/Enzymatic Reaction)	Pathway
	Literature summaries	Named Chemical	Named Chemical Interaction	

(Table header corner: Types of Information / Applications)

Figure 7. Information can be represented at different levels of specificity in the SigPath ontology. These different levels support different uses of the information (applications).

that represent molecules with only a name and a few aliases for descriptors. When a Named Chemical instance is used in SigPath, it is not clear if the component represents a protein, small molecule, or other type of component, and accordingly, the component may not have an accession code to link it to another database or to an organism. These limitations notwithstanding, the Named Chemical concept is useful to represent components with the level of detail that appear in an abstract or complete article. Together with Named Chemical Interaction, the Named Chemical concept can encode the level of information typically found in natural language in abstract or full text articles (although the names of the molecules are present in this material, obtaining the type of the molecule or the organism in which the interaction was studied may require consulting other sources of information than the original article). When represented in this limited way in SigPath, the information becomes searchable, and is sufficient to build diagrams or navigate interactions interactively (*see* section 6.3). Furthermore, the data can be further annotated and serve as the basis to create a less ambiguous representation.

6.2.2. Qualitative Level

At the qualitative level, SigPath can represent interactions (interactions are directed and have species on their left and right), possibly under the control of modifier species (e.g., species that can modulate the interaction by their presence or absence). Publications and individual user comments can be attached to interactions. Interactions are searchable by their interacting molecules. Molecules can in turn be searched by a variety of accession codes and names. Pathways are the qualitative representation of processes. A SigPath pathway consists of a set of interactions. Pathways are rendered automatically into diagrams to provide a graphical view of the information. Diagrams are linked on species to the page that describes the molecule.

6.2.3. Quantitative Level

This level requires that a kinetic mechanism and rate parameters be attached to an interaction. Kinetics are a way to describe how the concentrations of the species involved in an interaction control the rate of the interaction. Similarly to SBML, SigPath lets users associate a rate law to a kinetic (rate laws are mathematical expressions that evaluate to the rate of the reaction in the left to right direction). However, SigPath allows users to define Kinetic objects and document their variables (concentrations of species or constant parameters) so that other users of SigPath can reuse the same kinetic and attach them to other interactions. In SigPath, a Kinetic is endowed with a SigPath accession code. SigPath accession codes are called *spid* (for SigPath identifiers) and lets users identify and retrieve information in SigPath in a nonambiguous manner (they can be listed in articles as accession codes are for genes and proteins). Models are the quantitative representation of processes in SigPath. Models are a set of interactions described at the quantitative level, with initial conditions for the molecules involved in the interactions of the models. Models can be exported to files in SBML (L1 v1 to L2 v2) or Kinetikit format (42,43).

6.2.4. Navigating Through Pathway Information

Because of its graph-like structure, pathway information can be easier to understand when rendered in a graphical way. However, static diagrams of pathway information have the disadvantage that they rarely match exactly the amount of detail that a user needs for a certain task (such as viewing all the components and/or interactions that are relevant to a task). For this reason, we have extended SigPath with a tool to navigate pathway information (interactions, processes or components) and organize subsets of this information in interactive diagrams. SigPath Navigator is a Java Web Start application that provides a graphical metaphor for data in SigPath and lets users manipulate the data remotely. The Navigator is able to connect to and communicate with SigPath systems that are installed in laboratories from different geographical locations (provided the SigPath instances are reachable through the network where SigPath Navigator is started). This ability enables transfer of data from one SigPath system to another. SigPath Navigator also facilitates data curation, as will be described in the following worked example. SigPath Navigator leverages the SigPath XML exchange format to obtain or submit data from SigPath instances.

6.3. Worked Examples

6.3.1. Transfer of Data from One Instance to Another

In this example, we assume that two installations of SigPath are available to the user. These distinct installations of the SigPath system will be referred to as SigPath instance A and B. SigPath instance A contains a set of interactions that is not available in SigPath instance B. In this example, the user wants to transfer this set of data from instance A to B.

1. To connect to SigPath instance A, start SigPath Navigator and select "Connect to a SigPath instance" from the "SigPath" menu. The user needs to know the URL of the source SigPath instance. URLs for public versions of SigPath are predefined. If the URL to the public Web interface of SigPath is known, the user can determine the URL of the source SigPath instance by following these templates or contacting the administrators of the source SigPath instance.

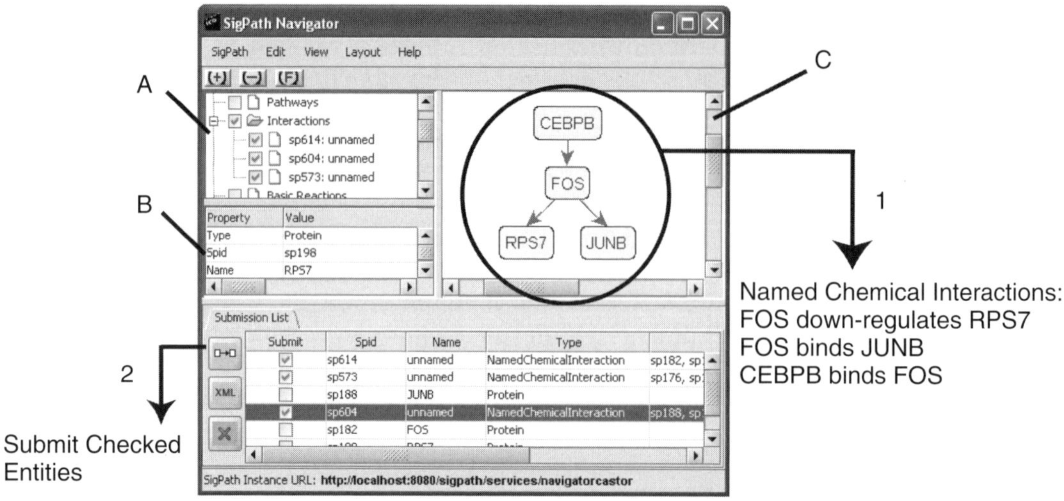

Figure 8. The SigPath Navigator user interface. (A) Tree panel displaying a list of components retrieved from the database. (B) Property panel displaying information about a selected component. (C) Graphics panel displaying a cartoon representation of 3 interactions. Arrow 1 shows the textual representation of 3 Named Chemical Interactions. Arrow 2 illustrates usage of the button.

2. The interactions can be retrieved by entering the *spid* values that uniquely identify them in the instance A. The Web interface of SigPath instance A can be used to search interaction data and obtain the *spid* values. In the navigator, enter the *spid* values one by one by selecting "Add spid" from the "Edit" menu.

3. At this stage, the user switches SigPath Navigator to instance B by selecting the "Connect to a SigPath instance" option from the "SigPath" menu. (A predefined instance can be selected or, again, a different URL can be entered; refer to point 1.)

4. Once connected to instance B, select "Submit Entities to SigPath" from the "SigPath" menu. The user should see a Submission List panel at the bottom of the application. Select the Named Chemical Interaction items in the list and click the "Submit Checked Entities" button (Arrow 2 in Figure 8). Only those items of information selected will be transferred to SigPath instance B.

5. If the submission is successful, the user should see a message dialog "Submission Successful." A detailed error message is provided if the transfer failed for any reason. The submission may fail if the items of information selected at step 4 are not self-contained. For instance they could refer to molecules not present in SigPath instance B. In this case, the SigPath Navigator can be used to identify the missing molecules or other elements of information, and these can be selected before initiating the data transfer. Transfer may also fail if any of the data elements transferred to instance B already exist in instance B. The interactive features of SigPath Navigator can help users pinpoint the source of these problems and remedy them.

6.3.2. Deletion of Data

There are rules and restrictions in SigPath to ensure deletion of data does not cause unintentional impact on other data. For example, if component JUNB participates with component FOS in an interaction, the system will not allow component JUNB to be deleted. This is because component JUNB is referenced in the representation of its interaction with FOS. SigPath Navigator reports these references as backward references. Reciprocally, the references from the interaction to components FOS and JUNB is said to be a forward reference. Forward references are simple enough to account for, but backward references can be difficult to find using the SigPath Web interface. In this example, deletion is only allowed if component JUNB is not referenced by another data element (any of the concepts shown in Figure 6). In addition, a user can only delete the data that he or she has submitted previously, not the ones submitted by other users. As a continuation of the previous example, the user now wants to delete the component RPS7 from the currently connected SigPath instance, as follows:

1. Select the component labeled RPS7, right-click and select the "Delete from SigPath Database" option. The user is prompted with 3 options: "Save Before Deleting," "Delete Now," or "Cancel." Click the "Delete Now" button. The user will be prompted with a message "Deletion denied: You may not delete entities that have backward references outside your selected list." In this case, the backward reference is the interaction with FOS (*see* Graphics panel in Figure 8). Thus, before deleting RPS7, the user should delete the interaction first.
2. Change the view by selecting the "Expand" option from the "View" menu. Select the interaction symbol labeled "FOS down-regulates RPS7," then right-click to select the "Delete from SigPath Database." Click the "Save before deleting" button.
3. The user can now proceed to delete RPS7 by repeating Step 1.

Electronic resources listed

HPRD	**http://www.hprd.org**	**Commercial, free for research**
Ingenuity Pathway database	**http://www.ingenuity.com**	**Commercial**
System Biology Markup Language	**http://www.sbml.org**	**Free, standard driven by community needs**
BioPax	**http://www.biopax.org**	**Free.**
Protégé	**http://protege.stanford.edu**	**Free, open-source**
Virtual Cell	**http://www.nrcam.uchc.edu**	**Data are freely available if shared by submitter.**
BioModels	**http://www.ebi.ac.uk/biomodels**	**Data are freely available.**
SigPath	**http://www.sigpath.org**	**Data and code are freely available.**
Twease	**http://www.twease.org**	**Open source**

Acknowledgments: The authors thank the colleagues and staff who contributed to the SigPath project over the years. Drs. Harel Weinstein and Ravi Iyengar provided vision, guidance and support. Without them, the project would not have been possible. Ethan Cerami and Marko Srdanovic helped with the infrastructure needed to support large-scale software development. Manuel Martin, Manda Wilson, Anat Maoz and Francois Le Fevre made significant contributions to the source code of the project. We also thank the many scientists who tested SigPath and provided feedback. They helped shape the information management scheme described in this chapter.

References

1. Max M, Shanker YG, Huang L, et al. Tas1r3, encoding a new candidate taste receptor, is allelic to the sweet responsiveness locus Sac. *Nat Genet* 2001;28(1):58–63.
2. Cooper KE. Some historical perspectives on thermoregulation. *J Appl Physiol* 2002;92(4):1717–1724.
3. Stewart S, Guan KL. The dominant negative Ras mutant, N17Ras, can inhibit signaling independently of blocking Ras activation. *J Biol Chem* 2000; 275(12):8854–8862.
4. Davis R, Szolovits P. What is Knowledge Representation? *AI Magazine* 1993;14(1):17–33.
5. Zeeberg BR, Feng W, Wang G, et al. GoMiner: a resource for biological interpretation of genomic and proteomic data. *Genome Biol* 2003;4(4): R28.
6. Dennis G, Jr., Sherman BT, Hosack DA, et al. DAVID: Database for Annotation, Visualization, and Integrated Discovery. *Genome Biol* 2003;4(5):P3.
7. Hosack DA, Dennis G, Jr., Sherman BT, Lane HC, Lempicki RA. Identifying biological themes within lists of genes with EASE. *Genome Biol* 2003;4(10): R70.
8. Ford G, Xu Z, Gates A, Jiang J, Ford BD. Expression Analysis Systematic Explorer (EASE) analysis reveals differential gene expression in permanent and transient focal stroke rat models. *Brain Res* 2006;1071(1):226–236.
9. Shen-Orr SS, Milo R, Mangan S, Alon U. Network motifs in the transcriptional regulation network of *Escherichia coli*. *Nat Genet* 2002;31(1):64–8.
10. Artzy-Randrup Y, Fleishman SJ, Ben-Tal N, Stone L. Comment on "Network motifs: simple building blocks of complex networks" and "Superfamilies of evolved and designed networks." *Science* 2004;305(5687):1107.
11. Albert R. Scale-free networks in cell biology. *J Cell Sci* 2005;118(Pt 21): 4947–4957.
12. Ma'ayan A, Jenkins SL, Neves S, et al. Formation of regulatory patterns during signal propagation in a Mammalian cellular network. *Science* 2005; 309(5737):1078–1083.
13. Neves SR, Iyengar R. Modeling of signaling networks. *Bioessays* 2002; 24(12):1110–1117.
14. von Dassow G, Meir E, Munro EM, Odell GM. The segment polarity network is a robust developmental module. *Nature* 2000;406(6792):188–192.
15. Alon U, Surette MG, Barkai N, Leibler S. Robustness in bacterial chemotaxis. *Nature* 1999;397(6715):168–171.
16. Urban S, Lee JR, Freeman M. A family of Rhomboid intramembrane proteases activates all Drosophila membrane-tethered EGF ligands. *EMBO J* 2002;21(16):4277–4286.
17. Jenssen TK, Laegreid A, Komorowski J, Hovig E. A literature network of human genes for high-throughput analysis of gene expression. *Nat Genet* 2001;28(1):21–28.

18. Friedman C, Kra P, Yu H, et al. GENIES: a natural-language processing system for the extraction of molecular pathways from journal articles. *Bioinformatics* 2001;17 Suppl 1:S74–S82.
19. Shi L, Campagne F. Building a protein name dictionary from full text: a machine learning term extraction approach. *BMC Bioinformatics* 2005; 6(1):88.
20. Peri S, Navarro JD, Amanchy R, et al. Development of human protein reference database as an initial platform for approaching systems biology in humans. *Genome Res* 2003;13(10):2363–2371.
21. Mishra GR, Suresh M, Kumaran K, et al. Human protein reference database–2006 update. *Nucleic Acids Res* 2006;34(Database issue):D411–D414.
22. Kohn KW. Molecular interaction map of the mammalian cell cycle control and DNA repair systems. *Mol Biol Cell* 1999;10(8):2703–2734.
23. Kitano H, Funahashi A, Matsuoka Y, Oda K. Using process diagrams for the graphical representation of biological networks. *Nat Biotechnol* 2005;23(8):961–966.
24. Cooper K, Torczon L. Engineering a Compiler. San Francisco: Morgan Kaufmann Publishers; 2004.
25. Hermjakob H, Montecchi-Palazzi L, Bader G, et al. The HUPO PSI's molecular interaction format–a community standard for the representation of protein interaction data. *Nat Biotechnol* 2004;22(2):177–183.
26. Lloyd CM, Halstead MD, Nielsen PF. CellML: its future, present and past. *Prog Biophys Mol Biol* 2004;85(2–3):433–450.
27. Finney A, Hucka M. Systems biology markup language: Level 2 and beyond. *Biochem Soc Trans* 2003;31(Pt 6):1472–1473.
28. Gruber TR. A Translation Approach to Portable Ontology Specification. *Knowledge Acquisition* 1993;5:199–220.
29. Harris MA, Clark J, Ireland A, et al. The Gene Ontology (GO) database and informatics resource. *Nucleic Acids Res* 2004;32(Database issue):D258–D261.
30. Keseler IM, Collado-Vides J, Gama-Castro S, et al. EcoCyc: a comprehensive database resource for Escherichia coli. *Nucleic Acids Res* 2005;33 Database Issue:D334–D337.
31. Eilbeck K, Lewis SE, Mungall CJ, et al. The Sequence Ontology: a tool for the unification of genome annotations. *Genome Biol* 2005;6(5): R44.
32. Noy NF, McGuinness DL. Ontology Development 101: A Guide to Creating Your First Ontology; 2001. Report No.: TR #SMI-2001-0880.
33. Skrabanek L, Campagne F. TissueInfo: high-throughput identification of tissue expression profiles and specificity. *Nucleic Acids Res* 2001;29(21): E102–E102.
34. Campagne F, Neves S, Chang CW, et al. Quantitative information management for the biochemical computation of cellular networks. *Sci STKE* 2004;2004(248):l11.
35. Resource Description Framework (RDF) Schema Specification 1.0, Candidate recommendation, World Wide Web Consortium (Mar. 2000). URL http://www.w3.org/TR/2000/CR-rdf-schema-20000327. 2000. (Accessed at http://www.w3.org/TR/2000/CR-rdf-schema-20000327.)
36. Dean M, Connolly D, van Harmelen F, et al. OWL web ontology language 1.0 reference, 2002.
37. Noy NF, Crubezy M, Fergerson RW, et al. Protege-2000: an open-source ontology-development and knowledge-acquisition environment. *AMIA Annu Symp Proc* 2003:953.

38. Stromback L, Lambrix P. Representations of molecular pathways: an evaluation of SBML, PSI MI and BioPAX. *Bioinformatics* 2005;21(24):4401–7.
39. Srdanovic M, Schenk U, Schwieger M, Campagne F. Critical evaluation of the JDO API for the persistence and portability requirements of complex biological databases. *BMC Bioinformatics* 2005;6(1):5.
40. Slepchenko BM, Schaff JC, Macara I, Loew LM. Quantitative cell biology with the Virtual Cell. *Trends Cell Biol* 2003;13(11):570–576.
41. Le Novere N, Bornstein B, Broicher A, et al. BioModels Database: a free, centralized database of curated, published, quantitative kinetic models of biochemical and cellular systems. *Nucleic Acids Res* 2006;34(Database issue): D689–D691.
42. Vayttaden SJ, Bhalla US. Developing complex signaling models using GENESIS/Kinetikit. *Sci STKE* 2004;2004(219):pl4.
43. Bhalla US. Use of Kinetikit and GENESIS for modeling signaling pathways. *Methods Enzymol* 2002;345:3–23.

Part IV

Methods and Software Platforms for Systems Biology

20

SBML Models and MathSBML

Bruce E. Shapiro, Andrew Finney, Michael Hucka, Benjamin Bornstein, Akira Funahashi, Akiya Jouraku, Sarah M. Keating, Nicolas Le Novère, Joanne Matthews, and Maria J. Schilstra

Summary

MathSBML is an open-source, freely downloadable *Mathematica* package that facilitates working with Systems Biology Markup Language (SBML) models. SBML is a tool-neutral, computer-readable format for representing models of biochemical reaction networks, and it is applicable to metabolic networks, cell signaling pathways, genomic regulatory networks, and other modeling problems in systems biology that is widely supported by the systems biology community. SBML is based on XML, which is a standard medium for representing and transporting data that is widely supported on the Internet, as well as in computational biology and bioinformatics. Because SBML is tool-independent, it enables model transportability, reuse, publication, and survival. In addition to MathSBML, a number of other tools that support SBML model examination and manipulation are provided on the http://sbml.org Web site, including libSBML, which is a C/C++ library for reading SBML models; an SBML Toolbox for MATLAB; file conversion programs; an SBML model validator and visualizer; and SBML specifications and schemas. MathSBML enables SBML file import to and export from *Mathematica*, as well as providing an API for model manipulation and simulation.

Key Words: SBML; libSBML; MathSBML; systems biology; XML; BioModels.

1. Motivation

The SBML is a tool-neutral, computer-readable, text file (XML) format for representing models of biochemical reaction networks. It is especially applicable to descriptions of cell signaling pathways, metabolic networks, genomic regulatory networks, and other modeling problems in systems biology (1,2). SBML is based on XML (the eXtensible Markup Language), which is a standard medium for representing and transporting data that is widely supported on the Internet (3) as well as in computational biology and bioinformatics (a recent PubMed search on "XML" returned 612 hits; INSPEC 4,200 hits; and Web of Science 3,009 hits; scholar.google.com, 810,000 hits) (4). The central goal of SBML is model portability. By encoding models in SBML, they can be freely interchanged between users, regardless of which software tool, hardware platform, or operating system each uses. So long as each modeler uses SBML-compliant software, they will both be able to run simulations from the same model, without modification, on their own platform, and compare results.

The benefits of this interoperability are enormous. Not only can users share models but they can also use multiple simulation tools and techniques within a single research project without rewriting their models (5). Say, for example, that a modeler wants to perform a combination of discrete stochastic and continuous dynamic simulations. Usually this means that he will need to use two different simulation tools. Typically, each software program has a unique model description format that is incompatible with other programs. If both tools are SBML compliant, then the model only needs to be encoded once.

A second benefit of this standardization is model publication and dissemination in the peer-reviewed literature. Published models are described in a variety of formats: differential equations, algebraic equations, reactions, pathway diagrams, event rules, etc. If, in addition, the author encodes his model in SBML and makes it available to the publisher (and eventually, the journal's readers) via an auxiliary Web site, then computationally astute peer-reviewers can test the models and verify the purported results independently of the authors. Furthermore, when the final paper is published, readers can easily reproduce the same results and incorporate them into and/or compare/contrast them with their own simulations. Journal editors appear to agree; for example, the instructions for authors of *Nature Molecular Systems Biology* include the statement "Where relevant and possible, authors are encouraged to submit datasets in SBML format" (6).

Finally, SBML can help ensure model survivability (7). When models are described in unique data formats, particularly when their authors code their own simulation engines, the software model survives only as long as the program is being used. Typically, this means that once a student graduates or a postdoctoral researcher moves on to a more permanent position this technology is lost by the original host institution. On the other hand, if commercial or widely available tools from the public domain are used, models typically survive only until a new version or software release requires a new data format, or more commonly, the program stops being supported on the modeler's preferred hardware/software environment. Although there is no guarantee that SBML will always be around, the designers of nearly a hundred different tools have already made their software SBML compliant or announced an intention to do so in the near future (the growing list is regularly updated on http://sbml.org).

2. The Evolution of SBML

SBML has been developed through an evolving international collaboration that reflects the wide variety of research being performed in systems

Table 1. SBML workshops and hackathons.

Meeting	Date	Location
1st Workshop	April 2000	Pasadena, CA, USA
2nd Workshop	Nov. 2000	Tokyo, Japan
3rd Workshop	June 2001	Pasadena, CA,USA
4th Workshop	Dec. 2001	Pasadena, CA, USA
5th Workshop	July 2002	Hatfield, UK
6th Workshop	Dec. 2002	Stockholm, Sweden
7th Workshop	May 2003	Ft. Lauderdale, FL, USA
1st Hackathon	July 2003	Blacksburg, VA, USA
8th Workshop	Nov 2003	St. Louis, MO, USA
2nd Hackathon	May 2004	Hinxton, UK
9th Workshop	Oct. 2004	Heidelberg, Germany
3rd Hackathon	May 2005	Tokyo, Japan
10th Workshop	Oct. 2005	Boston, MA, USA
4th Hackathon	April 2006	Nove Hrady, Czech Rep.
11th Workshop	Oct. 2006	Tokyo, Japan

biology. Owing to both the geographical diversity and the size of this group, most discussions have taken place electronically. Moderated (to edit out spam) discussion lists (sbml-discuss, libsbml-discuss) are maintained and archived at http://sbml.org/forums. Over 2,500 messages were posted to these lists between October 2002 and October 2005; another thousand or so were posted on earlier discussion lists that were combined with Systems Biology Workbench development, and countless other private messages have been sent between list members. These lists currently contain over 200 members coming from academic, commercial, and private environments, and from all continents.

The ideas developed and discussed through these forums were crystallized through a series of open workshops and working groups starting in the year 2000, many of which were followed by detailed specification documents or proposals. To date, 10 workshops and 3 "hackathons" have been held in Japan, the US, the UK, Sweden, and Germany (Table 1). Workshops provide a forum for users to become aware of new developments in SBML software, to discuss proposed SBML features so that consensus decisions can be made, and to maximize software interoperability by discussing issues that have arisen in the various implementations. Hackathons, on the other hand, provide a forum for software developers to gather and work simultaneously to solve interoperability issues. The minutes of all workshops and hackathons are available at http://sbml.org/.

The specification of the original language, called SBML Level 1, was released on 2 March 2001. Minor deficiencies and corrections were incorporated in the next release, SBML Level 1, Version 2 (L1V2), on 28 August 2003. Level 1, Version 2 replaces Level 1, Version 1, as it primarily corrects errors in the original document (8). A major revision, SBML Level 2, Version 1 (L2V1), was released on 28 June 2003 (9). Initials drafts of SBML Level 2, Version 2 (L2V2) and SBML Level 2, Version 3 (L2V3) were released on March 26, 2005 and March 20, 2007, respectively (10,11). All specification documents and related resources are maintained on the Web site at http://sbml.org.

The http://sbml.org Web site aims to provide for SBML what http://www.w3.org provides to the World Wide Web. While formal membership in an equivalent consortium is not currently required, the intent is to provide a formal, nonbiased (from the perspective of individual modeling tools) location where specifications, schemas, technical reports, discussion lists, a Wiki, and various online tools can be maintained. The tools provided, such as a validator, visualizer, conversion libraries, and libraries for reading and maintaining SBML files, are designed to aid all modelers in their SBML implementations, and will be the subject of section 5 of this chapter. The group that maintains http://sbml.org, the "SBML Team," is an international research team distributed at institutions around the world. The SBML Team is not the "keeper" of SBML—that role is for the systems biology community—merely organizers, editors, and fellow tool developers.

3. SBML Level 2 Models

In this section, we will review the format of SBML Level 2 Version 1 Models. Level 1 is omitted due to space limitations, as well as the overwhelming (and growing) prevalence of SBML Level 2; the interested reader should consult the references for additional information. The overall structure of an SBML model is

```
beginning of model definition
  list of function definitions
  list of unit definitions
  list of compartment definitions
  list of species
  list of parameters
  list of rules
  list of reactions
  list of events
end of model definition
```

The order of the lists cannot be modified, e.g., *species* must precede *parameters*, etc. SBML models are encoded as XML files; each XML file contains a single "model" object, which is itself enclosed within an `sbml` object (Figure 1). Each of the "lists" is optional; when present it must contain a nonzero number of object definitions of the following general form:

```
<listOfFoos>
  <foo . . .> . . . </foo>
  <foo . . .> . . . </foo>
  . . .
</listOfFoo>
```

where `foo` is one of the `functions`, `units`, `compartments`, `species`, `parameters`, `rules`, `reactions`, or `events`. With the exception of

```
<?xml version="1.0" encoding="UTF-8"?>
<sbml xmlns="http://www.sbml.org/sbml/level2"
    level="2" version="1">
   <model id="My_Model">
      <listOfFunctionDefinitions>
      ...
      </listOfFunctionDefintions>
      <listOfUnitDefinitions>
      ...
      </listOfUnitDefinitions>
      <listOfCompartments>
      ...
      </listOfCompartments>
      <listOfSpecies>
      ...
      </listOfSpecies>
      <listOfParameters>
      ...
      </listOfParameters>
      <listOfRules>
      ...
      </listOfRules>
      <listOfReactions>
      ...
      </listOfReactions>
      <listOfEvents>
      ...
      </listOfEvents>
   </model>
</sbml>
```

Figure 1. A skeleton SBML Level 2 model.

species, the final "s" is omitted in the individual object definitions; in all cases, the first letter of the object is uppercase in the listOf definition, and lowercase in the individual object definition (hence, listOfFoos has an uppercase "F" and is plural, and foo is singular and entirely lowercase).

3.1. SBML Object Hierarchy

All SBML objects derive from a class SBASE (Figure 2). Class SBASE (and hence all other SBML objects) contains three optional fields: a metaid, notes, and annotation. The metaid field is present for supporting metadata annotations using RDF, and has a data type of ID as defined by XML. Other tool users may also choose to use this metaid field. The notes field is a container for XHTML content. There are no restrictions on what a user may include in this content; however, unlike other fields, which are designed to be read by machines, the notes field is intended to provide a place to store information that can be read easily by humans. Furthermore, when a Web browser that does not support non-HTML XML display is used to view an SBML model, it is usually only the notes field that will be visible. Finally, the annotation field is a container for software generated information that is not intended to

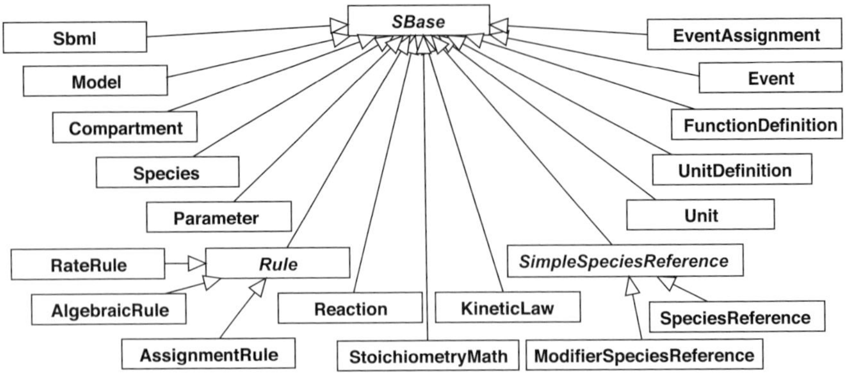

Figure 2. UML diagram of the SBML inheritance hierarchy showing the major data types in SBML.

be read by humans, but, nevertheless, contains information that cannot otherwise be encoded in SBML that is needed by particular software tools.

Nearly all SBML objects contain the following two fields: `id` and `name`. Most objects that have an `id` field will require that field, which is used to identify the particular instantiation of that object from other instantiations. The value of the id field must be an identifier that begins with a letter and contains only letters, numbers, and the underscore character. SBML is case sensitive, so that "x" and "X" represent two different identifiers. No two identifiers in the same scope may have the same name; thus, no species can have the same name as any compartment. Units are kept in a separate scope, and (as will be seen below) reactions may (optionally) use locally defined parameters that have a local scope. The `name` field is always optional, and its value may be any string of Unicode characters.

3.2. Mathematical Expressions in MathSBML

Several SBML objects allow (or require) mathematical expressions, notably `kineticLaw` (for a reaction), `stoichiometry` (of a species in a reaction), event triggers, event assignments, and rules. All mathematical expressions and formulas are expressed using a subset of MathML (12). MathML is an XML standard for encoding such expressions in a machine-readable format. MathML contains two flavors: presentation MathML, and content MathML. Presentation MathML is typically used to describe the placement of symbols on a page or a screen, whereas content MathML is used to describe the mathematical structure of an equation. For example, the following expresses $E = mc^2$ in content MathML,

```
<math xmlns='http://www.w3.org/1998/Math/MathML'>
  <apply>
    <eq/>
    <ci>E</ci>
    <apply>
      <times/>
```

Table 2. The subset of MathML that is allowed in SBML Level 2.

Object	Elements* Allowed
Token	cn**, ci, csymbol***, sep
Basic content	apply, piecewise, piece, otherwise
Relational operators	eq,neq, gt, lt, geq, leq
Arithmetic operators	plus, minus, times, divide, power, root, abs, exp, ln, log, floor, ceiling, factorial
Logical operators	and,or, xor, not
Qualifiers	degree, bvar, logbase
Trigonometric	sin, cos, tan, sec, csc, cot, sinh, cosh, tanh, sech, csch, coth, arcsin, rcos, arctan, arcsec, arccsc, arccot, arcsinh, arccosh, arctanh, arcsech, arccsch, arccoth
Constants	true, false, notanumber, pi, infinity, exponentiale
Annotation	semantics, annotation***, annotation-xml***

* The attributes style, class, and id may be used on any element.
** The attribute type may only take on one of the following: "e-notation," "real," "integer," or "rational."
*** encoding and definitionURL attributes are allowed are csymbol elements, and encoding is permitted on annotation and annotation-xml elements.

```
      <apply>
        <power/>
        <ci>c</ci>
        <cn type='integer'>2</cn>
      </apply>
      <ci>m</ci>
    </apply>
  </apply>
</math>
```

For the remainder of this chapter, whenever we refer to MathML, we will implicitly be referring to that subset of content MathML that is implemented in SBML Level 2 (Table 2). Like SBML, MathML is intended to be both generated and read by computers, and not by humans. Although short pieces of MathML are readable, the language's verbosity quickly makes it difficult to follow longer expressions. Fortunately, there are tools available to perform this translation; for example, in MathSBML (discussed in greater detail in later sections) there are two functions, InfixToMathSBML[*infix-expression*] and MathMLToInfix [*MathML-string*], which perform the conversion immediately.

3.3. Functions

A *function definition* associates a named identifier with a MathML lambda object that represents a mathematical function. For example,

```
<functionDefinition id="cube">
  <math xmlns="http://www.w3.org/1998/Math/MathML">
    <lambda>
```

```
      <bvar><ci> x </ci></bvar>
      <apply>
        <power/>
        <ci> x </ci>
        <cn> 3 </cn>
      </apply>
    </lambda>
  </math>
</functionDefinition>
```

defines a function cube that represents the mathematical expression x^3. A later MathML expression could then refer to the function cube via the apply command.

```
<math xmlns='http://www.w3.org/1998/Math/MathML'>
  <apply>
    <ci>cube</ci>
    <ci>x</ci>
  </apply>
</math>
```

3.4. Units

A *unit definition* defines physical units that can be applied to model objects in terms of a default set basic SI units (such as gram, liter, volt, etc.) For example, the user may define a unit "mmls" as millimoles per liter per second:

```
<unitDefinition id="mmls">
  <listOfUnits>
    <unit kind="mole" scale="-3"/>
    <unit kind="liter" exponent="-1"/>
    <unit kind="second" exponent="-1"/>
  </listOfUnits>
</unitDefinition>
```

and then give the value of a rate constant, later in the model, in units of mmls, e.g.,

```
<parameter id="K" value="0.007" units="mmls"/>.
```

3.5. Compartments

*Compartment*s are finite-sized containers for species. In SBML Level 1, a compartment may be a hierarchy of a topological enclosure with volume, but no geometric qualities. For example,

```
<compartment id="Membrane" spatialDimensions="2"
  constant="False"/>
<compartment id="Cell" outside="Membrane" size="1"/>
```

defines a compartment "Cell" surrounded by a second compartment "Membrane." In this example, Membrane is a two-dimensional surface surrounding a three-dimensional cell. The variable represents the compartment size, which may either be held fixed or allowed to change dynamically in a rule (by setting `constant='False'` in the `compartment` declaration). Besides the topological nesting, no other geometric information is normally encoded in SBML Level 2, although such information could be encapsulated in rules.

3.6. Species

Species are any chemical substances that can be measured by quantity or concentration that take part in a reaction. Examples include proteins, nucleic acids, and small molecules such as O_2 or ATP:

```
<species id="Glucose" compartment="cell"
  initialAmount="4" />
```

Other fields allow specifying initial concentration (instead of amount), units, charge, and whether or not the value should be kept constant, held as a boundary condition (allowed to be changed by rules but not by reactions), or variable.

3.7. Parameters

`Parameters` are constants or variables that do not represent substances. Parameters may be either global or locally specified within reactions (*see* section 3.9); an example was given in section 3.4. A parameter may be held fixed or allowed to change dynamically (by setting `constant='False'`). The values of dynamic *parameters* may be changed by *rules*, but not by *reactions*. Examples of parameters are rate constants, mass, and physical constants such as Avogadro's number. Rate constants and parameters that are referenced in multiple reactions must be defined globally; a rate constant that is used only in a single reaction can be defined as a local parameter.

3.8. Rules

`Rules` are mathematical expressions that describe the dynamics or values of variables. In SBML Level 2, there are three types of rules: assignment rules, rate rules, and algebraic rules. Assignment rules define the value of a parameter (or species) as a mathematical function of other variables in the system. Rate rules define the rate of change (derivative with respect to time) of a variable as a function of other system variables. Algebraic rules express algebraic constraints that should be satisfied by the system, such as $x + y - 7 = 0$. For example, the following defines a rate rule $dk_1/dt = A/(1 + A)$, followed by an assignment rule $k = k_1/k_2$ and an algebraic rule $0 = k_1 + k_2 + k_3$

```
<rateRule variable="k1">
  <math xmlns="http://www.w3.org/1998/Math/MathML">
  <apply>
   <divide/>
   <ci>A</ci>
   <apply>
```

```
      <plus/>
      <ci>A</ci>
      <cn type="integer">1</cn>
     </apply>
    </apply>
   </math>
  </rateRule>
  <assignmentRule variable="k">
   <math xmlns="http://www.w3.org/1998/Math/MathML">
    <apply>
     <divide/>
     <ci>k1</ci>
     <ci>k2</ci>
    </apply>
   </math>
  </assignmentRule>
<algebraicRule>
 <math xmlns="http://www.w3.org/1998/Math/MathML">
  <apply>
   <plus/>
   <ci>k1</ci>
   <ci>k2</ci>
   <ci>k3</ci>
  </apply>
 </math>
</algebraicRule>
```

The ordering of assignment rules is critical: a program is expected to evaluate them *in the order listed in the model.* Furthermore, (a) no more than one assignment or rate rule can be defined for any given identifier; (b) assignment rules override any initial conditions for that variable; (c) the math field of a rule can contain only identifiers that have been previously defined, and (d) an assignment rule cannot contain either the identifier for which the rule is defined or any element for which there is a subsequent assignment rule.

3.9. Reactions

A *reaction* is a statement describing a transformation, transport, or binding process that can change the amount of one or more species. For example,

```
<reaction id="R1">
  <listOfReactants>
    <speciesReference species="X" stoichiometry="1"/>
  </listOfReactants>
  <listOfProducts>
    <speciesReference species="Y"
      stoichiometry="2"/>
```

```
    <speciesReference species="Z"
      stoichiometry="1"/>
  </listOfProducts>
  <listOfModifiers>
    <modifierSpeciesReference species="A" />
  </listOfModifiers>
  <kineticLaw>
    <math xmlns="http://www.w3.org/1998/Math/MathML">
      <apply>
        <times/><ci>k</ci><ci>A</ci><ci>X</ci>
      </apply>
    </math>
    <listOfParameters>
      <parameter id="k" value="0.1" />
    </listOfParameters>
  </kineticLaw>
</reaction>
```

represents the reaction $X \xrightarrow{kA} 2Y + Z$. The parameter k defined in this example is only defined locally; its existence is unknown outside of the reaction definition (specifically, a separate parameter namespace is defined for each reaction to contain its local parameters). Local parameters may have the same id as global parameters and local parameters in different reactions are permitted to have the same id. When a local parameter has the same id as a global parameter, then any reference using that id within the reaction refers to the local parameter and not the global parameter. Stoichiometries can also be specified with MathML expressions.

3.10. Events

Events are explicit, instantaneous, discontinuous state changes that are triggered as a result of changing conditions within a model. Events specify a *trigger*, which is the condition that causes the event to occur (e.g., $mass > 1$ and $A < 2$); an *eventAssignment*, which is an action that occurs as a result of the event's triggering (e.g., set $mass = mass/2$); and a time *delay* (and associated *timeUnits*) between the occurrence of the trigger and the application of the eventAssignment. For example,

```
<event>
  <trigger>
    <math xmlns="http://www.w3.org/1998/Math/
      MathML">
      <apply>
        <and/>
        <apply><gt/><ci>mass</ci><cn>1</cn></apply>
        <apply><lt><ci>A</ci><cn>2</ci></apply>
      </apply>
      <apply><leq/><ci> P1 </ci> <ci> t </ci>
        </apply>
```

```
    </math>
  </trigger>
  <listOfEventAssignments>
    <eventAssignment variable="mass">
      <math xmlns="http://www.w3.org/1998/Math/
        MathML">
        <apply><divide/><ci>mass</ci><cn>2</cn></apply>
     </math>
     </eventAssignment>
  <listOfEventAssignments>
</event>
```

sets $mass = mass/2$ when the Boolean expressions $((mass > 1) \wedge (A < 2))$ change from false to true. The event will trigger only when the condition changes from false to true. If the condition later becomes false, and then true again, the event will trigger a second time.

4. Proposed Modifications to SBML

SBML is intended to meet the evolving needs of the systems biology community. Consequently SBML is being developed in levels, where each higher level adds additional features to the model definitions. These separate levels of SBML are intended to coexist; SBML Level 2 does not render SBML Level 1 obsolete. Software tools that do not need or cannot support higher levels may continue to use lower SBML levels; tools that can read higher levels are assured of also being able to interpret models defined in the lower levels. Minor changes in SBML are called versions; versions within the same level reflect minor changes within that level that were omitted from the earlier version, clarifications of intent and syntax, and typographical corrections to the model specifications.

As errors and omissions are discovered in the specifications, they are posted on an errata page at http://sbml.org. These corrections are then added to the next version of the specification. No new major features were added in SBML Level 1, Version 2; however, it did introduce several variant spellings (e.g., allow both meter and metre; species instead of specie), and a number of typographical errors, earlier omissions, and clarifications were introduced. SBML Level 2, Version 1 did introduce a number of features, notably: events; functions; the use of MathML rather than C-style infix expressions for formula strings; id and name fields for most objects; the removal of predefined rate laws; spatial dimensions; simplification of rule structure; and the addition of modifiers to a reaction definition.

Several minor language extensions have been proposed for SBML Level 2, Version 2 (13). (1) *Nested unit definitions* will allow new units to be defined in terms of other units defined in the same model, rather than merely in terms of the base SI units listed in the specification. (2) A new *list of species types* will represent classes of chemical entities independent of their locations; for example, two different species, one in the cytosol and one in the extracellular medium, can both be labeled as calcium ions. (3) A new Boolean *constraint rule* will define conditions (e.g., $A + B < C$) under which a model is valid. If a specified constraint is violated, then a simulator should halt and print a message indicating that the constraint was violated.

A substantial number of additional changes have been proposed for Level 3. Because of their greater specialization, and the fact that not all modelers will have need for all of these features, it is likely that Level 3 will be modular, in the sense that users will be able to specify at the beginning of a model which Level 3 features the model uses. These features, which are summarized below in alphabetical order, are described in detail on the SBML Wiki at http://sbml.org/wiki.

Alternative reaction extensions would provide the additional data structures that might be required to describe reactions nondeterministically through such features as probability models, Markov chains, Petric nets, pi-calculus, grammar rules, etc. The present implementation of SBML is based on chemical reactions and rate laws, and lends itself quite well to differential equation formalisms, but does not provide the proper set of information required for nondeterministic modeling. These reaction extensions could be closely related to *hybrid model* extensions.

Array and *set* extensions would describe collections of elements (bunches of things that are treated identically in some way) in terms of standard computational data structures, such as arrays, vectors, lists, sets, etc. These extensions can interact with the model composition extensions, in that arrays or lists of models could be described. For example, the *Arabidopsis* shoot apical meristem, which is the hemispherical tip of the growing plant shoot that consists of approximately 500 cells, could be described by a dynamic array or list of models (or compartments), with new models (or compartments) being instantiated when cells divide and old models being removed when cells die.

Complex species extensions would allow models to describe a single species in terms of its different states, such as phosphorylated/nonphosphorylated, or having different numbers or types of ligands bound to different sites. It is related to the Level 2 Version 2 extension of *species types*, but takes the idea further, in that a species can have both a *type* (e.g., MAPK) and a *state* (double-phosphorylated).

Controlled Vocabulary extensions would provide common terms to describe multiple aspects of the same thing. Different models might use different controlled vocabularies. For example, a reaction might be labeled as "Michaelis–Menten" or "Bi-Uni-Uni-Bi-Ping-Pong-Ter-Ter," or it might be described as "transcriptional," "transport," "activation," etc; a species might be labeled "substrate" or "catalyst." A controlled vocabulary could also include a mechanism for synonyms, indicating that "Km" and "Michaelis_Constant" represent the same parameter. This will most likely be developed in conjunction with the Systems Biology Ontology (SBO) effort, currently being led by three of the present authors (Nicolas Le Novère, Michael Hucka, and Andrew Finney). SBO consists of a taxonomy of the roles of reaction participants (e.g., "substrate," "inhibitor," "competitive inhibitor"); a controlled vocabulary for parameter rules in quantitative models ("Hill Coefficient"); and a classification of rate laws ("Mass-Action," "Henri-Michaelis-Menten").

Diagramming or *layout* extensions would allow a model to include specific descriptions of diagrams that describe the model. It would contain lists of graphical representations, or glyphs, of SBML model elements such as compartments, species, and reactions, and information about where to place the different glyphs on a diagram (digital or paper). The actual form of a specific *glyph*, e.g., whether a species should be represented by a simple black character string or by filled green oval, would be left up to the individual tool.

Dynamic model extensions provide ways to enable model structures to vary during a simulation. For example, a dynamic event might trigger cell division and add an additional compartment to the model. Dynamic extensions are closely related to array and model composition extensions.

Hybrid model extensions would allow different parts of the same model to be described by different formalisms. For example, one process could be described by a continuous differential equation, and another could be a discrete Markov process. Hybrid models could also involve alternative reaction formalisms and rules that allow dynamic switching between the formalisms for specific processes, constraints that need to be enforced during a simulation, and instantiations of submodels via model compositions.

Model Composition would provide the capability to define one SBML model in terms of other models (either in the same file or linked to another file), and include mechanisms for creating a hierarchy of submodels as "instances" of these models. For example, a model of a cell may contain multiple instances of a model of a mitochondria, with different parameter values, initial conditions, etc., or a tissue model may include various instances of a cell model.

Parameter Set extensions would facilitate the separation of initial conditions and parameter values of a model from the model structure itself. Some aspects are related to the idea of model composition. In its most basic form, a parameter set is a collection of key value pairs, where the key refers to an SBML object attribute; a specific parameter set could then be applied to an existing model, with the appropriate name/value pair substitutions made.

Spatial feature extensions would add geometric characteristics to a model. The only geometric aspects in SBML Level 2 are hierarchies of compartments that are described as being inside or outside of one another, and some aspect of area or volume. Spatial feature extensions could add information ranging from location, and adjacency lists to finite element or spline models describing the surface shape and features of a compartment.

5. Resources at http://sbml.org

In the following paragraphs, we briefly describe the tools at http://sbml.org that have been designed to support SBML development and that could be of use to nearly all SBML modelers. All of the tools described here are freely accessible at http://sbml.org.

5.1. Online Tools

The *online tools* enable the user to validate and visualize models in any version or level of SBML, and to convert Level 1 models to Level 2. The validator checks the model against the SBML XML schema and does limited consistency checks. It is possible for a model that is not valid SBML to be passed by the tool (because it does not include complete consistence checking), but it will invalidate any model that does not follow the SBML XML schema and a comprehensive set of rules that encode consistency constraints that are not expressible in the XML schema language. When a model is validated, the user will be provided with a model summary (e.g., number of each class of SBML object) and given the option to visualize the model or convert it to Level 2 (if the model is in Level 1). Errors are indicated by line number in the original model. The validator and converter are based on libSBML (*see* section 5.2). Visualization is provided utilizing Graphviz dot combined with an XSLT script, and displays the visualization as a .gif image in the Web browser. Because of server limitations, the online visualization is limited to models containing 100 or fewer reactions; however, the downloadable XSLT script can support models containing any number of reactions.

5.2. LibSBML

LibSBML is a C/C++ library providing an application programming interface (API) for reading, writing, and manipulating data expressed in SBML. LibSBML is a library designed to help read, write, manipulate, translate, and validate SBML files and data streams. It is not an application itself (although it does come with many example programs), but rather a library you can embed in your own applications. Although it is implemented in C/C++, it includes Java, Python, Perl, Lisp, and MATLAB language bindings in the distribution, and is written in such a way that users can write bindings from virtually any computer language implementation that allows cross-language bindings. The code is very portable and is supported on Linux, native Windows, and Mac OS-X operating systems.

The API provides an exhaustive list of getters (e.g., species_getInitialAmount), setters/unsetters (species_unsetSpatialSizeUnits), field state Booleans (species_isSetCompartment), object getters and creators (UnitDefiniton_addUnit, UnitDefinition_getUnit), enumerators, abstract classes corresponding to every SBML object, including full SBML field inheritance, and so forth. It also provides facilities for reading and writing SBML files, parsing models into abstract syntax trees, and SBML validation.

LibSBML understands all versions of SBML including Level 1, Versions 1 and 2, Level 2 version 1, and the draft SBML Layout Proposal. It is written in portable, pure ISO C and C++, and can be easily ported to nearly any operating system. LibSBML can be built using either GNU or MSVC tools. The library is linked with standard XML libraries and can be built with either Expat or Xerces. Finally, LibSBML provides full XML schema validation (Xerces only).

5.3. SBML Tools for MATLAB

The SBML Toolbox is a package for working with SBML models in MATLAB. Rather than providing a simulator, the SBML Toolbox provides facilities for converting an SBML model into a MATLAB-accessible format, so both the standard MATLAB solvers and/or user-developed simulators and libraries can be applied. The toolbox currently includes functions for reading and writing SBML models, converting SBML models into MATLAB data structures, viewing and manipulating those structures, converting them to MATLAB symbolic format, and simulating them using MATLAB's ODE solvers. At present, the toolbox includes functions to translate an SBML document into a MATLAB_SBML structure, save and load these structures to/from a MATLAB data file, validate each structure (e.g., reaction structure), view the structures using a set of graphical user interfaces, and to convert elements of the MATLAB_SBML structure into symbolic form, thus allowing access to MATLAB's Symbolic Toolbox. There are a small number of functions to facilitate simulation and a function that will output an SBML document from the MATLAB_SBML structure definition of a model. The toolbox is based on libSBML and requires a prior MATLAB installation. It has been tested in Windows, Linux, Unix, Cygwin, and MacOSX. Unix versions require a prior installation of libSBML; this is not required for the Windows version.

5.4. MathSBML

MathSBML provides facilities for reading and writing SBML models, converting them to systems of differential equations for simulation and plotting in *Mathematica*, and translating them to other formats. As with the SBML-Toolbox, its main purpose is to get models in and out of *Mathematica*, so that the user can apply them and/or all of the standard features of that language to the SBML model. MathSBML requires a prior installation of *Mathematica*, and is fully platform independent. MathSBML is the subject of section 7 of this chapter.

5.5. SBML Conversion Utilities

SBML Conversion utilities provide the ability to convert models described in other modeling languages into SBML. So far we have implemented two different model conversion utilities: KEGG2SML and CELLML2SBML. In addition, the online tools provide conversion form SBML Level 1 models to SBML Level 2 models.

KEGG2SBML is a Perl script that converts Kyoto Encyclopedia of Genes and Genomes (KEGG, http://www.genome.jpg/kegg) pathway database files to SBML files using LIGAND database files (14,15). It is compatible with all levels and versions of SBML, and includes support for <annotations> tags for CellDesigner. KEGG is a suite of databases and associated software for describing high-order functional behaviors of cells, systems, and organisms, and for relating those behaviors to the organisms' genomes. It includes several databases that describe protein interaction networks (the pathway database); chemical reactions (the ligand database); full-organism networks (gene and SSDB); and functional genomic (expression) and proteomic (BRITE) references. Despite the large amount of information (nearly one million different proteins and/or genes are in the gene database, for example) and extensive synonym and cross-links, it provides little or no information for actual reaction mechanisms or rate constants. KEGG2SBML requires Perl 5.6.1, expat, the Perl XML parser (XML::Parser), and libxml-perl, all of which are publicly available, as well as KEGG pathway database, KGML, and ligand database files that are available at the KEGG Web site. It has been tested on FreeBSD and Linux platforms, as well as Cygwin under Microsoft Windows.

CellML2SBML converts CellML models (16) to SBML (17). Like SBML, CellML is an XML-based modeling language used for storage and exchange of biological models. Although there are some common facilities in both languages, the two languages have slightly different goals. In particular, CellML is closely affiliated with anatomy and finite-element modeling languages (AnatML and FieldML). The CellML developers have been involved in the development of the SBML standard, and are currently developing a second tool (SBML2CellML) that will perform the conversion in the opposite direction. CellML2SBML is available for Windows and Linux systems and requires an XSLT processor to run. It consists of four XSLT style sheets.

5.6. Schemas, Specifications, and Test Suites

Full XML schemas (.xsd documents) have been defined for all versions of SBML, and are included in the download of libSBML, as well as in the specification documents.

The *SBML Test Suite* is a collection of models and associated automation scripts intended to serve as a test set for developers of SBML-enabled software. It also includes sample models in SBML Level 1 and Level 2 format. Syntactic testing determines if a tool accepts only well-formed SBML and rejects any syntactically incorrect SBML input. This may be accomplished by validation against a full XML schema. Semantic testing determines if the tool interprets well-formed SBML correctly, i.e., whether the software constructs the correct model and whether that model behaves correctly. This is usually tested by simulation and comparison with tabular output. The *semantic test suite* at http://sbml.org includes over 100 tests, including annotated SBML models and tabulated output, as well as an automated script for running the tests against your simulator, so long as the simulator can be invoked either from Windows Cygwin or a Unix command line.

6. BioModels Database

BioModels database (18,19), developed through an international collaboration between the SBML team (US/UK/Japan), EMBL-EBI (UK), DOQCD (IN) the Keck Graduate Institute (US), Systems Biology Institute (Japan), and JWS Online (South Africa) provides access to published peer-reviewed quantitative biological models. The peer review is provided by the publication process; a model must be published in some peer-reviewed form (e.g., a journal article) before it can be encoded in the database. The original paper's authors do not have to generate the SBML version of the model themselves. The model can be described in any language (e.g., differential equations, stochastic, lists of chemical reactions, etc.) within the paper, but only SBML (or CellML) models are incorporated within the database. Anybody can submit a model to the database, so long as it has been published and has the appropriate references, but it will not be propagated in the public version of the resource until the model has been verified by a database curator. Curators verify that the SBML model is valid, well formed, syntactically correct, and correctly represents the referenced publication, and that simulations based on these models reproduce (at least some of) the published results. Curators also annotate the components of the models with terms from controlled vocabularies such as GO (Gene Ontology of ChEBI) and links to other databases (such as UniProt, KEGG, and Reactome). This allows the users to search accurately for the models they need, but also to identify precisely every component of a model. Models can be retrieved in SBML, CellML, and various simulator-specific formats, such as XPP-Aut or SciLab.

7. Managing SBML with MathSBML

MathSBML (20) is an open-source *Mathematica* package that facilitates working with SBML models. Its primary purpose is to import SBML files into a *Mathematica* data structure so that users can manipulate the models within *Mathematica* without having to worry about the details of SBML structure. *Mathematica* is one of several platforms widely used by biological modelers, and it is available in many academic and commercial environments (e.g., over 500 US colleges and universities have site licenses). *Mathematica* is a symbolic computation environment that includes a wide range of features that are of use to computational biologists, notably numerical computation, graphics, and a programming language. Symbolic computation environments, also known as computer-algebra systems, allow the users to process equations symbolically, using formats that are similar to mathematical equations. From the perspective of computational biologists, this means that reactions and kinetic laws can be expressed in these they are used to, such as $A + B \rightleftarrows C$ or $C'[t] = k_1 A[t]B[t] - k_2 C[t]$. Besides the import feature, MathSBML also includes functions for simulation and plotting of SBML models, including differential-algebraic equations and events; a complete API (Tables 3 and 4) for manipulating SBML Level 2 models; the ability to display models in human-readable form as annotated html (or within *Mathematica* notebooks); and the ability to export new or modified models back to XML format. A summary of MathSBML commands is given in Table 5.

MathSBML provides full model interoperability with *Mathematica*, as well as a candidate reference implementation of SBML. MathSBML will run on any platform that has *Mathematica* 4.1 or higher installed. The solution of differential-algebraic systems (SBML models that have algebraic rules) requires *Mathematica* 5.0 or higher; purely differential systems (SBML without algebraic rules) can be solved on *Mathematica* 4.1. MathSBML is compatible with all levels and versions of SBML released to date, as well as several features proposed for future releases.

Table 3. Summary of the MathSBML API.

	Model	Compartment	Event	Function	Parameter	Reaction	Rule	Species	Unit	annotation
add_		✓	✓	✓	✓	✓	✓	✓	✓	
_ToSBML		✓	✓	✓	✓	✓	✓	✓	✓	✓
ToSymbolciSBML		✓	✓	✓	✓	✓	✓	✓	✓	✓
get_		✓	✓	✓	✓	✓	✓	✓	✓	
modify_		✓	✓	✓	✓	✓	✓	✓	✓	
remove_		✓	✓	✓	✓	✓	✓	✓	✓	
create_	✓									
createSymbolic_	✓									

The "_" in the name can be replaced with any checked object, e.g., `addFunction` or `modifyRule`. Controllable options are summarized in Table 4.

Table 4. SBML attributes that can be controlled via the API commands in Table 3.

API commands for:	Options*
species	id, name, compartment, initialAmount, initialConcentration, substanceUnits, spatialSizeUnits, hasOnlySubstanceUnits, boundaryCondition, charge, constant
compartment	id, name, constant, outside, spatialDimensions, size, units
event	id, name, trigger, delay, timeUnits, eventAssignment
function	id, name, math
parameter	id, name, annotation, notes, value, units, constant
reaction	id, name, fast, kineticLaw, modifiers, name, products, productStoichiometry, reactants, reactantStoichiometry, reaction, reversible, parameters (sub-options: value, name), timeUnits, substanceUnits;
rule	type, variable, math
species	id, name, compartment, initialAmount, initialConcentration, substanceUnits, hasOnlySubstanceUnits, boundaryCondition, charge, constant
unit	id, name, unit (sub-options: exponent, scale, multiplier, offset)
model	id, name; also: comments (XML Comments)

* All objects have modifiable annotation, notes, and metaid fields. Some options are mutually exclusive.

The "options" shown generally have a 1 : 1 correspondence with SBML attributes, although sometimes the spelling is different. For example, reaction products option refers to the SBML `speciesReferences` within the SBML `listOfProducts`; however, for the most part, the correspondence is clear.

Table 5. Summary of MathSBML commands excluding the API.

Function	MathSBML Entry Points
Algebraic/MathML conversion	`InfixToMathML, MathMLToInfix`
Convert model file format	`SBMLCopy`
Plot results of a simulation	`SMBLPlot, SBMLGridPlot, SBMLListPlot`
Simulation	`dataTable, SBMLNDSolve`
Import a model	`SBMLRead`
Export a model	`SBMLWrite, createModel`
Annotation control	`setAnnotationPackage, setAnnotationURL, setModelAnnotation, setSBMLAnnotation`
Model display	`showModel`

7.1. Model Import

Model import is performed using `SBMLRead`. Suppose, for example, that we are interested in modeling the cell cycle, and download the model "Novak1997_CellCycle" from the biomodels database into a local file `BIOMD0000000007.xml`. This file implements a model of DNA replication in the fission yeast *Schizosaccharomyces pombe* (21). We can read the model into the *Mathematica* computing environment with the command

```
m = SBMLRead ["BIOMD0000000007.xml", context → None],
```

which returns a *Mathematica* rule list (a standard technique used in *Mathematica* to describe complex data structures), as shown in Figure 3. This type of data structure allows the user to access all features of the model directly with *Mathematica*; a more SBML-oriented approach would be to use the model builder, which is described in a later section. A user could get a list of all of the assignment rules in the model, for example, by entering

```
r = SBMLAssignmentRules/.m
```

which will return

```
{IEB[t]==1-IE[t], UbEB[t]==1-UbE[t], UbE2B[t]==1-UbE2[t],
  Wee1B[t]==1-Wee1[t], Cdc25B[t]==1-Cdc25[t],
  Rum1Total[t]==G1R[t]+G2R[t]+PG2R[t]+R[t],
  Cdc13Total[t]==G2K[t]+G2R[t]+PG2[t]+PG2R[t],
  Cig2Total[t]==G1K[t]+G1R[t],
  k2[t]==0.0075 (1-UbE[t])+0.25 UbE[t],
  k6[t]==0.0375 (1-UbE2[t])+7.5 UbE2[t],
  kwee[t]==0.035 (1-Wee1[t])+0.35 Wee1[t],
  k25[t]==0.025 (1-Cdc25[t])+0.5 Cdc25[t],
  MPF[t]==G2K[t]+0.05 PG2[t], SPF[t]==0.25 G1K[t]+MPF[t]}
```

The third rule can be obtained as `rule3 = r[[3]]`, which would return

```
Rum1Total[t]==G1R[t]+G2R[t]+PG2R[t]+R[t]
```

as the value of the variable `rule3`.

```
{SBMLAlgebraicRules → {},
 SBMLAssignmentRules → {IEB[t] == 1 - IE[t], ⋯},
 SBMLCompartments → {Cell},
 SBMLConstants → {Cell → 1, ⋯},
 SBMLEvents → {
   "Start" → {
     "trigger" → "SPF[t] >= 0.1", "delay" → "60",
     "events" → {"kp[t]->kp[t]/2"}},
   ⋯},
 SBMLFunctions → {},
 SBMLIC → {UbE[0] == 1, ⋯},
 SBMLLevelVersion → 2.1`,
 SBMLMassActionEquations → {UbE′[t] == UbEB[t] v4[22], ⋯},
 SBMLMassActionVariables → {UbE[t],  },
 SBMLMassBalanceEquations → {UbE′[t] == v4[22], ⋯},
 SBMLModelid → "NovakTyson1997CellModel",
 SBMLModelName → "Novak1997_CellCycle",
 SBMLModelVariables → {UbE[t], ⋯},
 SBMLNameIDAssociations → {"NovakTyson1997CellModel" → "Novak1997_CellCycle", ⋯},
 SBMLODES → {Cdc25′[t] == -(0.25 Cdc25[t])/(0.1 + Cdc25[t]) + (Cdc25B[t] MPF[t])/(0.1 + Cdc25B[t]), ⋯},
 SBMLParameters → {mu, ⋯},
 SBMLReactions → {∅ → G2K, ⋯},
 SBMLSpecies → {UbE[t], ⋯},
 SBMLSpeciesCompartmentAssociations → {UbE → Cell, ⋯},
 SBMLSpeciesTypeAssociations → {},
 SBMLStoichiometryMatrix → {{0, 0, ⋯}, ⋯},
 SBMLUnitAssociations → {Cell → Units`litre, ⋯},
 SBMLUnitDefinitions → {Units`time → 60 Units`second, ⋯}}
```

Figure 3. Abbreviated form of data structure returned by SBMLRead after importing the cell cycle model described in the text. The ellipsis is used to indicate that some parts of the data structure have not been illustrated to save space in the present book chapter; in actuality, MathSBML will display the entire data structure.

One particularly useful feature of `SBMLRead` is that it constructs the complete set of differential equations that describe the model by combining all of the kinetic laws and rate rules in the model. This set of differential equations is returned as the field `SBMLODES`. `SBMLRead` also returns the stoichiometry matrix as a separate field, and this can be used to simulate models that do not have complete sets of kinetic laws. The corresponding mass-action and mass-balance equations are also generated.

7.2. Variable Scoping and Names

MathSBML attempts to match all identifiers in the *Mathematica* version of the model as closely as possible to the name in the model. In addition, the hierarchies of variable scoping are preserved, e.g., units and reaction parameters are kept in their own namespaces. *Mathematica* represents the scope of a symbol by its context. The context of a variable is indicated by predicating it with a string of characters ending in the back-quote character (normally found to the left of the number 1 on American keyboards).

SBML model variables are defined in a local context; the name of the context is determined by the model "name" in SBML Level 1, and by the model "id" in SBML Level 2. Thus, if the SBML model `foo` contains species `A` and `B`, and global parameters `f` and `k`, they will be represented as `foo`A`, `foo`B`, `foo`f`, and `foo`k`, respectively. Local parameters `k` and `kf` defined in reactions `R1` and `R2` will become `foo`R1`k`, `foo`R1`kf`, `foo`R2`k`, and `foo`R2`kf`, respectively. The only character that is allowed in an SBML identifier that is not allowed in a *Mathematica* identifier is the underscore ("_") character. The underscore has a special meaning in *Mathematica* that is used for pattern matching. SBMLRead replaces the underscore character with the \[UnderBracket] character (Unicode bottom square bracket 9141), which looks like a bracket ("[") turned on its side, with the ends pointing up. The underbracket is translated back to an underscore when a model is written back out as an XML file.

Mathematica contains a number of standard contexts. In particular, any variables that you type in during a *Mathematica* session that do not explicitly include a context are placed in the `Global'` context. You do not have to explicitly include the context in Global' variables. Thus the identifiers `A` and `Global`A` represent the same variable. You can change the default context form `Global`` to something else by changing the value of the *Mathematica* identifier `$Context.`

In SBMLRead, the option `context` → `None` indicates that the model should be placed in the local context. Thus, in the example in the previous section, we had a variable `Cdc13Total` and a global parameter `mu`, would normally be represented as represented as `NovakTyson1997-CellModel`Cdc13Total` and `NovakTyson1997CellModel`mu`. The units of the compartment `Cell` are specified in `litre`, which is represented as `NovakTyson1997CellModel`Units`litres`, because units are kept in their own namespace. This particular model does not use any local parameters in the kinetic laws form reactions; however, if we were to add a parameter `k` to the reaction `Cdc25Reaction` it would become `NovakTyson1997CellModel`Cdc25Reaction`k`. By using the option `context` → `CellCycle` in our call to `SBMLRead`, these would become `CellCycle`Cdc13Total`, `CellCycle`Units`litre`, and `CellCycle`Cdc25Reaction`k`.

7.3. Simulation and Plotting

Suppose you are interested in running a deterministic simulation of the model that was imported in the previous section. This feat is accomplished with `SBMLNDSolve`, which is a wrapper for the *Mathematica* numerical solver `NDSolve`. To run a simulation of NovakTyson-1997CellModel for 400 minutes (the units of time are redefined as minutes in the model) you would enter

```
r=SBMLNDSolve [m, 400]
```

The result is returned as a list of interpolation sets that are compatible with *Mathematica* interpolation and plotting functions. If you wanted to write a table of values of the model variables `Rum1Total` and `Cdc13Total` at intervals of 1 min from t = 150 to t = 200 to a comma-separated value file "results.csv,"

```
dt = dataTable[{Rum1Total, Cdc13Total}, {t, 150,
  200, 1}, r];
Export["results.csv", dt, "csv"]
```

Other standard output file formats, including .dif, .fit, fits, .hdf, .h5, .mat, .mtx, .tsv, .txt, .xls, are also supported.

Suppose instead of generating a table of data, you want a plot of those same variables:

```
SBMLPlot[r, {Rum1Total, Cdc13Total}]
```

The plot will normally be displayed on the screen and remain embedded in the *Mathematica* notebook. `SBMLPlot` is a wrapper for the *Mathematica* function `Plot`. Any standard plotting options can be specified as follows:

```
p = SBMLPlot[r, {Rum1Total, Cdc13Total},
    PlotStyles→{
      {Dashing[{.02}], Thickness[.005], Blue},
      {Thickness[.002], RGBColor[1,0,0]}},
    ImageSize→600,
    TextStyle→{FontFamily→Times, FontSize→18},
    holdLegend→True]
```

will generate a 600-pixel-wide image (Figure 4) with `Rum1Total` as a thick dashed blue line and `Cdc13Total` as a thinner solid red line. Figures can be exported, e.g., via

```
Export["myplot.jpg",p,"jpg"]
```

many standard graphics types are supported, including .bmp, .dcm, .dic, .eps, .gif, .jpg, .pbm, .pcx, .pdf, .pgx, .pict, .pnm, .png, .ppm, .svg, .tif, .wmf, and .xbm file formats.

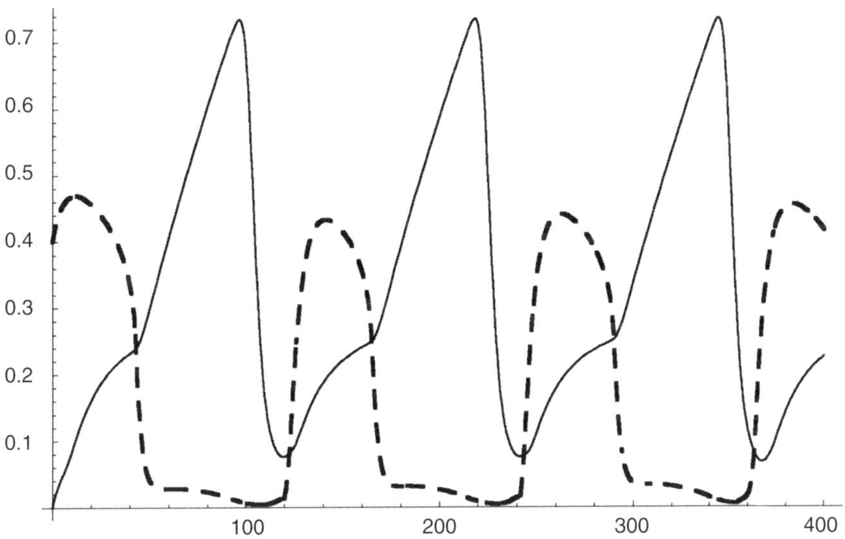

Figure 4. A plot of the two variables Rum1Total (dashed line) and Cdc13Total (solid line).

The MathSBML simulator, `SBMLNDSolve`, is a wrapper for *Mathematica's* `NDSolve`, which in turn evolved from the LSODA, IDA, and DASPK solvers. It incorporates a wide range of methods, including stiff and nonstiff integrators and switching methods and a framework for incorporating external solvers.

Events are implemented by the following algorithm, which ensures that events activate only when an event's trigger changes from false to true. Each event E in an SBML model in has a trigger expression T_E and assignments A_E. We replace the event E, with the following:

- a Boolean variable V_E with initial value false
- an event E_1 with trigger T_{E1} = (not V_E) $\wedge$ T_E and assignments A_E and $V_E = true$
- an event E_2 with trigger T_{E2} = (not T_E) $\wedge$ V_E and assignment $V_E = false$

The existence of the pseudoevents E_1 and E_2 and the new model variable V_E is completely transparent to the user, who is only aware of the existence of the events specified in the model. Events with delays are similarly handled by creating a pseudo-event that triggers once when the specified delay has elapsed. Our cell cycle model actually has two events, one with a delay, and one with multiple assignments:

- Event `start` (the start of S phase) occurs when `SPF` (S-phase promoting factor) crosses 0.1 from below; after a delay of 60 min, the model parameter `kp` is cut in half.
- Event `Division` (cell division) occurs when UbE crosses 0.1 from above. This triggers halving if the parameter `Mass` and doubling of the parameter `kp`.

Events are then detected in *Mathematica* 5.1 (and higher) by throwing and catching the event occurrence via the `NDSolve` `StepMonitor` option; in earlier versions, they are detected by the option `Stopping-Test`. In all cases, the precise event time is found by backward interpolation. If multiple events occur simultaneously, all are detected and processed.

7.4. The Model Builder: An SBML API

MathSBML contains a simple model editor, allowing users to create SBML models compatible with other simulators, as well as a *Mathematica* text-command based API that can be used to produce arbitrarily complex SBML files. The model editor contains a suite of commands to add, modify, or remove single SBML objects (such as a reaction, chemical species, or equation) from the current model (Tables 3, 4, and 5). The model may be either created *de novo* or read from a file. After building the model, the user can test it by running simulations, continue to modify it, or write the results as an SBML file, in no particular order.

There is a set of functions (`addX`, `modifyX`, and `removeX`) for each class of SBML model object: compartment, event, function, parameter, reaction, rule, species, and unit. Options allow users to specify specific object field values. For example, a partial list of the commands needed to create the Tyson cell-cycle model from scratch is illustrated in

```
<<mathsbml.m
newModel["CellCycle"];
addCompartment["Cell",size->1];
addUnit[id->"time",name->"minutes",
      unit->{"second"->{multiplier->60}}];
addParameter[id->"mu",value->0.00495];
addParameter[id->"Mass",value->1,constant->False];
addRule[type->"RateRule",variable->"Mass",math->Mass*mu];
      ⋮
addSpecies[IEB,boundaryCondition->True,initialAmount->0];
addSpecies[UbEB,boundaryCondition->True,initialAmount->0];
      ⋮
addRule[type->AssignmentRule,variable->IEB,math->1-IE];
addRule[type->AssignmentRule,variable->UbEB,math->1-UbE];
      ⋮
addSpecies[Rum1Total];
addSpecies[Cdc13Total];
      ⋮
addRule[type->AssignmentRule,variable->Rum1Total,
      math->R+G1R+G2R+PG2R];
addRule[type->AssignmentRule,variable->Cdc13Total,
      math->G2K+G2R+PG2+PG2R];
      ⋮
addEvent[id->"Start",trigger->SPF≥0.1,
      eventAssignment->{kp->kp2},delay->60];
addEvent[id->"Division",trigger->UbE≤0.1,
      eventAssignment->{kp->kp*2,Mass->Mass/2}];
      ⋮
addReaction[products->{G2K},id->"G2K_Creation",kineticLaw->k1];
      ⋮
createModel["cellCycle.xml"];
```

Figure 5. Model builder commands needed to create the cell cycle model. Because of space limitations, only a subset of the commands are shown; the vertical ellipses indicate that many commands were omitted. All of the omitted commands are of the form "addX." The last statement in the list, createModel, generates the SBML file cellCycle.xml.

Figure 5. The last step in the box creates an .xml file `cellCycle.xml`. A large number of consistency checks are made as the commands are typed to ensure that the SBML specification is satisfied. For example, every species must be associated with a compartment; if a compartment is not specified by the `addSpecies` statement, then the most recently referenced compartment is used. If no compartment has been defined yet, a new one is defined.

As the current model is built, it is stored internally by MathSBML. At any point in model development, either before or after the .xml file has been written, the model can be loaded into the simulator and tested via the `loadSimulator` command. The return value of `loadSimulator` is identical to the return value from `SBMLRead`, and therefore it is compatible with `SBMLNDSolve`. Similarly, `SBMLRead` will automatically load the model builder whenever a level-2 model is read in, so that it can be modified by add, remove, or modify commands.

Internally, an SBML model is stored as symbolic XML, a standard *Mathematica* data structure for handling XML files. The functions `getX[n]` return the nth object of class X in symbolic XML; the argument may be a number, an id, or a list of both. For example, `getReaction[2]` returns the second reaction in the model. The function `XMLOut` is used there to generate the corresponding XML fragment. Other functions `XtoSymbolicSBML` and `XtoSBML` allow one to generate the corresponding symbolic XML or XML fragment for any SBML object.

MathSBML is freely downloadable (LGPL license) from sourceforge at http://sf.net/projects/sbml. Full documentation with examples of all entry points, including the entire API is available, on the http://sbml.org Web site at http://sbml.org/software/mathsbml/. Instructions for downloading and installing MathSBML are also provided on that site. MathSBML will run under any operating system or platform on which *Mathematica* is already installed; a complete list of compatible systems is given at http://www.wolfram.com.

Acknowledgments: SBML was initially funded by a generous grant from the Japan Science and Technology Agency under the ERATO Kitano Symbiotic Systems Project. Additional support for the continued development of SBML and associated software and activities has come from the National Human Genome Research Institute (USA), the National Institute of General Medical Sciences (USA), the International Joint Research Program of NEDO (Japan), the ERATO-SORST Program of the Japan Science and Technology Agency (Japan), the Ministry of Agriculture (Japan), the Ministry of Education, Culture, Sports, Science and Technology (Japan), the BBSRC e-Science Initiative (UK), the DARPA IPTO Bio-Computation Program (USA), the Army Research Office's Institute for Collaborative Biotechnologies (USA), and the Air Force Office of Scientific Research (USA). Additional support is provided by the California Institute of Technology (USA), the University of Hertfordshire (UK), the Molecular Sciences Institute (USA), and the Systems Biology Institute (Japan).

References

1. Hucka M, Finney A, Sauro HM, et al. The systems biology markup language (SBML): a medium for representation and exchange of biochemical network models. *Bioinformatics* 2003;19:524–531.
2. Hucka M, Finney A, Bornstein BJ, et al. Evolving a lingua franca and associated software infrastructure for computational systems biology: the Systems Biology Markup Language Project. *IEE Sys Biol* 2004;1:41–53.
3. XML Core Working Group. Extensible Markup Language (XML). Available at http://www.w3.org/xml.
4. Achard F, Vaysseix G, Barillot E. XML, bioinformatics, and data integration. *Bioinformatics* 2001;17:115–125.2
5. Wanner BL, Finney A, Hucka M. Modeling the E. coli cell: The need for computing, cooperation, and consortia. In: Alberghina L, Westerhoff HV, eds. Systems Biology: Definitions and Perspective. Berlin: Springer-Verlag; 2005.
6. Nature Molecular Systems Biology. For Authors. Available at http://www.nature.com/msb/authors/index.html.
7. Finney A, Hucka M, Benjamin J, et al. Software infrastructure for effective communication and reuse of computational models. In: Szallasi Z, Stelling J, Periwal V, eds. System Modeling in Cellular Biology: From Concepts to Nuts. MIT Press; 2006.
8. Hucka M, Finney A, Sauro H, Bolouri H. Systems Biology Markup Language (SBML) Level 1: Structures and facilities for Basic Model Definitions. SBML Level 1, Version 2 (Final), 28 August 2003. Available at http://sbml.org/specifications/sbml-level-1/version-2/html/sbml-level-1.html.

9. Finney A, Hucka M. Systems Biology Markup Language (SBML) Level 2: Structures and Facilities for Model Definitions. SBML Level 2, Version 1 (Final), June 28, 2003. Available at http://sbml.org/specifications/sbml-level-2/version-1/html/sbml-level-2.html.
10. Finney A, Hucka M. Systems Biology Markup Language (SBML) Level 2: Structures and Facilities for Model Definitions, SBML Level 2, Version 2. Available at http://sf.net/projects/sbml. Assessed March 26, 2005.
11. Hucka M, Finney AM, Hoops S, Keating SM, Le Novere N. Systems Biology Markup Language (SBML) Level 2: Structures and Facilities for Model Definitions, SBML Level 2, and Version 3 (Draft). Available at http://sf.net/projects/sbml.
12. Asbrooks R, Buswell S, Carlisle D, et al. Mathematical Markup Language (MathML) Version 2.0 (Second Edition), http://www.w3.org/TR/2003/REC-MathML2-20031021/.
13. Finney A. Developing SBML Beyond Level 2: Proposals for Development. In: Danos V, Schachter V, eds. Lecture Notes in Computer Science. Computational Methods in Systems Biology. Berlin/Heidelberg: Springer; 2005:242–247.
14. Goto S, Nishioka T, Kanehisa M. LIGAND: chemical database for enzyme reactions. *Bioinformatics* 1998;14:591–599.
15. Goto S, Okuno Y, Hattori M, et al. LIGAND: database of chemical compounds and reactions in biological pathways. *Nucleic Acids Res* 2002;30:402–404.
16. Lloyd C, Halstead MDB, Nielson PF. CellML: its future, present, and past. *Prog Biophys Mol Biol* 2004;85(2–3):433–450.
17. Schilstra MJ, Lu L, Matthews J, Finney A, Hucka M, LeNovere N. CELLML2SBML: Conversion of CELLML inot SBML. *Bioinformatics* 2006;22(8):1018–1020.
18. Le Novère N, Finney A, Hucka M, et al. Minimal information requested in the annotation of models (MIRIAM). *Nature Biotechnol* 2005;23(12):1509–1515.
19. Le Novère N, Bornstein B, Broicher A, et al. BioModels database: a free, centralized database of curated, published, quantitative kinetic models of biochemical and cellular systems. *Nucleic Acids Res* 2006;34(Database issue):D689–D691.
20. Shapiro BE, Hucka M, Finney A, et al. MathSBML: a package for manipulating SBML-based biological models. *Bioinformatics* 2004;20(16):2829–2831.
21. Novak B, Tyson JJ. Modeling the control of DNA replication in fission yeast. *PNAS* 1997;94:9147–9152.

21

CellDesigner: A Graphical Biological Network Editor and Workbench Interfacing Simulator

Akira Funahashi, Mineo Morohashi, Yukiko Matsuoka, Akiya Jouraku, and Hiroaki Kitano

Summary

Understanding the logic and dynamics of gene-regulatory and biochemical networks is a major challenge of systems biology. To facilitate this research topic, we have developed CellDesigner, a modeling tool of gene-regulatory and biochemical networks. CellDesigner supports users to easily create such networks, using solidly defined and comprehensive graphical representation (SBGN, systems biology graphical notation). CellDesigner is systems biology markup language (SBML) compliant, and has Systems Biology Workbench(SBW)-enabled software so that it can import/export SBML-described documents and integrate with other SBW-enabled simulation/analysis software packages. CellDesigner also supports simulation and parameter search, which is supported by integration with SBML ordinary differential equation (ODE) Solver, enabling us to simulate through our sophisticated graphical user interface. We can also browse and modify existing SBML models with references to existing databases. CellDesigner is implemented in Java; thus, it runs on various platforms such as Windows, Linux, and MacOS X. CellDesigner is freely available from our Web site at http://celldesigner.org/.

Key Words: Pathway editor; biochemical simulation; SBML; SBW; systems biology; SBGN; XML.

1. Introduction

Systems biology is characterized by the synergistic integration of theory, computational modeling, and experimentation (1). Although software infrastructure is one of the most crucial components of systems biology research, there has been no common infrastructure or standard to enable integration of computational resources. To solve this problem, the SBML (http://sbml.org) (2) and the SBW (http://sbw.kgi.edu) have been developed (3). SBML is an open, extensible markup language (XML)–based format for representing biochemical reaction networks, and SBW is a modular, broker-based, message-passing framework for simplified intercommunication between applications. More than 100 simulation and analysis software packages already support SBML, or are in the process of being able to support them.

Identification of logic and dynamics of gene-regulatory and biochemical networks is a major challenge of systems biology. We believe that the standardized technologies, such as SBML, SBW, and SBGN, play an important role in the software platform of systems biology. As one such approach, we have developed CellDesigner (4), which is a process diagram editor for gene-regulatory and biochemical networks.

In this chapter, we will introduce the main features of CellDesigner.

2. Features of CellDesigner

Broadly classified, the current version (3.1 at the time this was written) of CellDesigner has the following features:

- Representation of biochemical semantics
- Detailed description of state transition of proteins
- SBML compliant (SBML Level 1 and 2)
- Integration with SBW-enabled simulation/analysis modules
- Integration with native simulation library (SBML ODE Solver)
- Capability of database connections
- Extreme portability as a Java application

The aim in developing CellDesigner is to supply a process diagram editor with standardized technology (SBML in this case) for every computing platform, so that it could confer benefits to as many users as possible. By using the standardized technology, the model could be easily used with other applications, thereby reducing the cost to create a specific model from scratch. The main standardized features that CellDesigner supports could be summarized as "graphical notation," "model description," and "application integration environment." The standard for graphical notation plays an important role for efficient and accurate dissemination of knowledge (5), and the standard for model description will enhance the portability of models between software tools. Similarly, the standard for application integration environment will help software developers to provide the ability for their applications to communicate with other tools.

2.1. Symbols and Expressions

CellDesigner supports graphical notation and listing of symbols based on a proposal by Kitano et al. (5). The definition of graphical notation has now been developed as international community–based activities called SBGN (http://sbgn.org). Although several graphical notation systems have been already proposed (6–11), each has obstacles to becoming a standard. SBGN is proposed for biological networks designed to express sufficient information a in clearly visible and unambiguous way (5). We expect that these features will become part of the standardized technology for systems biology. The key components of SBGN, which we propose, are as follows:

1. To allow representation of diverse biological objects and interactions.
2. To be semantically and visually unambiguous.
3. To be able to incorporate notations.
4. To allow software tools to convert a graphically represented model into mathematical formulas for analysis and simulation.
5. To have software support to draw diagrams.
6. To have a freely available notation scheme.

To accomplish these requirements for the notation, Kitano et al. (5) first decided to define a notation by using a process diagram, which graphically represents state transitions of the molecules involved. In the process diagram representation, each node represents the state of molecule and complex, and each arrow represents state transition among states of a molecule. In the conventional entity-relationship diagrams, an arrow generally means "activation" of the molecule. However, it confuses the semantics of the diagram, as well as limiting possible molecular processes that could be represented. A process diagram is more intuitively understandable than the entity-relationship diagram; one of the reasons for this is that the process diagram could be explicitly represented as a temporal sequence of events, which entity-relationship cannot. For example, during the process of maturation-promoting factor (MPF) activation in the cell cycle, kinases such as Wee1 phosphorylates residues of Cdc2, which is one of the components of MPF (Figure 1). However, MPF

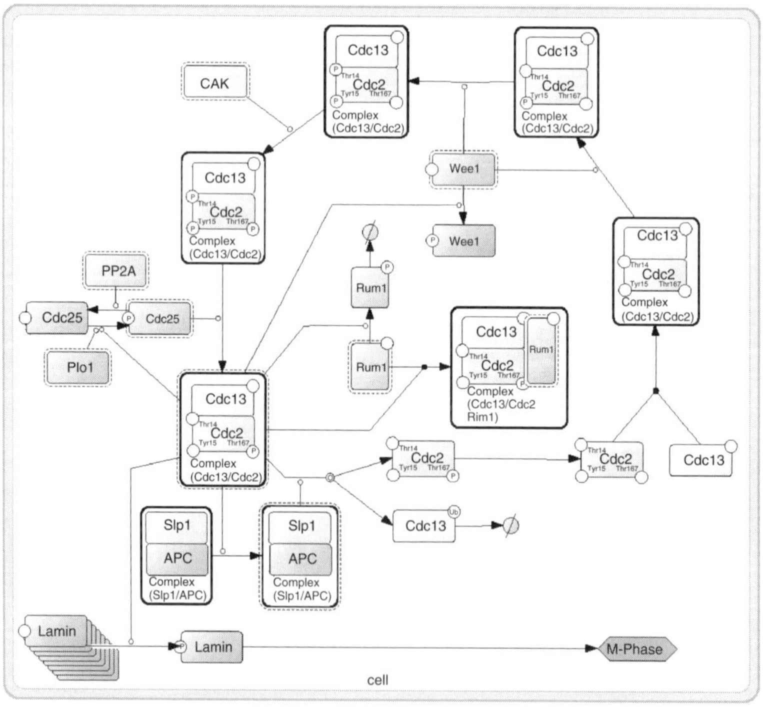

Figure 1. A process diagram representation of the MPF cycle.

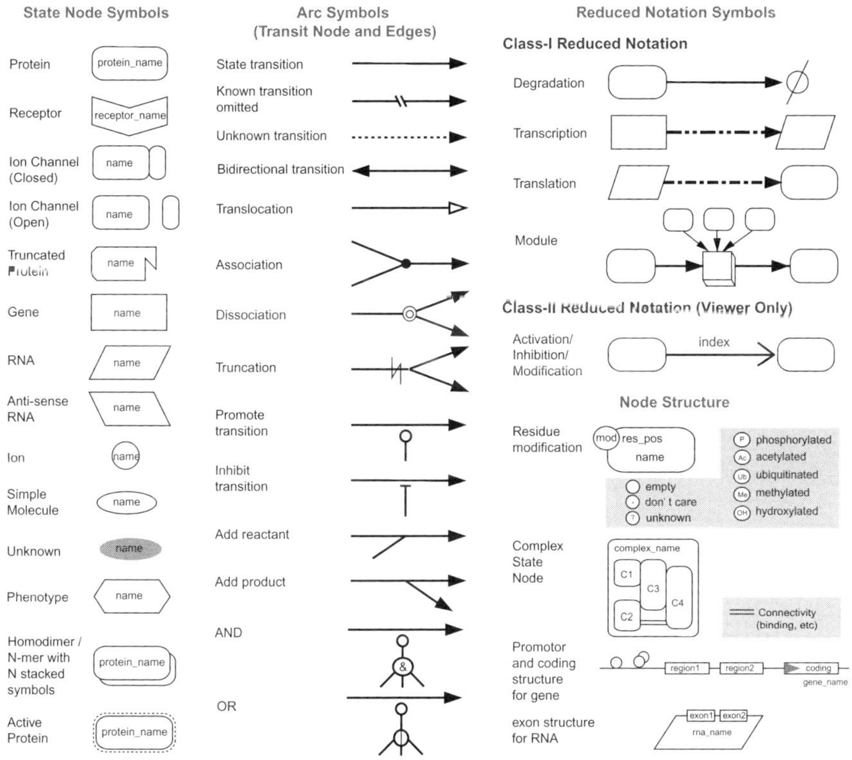

Figure 2. Proposed set of symbols for representing biological networks with process diagrams.

is not yet activated by this phosphorylation. If we use an arrow for activation, we cannot properly represent the case. In the process diagram, on the other hand, whether a molecule is "active" or not is represented as a state of the node, instead of by an arrow. The promotion and inhibition of catalysis are represented as a modifier of state transition, using a circle-headed line and a bar-headed line, respectively.

Although a process diagram is suitable for representing temporal sequence, either the process diagram or the entity-relationship approach could be used, depending upon the purpose of the diagram. Both notations could actually maintain compatible information internally, but differ in visualization (5).

We propose, as a basis of SBGN, a set of notations that enhance the formality and richness of the information represented. The symbols used to represent molecules and interactions are shown in Figure 2.

The goal of SBGN is to define a comprehensive system of notation for visually describing biological networks and processes, thereby contributing to the eventual formation of a standard notation. For such a graphical notation to be practical and to be accepted by the community, it is essential that software tools and data resources be made available. Even if the proposed notation system satisfies the requirements of biologists, lack of software support will drastically decrease its advantages. CellDesigner currently supports most of the process diagram

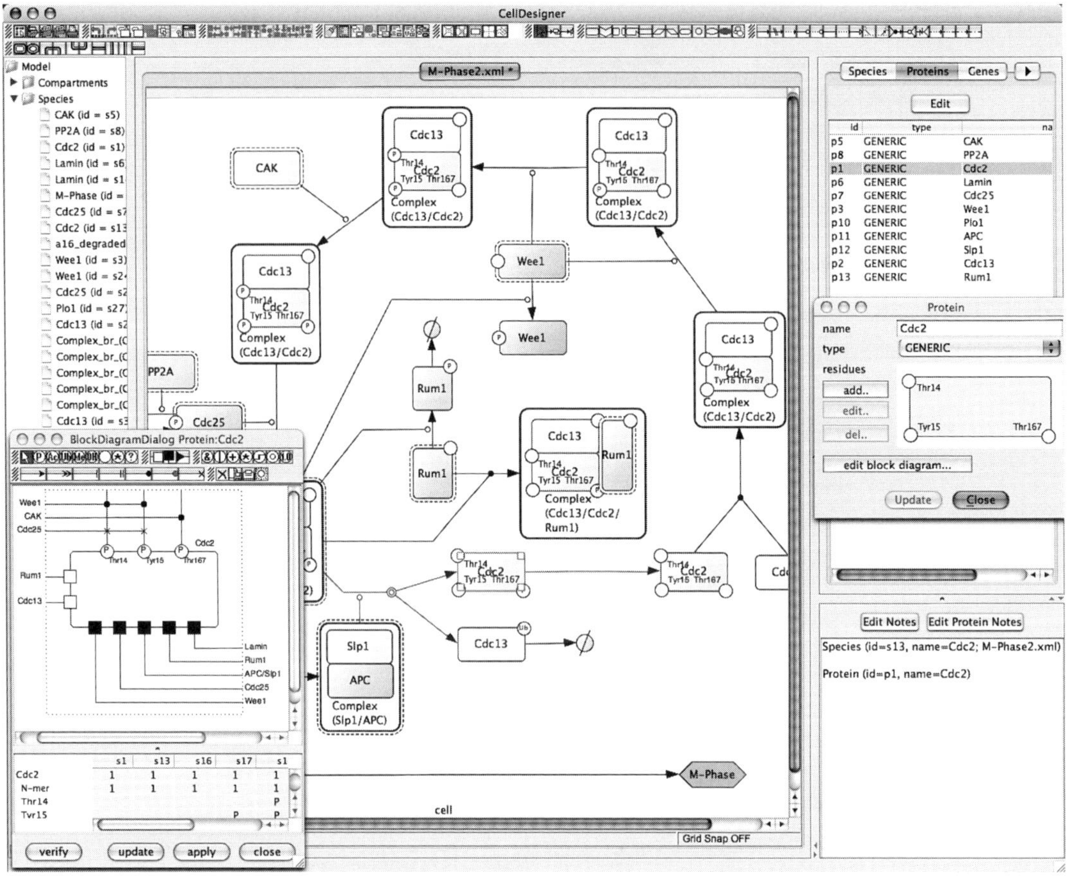

Figure 3. Screenshot of CellDesigner.

notation proposed, and will fully implement the notation in the near future (Figure 3).

2.2. SBML Compliant

CellDesigner is an SBML-compliant application—it supports SBML reading and writing capabilities. SBML is a tool-neutral, computer-readable format for representing models of biochemical reaction networks, and it is applicable to metabolic networks, cell signaling pathways, gene regulatory networks, and other modeling problems in systems biology. SBML is based on XML, which is a simple, flexible text format for exchanging a wide variety of data (11). The initial version of the specification was released in March 2001 as SBML Level 1. The most recently released version of SBML is Level 2 Version 1, and Level 2 Version 2 was released in January 2006. Currently, SBML is supported by over 100 software systems and is widely used. CellDesigner uses SBML as its native model description language, and thus, once a user creates a model by CellDesigner, all information inside the model will be stored in SBML and the model could be used by other software systems without any conversion of the model. As mentioned, CellDesigner draws a pathway with its specialized graphical notation. Because such layout information has not been supported by SBML, CellDesigner stores its layout information under an "annotation" tag, which does not conflict with current SBML specification. There is a working group of layout extensions for SBML, and they will be incorporated to SBML Level 3. We are currently working to implement a conversion module to export SBML layout extension from CellDesigner. CellDesigner has an auto layout function so that it can read all SBML Level 1 and 2 documents, whether the model contains layout information or not. By using this function, users can use existing SBML models, such as KEGG, BioModels database, etc. We have converted more than 12,000 metabolic pathways of KEGG to SBML (the pathways are available from http://systems-biology.org/). Other SBML models are available from the BioModels Database (http://www.ebi.ac.uk/biomodels/). We could also use our own SBML models created by CellDesigner on other SBML-compliant applications (http://systems-biology.org/001/).

2.3. SBW Enabled

CellDesigner is an SBW-enabled application. With SBW installed, CellDesigner could integrate all SBW-enabled modules (Figure 4). For example, users could browse or modify a model converted from an existing database with CellDesigner, and launch a simulator from CellDesigner (by selecting Simulation Service or Jarnac Simulation Service from the SBW menu) to run simulations in real time. There are many other SBW-enabled modules, such as ODE-based simulator, stochastic simulator, MATLAB, FORTRAN translator, bifurcation analysis tool,

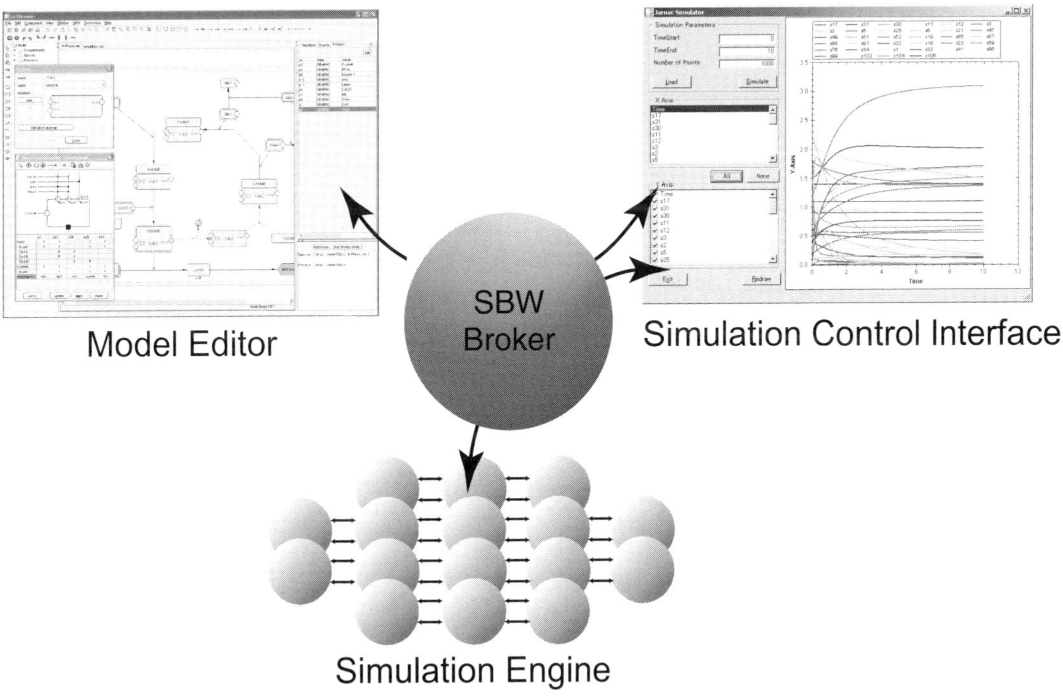

Figure 4. Illustration of the relationship between SBW Broker and SBW modules.

and optimization module. These SBW-enabled modules are freely available from http://sbw.kgi.edu/.

2.4. Supported Environment

CellDesigner is implemented in Java, and could run on many platforms that support Java Runtime Environment (JRE). Currently, CellDesigner runs on the following platforms:

- Windows (98SE or later)
- MacOS X (10.3 or later)
- Linux (Fedora Core 4 or later)

The current version of CellDesigner requires JRE1.4.2 or higher and X Window System for UNIX platforms.

2.5. Exporting Capability

Because CellDesigner is supposed to be a "design tool" for representing gene regulatory and biochemical networks, the pathways described by CellDesigner should be easily used in various situations (e.g., figures in a manuscript). CellDesigner could thus export the pathways in various formats; it can currently export in JPEG, PNG, and SVG formats.

2.6. Simulation Capability

One of our aims is to use CellDesigner as a simulation platform; thus, integration capability with native simulation library has been implemented. SBML ODE Solver (12) can be invoked directly from CellDesigner, which enables us to run ODE-based simulations.

The SBML ODE Solver Library is a programming library for symbolic and numerical analysis of chemical reaction network models encoded in SBML. It is written in ISO C and distributed under the open source LGPL license. SBML ODE Solver can read SBML models by using libSBML (13), and then construct a set of ODEs and their Jacobian matrix, and so forth. SBML ODE Solver uses SUNDIALS' CVODES (http://www.llnl.gov/CASC/sundials/) for numerical integration and sensitivity analysis. The performance of simulation engine is a critical issue for a simulation platform, so we have wrapped the C API of SBML ODE Solver from Java by using Java Native Interface. This resulted in a small overhead of simulation execution time compared with native library, and still retained the broad support of multiple operating systems. The simulation engine itself is executed by native library, and the results are shown in a GUI window written in Java (Figure 5). The simulation results can be exported to CVS, JPEG, and PNG formats, and to various bitmap files.

2.7. Database Connection Capability

To efficiently conduct network analysis, connection with databases is significant, as users may want to further examine network characteristics. We have added this capability, enabling direct connection with the following databases:

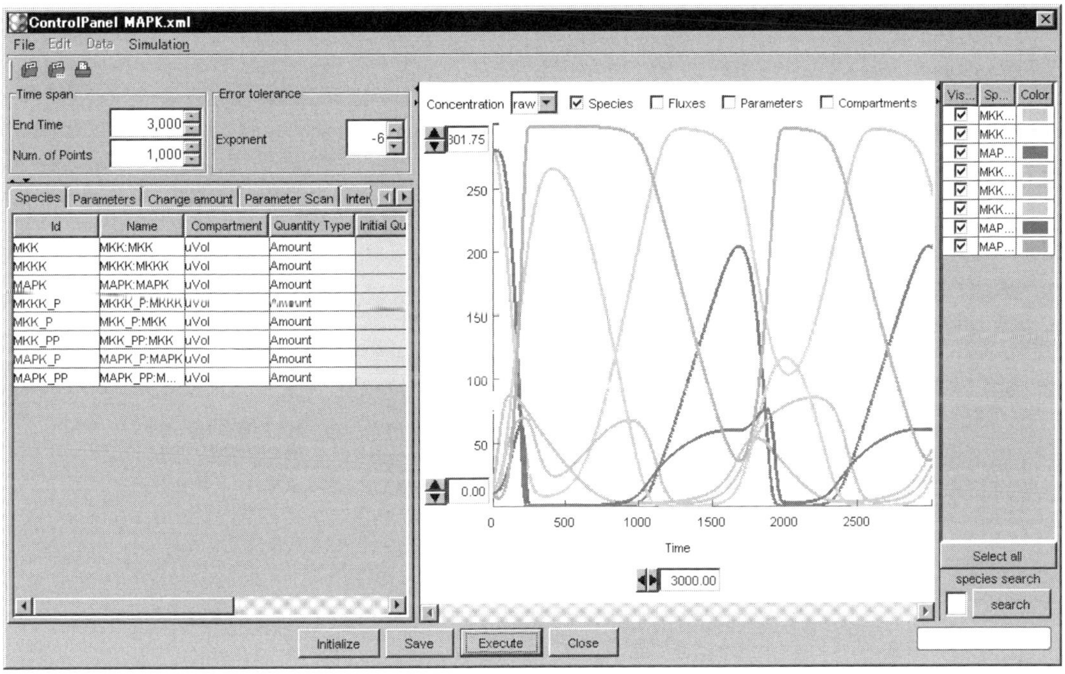

Figure 5. Snapshot of a simulation result obtained by integration with SBML ODE Solver.

- SGD (*Saccharomyces* Genome Database, http://www.yeastgenome.org)
- DBGET (Database retrieval system for a diverse range of molecular biology databases, http://www.genome.ad.jp/dbget/)
- iHOP (Information Hyperlinked Over Proteins, http://www.ihop-net.org/UniPub/iHOP/)
- PubMed (http://www.pubmed.gov)
- BioModels (Database of annotated computational models, http://www.biomodels.net/)

Once a species is selected, users can select the "Database" menu, from which those databases can be chosen to query according to the name of the species. For PubMed connection, a PubMed search is conducted according to the ID written in the Notes of the components. From the BioModels database, users can import (and not query) SBML-based models, which are those curated computational models prepared for simulations or further various analyses.

2.8. Collaboration with Worldwide Groups

Our approach seems to have attracted quite a lot of attention. We have been collaborating with several groups. One of the groups is Applied Biosystems, a biotech company. In their application, the engine of CellDesigner is used in the Web-based front end of the PANTHER pathway system to represent protein networks ([14,15] http://www.pantherdb.org/; Figure 6).

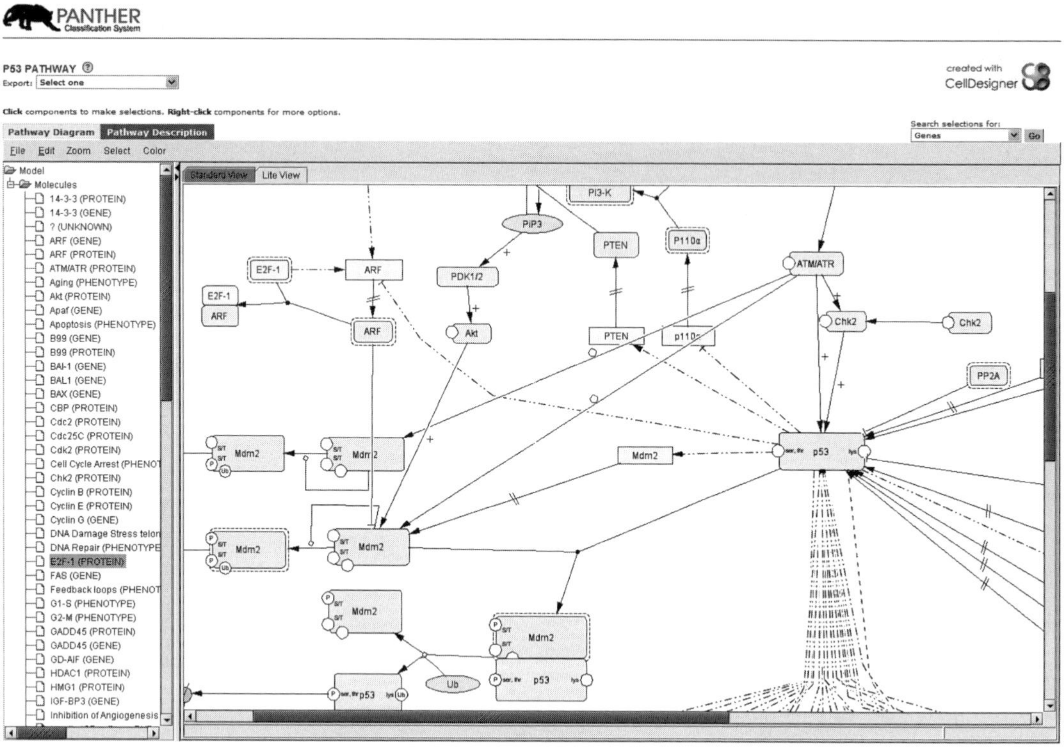

Figure 6. Screenshot of PANTHER.

Similar to the case of PANTHER, CellDesigner will be used as a pathway viewer on the BioModels database (http://www.biomodels.net). Given a curated computational model-based database, we developed an Applet-based CellDesigner so that it could be embedded in the front end of the BioModels database.

On the other hand, we have recently joined BioPAX-DX (data exchange), which aims to facilitate data exchange in the short term, by providing a data schema. BioPAX is a worldwide collaborative effort to create a data exchange format for biological pathway data.

Other collaborative efforts include SBML ODE Solver development from the University of Vienna, and Taverna from the Taverna project.

3. How Does it Work?

Building models with CellDesigner is quite straightforward. To create a model, the user selects "New" from "File" menu, inputs the name of an SBML document, and a new canvas will then appear. The user could then add a species, such as a protein, gene, RNA, ion, simple molecule, and so forth. A new window will appear asking the name of the species. The size of each species can be changed by clicking and dragging the corner of species. The user can also define the default size of each species using the "Show Palette" option from the "Window" menu. Species' can be moved by dragging and dropping.

To draw reactions, a type of reaction should first be selected from the UI buttons, and a reactant species then clicked, followed by a product species. To add more reactants, the user can click the "Add reactant" button, and then choose a species and a reaction.

As we briefly mentioned, the modeling process with CellDesigner is in straightforward steps that should not cause users any confusion.

CellDesigner can also represent common types of reactions, such as catalysis, inhibition activation, and so forth. The procedure for representing such reactions is just the same as adding reactants or products to an existing reaction; i.e., to select a species (modifier), followed by a reaction. The user can also easily edit the symbols for proteins with modification residues, and hence, can describe detailed state transitions between species of an identical protein by adding different modifications.

The models are stored in an SBML document, which contains all the necessary information referring to species, reactions, modifiers, layout information (geometry), state transitions of proteins, modification residues, etc. These SBML models could be used on other SBML-compliant applications.

If users want to run a simulation based on the SBML model, select Simulation menu, which, in turn, calls on the SBML ODE Solver directly. The control panel appears, enabling users to specify the details of parameters, to change the amount of a specific species, to conduct a parameter search, and to run a simulation interactively. To conduct a time-evolving simulation, users may need to know the basics of the SBML specification (*see* http://sbml.org for details).

If users select the SBW menu, on the other hand, CellDesigner passes the SBML data to the SBML-compliant tools via SBW, whereas you need to set up SBW before you invoke SBW connection.

4. What Distinguishes CellDesigner's Technology from Others Currently Available?

Currently, many other applications exist that include pathway design features. The advantages of CellDesigner over other pathway design tools could be summarized as follows:

- Based on standard technology (i.e., SBML compliant and SBW enabled),
- Supports clearly expressive and unambiguous graphical notation systems (SBGN), which is aimed at contributing to eventual standard formation
- Runs on many platforms (e.g., Windows, MacOS X, Linux)

As described above, the aim of the development of CellDesigner is to supply a process diagram editor with standardized technology for every computing platform, so that it will benefit as many biological researchers as possible. Some of the existing applications are SBML-compliant, and some run on multiple platforms

These tools are powerful in some aspects and they are not intended to support the features as CellDesigner. Some of them have the facility to create pathways, and some also include a simulation engine or database integration module. CellDesigner does include a simulation engine provided by SBML ODE Solver development team, and it can also connect to other SBW-enabled applications so that users could switch the simulation engine on the fly. Furthermore, we have been converting existing databases to SBML (e.g., KEGG), and all SBML-compliant applications could easily be browsed, edit the models, and even simulate via CellDesigner.

The overriding advantage of CellDesigner is that it uses open and standard technologies. The models created by CellDesigner could be used on many other (over 100) SBML-compliant applications, and its graphical notation system will make the representation of models in a more efficient and accurate manner.

5. Future Work

In future versions of CellDesigner, we plan to implement more capabilities. Improvement of the autolayout function is a big issue; the bigger (e.g., more than a few hundred nodes) the network diagram becomes, the slower the performance of CellDesigner becomes, which causes our current version not to align each node and edge well. Integration with other modules is also underway, such as other simulation, analysis, and database modules. Current version of CellDesigner has been implemented as a Java application, although we are developing a JWS (Java Web Start) version of CellDesigner so that it could be used as a Web-based application as well.

To be widely used from biologists to theorists, we believe that it is essential to meet the standard. We are thus actively working as SBML and SBGN working group members, with aims to establish de facto standards in the systems biology field; SBML seems to have already become de facto as model description language. SBML Level 3 will include layout extensions, and we will incorporate the functions in our new release of CellDesigner. BioPAX (http://ww.biopax.org) is another big activity, which tries to connect widely distributed data resources seamlessly. We also plan to connect CellDesigner with the BioPAX data format so that users could use CellDesigner from BioPAX platform and vice versa.

From software development perspectives, providing API, plug-in interface, or open-source strategy might be a solution to speed up the development, and enable users to customize the software depending on their needs. Although we have been providing binary programming of CellDesigner so far, we are now working to extend our development scheme.

We want CellDesigner to be used by anyone who is working in a biology-related field. As described throughout this chapter, CellDesigner is designed to be as user-friendly as possible, thus allowing users to draw pathway diagrams as easily as drawing with other drawing tools, such as Microsoft Visio or Adobe Illustrator. Because our proposed notation itself is rigidly defined, the diagrams could be used for presentation, or even as a knowledge base; the diagrams could be used as figures in a manuscript or a pathway representation of databases. Because pathway diagram notation is now getting more attention, which has resulted in the formation of an SBGN working group (http://sbgn.org), we hope the notation will be more refined as a de facto standard representation, which will be reflected in the presentation of CellDesigner.

Our concept for developing CellDesigner is to make it "easy to create a model, to run a simulation, and to use analysis tools." This will be achieved by extending the development of corresponding native libraries or SBW-enabled modules. Improvement of the graphical user interface is also required, including the mathematical equation editor, so the user can easily write equations by selecting and dragging a species.

6. Conclusion

We have introduced CellDesigner, which is a process diagram editor for gene-regulatory and biochemical networks based on standardized technologies and with wide transportability to other SBML-compliant applications and SBW-enabled modules. Since the release of CellDesigner, there have been 12,000 downloads. CellDesigner also aims to support standard graphical notation. Because the standardization process is still under way, our technologies are still changing and evolving. As we are in partnership with SBML, SBW, and SBGN working groups, we will go through with these standardization projects, thereby improving the quality of CellDesigner.

The current version of CellDesigner is 3.1-RELEASE (as of March 2006), and it runs on multiple platforms, such as Windows, Linux, and MacOS X, and is freely available from http://celldesigner.org/.

Acknowledgments: We would like to thank Noriko Hiroi (Kitano Symbiotic Systems Project, JST, Japan), Norihiro Kikuchi (Mitsui Knowledge Industry Co., Ltd., Japan), and Naoki Tanimura (Mizuho Information & Research Institute, Inc., Japan) for fruitful discussion.

This research is supported in part by the ERATO-SORST program (Japan Science and Technology Agency), the International Standard Development area of the International Joint Research Grant (NEDO, Japanese Ministry of Economy, Trade, and Industry), and through the special coordination funds for promoting science and technology from the Japanese government's Ministry of Education, Culture, Sports, Science, and Technology.

References

1. Kitano H. Systems biology: A brief overview. *Science* 2002;295;1662–1664.
2. Hucka M, Finney A, Sauro HM, et al. The systems biology markup language (SBML): a medium for representation and exchange of biochemical network models. *Bioinformatics* 2003;19:524–531.

3. Sauro HM, Hucka M, Finney A, et al. Next generation simulation tools: the Systems Biology Workbench and BioSPICE integration. *OMICS* 2003;7: 355–372.
4. Funahashi A, Tanimura N, Morohashi M, Kitano H. CellDesigner: a process diagram editor for gene-regulatory and biochemical networks. *BIOSILICO* 2003;1:159–162.
5. Kitano H, Funahashi A, Matsuoka Y, et al. The process diagram for graphical representation of biological networks. *Nat Biotechnol* 2005;23:961–966.
6. Kohn K. Molecular Interaction Maps as information organizers and simulation guides. *Chaos* 2001;11:84–97.
7. Kohn K. Molecular interaction map of the mammalian cell cycle control and DNA repair systems. *Mol Biol Cell* 1999;10:2703–2734.
8. Pirson I, Fortemaison N, Jacobs C, et al. The visual display of regulatory information and networks. *Trends Cell Biol* 2000;10:404–408.
9. Cook DL, Farley JF, Tapscott SJ. A basis for a visual language for describing, archiving and analyzing functional models of complex biological systems. *Genome Biol* 2001:2.
10. Maimon R, Browing S. Diagrammatic Notation and Computational Structure of Gene Networks. In: Proceedings of the Second International Conference on Systems Biology 2001:311–317.
11. XML Core Working Group. Extensible Markup Language (XML). Available at http://www.w3.org/XML/
12. Machne R, Finney A, Muller S, et al. The SBML ODE Solver Library: a native API for symbolic and fast numerical analysis of reaction networks. *Bioinformatics* 2006;22:1406–1407.
13. Hucka M, Finney A, Bornstein BJ, et al. Evolving a Lingua Franca and Associated Software Infrastructure for Computational Systems Biology: The Systems Biology Markup Language (SBML) Project. *Sys Biol* 2004;1:41–53.
14. Mi H, Gupta A, Gok, MA, et al. The PANTHER database of protein families, subfamilies, functions and pathways. *Nucleic Acids Res* 2005;33:284–288.
15. Thomas PD, Campbell MJ, Kejariwal A, et al. PANTHER: a library of protein families and subfamilies indexed by function. *Genome Res* 2003;13: 2129–2141.

22

DBRF-MEGN Method: An Algorithm for Inferring Gene Regulatory Networks from Large-Scale Gene Expression Profiles

Koji Kyoda and Shuichi Onami

Summary

The difference-based regulation finding–minimum equivalent gene network (DBRF-MEGN) method is an algorithm for inferring gene regulatory networks from gene expression profiles corresponding to gene perturbations. In this method, gene regulatory networks are modeled as signed directed graphs, and the most parsimonious graphs consistent with gene expression profiles are deduced by using a graph theoretical procedure. The method is applicable to large-scale gene expression profiles, and gene regulatory networks deduced by the method are highly consistent with gene regulations identified through classic small-scale experiments in genetics and cell biology. Free software for the method is available and runs under Windows or Linux platforms on a typical IBM-compatible personal computer. The DBRF-MEGN method will provide invaluable information for basic biology and drug discovery.

Key Words: Gene network inference; signed directed graph; microarray; gene expression profiles; perturbation; deletion mutant; overexpression mutant.

1. Introduction

The DBRF-MEGN method is an algorithm for inferring gene regulatory networks (hereafter called *gene networks*) from gene expression profiles corresponding to gene perturbations. Combination of technologies for perturbing (i.e., down- or up-regulating) the activity of genes (1–6), along with those for systematic measurement of expression of genes (7,8), enables us to obtain large-scale gene expression profiles corresponding to gene perturbations (9,10). Inference of gene networks from such gene expression profiles will greatly contribute to both basic biological advances and drug discovery.

Many procedures have been developed for inferring gene networks from gene expression profiles corresponding to gene perturbations. In these procedures, gene networks are modeled by using various mathematical frameworks. Ideker et al. modeled gene networks as acyclic Boolean networks and deduced a network consistent with profiles by using a combinatorial optimization technique (11). Pe'er et al. modeled gene networks as Bayesian networks and estimated gene networks by using machine learning technology (12). Wagner modeled gene networks as directed acyclic graphs and deduced the most parsimonious graph consistent with profiles by using a graph theoretical procedure (13).

In the DBRF-MEGN method, gene networks are modeled as signed directed graphs (SDGs), and the most parsimonious SDGs consistent with gene expression profiles are deduced by using a graph theoretical procedure (14). In the SDGs, regulation between two genes is represented as a signed directed edge whose sign (positive or negative) represents whether the effect of the regulation is activation or inhibition, and whose direction represents what gene regulates what other gene. An outstanding feature of the DBRF-MEGN method is the utility of gene networks deduced by this method. These deduced gene networks can be compared directly with those identified through classic small-scale experiments in genetics and cell biology, and the deduced networks can be interpreted in the same way as those identified through small-scale experiments (14). This utility results from the fact that the SDG used in the DBRF-MEGN method is the most common representation of gene networks in genetics and cell biology, and also from the fact that the algorithm of the DBRF-MEGN method is based on logic that is most commonly used in genetics and cell biology to infer gene networks from small-scale gene-perturbation experiments (14).

In this chapter, we introduce the DBRF-MEGN method. First, we briefly describe the algorithm of this method. Second, we provide examples of application of the method to real large-scale gene expression profiles. Third, we show how to install and use the free software for the DBRF-MEGN method. Finally, we discuss the perspectives of the method.

2. Algorithm

The DBRF-MEGN method consists of five processes: i) difference-based deduction of initially deduced edges; ii) removal of nonessential edges from the initially deduced edges; iii) selection of the uncovered edges in main components from the nonessential edges; iv) separation of the uncovered edges in main components into independent groups; and v) restoration of the minimum number of edges from each independent group. The method deduces the most parsimonious SDGs consistent with gene expression profiles. Those graphs are called MEGNs.

2.1. Difference-Based Deduction of Initially Deduced Edges

The first process of the DBRF-MEGN method deduces signed directed edges (hereafter called *edges*) using an assumption that is commonly used in genetics and cell biology (Figure 1A); i.e., there exists a positive (negative) regulation from gene A to gene B when the expression level of gene B in the condition where the activity of gene A is down-regulated is lower (higher) than in the control condition. For each possible pair of genes in the profiles, the process determines whether positive or negative regulation between those two genes exists and deduces all edges consistent with both the assumption and the profiles by detecting the difference in expression level between the control condition and a condition where activity of a gene is perturbed (Figure 1B). These edges are called *initially deduced edges*.

2.2. Removal of Nonessential Edges from the Initially Deduced Edges

The initially deduced edges include not only those representing direct gene regulations but also those representing indirect gene regulations. We define the regulation from gene A to gene B as direct when gene A regulates gene B independently of other gene regulations; e.g., a transcription factor A binds to upstream regulatory regions of gene B and increases the transcription of gene B. On the other hand, we define gene regulation as indirect when gene A regulates gene B as a result of other regulations; e.g., a transcription factor A increases the transcription of transcription factor C, which then increases the transcription of gene B. A desirable gene network consists only of direct gene regulations because indirect regulations do not correspond to molecular mechanisms of gene regulation. To choose edges representing direct gene regulations from the initially deduced edges, the second process of the DBRF-MEGN method removes all edges that are deductively explained by two other initially deduced edges (Figure 1C) because an indirect regulation is deductively explained by direct regulations. The resulting edges are called *essential edges* and the removed edges are called *nonessential edges*.

2.3. Selection of Uncovered Edges in Main Components from Nonessential Edges

The essential edges sometimes fail to deductively explain all initially deduced edges (Figure 1D). Some edges represent direct gene regulations even when they are deductively explained by two other edges. Therefore, the second process sometimes removes edges representing direct gene regulations, resulting in excess removal of edges. It is difficult to know whether a nonessential edge represents direct or indirect gene regulation when only expression profiles corresponding to single-gene perturbations are available. Therefore, instead of looking for edges representing direct regulations among the nonessential edges, the DBRF-MEGN method compensates for excessively removed edges by restoring a minimum number of nonessential edges so that the resulting edges (the essential edges and the restored nonessential edges) can explain all initially deduced edges. Often, several sets of such nonessential edges exist, and the method deduces all sets. An SDG consisting of essential edges and the restored nonessential edges is a most parsimonious SDG consistent with given profiles; i.e., a MEGN.

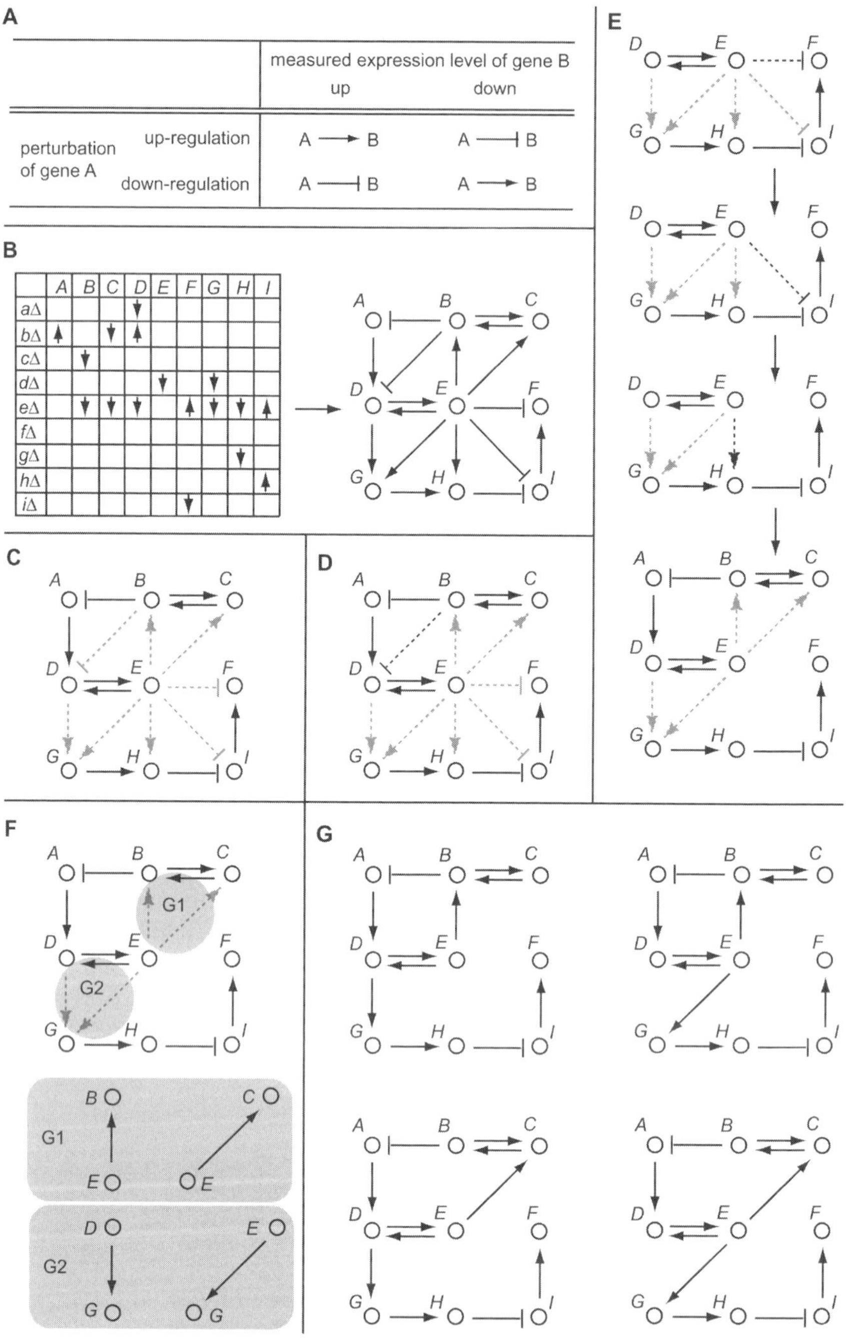

Figure 1. Example of the deduction of MEGNs from the gene expression profiles of gene deletion mutants. (A) Assumptions used in the DBRF-MEGN method. (B) Deduction of initially deduced edges. The matrix represents a set of expression profiles and the schematic represents a set of initially deduced edges. (C) Essential edges. Nonessential edges are gray-dotted. (D) Uncovered edges. Uncovered edges are gray-dotted and covered edges are black-dotted. (E) Exclusion of uncovered edges in peripheral components. (F) Independent groups of uncovered edges in main components. For each group, the minimum number of edges with which essential edges can explain all edges in the group are shown. (G) MEGNs of the profiles. Combinations of the minimum numbers of edges of independent groups produce all MEGNs.

The computation of the process described for deducing MEGNs is bounded by $n^3\sum_{i=0}^{m} {}_{(I-S)}C_i \cdot (I-S-i)$, where n is the number of genes in the profiles, m is the number of nonessential edges to be restored, I is the number of initially deduced edges, and S is the number of essential edges. This computation is impractical, however, because ${}_{(I-S)}C_m$ increases rapidly as $I - S$ and/or m increase. To reduce the computational cost, the third process of the DBRF-MEGN method distinguishes nonessential edges that have a chance to be included in the MEGNs from those that do not, before selecting the sets of nonessential edges to be restored. This process consists of two subprocesses: (a) selection of uncovered edges and (b) selection of uncovered edges in main components. The resulting nonessential edges are called uncovered edges in main components. From these edges, the later processes of the method select edges that are included in the MEGNs. The process reduces the computational cost of deducing MEGNs to $n^3\sum_{i=0}^{m} {}_{U}C_i \cdot (U-i)$, where U is the number of uncovered edges in main components.

2.3.1. *Selection of Uncovered Edges*

The first subprocess distinguishes the nonessential edges that are deductively explained by the essential edges from those that are not. Those edges are called *covered edges* and *uncovered edges*, respectively.

2.3.2. *Selection of Uncovered Edges in Main Components*

The second subprocess distinguishes the uncovered edges that have a chance to be included in the MEGNs from those that do not (Figure 1 E). Those edges are called *uncovered edges in main components* and *uncovered edges in acyclic components*, respectively. The uncovered edges in peripheral components are selected as follows (Figure 1E): (i) select uncovered edges that do not deductively explain any uncovered edges; (ii) select uncovered edges that deductively explain uncovered edges that have been selected, but that do not deductively explain those that have not; and (iii) repeat (ii) until no new uncovered edges are selected. The selected uncovered edges are uncovered edges in peripheral com-ponents, and those that are unselected are uncovered edges in main components.

2.4. Separation of Uncovered Edges in Main Components into Independent Groups

To further reduce the computational cost, the fourth process of the DBRF-MEGN method separates uncovered edges in main components into independent groups so that edges to be restored can be deduced independently for each group (Figure 1F). The independent groups are generated so that the edges in one group do not deductively explain those in other groups. The process reduces the computational cost of deducing MEGNs to $\sum_{j=1}^{t}\sum_{i=1}^{m_j} {}_{D_j}C_i \cdot (D_j - i) \cdot n_j^3$, where t is the number of independent groups, n_j is the number of genes in the jth independent group, D_j is the number of edges in the jth independent group, and m_j is the number of edges in the jth independent group to be included in a MEGN.

2.5. Restoration of Minimum Number of Edges from Each Independent Group

For each independent group, the fifth process of the DBRF-MEGN method deduces the minimum number of edges with which essential edges can deductively explain all edges in the group. All sets of such edges are deduced for each group. The essential edges and any possible combination of these sets from each group generate a MEGN of the profiles (Figure 1G).

3. Application

In this section, to show the applicability and validity of the DBRF-MEGN method, we provide examples of application of the DBRF-MEGN method to real large-scale gene expression profiles (14). In these examples, a subset of large-scale gene expression profiles obtained from *Saccharomyces cerevisiae* (9) was used. The set of profiles comprises the expression levels of 265 genes measured in 265 gene deletion mutants corresponding to those genes. Each expression level accompanies a *p*-value, which corresponds to the significance of the difference from the expression level in the wild type (9). We considered the expression level in the deletion mutants to be increased (decreased) when the level significantly differed from that in the wild type at a *p*-value less than a predefined threshold.

3.1. Applicability to Large-Scale Gene Expression Profiles

The computational cost of use of the DBRF-MEGN method depends on the *p*-value threshold because the numbers of initially deduced edges, essential edges, and edges in MEGNs depend on the *p*-value threshold. An optimal *p*-value threshold for the method is 0.01 (14). To show the applicability of the DBRF-MEGN method to large-scale gene expression profiles, we provide examples of the computational costs of use of the DBRF-MEGN method using various *p*-value thresholds (14). The DBRF-MEGN method deduced 829 initially deduced edges, 675 essential edges, and a unique MEGN consisting of those 675 essential edges from the expression profiles for the 265 *S. cerevisiae* genes when the *p*-value threshold was 0.01 (Figure 2). The computation took approximately 0.02 s on an Intel Pentium 4 PC (2.8 GHz, 1 GB RAM).

Essential edges failed to explain the initially deduced edges when the *p*-value threshold was 0.03 or 0.05. In these two cases, the method successfully deduced 2 and 16,384 MEGNs, consisting respectively of 964 and 1,090 edges, where 1,341 and 1,666 initially deduced edges and 963 and 1,076 essential edges, respectively, were deduced. The computation took approximately 0.02 s and 0.75 s, respectively. These examples show the applicability of the DBRF-MEGN method to real large-scale gene expression profiles.

3.2. Validity of MEGNs

MEGNs deduced by the DBRF-MEGN method can be directly compared with gene networks identified through classic small-scale

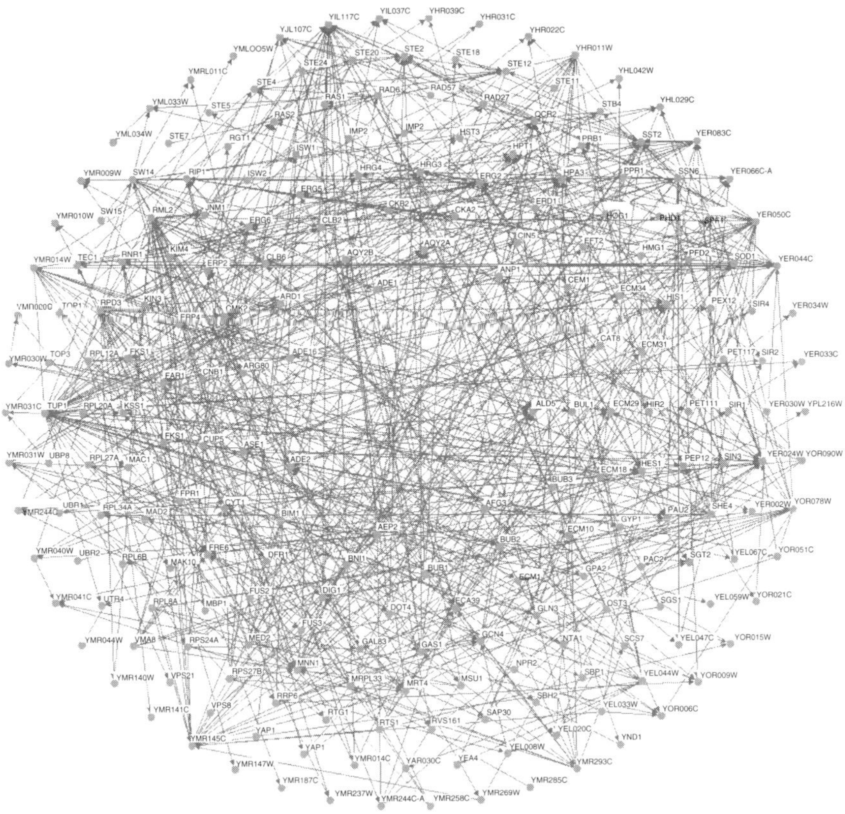

Figure 2. MEGN deduced from the expression profiles of 265 *Saccharomyces cerevisiae* genes. The network consists of 265 genes and 675 edges, of which 230 are positive and 445 negative. Graphical display of the gene network was created using the Osprey program (15).

experiments in genetics and cell biology. To show the validity of the DBRF-MEGN method, we will compare the MEGN deduced from the expression profiles for the 265 *S. cerevisiae* genes and the known gene network in a pheromone response pathway reported in the literature (14). The pheromone response pathway is one of the most thoroughly identified cellular cascades in *S. cerevisiae*.

First, we focus on transcriptional regulation by Ste12p, which is the central transcription factor in the pheromone response pathway. Because the expression profiles used in the DBRF-MEGN method were a collection of mRNA levels in gene deletion mutants, an edge directing from *STE12* was expected to represent a transcriptional regulation by Ste12p. Among the 265 genes in the profiles, 6 are transcriptional targets of Ste12p (Figure 3A) (16–21). The method cannot deduce self-regulations because of its assumption (*see* Figure 1A for a schematic of this assumption). Therefore, we expected that five positive edges directing from *STE12* to *FAR1*, *FUS3*, *SST2*, *STE2*, and *TEC1* would be deduced (Figure 3B). As expected, the method deduced five edges directing from *STE12*, all five of which were positive edges directing to each of those five genes (Figure 3C).

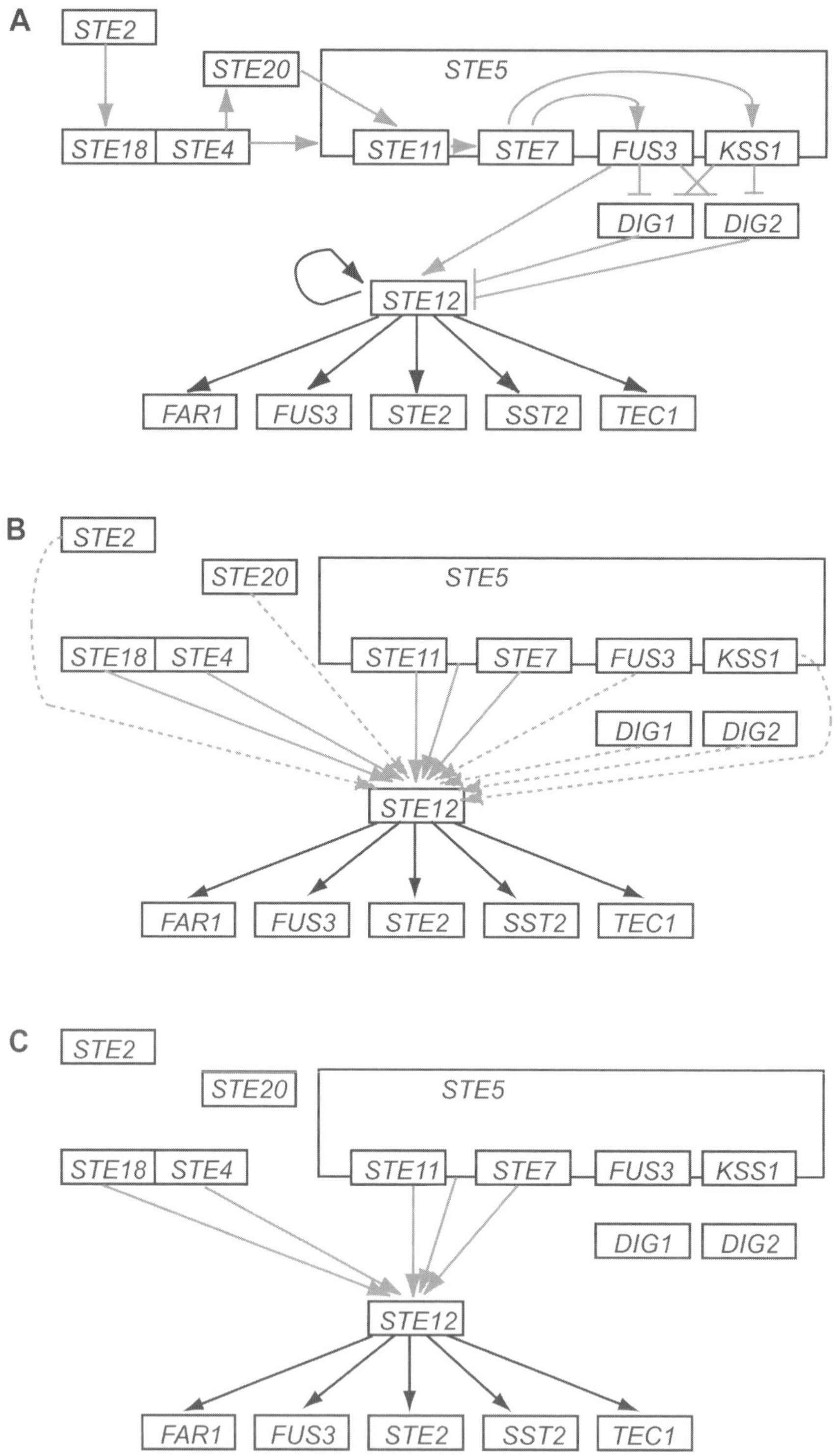
A
STE2
STE20
STE5
STE18
STE4
STE11
STE7
FUS3
KSS1
DIG1
DIG2
STE12
FAR1
FUS3
STE2
SST2
TEC1
B
STE2
STE20
STE5
STE18
STE4
STE11
STE7
FUS3
KSS1
DIG1
DIG2
STE12
FAR1
FUS3
STE2
SST2
TEC1
C
STE2
STE20
STE5
STE18
STE4
STE11
STE7
FUS3
KSS1
DIG1
DIG2
STE12
FAR1
FUS3
STE2
SST2
TEC1

Figure 3. Validation of MEGN in the *Saccharomyces cerevisiae* pheromone response pathway. (A) Known pheromone response pathway. Six transcriptional regulations (black edges) and 14 posttranscriptional regulations (gray edges) were reported previously (16–23). (B) Expected edges in the pheromone response pathway. Five edges from *STE12* to transcriptional targets (black edges) and five edges from posttranscriptional regulations to *STE12* (gray edges) were expected. Six edges from posttranscriptional regulators to *STE12* (dotted gray edges) were not expected because of the experimental conditions or the redundancy of gene regulation. (C) MEGN in the pheromone response pathway. Five expected edges from *STE12* (black edges) and five expected edges to *STE12* (gray edges) were deduced.

Next, we focus on the posttranscriptional regulation cascade that regulates Ste12p activity. Deletion of a single gene in this cascade increases (decreases) Ste12p activity, which then increases (decreases) the *STE12* mRNA level, because Ste12p self-increases its own transcription (20). The expression profiles used were a collection of mRNA levels in gene deletion mutants. Therefore, an edge directing from a gene to *STE12* was expected to indicate the existence of a posttranscriptional regulation cascade from this gene to Ste12p, unless the gene was a transcriptional regulator. Among 265 genes, 11 are involved in the posttranscriptional regulation cascade that regulates Ste12p activity (Figure 3A) (21–23). However, deletion of 6 of those 11 genes was not expected to affect the *STE12* mRNA level for the following reasons: *STE2* encodes the receptor for α-factor (24); the receptor would not have been activated in any of the experiments in which gene expression profiles were measured because MATa cells, which do not secrete α-factor (25), were used in these experiments; deletion of *STE20* does not completely block pheromone-induced Ste12p activation, suggesting that an unidentified pathway bypasses Ste20p activity (26); and *FUS3* and *KSS1* (27) and *DIG1* and *DIG2* (22) are functionally redundant. Therefore, we expected that five positive edges directing to *STE12* from *STE4*, *STE5*, *STE7*, *STE11*, and *STE18* would be deduced by the DBRF-MEGN method (Figure 3B). As expected, the method deduced five edges directed to *STE12*, all five of which were positive and directed from each of these five genes (Figure 3C). These results show the validity of the DBRF-MEGN method.

The validity of the DBRF-MEGN method is supported by analyses of two other cellular pathways, i.e., the general amino acid control system and the copper and iron homeostasis system (14).

4. Software

In this section, we briefly describe how to install and use the free software for the DBRF-MEGN method. The section is divided into four subsections: i) hardware and software; ii) format of the input file; iii) running the software; iv) format of the output file. Information on obtaining the software is available at our Web site (http://www.so.bio.keio.ac.jp/dbrf-megn/).

4.1. Hardware and Software

The program for the DBRF-MEGN method runs under Windows or Linux platforms on a typical IBM-compatible personal computer. The code for the DBRF-MEGN method is written in ANSI C++. Therefore, libraries for C++ are needed to run the program. Although we have not confirmed whether the program runs on other platforms, it should run on platforms based on UNIX that have a C++ compiler (e.g., Mac OS X or Solaris).

4.2. Format of Input File

The program requires an input file that is in tab-delimited text (Figure 4A). The first row lists the genes whose activities are perturbed (up- or down-regulated). The first column lists the genes whose expression levels are measured. The number and order of genes in the two lists must be the same. Each element of the remaining rows or columns represents the log-ratio of the gene expression level under the condition where the activity of the gene is perturbed relative to that in the control condition.

A

	gene_a	gene_b	gene_c	gene_cd	gene_e	gene_f	gene_g
GENE_A	0	−1.1	−1.2	0	0	0	0
GENE_B	−1.1	0	−1.4	0	1.5	0	0
GENE_C	−1.5	−1.9	0	0	0	0	1.7
GENE_D	0	0	0	0	−1.2	−1.3	0
GENE_E	0	0	0	−1.1	0	−1.7	0
GENE_F	0	0	0	−1.2	−1.2	0	0
GENE_G	0	0	0	0	0	0	0

B

from	to	effect
GENE_E	GENE_B	N
GENE_G	GENE_C	N

C

num. of independent groups: 2

group 1 (1 out of 2 sets)

set	from	to	effect	from	to	effect	from	to	effect
1	GENE_A	GENE_B	P	GENE_B	GENE_C	P	GENE_C	GENE_A	P
2	GENE_A	GENE_C	P	GENE_B	GENE_A	P	GENE_C	GENE_B	P

group 2 (1 out of 2 sets)

set	from	to	effect	from	to	effect	from	to	effect
1	GENE_D	GENE_E	P	GENE_E	GENE_F	P	GENE_F	GENE_D	P
2	GENE_D	GENE_F	P	GENE_E	GENE_D	P	GENE_F	GENE_E	P

Figure 4. Format of data files of the DBRF-MEGN method. (A) Example of the input file for gene deletion mutants. (B) Output file for essential edges. (C) Output file for restored edges.

When the expression level under the condition where activity of the gene is down-regulated is significantly higher (lower) than in the control condition, the corresponding element has a plus (minus) value. When there is no significant difference in gene expression level between the control condition and the condition of gene perturbation, the corresponding element has the value of zero.

4.3. Running the Software

The program runs by using the command "*dbrf_megn [OPTION] <data file>*" from the command line. The *[OPTION]* will be "*-D*" or "*-U*," which indicates that the following *<data file>* is a set of gene expression profiles corresponding to the down- or up-regulation of genes, respectively. The *<data file>* is the input file whose format is described in the previous section. When the MEGN consists only of the essential edges, the program generates only one file, named *essential_edges-<data file>*, which includes a list of the essential edges. When a MEGN consists of the essential edges and the restored edges, the program generates two files, one named *essential_edges-<data file>*, which includes a list of the essential edges, and another named *restored_edges-<data file>*, which includes sets of the restored edges in independent groups.

4.4. Format of Output File

Each row of the *essential_edges-<data file>* represents a deduced essential edge, which is directed from the gene in the first column to that in the second column (Figure 4B). The third column represents the sign of the edge (P, positive; N, negative).

Each section of the *restored_edges-<data file>* corresponds to an independent group (Figure 4C). Each row of each section represents a set of the minimum number of restored edges in the corresponding group. Each MEGN consists of the essential edges and a combination of one set from each of the independent groups.

5. Perspectives

MEGNs deduced by the DBRF-MEGN method allow us to obtain invaluable information for understanding cellular function. This is because the SDG used in the DBRF-MEGN method is the most common representation of gene networks in genetics and cell biology, and the algorithm of the DBRF-MEGN method is based on the logic most commonly used in genetics and cell biology to infer gene networks from small-scale gene perturbation experiments. For example, a procedure has been proposed for predicting transcriptional targets and modulators of the transcriptional activity of transcription factors from MEGN (14). MEGNs can be applied to various analyses of network structures (28,29). These analyses will provide invaluable insight into the nature of gene networks.

The DBRF-MEGN method is a powerful analytical tool, not only for large-scale gene expression profiles but also for small- or medium-scale gene expression profiles. The algorithm of the DBRF-MEGN method without the processes of *Selection of the uncovered edges in cyclic components from the nonessential edges* and *Separation of the uncovered edges in main components into independent groups* is the procedure most commonly used in genetics and cell biology to manually infer gene networks from small-scale gene-perturbation experiments. However, the computation of the DBRF-MEGN method without these two processes is bounded by $n^3\sum_{i=0}^{m} {}_{(I-S)}C_i \cdot (I-S-i)$, and the computation increases rapidly as $I - S$ and/or m increase. This computational cost indicates that manual inference of gene networks is impractical even for small- or middle-scale gene expression profiles. The DBRF-MEGN method deduces all of the exact solutions for the most parsimonious SDGs consistent with given expression profiles. Therefore, use of the DBRF-MEGN method ensures accurate analysis of small- and medium-scale gene expression profiles.

Deduction of MEGNs from more than one set of gene expression profiles is a new research strategy in systems biology. Both positive and negative edges are sometimes deduced in the same direction between the same pair of genes when more than one set of gene expression profiles is analyzed by the DBRF-MEGN method; e.g., positive and negative edges from gene A to gene B are deduced from gene expression profiles of deletion mutants and of overexpression of genes, respectively. Interestingly, these edges may reflect nonlinearity of gene regulation. Comparisons of MEGNs deduced from different sets of gene expression profiles will provide new opportunities for understanding the nonlinear dynamics of gene regulatory networks in cells.

Integration of diverse functional genomic data is a major approach in current gene network inference (30,31). Various functional genomic data are available. For example, the chromatin immunoprecipitation assay provides protein–DNA associations (20,28). Two-hybrid systems and *in vivo* pull-down assay provide protein–protein interactions (32–35). Fluorescent labeling technology provides information on protein localization (36). Integration of MEGNs and these functional genomic data will provide information invaluable to basic biology and drug discovery.

References

1. Liu H, Krizek J, Bretscher A. Construction of a GAL1-regulated yeast cDNA expression library and its application to the identification of genes whose overexpression causes lethality in yeast. *Genetics* 1992;132:665–673.
2. Baudin A, Ozier-Kalogeropoulos O, Denouel A, et al. A simple and efficient method for direct gene deletion in *Saccharomyces cerevisiae. Nucleic Acids Res* 1993;21:3329–3330.
3. Wach A, Brachat A, Pohlmann R, et al. New heterologous modules for classical or PCR-based gene disruptions in *Saccharomyces cerevisiae. Yeast* 1994;10:1793–1808.
4. Lorenz MC, Muir RS, Lim E, et al. Gene disruption with PCR products in *Saccharomyces cerevisiae. Gene* 1995;158:113–117.
5. Gari E, Piedrafita L, Aldea M, et al. A set of vectors with a tetracycline-regulatable promoter system for modulated gene expression in *Saccharomyces cerevisiae. Yeast* 1997;13:837–848.

6. Fire A, Xu S, Montgomery MK, et al. Potent and specific genentic interference by double-stranded RNA in *Caenorhabditis elegans. Nature* 1998; 391:806–811.
7. Schena M, Shalon D, Davis RW, et al. Quantitative monitoring of gene expression patterns with a complementary DNA microarray. *Science* 1995;270:467–470.
8. Lockhart DJ, Dong H, Byrne MC, et al. Expression monitoring by hybridization to high-density oligonucleotide arrays. *Nat Biotechnol* 1996;14:1675–1680.
9. Hughes TR, Marton MJ, Jones AR, et al. Functional discovery via a compendium of expression profiles. *Cell* 2000;102:109–126.
10. Mnaimneh S, Davierwala AP, Haynes J, et al. Exploration of essential gene functions via titratable promoter alleles. *Cell* 2004;118:31–44.
11. Ideker TE, Thorsson V, Karp RM. Discovery of regulatory interactions through perturbation: inference and experimental design. *Pac Symp Biocomput* 2000;305–316.
12. Pe'er D, Regev A, Elidan G, et al. Inferring subnetworks from perturbed expression profiles. *Bioinformatics* 2001;17 Suppl:S215–S224.
13. Wagner A. How to reconstruct a large genetic network from n gene perturbations in fewer than n^2 easy steps. *Bioinformatics* 2001;17:1183–1197.
14. Kyoda K, Baba K, Onami S, et al. DBRF-MEGN method: an algorithm for deducing minimum equivalent gene networks from large-scale gene expression profiles of gene deletion mutants. *Bioinformatics* 2004;20:2662–2675.
15. Breitkreutz BJ, Stark C, Tyers M. Osprey: a network visualization system. *Genome Biol* 2003;4:R22.
16. Errede B, Ammerer G. STE12, a protein involved in cell-type-specific transcription and signal transduction in yeast, is a part of protein-DNA complexes. *Genes Dev* 1989;3:1349–1361.
17. Sprague GFJr, Thorner JW. Pheromone response and signal transduction during the mating process of *Saccharomyces cerevisiae*. In: Jones EW, Pringle JR, Broach JR eds. The molecular and cellular biology of the yeast *Saccharomyces.* New York: Cold Spring Harbor Laboratory Press; 1992:657–744.
18. Oehlen LJ, McKinney JD, Cross FR. Ste12 and Mcm1 regulate cell cycle-dependent transcription of *FAR1*. *Mol Cell Biol* 1996;16:2830–2837.
19. Oehlen L, Cross FR. The mating factor response pathway regulates transcription of *TEC1*, a gene involved in pseudohyphal differentiation of *Saccharomyces cerevisiae*. *FEBS Lett* 1998;429:83–88.
20. Ren B, Robert F, Wyrick JJ, et al. Genome-wide location and function of DNA binding proteins. *Science* 2000;290:2306–2309.
21. Roberts CJ, Nelson B, Marton MJ, et al. Signaling and circuitry of multiple MAPK pathways revealed by a matrix of global gene expression profiles. *Science* 2000;287:873–880.
22. Tedford K, Kim S, Sa D, et al. Regulation of the mating pheromone and invasive growth responses in yeast by two MAP kinase substrates. *Curr Biol* 1997;7:228–238.
23. Elion EA. The Ste5p scaffold. *J Cell Sci* 2001;114:3967–3978.
24. Jenness DD, Burkholder AC, Hartwell LH. Binding of α-factor pheromone to yeast a cells: chemical and genetic evidence for an α-factor receptor. *Cell* 1983;35:521–529.
25. Herskowitz I, Rine J, Strathern J. Mating-type determination and mating-type interconversion in *Saccharomyces cerevisiae.* In: Jones RW, Pringle JR, Broach JR, eds. The molecular and cellular biology of the yeast *Saccharomyces.* New York: Cold Spring Harbor Laboratory Press, 1992:319–414.

26. Ramer SW, Davis RW. A dominant truncation allele identifies a gene, *STE20*, that encodes a putative protein kinase necessary for mating in *Saccharomyces cerevisiae. Proc Natl Acad Sci USA* 1993;90:452–456
27. Elion EA, Brill JA, Fink GR. FUS3 represses CLN1 and CLN2 and in concert with KSS1 promotes signal transduction. *Proc Natl Acad Sci USA* 1991;88:9392–9396.
28. Lee TI, Rinaldi NJ, Robert F, et al. Transcriptional regulatory networks in *Saccharomyces cerevisiae. Science* 2002;298:799–804.
29. Farkas I, Jeong H, Vicsek T, et al. The topology of the transcriptional regulatory network in the yeast, *Saccharomyces cerevisiae. Physica A* 2003; 318:601–612.
30. Ideker T, Thorsson V, Ranish JA, et al. Integrated genomic and proteomic analysis of a systematically perturbed metabolic network. *Science* 2001;292: 805–817.
31. Gunsalus KC, Ge H, Schetter AJ, et al. Predictive model of molecular machines involved in *Caenorhabditis elegans* early embryogenesis. *Nature* 2005;436:861–865.
32. Uetz P, Giot L, Gagney G, et al. A comprehensive analysis of protein-protein interactions in *Saccharomyces cerevisiae. Nature* 2000;403:623–627.
33. Ito T, Chiba T, Ozawa R, et al. A comprehensive two-hybrid analysis to explore the yeast protein interactome. *Proc Natl Acad Sci USA* 2001;98: 4569–4574.
34. Gavin AC, Bosche M, Krause R, et al. Functional organization of the yeast proteome by systematic analysis of protein complexes. *Nature* 2002;415: 141–147.
35. Ho Y, Gruhler A, Helibut A, et al. Systematic identification of protein complexes in *Saccharomyces cerevisiae* by mass spectrometry. *Nature* 2002;415: 180–183.
36. Huh WK, Falvo JV, Gerke LC, et al. Global analysis of protein localization in budding yeast. *Nature* 2003;425:686–691.

23

Systematic Determination of Biological Network Topology: Nonintegral Connectivity Method (NICM)

Kumar Selvarajoo and Masa Tsuchiya

Summary

The understanding of dynamic behavior of biological networks for a given stimulus is a huge challenge. The usage of static pathway maps under these circumstances is clearly insufficient to address key issues like regulatory behavior or oscillatory features of the network. We have devised a computational methodology called the *Nonintegral Connectivity Method* (*NICM*), which does not rely on kinetic parameters and is not based on differential equations. *NICM* systematically analyzes the dynamic phenotype of a set of network species, to a given external perturbation, and determines their local reaction connectivity (network motif). As a proof-of-concept, we show that *NICM* successfully detects several network motifs by analyzing the response profiles of constituent reactants constructed using mass action kinetics with pulse perturbation. To test the applicability on a biological system, we analyzed published phenotypes of yeast glycolytic metabolites. Our simulation suggests that a previously unattended step, the downstream reaction of fructose 1,6-bisphosphate (FBP), becomes the rate-limiting step in glucose pulse experiments. The discovery of such key regulatory steps could pave the way for systematic identification of novel targets for drug development.

Key Words: Biological network; connectivity method; linear response; glycolysis; *Saccharomyces cerevisiae*.

1. Introduction

Despite the numerous integrated studies (genomics, proteomics, and metabolomics) of cellular systems, we have not made significant progress in the understanding of even the simplest dynamic behavior of organisms, e.g., cell growth. One of the key reasons behind this slow progress is deeply rooted in the predominant application of steady-state static interaction maps of biological reactions to model the constantly evolving *in vivo* cellular system. To circumvent the problem, it becomes increasingly important to study the temporal topological changes of cellular networks under various kinds of perturbations, and to systematically identify the modifications in cellular interactions (1–5).

One particular area in biology that has received elaborate theoretical foundation is the conceptualization of biochemical network connectivity and regulation. However, as mentioned above, most, if not all, of these approaches use static interaction maps, steady-state conditions, and *in vitro* parameters to infer the properties of the network (6–10). Although these approaches have been instrumental to reveal biological properties, such as robustness, oscillatory behavior, and optimal growth rates, the understanding of cellular adaptation to environmental or physical changes (perturbations) has not been defined successfully. One of the main reasons for this situation has been the general lack of quantitative data of biological entities in a dynamic fashion. Recently, with the progress in the development of experimental methodologies, we are now presented with renewed opportunity to develop and analyze novel computational and mathematical approaches (4,15,16). For instance, we can now measure *in vivo* dynamics of numerous metabolites within cells at the same time, to a given external perturbation (17–19). This kind of measurement presents us with better insights into the molecular response of the cells in regulating any external influence.

In this report, we discuss the development of a novel computational methodology, *NICM*, which promises to systemically detect the dynamic connectivity of unknown or partially known biological/chemical network in a pulsed, perturbed system (22,23). *NICM* does not require *a priori* knowledge of the connectivity of reactants in a system. *NICM* requires only the reactants' dynamic concentration profile, and, utilizing formalized mathematical rules (called connectivity rules), it determines the relationship between all the reactants in the system. The *NICM* expressions constitute different combinations of depletion and formation wave terms to represent the propagation of pulse perturbation in various kinds of network motifs. As proof-of-concept, we used *NICM* to re-infer the original reaction network set up using pulse perturbation on a simple mass action kinetic system. In all cases, we were able to successfully reconstruct the original topology of the network motif. We then described the ability of *NICM* to analyze the dynamics of yeast glycolytic metabolites to an external perturbation.

2. Methods

2.1. NICM

The *NICM* is based on devising a basic principle for the response profile of species in a biological network that undergoes an upstream impulse perturbation. To illustrate the concept, let us give an impulse α to a first-order reaction constituting reactants A and B. The rate of change of A and B are:

$$\frac{dA}{dt} = -k_1 A \tag{1}$$

$$\frac{dB}{dt} = k_1 A, \tag{2}$$

where k_1 is the rate constant with initial conditions ($t = 0$) $A = A_o + \alpha$, $B = B_o$. The dynamic concentration profiles become:

$$A = \alpha e^{-k_1 t} + A_o \tag{3}$$

$$B = \alpha(1 - e^{-k_1 t}) + B_o \tag{4}$$

We notice from equations (3) and (4) that A possesses a depletion wave term, and B contains a formation wave term exclusively (by the observation of their exponential terms), and both reactants have the same rate constant.

Suppose we now include another reaction for B (with rate constant k_2). The rate of change of B becomes (A remains unchanged):

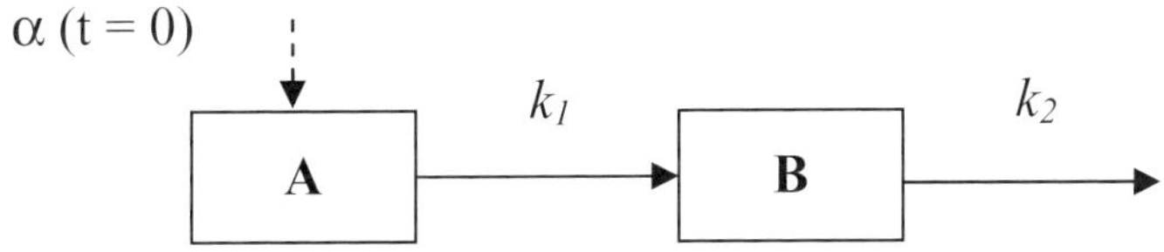

Insert 1

$$\frac{dB}{dt} = k_1 A - k_2 B. \tag{5}$$

Integrating equations (1) and (5) yields:

$$B = \frac{k_1}{k_2 - k_1}\left(e^{-k_1 t} - e^{-k_2 t}\right) + B_o. \tag{6}$$

We factorize (6) in respect to $e^{-k_1 t}$ if $k_2 > k_1$ (or $e^{-k_2 t}$ if $k_1 > k_2$) and obtain:

$$B = \frac{k_1}{k_2 - k_1}\left(1 - e^{-(k_2 - k_1)t}\right)e^{-k_1 t} + B_o. \tag{7}$$

We use equation (7) as a foundation to set up a generalized expression for the concentration profile of a reactant X, acting along a pathway, to constitute both the formation and depletion wave terms:

$$X = \alpha(1 - e^{-p_1 t})e^{-p_2 t} + X_o \tag{8}$$

perturbation coefficient (α) | formation coefficient (p_1) | depletion coefficient (p_2) | initial condition (X_o)

The perturbation coefficient, α, represents the strength of perturbation, the formation coefficient, p_1, represents the strength of formation wave, and the depletion coefficient, p_2, represents the strength of depletion wave. We named this method, for constructing mathematical expressions for a reactant's dynamic concentration profile based on detecting the formation and depletion wave terms, the *NICM*. This is distinct from the conventional way of integrating the velocity expressions to determine a reactant's concentration profile, as in equations (1) and (2).

Equation (8) can also be simplified into:

$$X = \alpha(t)e^{-p_2 t}, \tag{9}$$

where $\alpha(t) = \alpha(1 - e^{-p_1 t})$. If, however, X receives an instanteous α_o pulse perturbation, then $\alpha(t) = \alpha_o$ and equation (9) becomes:

$$X = \alpha_o e^{-p_2 t}. \tag{10}$$

Using the general form of equation (8), we will describe how we develop concentration profiles for connected systems (*NICM*). For example, let us assume A (zero initial concentration) has a depletion wave term for an α_o pulse perturbation as depicted in insert 2.

Using equation (10) we obtain:

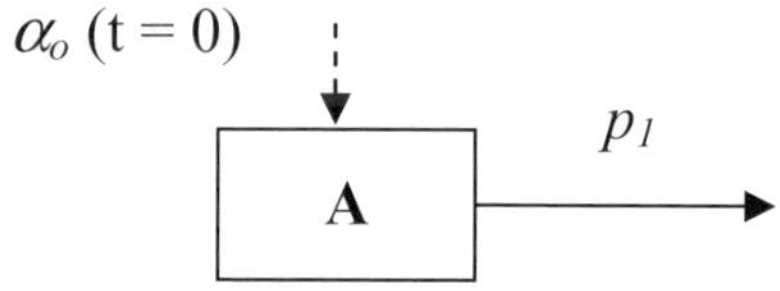

Insert 2

$$A = \alpha_o e^{-p_1 t}. \tag{11}$$

For the case of B, as in insert 3, using equation (8) we obtain:

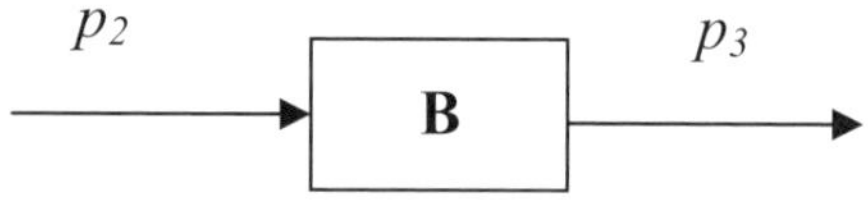

Insert 3

$$B = \beta_o(1 - e^{-p_2 t})e^{-p_3 t} \tag{12}$$

If A and B belong to a connected system, using conservation of mass we obtain $\alpha_o = \beta_o$ and $p_2 = p_1$. Therefore, equations (11) and (12) now become

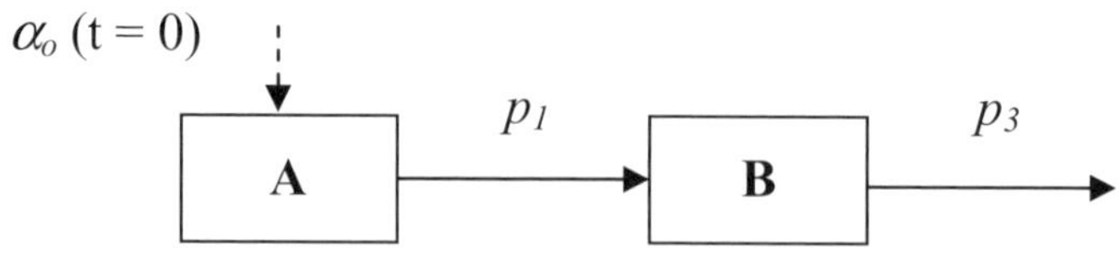

Insert 4

$$A = \alpha_o e^{-p_1 t} \tag{13}$$

$$B = \alpha_o(1 - e^{-p_1 t})e^{-p_3 t} \tag{14}$$

In this way, we see that the depletion coefficient of A (p_1) is equal to the formation coefficient of B.

2.2. Application of *NICM*

NICM requires time-series concentration profile of reactants in a system to a given stimulus. Using this information only, we identify whether a reactant constitutes a formation wave term, a depletion wave term (or a combination of both). By comparing the number of formation and/or depletion wave terms present and using reaction connectivity rules (*see* section 2.3), we construct a reaction network between the reactants. As proof-of-concept, we compare the performance of *NICM* with commonly used mass action kinetic analysis on various types of network motifs. We emphasize that when comparing mass action kinetics with pulse perturbation, our method is only an approximation to the true solution. However, by analyzing the dynamics of yeast glycolysis *in vivo*, which is a highly nonlinear pathway, we show that our method is not restricted to linear kinetics alone (*see* section 3). In short, our method is instrumental to the identification of local connectivity of each reactant and the connection of a series of reactants into a pathway (construction of network) (Figure 1).

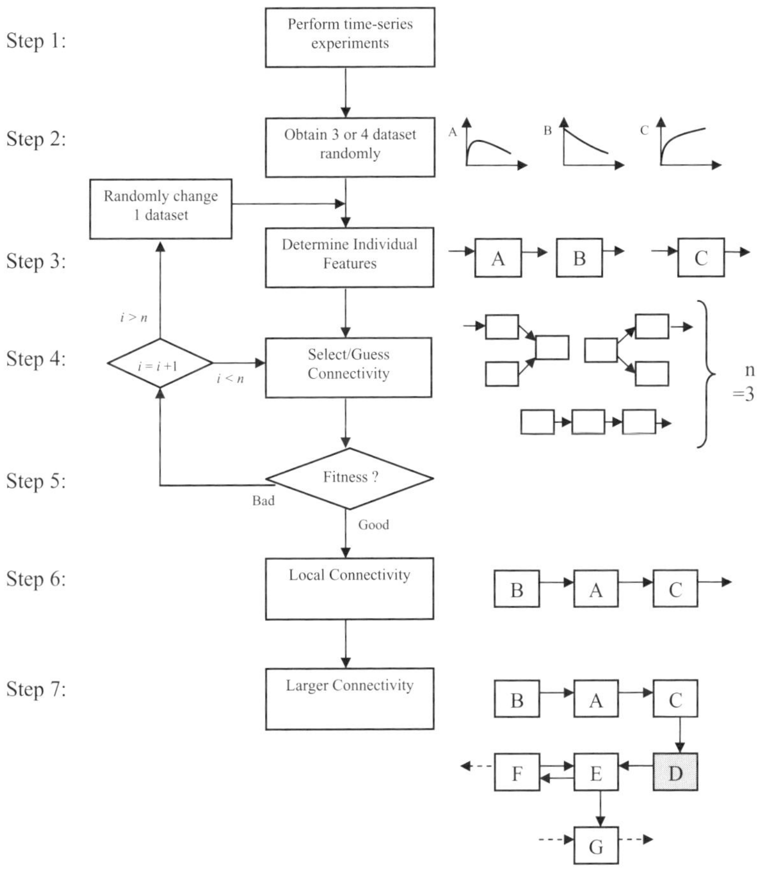

Figure 1. Flowchart for the identification of local network connectivity or motif using the *NICM* methodology. See text for details. *n* is the number of possible connectivity motifs.

To illustrate the *NICM* procedure, let us suppose we have a stimulated system and are able to obtain various time variant concentration profile of reactants whose connectivity is unknown (Figure 1, Step 1). We first select, randomly, 3 or 4 reactants' response profile at a time (Figure 1, Step 2). For these reactants, we determine for each reactant whether they possess formation wave term, depletion wave term, or a combination of both (Figure 1, Step 3). Next, using this information, we will connect the chosen reactants to form possible network motifs. Using each motif, we will derive the *NICM* mathematical expressions (e.g., linear-chain motif equations (15–17), diverging motif equations (22–25), etc.; *see* section 2.3). We then perform the fitting process (Figure 1, Step 5). The fitness evaluation is done using a genetic algorithm (26,27) and least square fit (28) (*see* Appendix 1). We perform fitness for each reactant profile and sum them together (for the selected 3 or 4 reactants). If the combined fitness is below a set tolerance as required (*see* Appendix 1), the selected motif (local connectivity) is accepted (Step 6), otherwise, a different motif is chosen (Step 4). If no acceptable result is obtained (fitness above tolerance limit) for all possible motifs, we move to Step 2 and exchange one dataset randomly. We perform steps 3–5 repeatedly until we obtain a good fitness result and select the motif (Step 6). To consider larger networks (e.g., 20 reactants), we will first search for another reactant that connects to the reference motif obtained from Step 6. We, again, evaluate the overall optimal fitness value including this new reactant (e.g., reactant D; Figure 1). We then anchor this new reactant and search for subsequent network motif where it "attaches." In other words, we build additional motifs attached to the reference motif (motif–motif connection) and form a bigger module. By continuing this process of connecting modules to modules, we can eventually obtain the entire connectivity (Figure 1, Step 7). In this manner, we are able to reduce the number of combinatorial complexity of network connectivity, and the entire process of Figure 1 can be implemented using an automated system, such as genetic programming (24,25).

To demonstrate *NICM* (Steps 1 to 6 only), we created a theoretical motif that contain 3 reactants and produced time-series plots of their concentration profile, using mass action kinetics with pulse perturbation and predefined rate constants (Figure 2). By simply using this information and applying *NICM*, we successfully reconstructed the original network (linear-chain motif with reversible step, *see* section 2.3). However, when we deliberately used an incorrect connectivity (linear-chain motif without reversible step) to fit the data, the accuracy of the result is very poor (Figure 2).

We now show the development of *NICM* connectivity rules for each reactant in various types of reaction motifs (29), such as linear pathways, diverging pathways, reversible steps, and feedback/forward network using *NICM*. Note that in our definition of motif, to reduce the total number of possible motifs, we include reversible steps as part of the parent motif (*see* sections 2.3.1 and 2.3.4). We perform proof-of-principle simulations and compare the results with mass action kinetic models using pulse perturbation for each network motif.

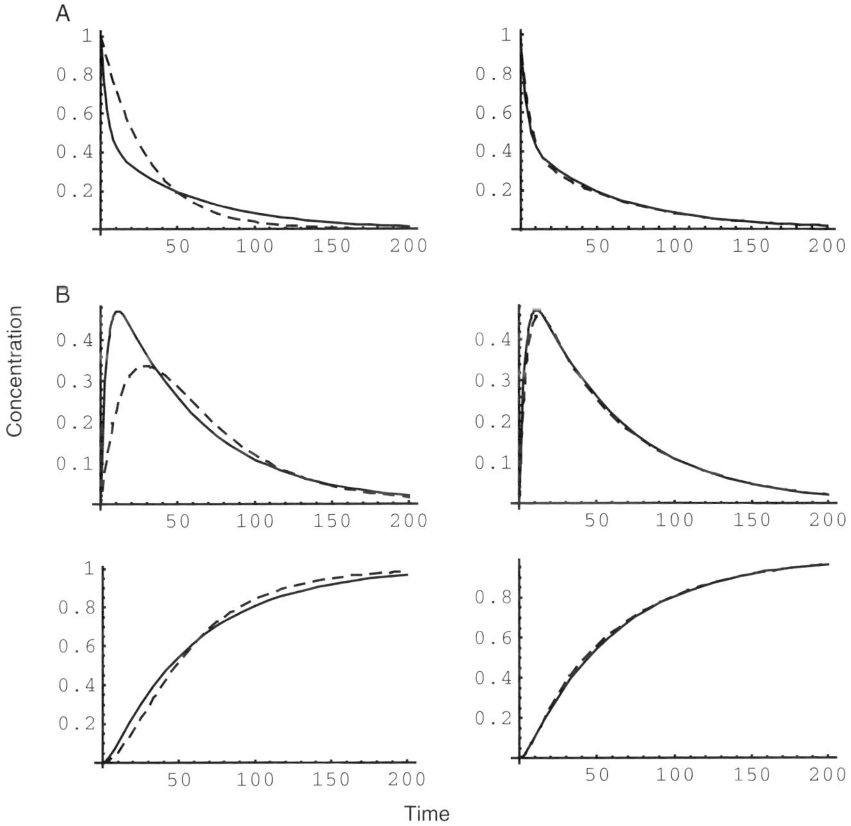

Figure 2. Generation of time-series plots for a linear-chain motif with reversible step using mass action kinetics with pulse perturbation (*see* insert 14). Unit pulse ($\alpha = 1$) is applied to A with rate constants $k_1 = 0.15$, $k_2 = 0.03$, and $k_3 = 0.1$. (A) The simulation results of *NICM* using linear-chain motif (*see* insert 5) ($p_1 = 0.033$ and $p_2 = 0.02$, obtained using GA, equations [15–17]). (B) The simulation results of *NICM* using linear-chain motif with reversible step (see insert 14) ($p_1 = 0.115$, $p_2 = 0.0165$, $p_3 = 0.074$, and $\beta = 0.43$, equations [30–32]). The maximum error for (A) is 42%, and for (B) is 4% (*see* Appendix 1 for the definition of error). The x axis represents time and y axis represents concentration, both in arbitrary units. For all subsequent plots we adopt the same representation of the axes (x, time; y, concentration of reactant). Solid lines indicate the mass action kinetics with pulse perturbation, and dotted lines indicate the result using *NICM*.

2.3. NICM Connectivity Rules

2.3.1. Linear-Chain Motif

Let us look at simple two-reaction linear pathway. Using *NICM* (with the aid of equations (13) and (14)) we can represent, for α_o pulse perturbation, the time-series concentration profiles of A, B, and C as:

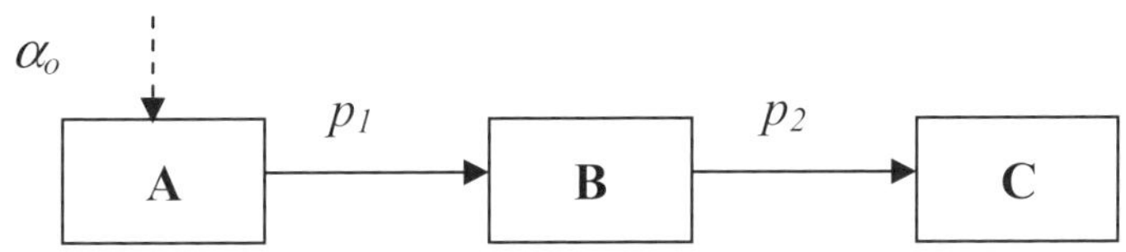

Insert 5

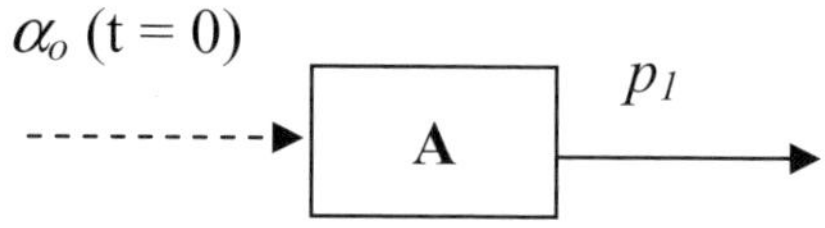

Insert 6

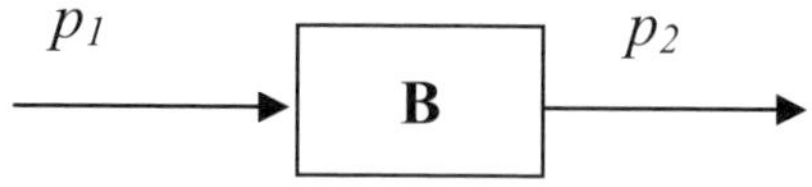

Insert 7

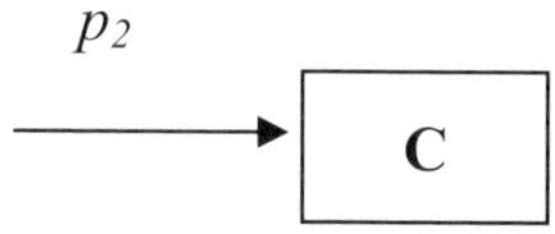

Insert 8

$$A = \alpha_o e^{-p_1 t} \tag{15}$$

$$B = \alpha(1 - e^{-p_1 t})e^{-p_2 t} \tag{16}$$

$$C = \alpha(1 - e^{-p_1 t})(1 - e^{-p_2 t}) \tag{17}$$

Because B and C are connected, the depletion coefficient of B is equal to the formation coefficient of C.

Although, equations (15) to (17) were easily derived using *NICM*, they do not represent the true solution (when compared to mass action kinetics). Only up to the simple case of equation (7) does our basic rule yield an exact solution; beyond this, the solution approximates the true solution (linear response solution). Nevertheless, the approximation generally yields marginal difference (the error is usually less than 10%; *see* Appendix 1) when compared to the true solution for a given set of rate constants. This discrepancy is not significant, as experimental variations are usually much larger (30). However, under certain circumstances, for example, when $k_1 = 2k_2$ or $k_2 = 2k_1$, the *NICM* expressions become equivalent to the solution of mass action kinetics with impulse perturbation. For example, if we take $\alpha = 1.0$, $k_1 = 0.1$, and $k_2 = 0.2$, we obtain p_1 and $p_2 = 0.1$ (using GA (26,27)) with no error between the two methods; i.e., exact solution (Figure 3).

2.3.2. *Merging Motif*

Let us look at a situation where two reactants react independently within a pathway.

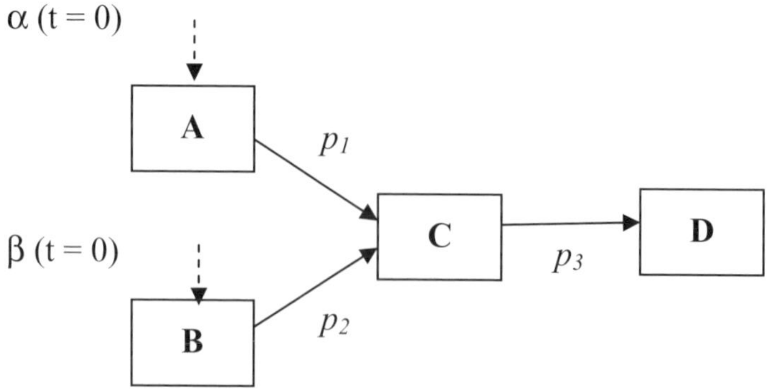

Insert 9

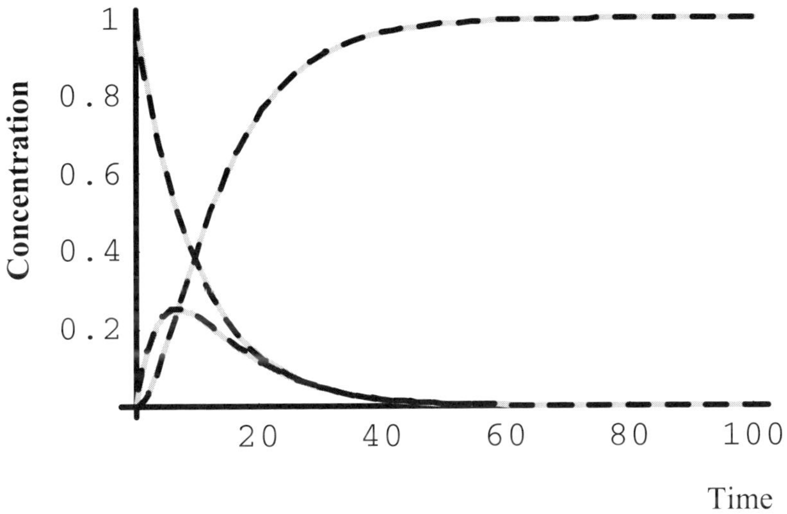

Figure 3. Comparison of mass action kinetics with pulse perturbation and *NICM* for linear-chain motif (insert 5). The time variant plots of *A*, *B*, and *C* were generated using mass action kinetics with unit pulse perturbation and rate constants $k_1 = 0.1$ and $k_2 = 0.2$. We then used the linear-chain motif expressions of *NICM* to fit the data. The fitting process involves choosing the relevant coefficients in the *NICM* expressions (equations [15–17]) that gives the lowest fitness value (*see* Appendix 1). We obtained $p_1 = 0.1$ and $p_2 = 0.1$. Solid lines indicate mass action kinetics solution, and dotted lines indicate the result obtained using *NICM*. Note that we cannot distinguish the two corresponding lines, as they overlap each other (in this case, *NICM* yields exact solution to mass action solution).

We notice that *A* and *B* have depletion terms only, *C* posseses both formation and depletion wave terms, and *D* has only depletion wave term. Now, individually perturbing *A* by α and *B* by β and applying *NICM*, we can represent each reactant's concentration as:

$$A = \alpha e^{-p_1 t} + A_o \quad (18)$$

$$B = \beta e^{-p_2 t} + B_o \quad (19)$$

$$C = [\alpha(1 - e^{-p_1 t}) + \beta(1 - e^{-p_2 t})]e^{-p_3 t} + C_o \quad (20)$$

$$D = [\alpha(1 - e^{-p_1 t}) + \beta(1 - e^{-p_2 t})](1 - e^{-p_3 t}) + D_o \quad (21)$$

where p_1, p_2, and p_3 are the depletion coefficients for reactions *A* to *C*, *B* to *C*, and *C* to *D*, respectively. A_o, B_o, C_o and D_o represent the initial equilibrium concentration. Note, we add the formation terms for *C* and *D*, as the two formative reactions are independent to each other.

Putting α and $\beta = 1$ and $A_o = B_o = C_o = D_o = 0$ in equations (18–21), we are only required to determine p_3 because, according to equations (18–19), $p_1 = k_1$ and $p_2 = k_2$. In Figure 4, we compared the *NICM* performance against mass action solutions and produced closely matched results.

2.3.3. Diverging Motif

Consider now reactant *B* producing *C* and *D* (assume *D* has an initial equilibrium concentration D_o), perturbing *A* by α and applying *NICM*:

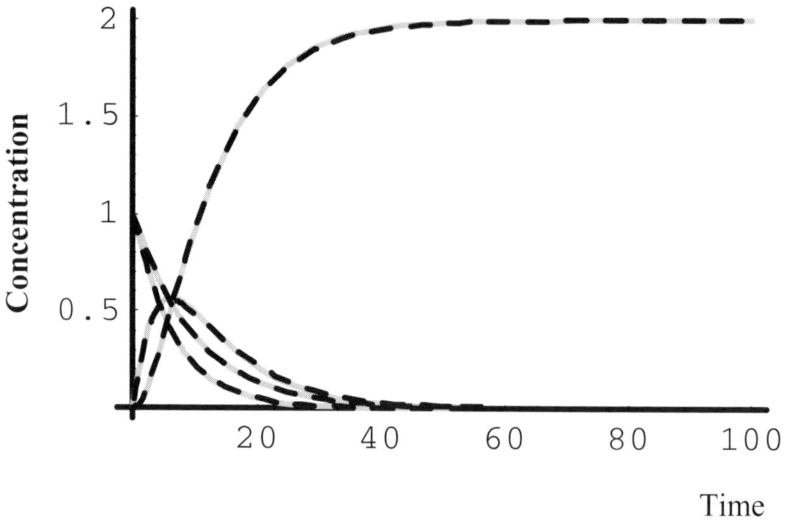

Figure 4. Comparison of mass action kinetics with pulse perturbation and *NICM* for a merging motif (insert 9). The time-variant plots of *A*, *B*, *C*, and *D* were generated using mass action kinetics with unit pulse perturbation and rate constants $k_1 = 0.15$, $k_2 = 0.1$, and $k_3 = 0.2$. Merging motif expressions of *NICM* (equations [18–20]) provided a good fit (maximum error is 3%) to the data when $p_1 = 0.15$, $p_2 = 0.1$, and $p_3 = 0.107$. Solid lines indicate mass action kinetics solution, and dotted lines indicate the result obtained using *NICM*.

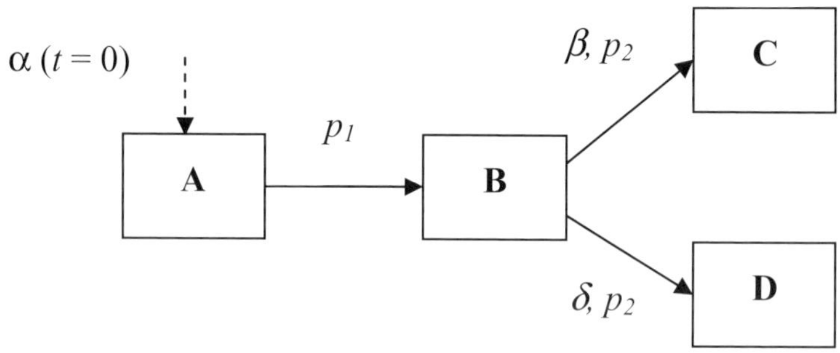

Insert 10

$$A = \alpha e^{-p_1 t} + A_o \quad (22)$$

$$B = \alpha(1 - e^{-p_1 t})e^{-p_2 t} + B_o \quad (23)$$

$$C = \beta(1 - e^{-p_1 t})(1 - e^{-p_2 t}) + C_o \quad (24)$$

$$D = \delta(1 - e^{-p_1 t})(1 - e^{-p_2 t}) + D_o \quad (25)$$

The perturbation coefficient, α, is shared by the reaction *B* to *C* and *B* to *D*, i.e., the actual perturbation is split by the diverging pathway. Therefore, we have introduced new perturbation coefficients for *C* and *D*, namely, β and δ where in a mass conserved system, $(\beta + \delta)$ must be equal to α. The proportion of α and β will be determined by the actual rate of reaction *B* to *C* and *B* to *D*. For example, if the rate for *B* to *C* is greater than *B* to *D*, then α must be greater than β. The formation coefficient remains the same between reaction *B* to *C* and *B* to *D* to satisfy the law of mass conservation.

Letting $\alpha = 1$ and $A_o = B_o = C_o = D_o = 0$ in equations (22–25), the maximum error between the two methods for all reactants is approxi-

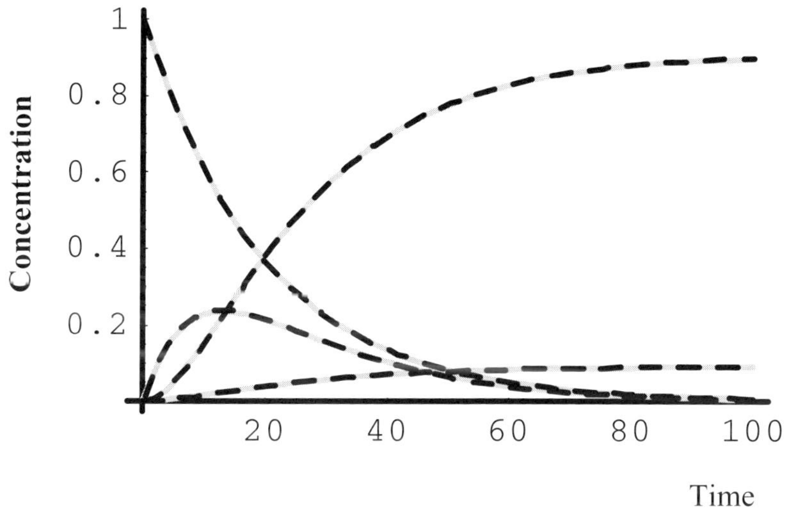

Figure 5. Comparison of mass action kinetics with pulse perturbation and *NICM* for a diverging motif (insert 10). The time variant plots of *A*, *B*, *C*, and *D* were generated using mass action kinetics with unit pulse perturbation and rate constants $k_1 = 0.05$, $k_2 = 0.01$, and $k_3 = 0.1$. The diverging motif expressions of *NICM* (equations [22–25]) provided a good fit (2% error) when $p_1 = 0.05$, $p_2 = 0.05$, $\alpha = 0.09$, and $\beta = 0.91$. Solid lines indicate mass action kinetics solution, and dotted lines indicate the result obtained using *NICM*.

mately 2% (Figure 5). In addition, by looking at just the perturbation coefficients, β and δ, we could decide which branching pathway is dominant (from Figure 5, pathway *B* to *D* is approximately 10 times more dominant than *B* to *C*).

2.3.4. Linear-Chain Motif with Reversible Step

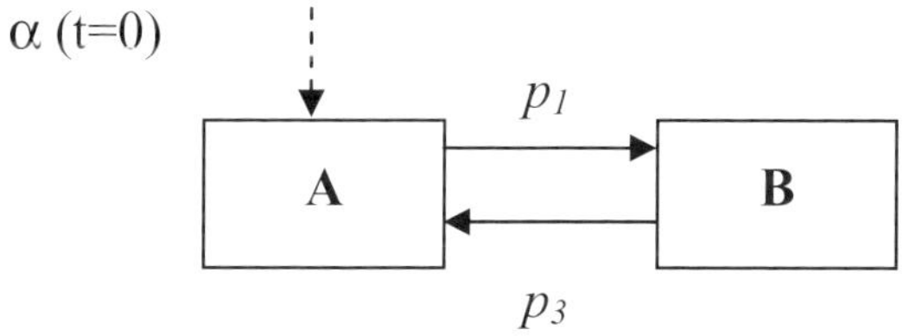

Insert 11

The development of equations through *NICM* for reversible reactions is generally not straightforward. It is best to describe the reactants' concentration profiles by first decomposing the reaction steps into a sequence of events that represents the propagation of the perturbation α:

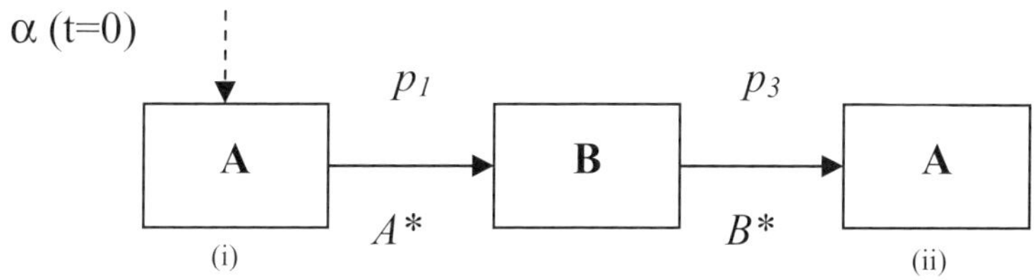

Insert 12

Consider first reactant A only. We split A into two parts, marked (i) and (ii). For each part, we could construct the following relations:

$$A(i) = \alpha e^{-p_1 t} \quad (26)$$

$$A(ii) = B^*(1 - e^{-p_3 t}) = \beta(1 - e^{-p_1 t})(1 - e^{-p_3 t}) \quad (27)$$

$$A = A(i) + A(ii) = \alpha e^{-p_1 t} + \beta(1 - e^{-p_1 t})(1 - e^{-p_3 t}) \quad (28)$$

Some portion of B (B^*) is converted back to A, $B^* = \beta(1 - e^{-p_1 t})$ so as to obey the law of mass conservation ($A^* + B^* = \alpha(1 - e^{-p_1 t})$), which is equivalent to the formation term of B in the absence of the reversible step.

For reactant B:

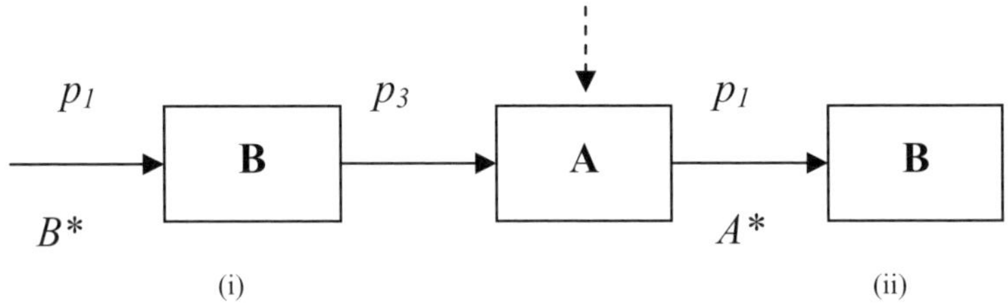

Insert 13

$$B(i) = B^* e^{-p_3 t} = \beta(1 - e^{-p_1 t})e^{-p_3 t} \quad (29)$$

$$B(ii) = A^* = (\alpha - \beta)(1 - e^{-p_1 t}) \quad (30)$$

$$B = B(i) + B(ii) = \beta(1 - e^{-p_1 t})e^{-p_3 t} + (\alpha - \beta)(1 - e^{-p_1 t}) \quad (31)$$

Now consider the time event propagation of pulse perturbation for A and B:

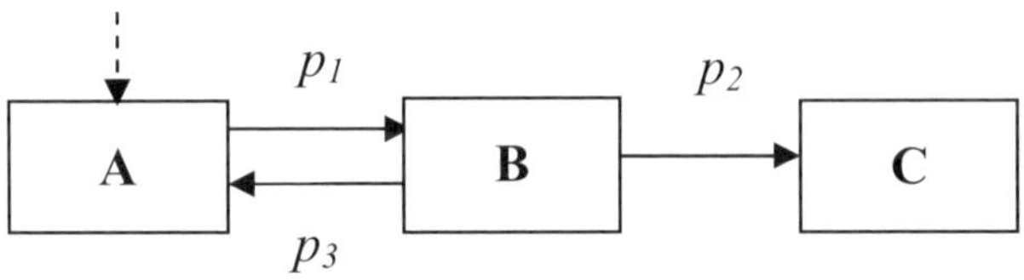

Insert 14

For A:

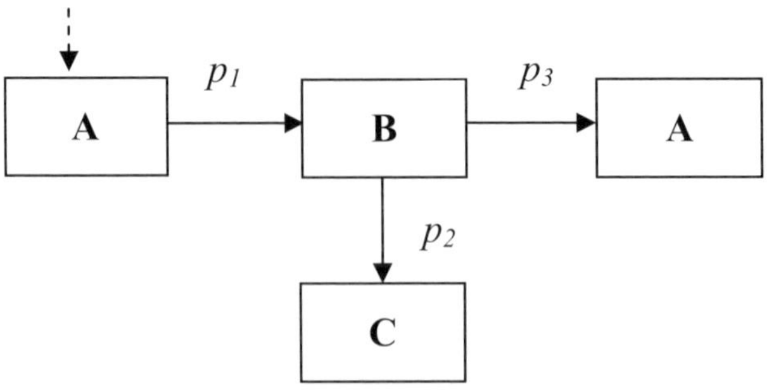

Insert 15

For B:

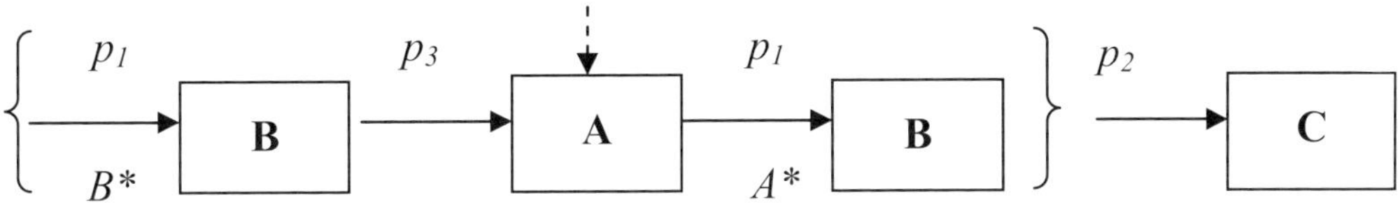

Insert 16

We can show that:

$$A = \alpha e^{-p_1 t} + \beta(1 - e^{-p_1 t})(1 - e^{-p_3 t})e^{-p_2 t} \tag{32}$$

$$B = \lfloor \beta(1 - e^{-p_1 t})e^{-p_3 t} + (\alpha - \beta)(1 - e^{-p_1 t}) \rfloor e^{-p_2 t} \tag{33}$$

$$C = \alpha(1 - e^{-p_1 t})(1 - e^{-p_2 t}) \tag{34}$$

To demonstrate the applicability of equations (32–34) for reversible reactions, we built a mass action kinetic model with pulse perturbation and rate constants $k_1 = 0.1$, $k_2 = 0.001$, $k_3 = 0.1$, which is a bottleneck situation; i.e., the flux to C is very low as compared to the flux to B. Figure 6 shows the comparison of the two simulation results. The maximum error between the two results for all reactants is approximately 9%.

2.3.4.1. Other Considerations: Our method can also be extended to include feedback/feedforward motifs and oscillatory behavior (when depletion and formation coefficient, p, becomes a complex number).

2.3.5. Feedback Motif

If the concentration of C increases the rate of reaction of A to B, we have a positive-feedback loop:

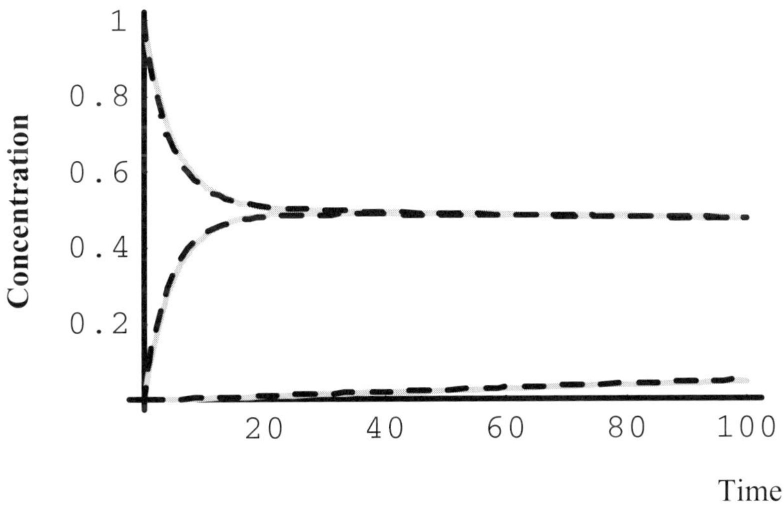

Figure 6. Comparison of mass action kinetics with pulse perturbation and *NICM* for a linear-chain motif with reversible step (insert 14). The time-variant plots of A, B, and C were generated using mass action kinetics with unit pulse perturbation and rate constants $k_1 = 0.1$, $k_2 = 0.001$, and $k_3 = 0.1$. The expressions of *NICM* (equations [32–34]) provided total error of approximately 9% when $p_1 = 0.115$, $p_2 = 0.0005$, $p_3 = 0.125$, $a = 1.0$, and $b = 0.5$. Solid lines indicate mass action kinetics solution, and dotted lines indicate the result obtained using *NICM*.

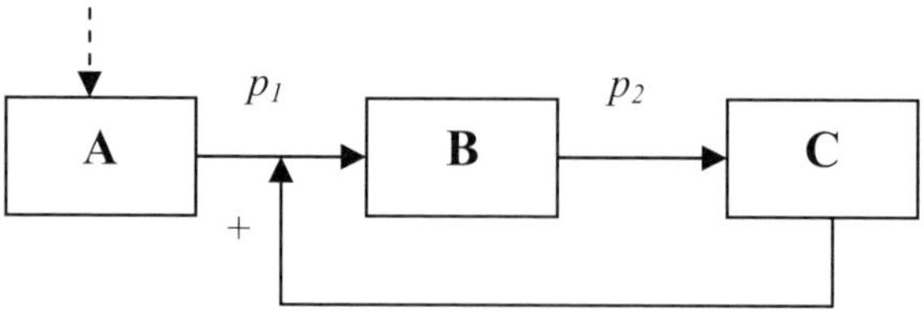

Insert 17

We could represent this scenario as following:

$$A = A_1 e^{-[p_1 * (C - C_o)]t} + A_o \tag{35}$$

$$B = A_1(1 - e^{-[p_1 * (C - C_o)]t}) e^{-p_2 t} + B_o \tag{36}$$

$$C = A_1(1 - e^{-[p_1 * (C - C_o)]t})(1 - e^{-p_2 t}) + C_o \tag{37}$$

The depletion coefficient p_1 of A (e.g., in equation (11)), in the absence of any feedback mechanism, has been replaced with $p_1*(C - C_o)$, that is, replacing a constant coefficient with a linear function coefficient (an assumed relation to demonstrate the feedback activation). We use this linear relation to show that as C increases, it increasingly activates the reaction A to B. However, in the situation that $0 < p_1 < 1$ (usual case) and $(C - C_o) < 1$, we need to multiply the term $p_1*(C - C_o)$ by a constant factor to make it >1. Of course, feedback regulation could be a complex process and may involve a nonlinear type of regulation. Here, we have stuck to a simple case to demonstrate the capability of *NICM* to consider feedback properties.

2.3.6. Feedforward Motif

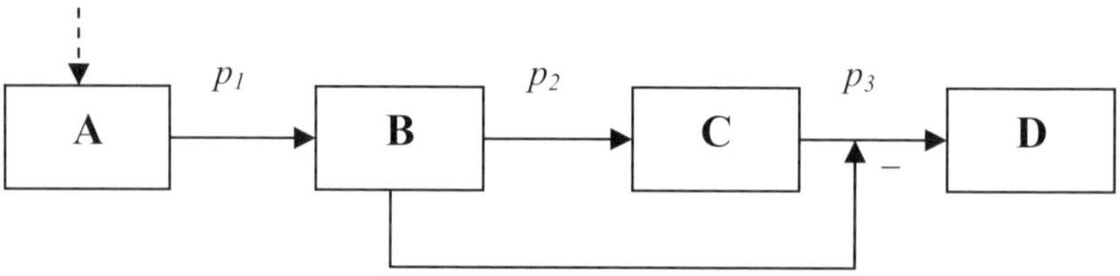

Insert 18

In a negative-feedback system, increasing concentration of B negatively regulates the conversion of C to D. For such a system, we represent the various reactants' concentration profile as:

$$A = A_1 e^{-p_1 t} + A_o \tag{38}$$

$$B = A_1(1 - e^{-p_1 t}) e^{-p_2 t} + B_o \tag{39}$$

$$C = A_1\left(1 - e^{-p_1 t}\right)\left(1 - e^{-p_2 t}\right) e^{-\frac{p_3}{(B - B_o)} t} + C_o \tag{40}$$

$$D = A_1\left(1 - e^{-p_1 t}\right)\left(1 - e^{-p_2 t}\right) 1 - e^{-\frac{p_3}{(B - B_o)} t} + D_o \tag{41}$$

In the event that $(B - B_o) = 0$, we will set $C = C_o$.

We next show the applicability of *NICM* to better understand the dynamic features of yeast glycolytic metabolism under glucose pulse perturbation.

3. Analysis of Yeast Glycolytic Network

With the recent advent of sophisticated time-course intracellular phenotype measuring tools, it now becomes possible to observe the global dynamic phenotype of biological networks *in vivo* (17–19). The generation of these data is indispensable, as it allows one to analyze algorithm over a range of period, rather than the traditional steady-state analysis.

We attempted to test our methodology on a section of the glycolytic pathway of *Saccharomyces cerevisiae*. We chose glycolysis because it is well studied and is known to contain highly nonlinear features, such as feedback and feedforward loops (Figure A2.1, Appendix 2). This provides us with the opportunity to test the applicability of *NICM* on biological systems.

We obtained the experimental time-course response of the glycolytic phenotype of *S. cerevisiae* to a glucose pulse from available literature (17). Although the dataset is not sufficient for us to reconstruct the entire glycolytic pathway connectivity using all of our theoretical motifs (*see* section 2.3), we intend to demonstrate the applicability of our technique to decipher some key properties of the biological regulation that occurs during the experimental perturbation. We wanted to know whether our technique is able to capture new knowledge of *S. cerevisiae* glycolysis that had been previously inferred.

Before we use our technique on the dataset, we ensured the reproducibility of the dataset by searching for other published sources that performed similar experiments. We found a paper from a different group that reported similar phenotype for a similar glucose pulse experiment (19).

Figure A2.2 (Appendix 2) shows some glycolytic phenotype of *S. cerevisiae*. By simple visual inspection or comprehending the article, one would not be able to fully understand the result of the study. We performed our methodology (Figure 1) on this dataset but in this paper; we restrict our analysis to the first 3 metabolites, glucose-6-phosphate (G6P), fructose-6-phosphate (F6P), and fructose 1,6-bisphosphate (FBP) only. As we do not have the information of other intercepting pathways to these metabolites, e.g., the pentose phosphate pathways (PPP), we started our analysis by assuming that our local network is of the linear chain motif (*see* section 2.3) that is, → G6P → F6P → FBP → etc., equation (12). We next performed the fitting process (Figure A2.3, Appendix 2) and obtained the following expressions with the best fitness value

$$G6P = 4.80(1 - e^{-0.230t})e^{-0.016t} + 0.90$$

$$F6P = 2.01(1 - e^{-0.130t})e^{-0.037t} + 0.17$$

$$FBP = 2.68(1 - e^{-0.258t})e^{-0.0003t} + 0.11$$

Despite restricting our analysis to linear motifs, we obtained a very interesting result. We notice for G6P, the depletion coefficient is much smaller than the formation coefficient of F6P (equation (8)). If we assume F6P is solely produced by G6P (as we have done through the application of linear-chain motif), then the formation coefficient value of F6P should be similar to that of the depletion coefficient of G6P (p_2 of equation (8)). This could imply that the faster formation of F6P could be due to other intercepting reactions that we have not factored into our initial motif assumption. This result is justified by the fact that F6P is also known to be produced by PPP, which we deliberately excluded in our motif selection.

Next, looking at FBP, the depletion coefficient is very small or insignificant. This clearly indicates that the metabolite of interest has reached a saturated value, perhaps due to downstream rate-limiting mechanism. This observation is strengthened by the lower than expected yield of downstream metabolites like glycerol and glyceraldehyde-3-phosphate (G3P) in (17).

To test our hypothesis, we localized the FBP network, as represented in Figure 7A, and performed a mass action kinetics analysis using pulse perturbation around this network. Our aim is to determine the various k values in the model, by making a close fit to the experimental data, and then comparing them to infer the presence or absence of any rate-limiting phenomenon that we suspect. We noticed that the value of k_2 is much larger than that of k_4 and k_5 ($k_2 = 0.05$, $k_4 = 0.0004$, $k_5 = 0.0002$) (Figure 7B), thus indicating a bottleneck scenario for FBP (Figure A2.2, Appendix 2). This positively implies that downstream of FBP, the enzyme aldolase activity may have reached a saturation point, thereby becoming the rate-limiting enzyme for glucose pulse experiments. If this result is proven to be true with subsequent wet experiments, there will be the hope to use *NICM* and yeast for industrial benefit, such as increased production of ethanol for beer brewery, by carefully targeting the enzyme aldolase.

4. Conclusion

We have developed a novel quantitative method, the *NICM*, for the dynamic analysis of biological pathways and networks, without the need to form differential equations. Starting with only the time-series profile of reactants in a system, the decomposition of each reactant into formation, depletion term waves or a combination of both, help us to connect a set of reactants into the frequently observed network motifs. When applied on published experimental work on yeast dynamics, we predict that yeast glycolysis under glucose pulse may constitute a novel rate-limiting step, aft of FBP. Furthermore, in a recent paper, the application of pulse perturbation on mass action kinetics on Toll-like receptor signaling in macrophages, predicts that the MyD88-independent pathway consists of novel intermediates (31). This strongly indicates that *NICM* is not just restricted to metabolic pathway analysis but also can be used to model signaling networks, as we demonstrate that *NICM* successfully identifies network motifs developed using mass action kinetics with pulse perturbation.

Acknowledgments: We thank Prof. Masaru Tomita of Keio University and Dr. Patrick Tan of Genome Institute of Singapore for valuable discussions and the Bioinformatics Institute, Singapore, for funding the work.

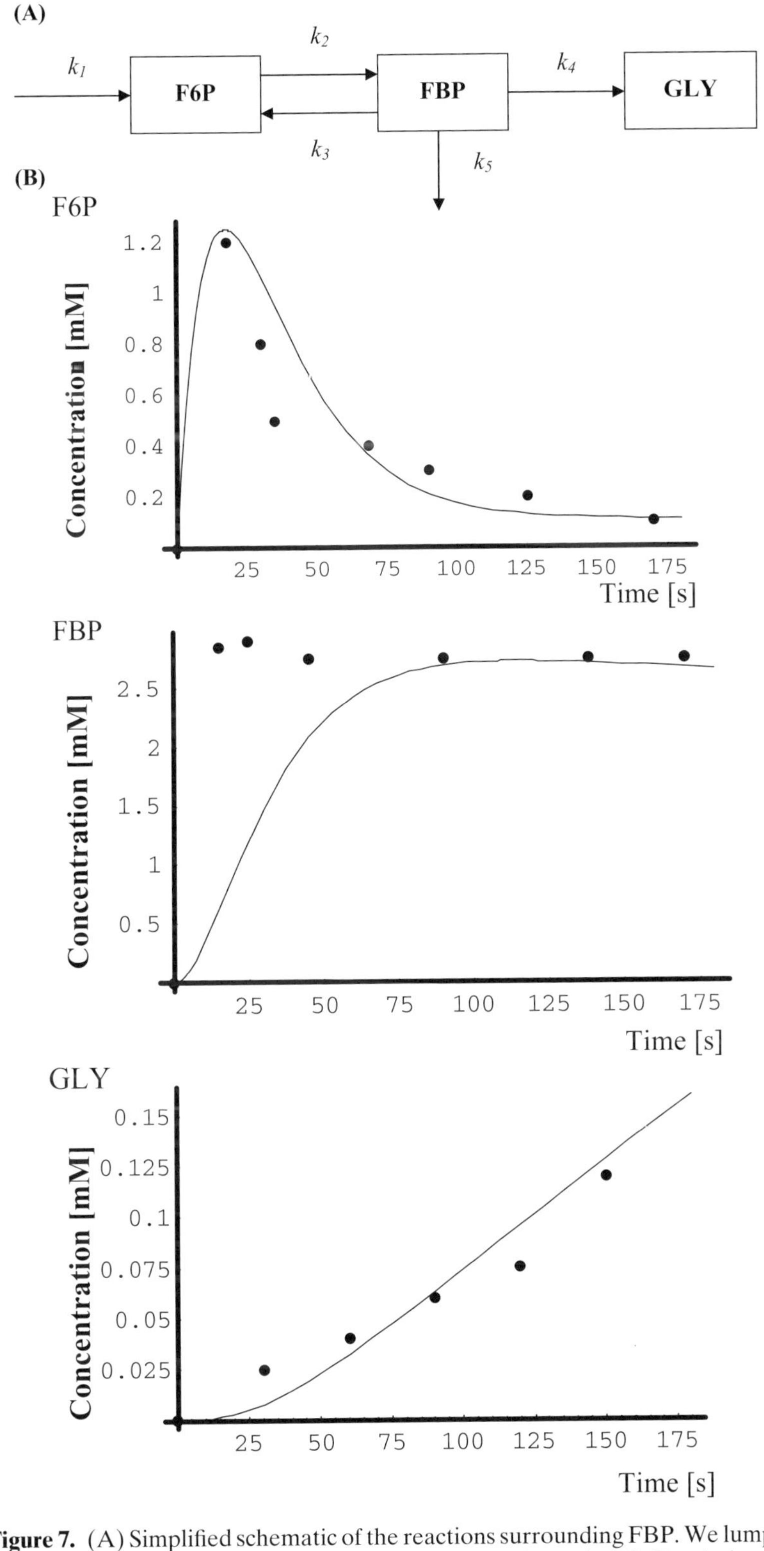

Figure 7. (A) Simplified schematic of the reactions surrounding FBP. We lumped the reactions of dihydroxyacetone phosphate and glycerol. (B) Simulations of F6P, FBP, and glycerol using pulse perturbation ($\alpha = 3$) on mass action kinetics with $k_1 = 0.64$, $k_2 = 0.05$, $k_3 = 0.002$, $k_4 = 0.0004$, and $k_5 = 0.0002$. Solid line represents solution of mass action kinetics, and dots represent experimental results adapted from (17). The x axis represents time (s) and the y axis represents the concentration (mM) of metabolites. Although the steady-state level for FBP is accurately simulated, the initial transient data (up to approximately 50) does not match. We believe this poor transient prediction could be caused by missing links (interactions) in our present understanding of FBP connectivity, as in (A).

References

1. Nilsson R, Bajic VB, Suzuki H, et al. Transcriptional network dynamics in macrophage activation. *Genomics* 2006;88(2):133–142.
2. Proulx SR, Promislow DE, Phillips PC. Network thinking in ecology and evolution. *Trends Ecol Evol* 2005;20(6):345–353.
3. Ihmels J, Levy R, Barkai N. Principles of transcriptional control in the metabolic network of *Saccharomyces cerevisiae. Nat Biotechnol* 2003;22:86–92.
4. Vance W, Arkin A, Ross J. Determination of causal connectivities of species in reaction networks. *PNAS* 2002;99(9):5816–5821.
5. Torralba AS, Yu K, Shen P, et al. Experimental test of a method for determining causal connectivities of species in reactions. *PNAS* 2003;100(4):1494–1498.
6. Cornish-Bowden A, Wharton CW. Enzyme Kinetics. Oxford: IRL Press; 1988.
7. Fell DA. Understanding the control of metabolism. Portland Press, 1996.
8. Stephanopoulos GN, Aristidou AA, Nielsen J. Metabolic Engineering. San Diego: Academic Press, 1998.
9. Edwards JS, Ibarra RU, Palsson BO. *In silico* predictions of Escherichia coli metabolic capabilities are consistent with experimental data. *Nat Biotechnol* 2001:19:125–130.
10. Voit EO. Computational Analysis of Biochemical Systems: A Practical Guide for Biochemists and Molecular Biologists. Cambridge University Press, 2000.
11. Heijnen JJ. Approximative kinetic formats used in metabolic network modeling. *Biotechnol Bioeng* 2005;91(5):534–545.
12. Palsson B. The challenges of *in silico* biology. *Nat Biotechnol* 2000;18: 1147–1150.
13. Bailey JE. Complex biology with no parameters. *Nat Biotechnol* 2001;19: 503–504.
14. Bailey JE. Mathematical modeling and analysis in biochemical engineering: Past accomplishments and future opportunities. *Biotechnol Prog* 1998; 14:8–20.
15. Tsuchiya M, Ross J. Application of Genetic Algorithm to Chemical Kinetics: Systematic Determination of Reaction Mechanism and Rate Coefficients for a Complex Reaction Network. *J Phys Chem A* 2001;105(16):4052–4058.
16. Arkin A, Shen P, Ross JA. A Test Case of Correlation Metric Construction of a Reaction Pathway from Measurements. *Science* 1997;277(5330): 1275–1279.
17. Theobald U, Mailinger W, Baltes M, et al. *In vivo* analysis of metabolic dynamics in *Saccharomyces cerevisiae*. I. Experimental observations. *Biotechnol Bioeng* 1997;55:305–316.
18. Lange HC, Eman M, van Zuijlen G, et al. Improved rapid sampling for *in vivo* kinetics of intracellular metabolites in *Saccharomyces cerevisiae. Biotechnol Bioeng* 2001;75(4):406–415.
19. Visser D, van Zuylen GA, van Dam JC, et al. Analysis of *in vivo* kinetics of glycolysis in aerobic *Saccharomyces cerevisiae* by application of glucose and ethanol pulses. *Biotechnol Bioeng* 2004;88(2):157–167.
20. Vlad MO, Morán FY, Ross J. Response theory for random channel kinetics in complex systems. Application to lifetime distributions of active intermediates. *Physica A* 2000;278:504–525.
21. Vlad MO, Arkin A, Ross J. Response experiments for nonlinear systems with application to reaction kinetics and genetics. *PNAS* 2004;101(19): 7223–7228.

22. Selvarajoo K. Computational Modelling of Biological Pathways. Nanyang Technological University PhD Thesis. 2004.
23. Tan P, Selvarajoo K. *In silico* Modelling of Biochemical Pathways. US National Application Number: 10/222 029 International Application Number: PCT/GB2002/003, 2002.
24. Swain AK, Morris AS. An evolutionary approach to the automatic generation of mathematical models. *Appl Soft Comput* 2003;3(1):1–21.
25. Sugimoto M, Kikuchi S, Tomita M. Reverse engineering of biochemical equations from time-course data by means of genetic programming. *Biosystems* 2005;80(2):155–164.
26. Carroll DA. Chemical laser modelling with genetic algorithms. *AIAA* 1996;34(2):338–346.
27. Tsuchiya M, Ross J. Advantages of external periodic events to the evolution of biochemical oscillatory reactions. *PNAS* 2003;100(17):9691–9695.
28. Lancaster P, Salkauskas K. Curve and Surface Fitting: An Introduction. Academic Press, 1986.
29. Milo R, Shen-Orr S, Itzkovitz S, et al. Network motifs: simple building blocks of complex networks. *Science* 2002;298(5594):824–827.
30. Molloy MP, Brzezinski EE, Hang J, et al. Overcoming technical variation and biological variation in quantitative proteomics. *Proteomics* 2003;3:1912–1919.
31. Selvarajoo K. Discovering differential activation machinery of the Toll-like receptor 4 signaling pathways in MyD88 knockouts. *FEBS Lett* 2006;580(5);1457–1464.

Appendix 1

Genetic Algorithm (GA)

The GA works by encoding each coefficient in a given *NICM* mathematical expression into a binary string. The algorithm evolves parameter sets by the operations of selection, crossover, and mutations into the strings from one generation to the next (26,27). A variable P, such as α, p_1, p_2 in *NICM* expressions, e.g., equations (15–17), adopts a simple form, $P_{in}(1 + P_{max} \cdot R/(2^{16}-1))$, where P_{in} is the initial value of P, P_{max} is the maximum change of the parameter over the entire generation of interest, and R is initially set to be 0; then a binary string corresponding to R is evolved within the range $0 \leq R \leq 2^{16}-1$ by the GA operations. In the GA, we assign strings for 24 individuals for the variables in the *NICM*, which are used to search for the minimum fitness value.

Fitness Value

Fitness, f, or error, for each set of motif is based upon least square fit (28). It is defined as the sum of the absolute difference of mass action kinetic solution, represented as X in equation (A1), and *NICM* mathematical expression, Y, divided by mass action kinetics solution, at all discretized time points (j); the fitness of each reactant is then summed in a motif.

$$f = \sum_{i}^{n} \sum_{j}^{m} \frac{\left(X_j^i - Y_j^i\right)}{X_j^i}, \qquad \text{(A1)}$$

where *i* and *j* represent each reactant and time point, respectively. In all of our cases (Figures 2–6), $n = 3$ or 4, depending on the kind of motif, and $m = 200$.

Tolerance

We set the tolerance for accepting a network motif if the fitness value, *f*, is less than or equal to 0.01 or 10% because choosing the incorrect motif usually yields error more than 25% (Figure 2).

Computational Implementation

The current form of *NICM* software application is performed using Mathematica Version 5.1, which runs on Intel Pentium 4 (3 GHz) and 1 GB of RAM.

Appendix 2

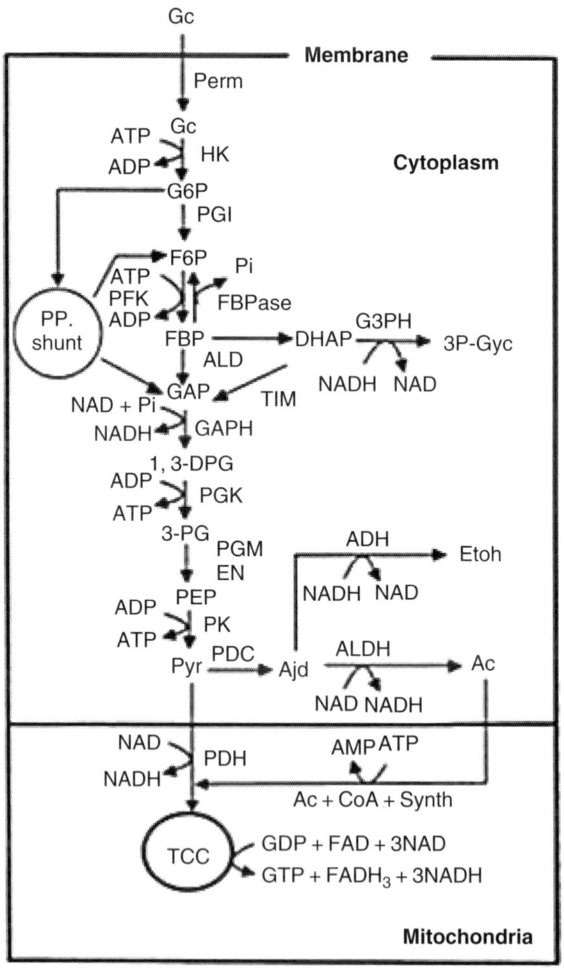

Figure A2.1. Schematic of glycolytic pathway in *S. cerevisiae*. Adapted from Theobald et al., 1997.

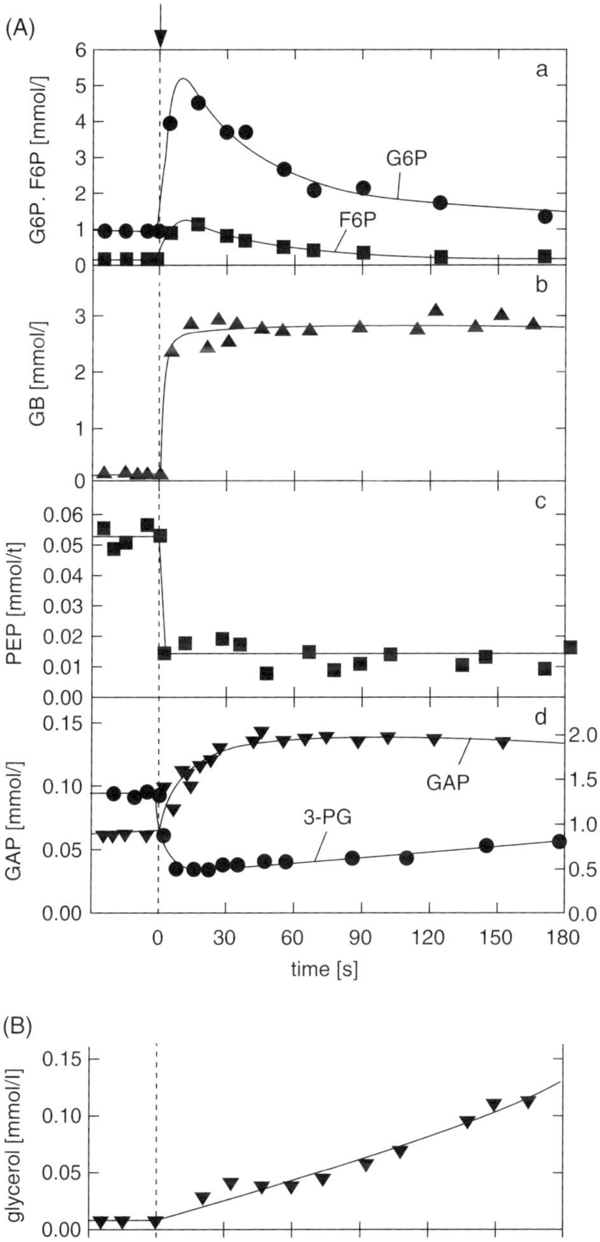

Figure A2.2. (A) The changes in the levels of G6P, F6P (a), FBP (b), phosphoenolpyruvate (c), and GAP/3-phosphoglycerate (d) from steady-state conditions after a glucose pulse perturbation. (B) The changes in the levels of glycerol from steady-state conditions after a glucose-pulse perturbation. We represent GAP as G3P throughout the text. Adapted from Theobald et al., 1997.

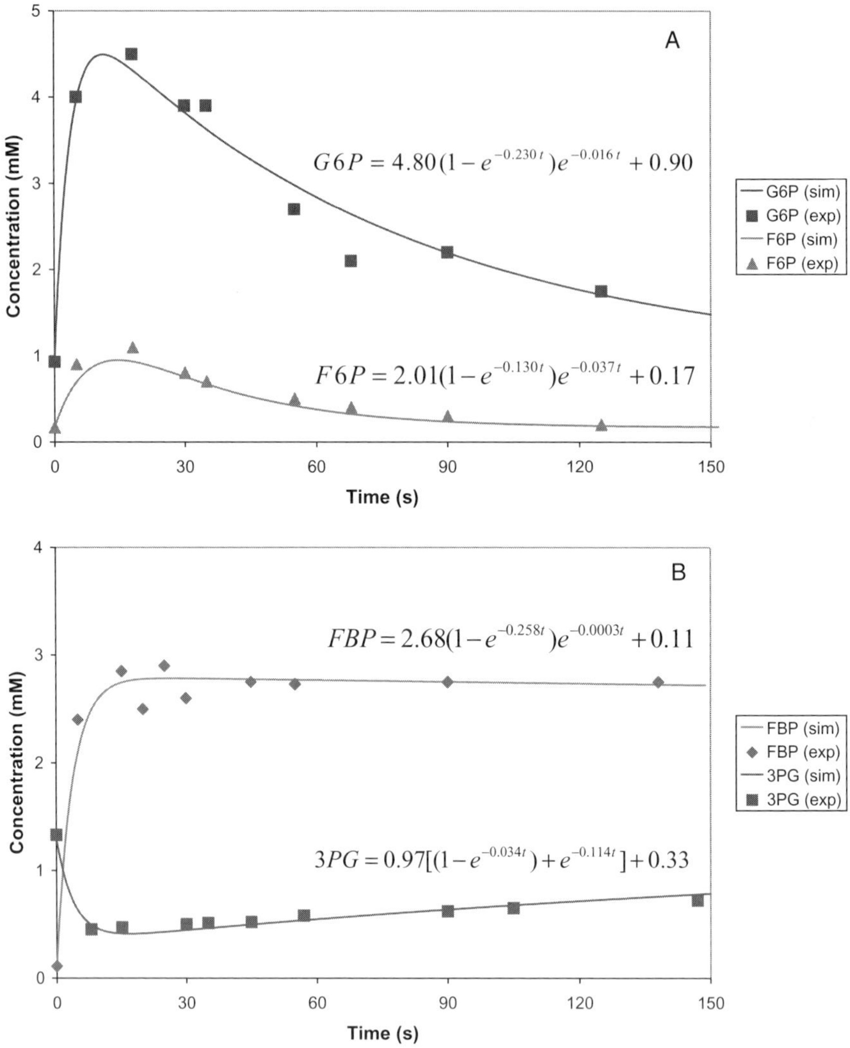

Figure A2.3. Using *NICM* expressions and GA, we fit the dynamic concentration profiles of (A) G6P, F6P, (B) FBP, 3PG, (C) G3P, phosphoenolpyruvate, and (D) pyruvate. The *NICM* expression with coefficient values for each metabolite is shown in each graph. The experimental data, denoted by (exp), were obtained from Theobald et al., 1997.

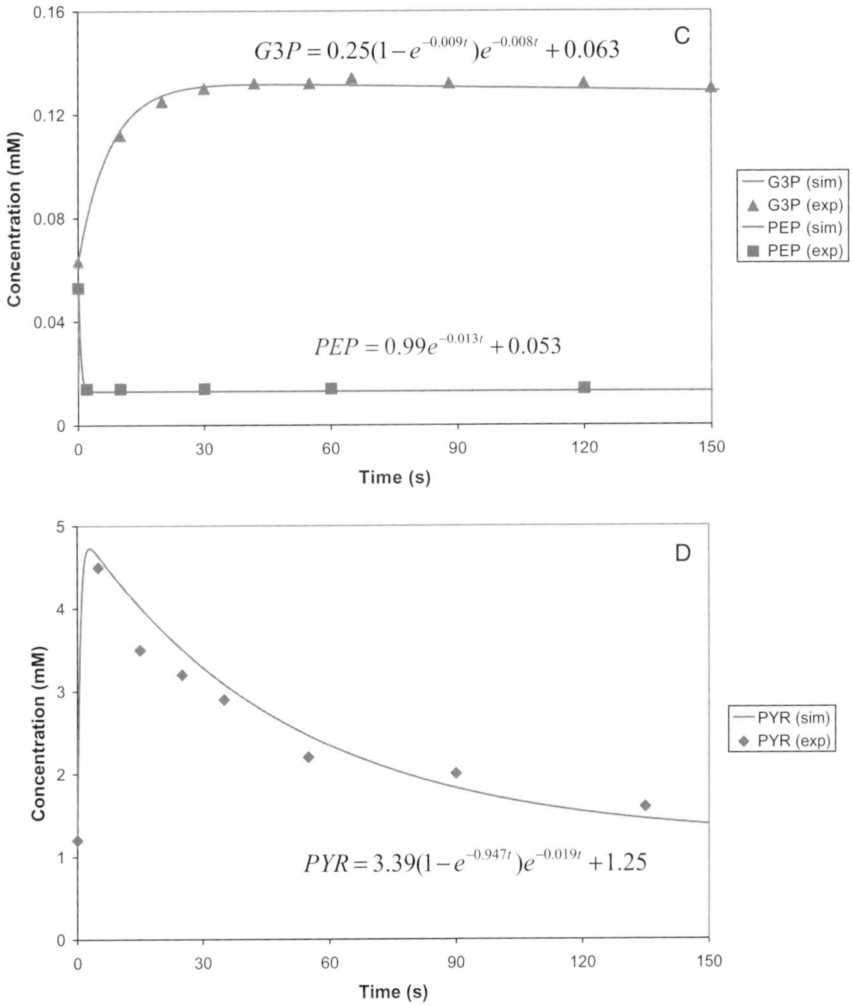

Figure A2.3. *Continued*

24

Storing, Searching, and Disseminating Experimental Proteomics Data

Norman W. Paton, Andrew R. Jones, Chris Garwood,
Kevin Garwood, and Stephen Oliver

Summary

The chapter introduces the challenges of storing, sharing, and querying proteomics data caused by the complexity of the experimental techniques, and the speed with which the techniques evolve. Public proteome databases are difficult to develop and populate because of the range of data types and queries that must be supported, and the quantity of metadata required to validate results. There are several data standards under development that should alleviate some of the challenges, and databases that utilize the standards are becoming more widely supported. The chapter describes a model of a complete proteomics pipeline, including the metadata that should be captured to allow confidence to be placed on the results. Software is also required, which can produce data conforming to the standards and that can be used to query proteomics data repositories. The chapter outlines the requirements for software and presents two exemplars developed at the University of Manchester. Finally, there is a description of the likely future developments in standardization for proteomics.

Key Words: Proteomics; mass spectrometry; data standard; database; PEDRo; Proteomics Standards Initiative.

1. Introduction

Experimental proteomics involves the identification and, in some cases, quantification of as many proteins as possible in a biological sample (1). Proteomics is rapidly evolving into a high-throughput technology, in which substantial and systematic studies are conducted on samples from a wide range of physiological, developmental, or pathological conditions. As a result, effective archiving and sharing of proteomic data is becoming increasingly important to enable comparison, validation, and further analysis of experimental results.

Figure 1 illustrates the steps in a typical proteomics experimental pipeline. Although the results of a proteomics experiment can be

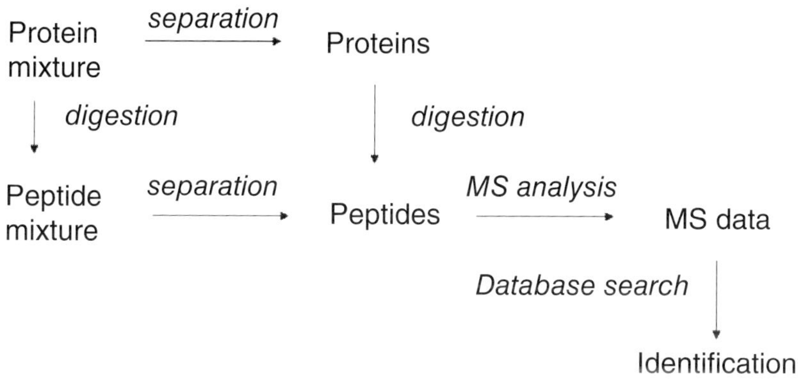

Figure 1. A typical proteomics experimental process.

summarized as an association between a description of a sample and the list of identifications made, many applications of a proteomics data repository require access to information on all aspects of the pipeline.

Establishing the most appropriate kinds of data to capture about a proteomics experiment is not straightforward, as this depends on the use that is to be made of the data. In a data repository, users may want to search for results based on widely varying criteria; for example, the proteins identified, the change in the level of a protein over time, the method by which a sample was extracted, etc. Furthermore, the users of a proteome data repository may, themselves, be diverse, including experimentalists with minimal direct experience of proteomics, but who are interested in proteins or organisms for which proteome studies have been conducted; proteome scientists who want to identify how successful specific techniques have been in different contexts; or mass-spectrometric analysts who want to compare their results with others.

A further challenge lies in the variety of proteomic technologies and practices. For example, samples may be subject to a wide range of prefractionation techniques (2), which affect the proteins that are found in a sample. Such techniques may be applied, for example, because low-abundance proteins are only likely to be detected using two-dimensional gel electrophoresis if steps are taken to remove highly abundant proteins before separation. As a result, the presence or absence of proteins in the result of an experiment is affected by the details of sample processing, and thus scrutiny of a result; for example, to understand whether or not a specific protein might have been expected to have been detected requires details on how the sample was processed throughout the experimental pipeline. The large number of different techniques for sample extraction, preparation, and separation, combined with the complexity of mass spectrometric techniques, and the subtleties of the software used to make identifications, means that capturing the data from a proteomics pipeline in a systematic manner is a challenging business.

The need for consistent and effective descriptions of proteomics experiments is reflected both in the presence of a standards body for proteomics: the Proteomics Standards Initiative (PSI; http://psidev.sourceforge.net/) of the Human Proteome Organisation, and of formal guidelines from journals for the description of proteomics experiments (3). Although the resulting standards and guidelines have yet to reach maturity, in the sense that they have become comprehensive in their coverage or universal in their adoption, their very presence suggests that what constitutes best practice for the storage, searching, and dissemination of proteomics experimental data is likely to become more widely agreed upon in the near future than it has been in the recent past.

The remainder of this chapter is structured as follows. Section 2 summarizes the major contributions in the field that have taken place over the last few years. Section 3 describes a model of the main components of a proteomics experiment. Section 4 focuses on the capture and dissemination of proteomics data, using two software applications as exemplars. Conclusions and future perspectives are provided in Section 5.

2. Previous Work

There are many different tasks that a proteomics data repository might be expected to support. These include: (i) search, e.g., to identify the experiments that found particular proteins, or to find the quantitative experiments that have been carried out for a designated species; (ii) browse, e.g., to view the gels that have been produced for an organism, or to compare a reference gel with one obtained in a specific experiment; (iii) validation, e.g., to assess the identifications claimed in an experiment, or to establish the confidence that should be put on the identifications made; and (iv) analysis, e.g., to repeat the running of identifications using different software or an updated database.

The earliest proteome databases mostly supported tasks (i) and (ii). For example, the SWISS-2DPAGE database was established by the Swiss Institute of Bioinformatics in 1993 to store images of two-dimensional gel electrophoresis (4). The database provides a Web interface that allows searching of proteins that have been identified on a gel, or browsing of gels by species or tissue type of the sample studied. Historically, SWISS-2DPAGE has not stored sufficient information about the methods used to identify a protein from a gel, such as the mass spectrometry data that was searched against a sequence database, to allow the verification of each result. A relatively recent development is the inclusion of an option that allows partial recording of such data (for example, giving the peak list produced by mass spectrometry), which could in theory allow an identification to be verified if the peak list had been searched using a standard set of database parameters. However, there is significant variation in the methods of mass spectrometry and database searches, such that storage only of a peak list is not sufficient to establish confidence in a protein identification.

More recently, databases have been produced that also support tasks (iii) and (iv); by focusing on the storage of the peak lists produced by mass spectrometry, the proteins identified from the peak lists, and the details of the methods used. The resulting databases, such as PRIDE (5) and GPM (6), have grown rapidly to include many protein identifications, as most of the data required can be captured automatically from mass spectrometry and proteomics software.

GPM contains data imported in the mzXML format defined by the Institute of Systems Biology (7). The mzXML format captures raw mass spectrometry data (the peak list) and metadata (such as machine parameters). A set of supporting tools is freely available that converts the output format from various types of instruments to mzXML. The format can also be used as input to a number of search applications that identify proteins from sequence databases. As such, mzXML allows data produced in one laboratory to be validated or re-analyzed by researchers who do not have access to the (often proprietary) software that produced the raw data or performed the initial database search. PRIDE, hosted at the European Bioinformatics Institute (www.ebi.ac.uk), imports data in the mzData format, developed by PSI. The mzData format has similar coverage to mzXML, and converters exist to transfer data between mzXML and mzData, such that either could be viewed as a viable standard for mass spectrometry data.

These databases contain rather few details on upstream sample processing, and thus support only certain forms of validation. For example, they allow validation that a protein has been correctly identified within a sample (checking for false positives). However, without sufficient information about the sample processing it is not possible to get an indication of false negatives; that is, the proteins that have not been identified because they were "lost" during separation. There are databases that model the complete proteomics investigation, such as PEDRoDB (8), but they are typically much more expensive to populate, in terms of time and effort, and are supported less directly by the early PSI standards.

As such, proteome data management should be seen as being in transition; some substantial resources now exist, but have yet to be fully supported by standards, and journal publication of proteomic experiments does not always require deposition of results in repositories. However, data standards and repositories are growing in importance, and can be expected to be a prominent part of the proteomics research landscape in the near future.

3. Modeling Proteomics Data

The PEDRo (Proteomics Experimental Data Repository) model was developed at the University of Manchester in 2003. PEDRo was intended to instigate community discussion about the requirements for a standard data format that would allow all four tasks described in section 2 to be supported. To achieve this goal, significant detail about the experimental procedures employed that produced the results must be captured. The proteins detected in a sample are influenced by the techniques used for extraction, separation, and identification. Furthermore, small differences in the genotype or environmental conditions of an organism can greatly affect the proteome detected in the sample.

The four sections of PEDRo (Figure 2) contain information about: (A) the sample being studied, (B) the protein separation technique employed, (C) the experimental procedure used to perform mass spectrometry, and (D) mass spectrometry data and analysis performed over the data.

Section A of the model allows the user to specify the hypothesis of the experiment and give a citation for the methodology employed in the experiment. The model also captures the sample on which proteomics was performed, allowing the user to enter the species of origin and the tissue, strain or cell type (if applicable). Such information is usually stored in a database to allow queries to find datasets of interest. The model does not contain a highly detailed structure for capturing all the potential variations in samples that could exist. It is a difficult challenge to develop a model for biological samples because there is so much

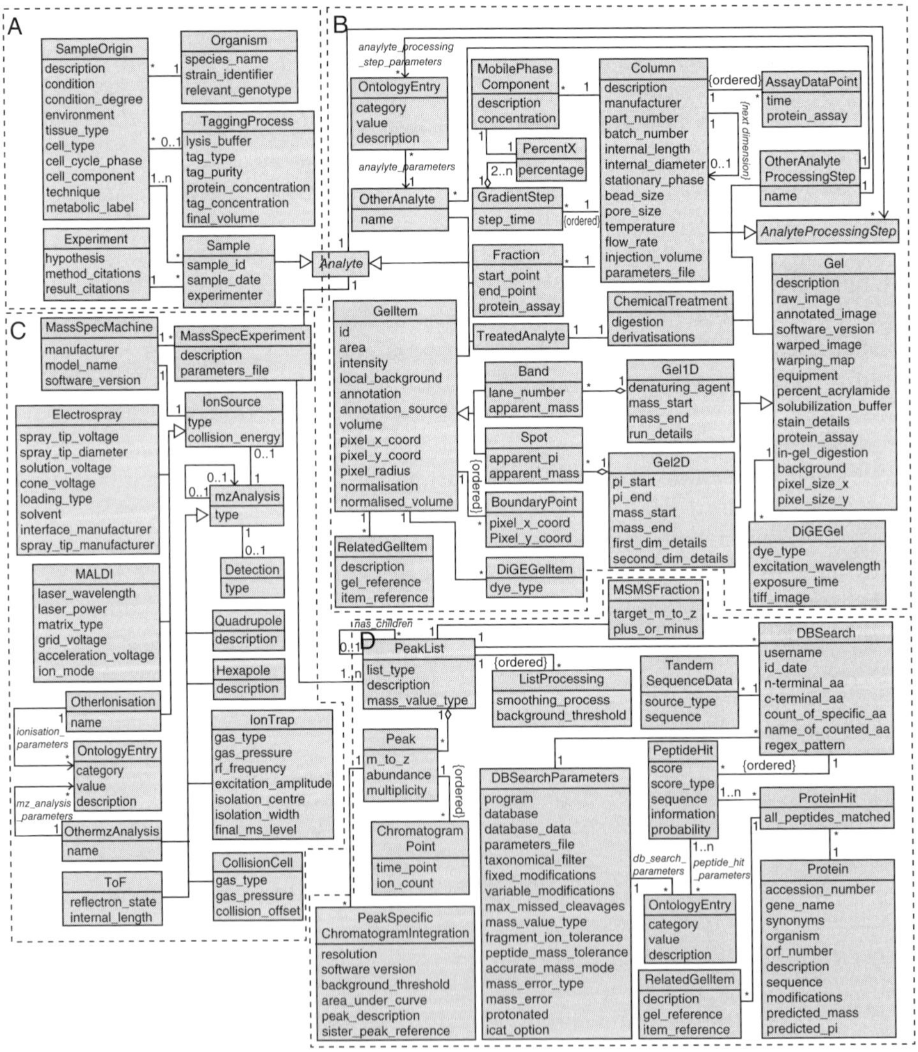

Figure 2. The PEDRo model of proteome data (9).

variation in the types of information that are relevant in a given context. For instance, the important information could be the description of a sample of river water or the case history of an individual with heart disease. The future development of proteome formats is likely to focus on defining relatively simple models of biological samples. The detail about the sample will be captured using definitions of samples obtained from controlled vocabularies (ontologies) that can grow over time. Experts in the biology of a particular organism or tissue type can develop the controlled vocabularies, whereas experts in proteomics will develop the underlying model.

Section B describes various techniques that could be employed to separate proteins. The model includes techniques, such as one- and two-dimensional gels, and column separations, such as liquid chromatography. The model captures a significant number of parameters, which is important because the parameter values may affect the set of proteins detected. One example is the pH range of a 2D gel. In one of the gel's dimensions, proteins are separated according to their net charge within a pH gradient, such as pH 4–7, until they reach their isoelectric point (pI), which is when they cease to migrate. Only soluble proteins with pI values in this range will be detected, leaving those proteins with pI values outside the chosen range undetected.

The model also describes the outputs of the separations, such as spots detected on a gel or fractions collected from a column. The spots or fractions can be used as input to the next part of the model, the mass spectrometry procedure.

Various methods of mass spectrometry are described in Section C of Figure 2. The model captures the method of ionization and of detection. Each of these processes has a significant number of parameters used with a particular type of instrument. The parameters will rarely be used for querying a database, but it is important that they are captured to allow confidence to be placed in the final results, or to allow a particular method to be repeated.

Section D captures the data produced from mass spectrometry (`Peak` and `Peaklist`) and searches performed with the data to identify proteins (`DBSearch` and `DBSearchParameters`). This part of the model contains many components that are vital for placing confidence in the results. It has previously been demonstrated that the set of proteins identified by mass spectrometry is affected by the algorithm used, the search parameters, and the quality of the sequence database searched (10). This level of detail must be stored because, if a protein has been identified in a sample, it is vital that the confidence of an identification be established, which can depend on various criteria specific to the software used for the search. Furthermore, the data produced by mass spectrometry can be searched with a new algorithm, or against a new version of the sequence database, to validate a particularly important result or to find additional proteins that were not identified in the initial search.

PEDRo has been used as a starting point from which a set of smaller modular formats has been developed, managed by PSI. The mzData format (http://psidev.sourceforge.net/ms/#mzdata) captures the machine parameters and data produced by mass spectrometry, similar in scope to mzXML. A format is also under development that handles the identification (and quantification) of protein data by database searches with mass spectrometry data. PSI will also manage formats describing gel electrophoresis, liquid chromatography, and other protein separation techniques. In essence, these formats contain large amounts of metadata describing the techniques used to extract and separate proteins from complex mixtures before identification by mass spectrometry. The actual data is the peak list produced by mass spectrometry and the proteins identified by a database search. Such a large quantity of metadata is also required to place these data within the context of a complete proteomics pipeline, to assess the confidence that can be put on a result, and to determine the validity of performing comparisons of data produced using different techniques.

4. Capturing and Disseminating

In this section, we describe two software applications called Pedro and Pierre, developed at the University of Manchester, which can perform capture and dissemination of proteomics data.

Pedro is a software application that presents a graphical interface enabling a user to enter the metadata and data for their domain of interest (11). One of the challenges with developing software for proteomics is the speed with which the experimental techniques evolve. It is difficult to design software that can accommodate changes without significant effort on the part of software developers following each new technique that must be described. Pedro's approach is to generate data-entry forms based on a model that is independent of the software. Pedro is not concerned with the nature of the model, only that it is represented in a standard framework (an XML Schema, http://www.w3.org/XML/Schema; both the Pedro model and the PSI models have representations in XML Schema). As such, any model can be loaded into Pedro for the purpose of generating data-entry forms. To support proteomics, Pedro has been developed in parallel with the PEDRo model described in section 3 but, importantly, there is no reliance, within Pedro, on any of the concepts in the PEDRo model. In practice, this means that if a novel kind of data is added to PEDRo, the new data type will be rendered in the Pedro interface without requiring any change to the software. This generic approach to form generation has allowed Pedro to be used directly in many other fields, including medical patient records, grid service descriptions, and genealogy (example models can be found at http://pedrodownload.man.ac.uk/models.html).

Pedro considers three types of users or clients: developers, data modelers, and end users. A developer is considered someone who develops software that is plugged into the Pedro system at one of several extensibility points, for example, to import data from a proprietary file format. A data modeler is responsible for creating the XML Schema, such as the PEDRo model, which will be used for the generation of data-entry forms. The data modeler need not be an expert in the domain that is being

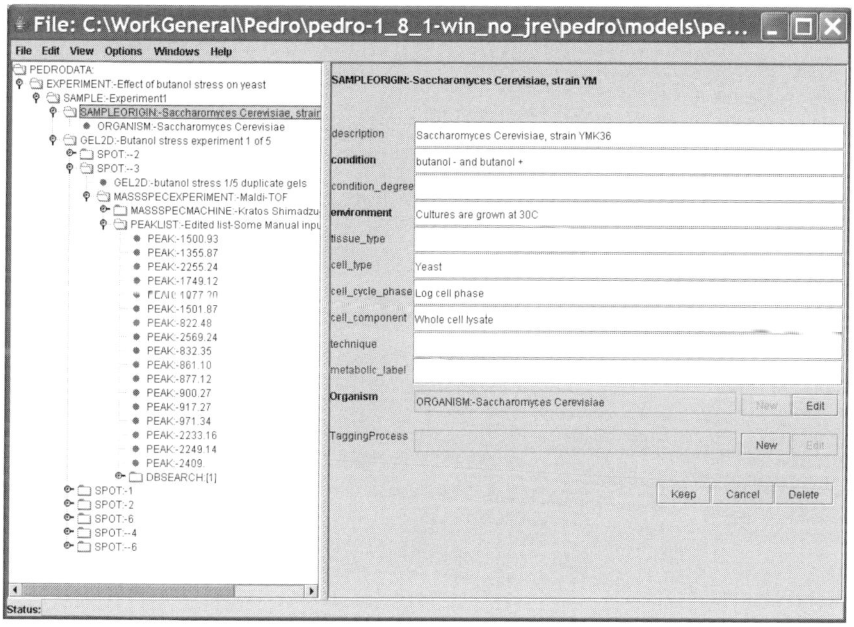

Figure 3. A screenshot of the Pedro form used for capturing the details of the biological sample used in a proteomics experiment.

modeled, but he would be expected to work in collaboration with such an expert. An end user is someone who uses the generated forms, as illustrated for the PEDRo model in Figure 3, for data entry. The end user is not expected to have any in-depth knowledge of computers or programming. In proteomics, the end user is a scientist, using the software to capture her data.

Among Pedro's features is the ability to import terms from externally controlled vocabularies (ontologies). For example, if the user needs to specify a "species name," a call is made to a relevant ontology resource, such as the NCBI taxonomy (http://www.ncbi.nlm.nih.gov/Taxonomy/). The user is shown a list of possible terms from which he may select the correct taxonomic name. This feature ensures that the name is entered correctly according to a standard definition, for instance as "*Homo sapiens*" rather than "*H. sapiens*" or "human." The use of controlled vocabularies is important because it facilitates the subsequent retrieval of data, as the user need only search using the term "*Homo sapiens*," rather than all of the possible synonyms, to find the datasets of interest.

Pedro also has facilities for creating templates (Figure 4). These are records that have been populated with data that may be used again multiple times, thus, saving the user from having to repeatedly re-enter the same data. For instance, the template feature is especially useful if the same experimental protocols have been employed in multiple settings, thus preventing unnecessary entry of redundant information. Pedro allows data modelers to set context-sensitive help pages for end users.

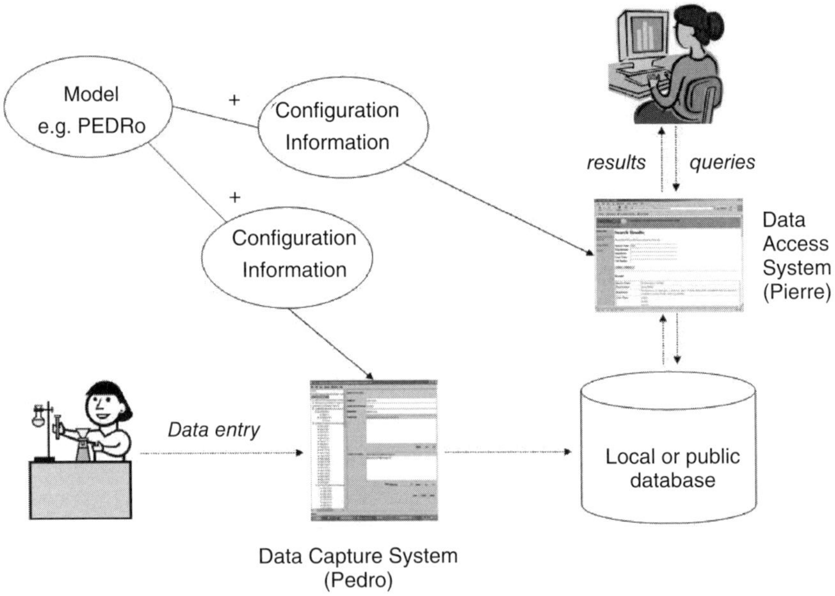

Figure 4. The user interaction with Pedro to submit data to a repository, and with Pierre to access the data.

These can be activated by the end user and can provide additional information about a particular field. For instance, a description could be provided to explain the type of data that should be entered in a field called "environment."

Pedro allows the proteomics research community to expand the capabilities of the tool by making use of plug-in technology. One such example would be importing the peak list produced by a particular type of mass spectrometer and converting the format to a standard, such as mzData.

The other application developed for the proteomics research community is called Pierre. Pierre's role is to provide users with an interface for browsing, searching, and querying data repositories. Like Pedro, the Pierre infrastructure allows any data model to be loaded and thus the infrastructure automatically accommodates changes as the requirements of the technology evolve.

Up to five auto-generated software products can be created when Pierre is run. Four of the products are interfaces that allow users to access the data repository in different ways. Pierre can create a Web interface that allows users to browse or search data from any location, as illustrated in Figure 5. Alternatively, a stand-alone application can be downloaded and installed locally, which offers more features than the Web interface. Pierre creates two other interfaces, intended for users who are comfortable with UNIX-style systems or who may not want Web or down-loadable application interfaces. The first is a command-line interface, which allows queries to be asked directly of the database, without

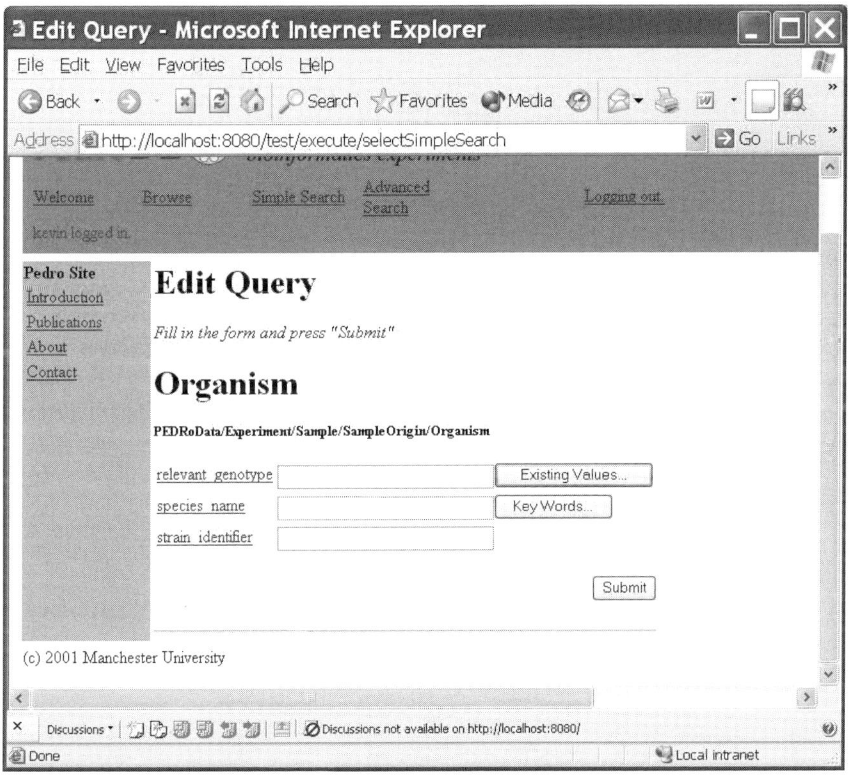

Figure 5. An example of a canned query in a Web interface generated by Pierre.

having to load a graphical application. The command line feature can also be useful for connecting a database with other pieces of software by means of piping commands. The text menu interface is similar to the command line, except that the user can navigate using a series of menu lists. Finally, Pierre can create an application programming interface (API) that allows the developers of other applications to embed services offered by Pierre within other software or a larger framework.

Pierre allows end users to browse and to conduct three kinds of searches. Simple search lets users interrogate a data repository based on preset queries. These so-called canned queries provide an easy means of querying because they are based on standard questions that have been identified by a designer as relevant to the user community. Example queries to a proteomics database could include "find all the experiments on mice," "find all the proteins that have been identified on gels," or "find the proteins that have been identified at more than one lab." Advanced search enables end users to construct their own queries from fields that occur in the schema. Advanced search allows Boolean queries to be constructed, such as "find experiments on mice AND find experiments where protein name = p53." Expert search allows users to directly enter a query in the query language offered by the underlying data repository.

To perform an expert search, the user must have knowledge of the query language and the data model.

There are many different methodologies that could be used to develop data repositories for proteomics. The Pedro/Pierre software is only one example; however, it differs significantly from many others in that the software architecture is independent of the model used to represent the data. In proteomics, and in the wider context of systems biology, the methodologies used to study genes, proteins and metabolites continue to evolve at a remarkable rate, with new experimental protocols published at frequent intervals (12,13). The approach taken with Pedro/Pierre has the advantage that the software can remain stable even if the data model continues to evolve. As new data models are produced, new interfaces can be provided to the database with minimal software development effort.

5. Conclusions and Future Perspectives

This chapter has reviewed various issues relating to the modeling, capture and sharing of proteomics experimental data. The overall lesson is that the position is evolving rapidly. At the time of writing, there is relatively little sharing of proteomics data, but repositories are being actively developed at several sites, and standards are emerging that will encourage the systematic capturing and sharing of such data. Given all this, proteomics data should soon be on a similar footing to other types of functional genomics data, where there is widespread sharing and support from standards for both microarray and protein interaction data.

Proteome data are intrinsically challenging to handle. Seemingly straightforward conclusions, such as that a protein is present in a sample, are based on several complex computational analysis steps, some of which, in turn, depend on external sequence databases. In addition, whether or not a protein is detected may depend on subtle features of the way the sample is processed, which may also affect quantitative results. All this is compounded by the wide and growing collection of techniques for sample processing and mass spectrometric identification. Furthermore, different users of proteome data may require access to very different levels of detail to support rather different tasks. For example, the data required to compare the effectiveness of different separation techniques are different from those required to compare the effectiveness of different identification algorithms. As such, there is still further work to be done to understand how best to make use of archives of proteome experimental data, and to establish how best to integrate such results with those from other high-throughput experimental methods.

Acknowledgments: Work in the authors' laboratories was funded by the PEDRo and COGEME grants of the UK Biotechnology and Biological Sciences Research Council, whose support we are pleased to acknowledge.

References

1. Liebler DC. Introduction to Proteomics. Totowa: Humana Press; 2002.
2. Righetti PG, Castagna A, Herbert B, et al. Prefractionation techniques in proteome analysis. *Proteomics* 2003 Aug;3(8):1397–1407.
3. Carr S, Aebersold R, Baldwin M, et al. Working Group on Publication Guidelines for Peptide and Protein Identification Data. The need for guidelines in publication of peptide and protein identification data: Working Group on Publication Guidelines for Peptide and Protein Identification Data. *Mol Cell Proteomics* 2004 Jun;3(6):531–533.
4. Hoogland C, Mostaguir K, Sanchez JC, Hochstrasser DF, Appel RD. SWISS-2DPAGE, ten years later. *Proteomic.* 2004 Aug;4(8):2352–2356.
5. Martens L, Hermjakob H, Jones P, et al. PRIDE: The proteomics identifications database. *Proteomics* 2005 Oct;5(13):3537–3545.
6. Craig R, Cortens JP, Beavis RC. Open source system for analyzing, validating, and storing protein identification data. *J Proteome Re.* 2004 Nov–Dec; 3(6):1234–1242.
7. Pedrioli PG, Eng JK, Hubley R, et al. A common open representation of mass spectrometry data and its application to proteomics research. *Nature Biotechnol* 2004 Nov;22(11):1459–1466.
8. Garwood K, McLaughlin T, Garwood C, et al. PEDRo: a database for storing, searching and disseminating experimental proteomics data. *BMC Genomics* 2004 Sep 17;5(1):68.
9. Taylor CF, Paton NW, Garwood KL, et al. A systematic approach to modeling, capturing, and disseminating proteomics experimental data. *Nature Biotechnol* 2003, Mar;21(3):247–254.
10. Kapp EA, Schutz F, Connolly LM, et al. An evaluation, comparison, and accurate benchmarking of several publicly available MS/MS search algorithms: sensitivity and specificity analysis. *Proteomics* 2005 Aug;5(13): 3475–3490.
11. Garwood KL, Taylor CF, Runte KJ, et al. Pedro: a configurable data entry tool for XML. *Bioinformatics* 2004 Oct 12;20(15):2463–2465.
12. Dunkley TP, Watson R, Griffin JL, et al. Localization of organelle proteins by isotope tagging (LOPIT). *Mol Cell Proteomics* 2004 Nov;3(11): 1128–1134.
13. Ross PL, Huang YN, Marchese JN, et al. Multiplexed protein quantitation in *Saccharomyces cerevisiae* using amine-reactive isobaric tagging reagents. *Mol Cell Proteomics* 2004 Dec;3(12):1154–1169.

25

Representing and Analyzing Biochemical Networks Using BioMaze

Yves Deville, Christian Lemer, and Shoshana Wodak

Summary

Systems biology aims at understanding the holistic behavior of biological systems. A very important step toward this goal is to develop a theoretical framework in which we can embed the detailed knowledge that biologists are accumulating at increasing speed, which will then allow us to compute the outcomes of the complex interplay between the myriad interactions that take place in the system. This chapter deals with important basic aspects of this theoretical framework that lie on the divide between systems biology and bioinformatics. In the first part, it discusses the conceptual models used for representing detailed knowledge on various types of biochemical pathways and interactions. As much of this knowledge deals with the complex networks of functional and physical interactions between the different molecular players, the second part of this chapter reviews the conceptual models and methods used to analyze various properties of these networks.

Key Words: Biochemical networks; network analysis; metabolic pathways; signal transduction; artificial intelligence; BioMaze.

1. General Introduction

The major challenge of the post-genomic era is the interpretation of the vast body of genomic sequence information in terms of the biological function of the gene products and the mechanisms of the cellular processes in which they are involved. This endeavor is driven in great part by the expectation that the gained understanding will lead to new ways of diagnosing and curing human diseases, and making our planet a better place to live.

But the task is daunting. The very notion of biological function is complex. The function of proteins, which are one type of gene product, essentially depends on the molecular interactions they make and on the cellular context in which they find themselves. Understanding function, thus, requires knowledge of how the different molecular players cooperate to produce the observed behavior of the living cell and of key processes therein. Acquiring this knowledge is the main object of systems biology, a field that has attracted renewed interest in recent years, and to which this volume is devoted.

A first key step in this endeavor is to acquire the information necessary to describe the system under study in a useful way. Major efforts are therefore being devoted worldwide to collecting such information by diverse means. Experimental procedures are used to measure gene expression profiles (1), transcription factor–gene interactions (2), and mRNA lifetimes (3) on the genome scale. Protein–protein interactions are characterized using high-throughput pull-downs or two-hybrid screens (4), and indirect "interactions" between genes are being probed by multiple gene deletions (5). In parallel, protein–protein interactions, sets of co-regulated genes (6,7), and metabolic pathways (8,9,10) are inferred using theoretical methods. These methods exploit information on protein and DNA sequences in related genomes, on protein three-dimensional (3D) structures, domain architecture, and gene order (11). Others use automatic procedures to extract links from texts of Medline abstracts (12).

All of these approaches yield very large bodies of valuable, but rather noisy, data, which systems biology research endeavors to exploit. Clearly, the bulk of the data pertains to the description of the circuitry of the cellular systems; the interaction, regulatory networks, and pathways (on gene regulation, metabolism and signal transduction), and provides limited information on the temporal sequence of events, or on their spatial organization. But obtaining detailed information on the circuitry is a key first step that can yield valuable clues on the system-level behavior (13), provided, however, that this information can be adequately validated and readily analyzed.

Currently, such analyses face various difficulties. The ability of accessing and manipulating the information is limited by the fact that it is distributed across heterogeneous databases. Also, our current knowledge of the various cellular processes (protein interaction, gene regulation, or signal transduction) is poorly structured and partial. The data can therefore be incomplete, inconsistent, or approximate. In addition, the size of the pathways and networks available for analysis can be very large, leading to problems of spatial and temporal computational complexity. All this makes the representation and analysis of pathways and interaction networks, which we denote here as *biochemical networks*, challenging problems in systems biology and bioinformatics.

This chapter describes strategies for addressing these challenges, with examples taken from our own work on the BioMaze system. Section 2 discusses data models for representing rich information on biochemical networks for archival and query purposes. That section starts with a short overview of existing models and proceeds with a description of an attractive integrative data model implemented in the BioMaze database. Section 3 deals with data models used for the purpose of performing computational analyses of biochemical networks. Section 4 reviews analysis methods that use standard graph-based techniques and presents some recent advances in the application of constraint satisfaction methods. A brief description of the BioMaze database system featuring the described data model and analysis methods is presented in Section 5.

2. Data Models for Representing Biochemical Networks

2.1. Overview of Existing Models

Building the theoretical frameworks for investigating the rapidly growing body of biological data and subjecting it to the corresponding systematic analyses requires that the data be appropriately structured and organized. Furthermore, the heterogeneity of the terms used by biologists must be reduced through the creation of controlled vocabularies, and standards must be developed to formalize the description of both experimental data and mathematical models of cellular and physiological processes.

Much of this has been happening in recent years. The development of specialized databases for representing information on cellular processes and interactions (14), has required the design of a new generation of data models that are much more complex than those previously used in databases representing information on gene/genome and protein sequences (15). Pioneering database projects such as EcoCyc (9,16), KEGG/Ligand (17), and WIT (18) initially focused on metabolic pathways, developing data models specifically tailored to this types of processes. With the exception of EcoCyc, which featured a rather sophisticated hierarchical data organization early on, the other data models were initially quite rudimentary, often representing pathways as collections of molecular functions, or unordered collections of catalyzed reactions, with the order of the reactions provided by the graphical representations (maps). However, these databases have now elaborated and extended their models to allow the representation of other processes, such as gene regulation, transport and signal transduction. In some databases, such as KEGG/Ligand, information on pathways from different organisms is integrated in the same data structure, whereas in others, such as EcoCyc and its sister databases, each organism has its processes represented in a separate database. Elaborate data structures capable of accommodating different types of processes are featured in more recent database efforts, such as the Reactome database, which focuses on human pathways (19), and Patika, which handles metabolic and signal transduction pathways of different organisms (20).

The data models underlying all these databases have many common features, which reflect a consensus view reached in the field. But they remain different enough to preclude easy integration. This has prompted the development of a community-wide standard for pathway data exchange, BioPax (21), which goes far beyond what XML (Extensible Markup Language) has to offer, as XML is a simple, flexible text format derived from SGML (ISO 8879), originally designed to meet the challenges of large-scale electronic publishing.

Another category of databases includes those specializing entirely on representing processes other than metabolism. These include databases, such as DIP (22), BIND (23), MINT (24), and IntAct (25), whose primary focus is the representation of data on protein–protein interactions. Their data models are usually much simpler than those of the pathway databases. They are limited to representing pairwise interaction events with their associated annotations, often varying widely in scope and coverage. But they differ sufficiently from one another to have prompted the development of a community-wide standard (PSI) for the data exchange format (26).

Two other types of databases with distinct data structures are those specializing in gene regulation and in signal transduction pathways. Gene regulation databases, such as TRANSFAC (27) or Regulon DB (28), represent information on transcription factor–gene association, on the transcription factor–binding sequences, associated sequence motifs, and relevant annotations. Only a few databases, mostly those already specializing in metabolic pathways, feature data models representing the regulatory networks that can be constructed by linking together different transcription factor–gene and transcription factor–protein interactions.

By far, the most complex data models are those for representing signal transduction pathways. Many aspects of these models are hierarchic. A good example is in TRANSPATH (29), where the data model combines hierarchies at the molecular, reaction, and pathway levels to yield the required description of the corresponding pathways. A molecular hierarchy is used to represent higher level information on orthology or protein family on the one hand, and lower level details on signaling interactions, like the protein chain and particular domain involved in the interaction, as well as the corresponding chemical modification (usually phosphorylation). Reaction and pathway hierarchies are used to combine the molecular events into pathways at various levels of granularity. Other databases, such as CSNDB (30) and INOH (31), feature different models.

In parallel, conventions for representing signal transduction networks have also been proposed (32). These comprise a set of rules and symbols for the visual representation of elementary biochemical processes, such as various types of protein–protein association events, enzymatic catalysis, inhibition, and protein modification, to more complex processes, such as degradation or gene expression.

Last, one should mention the so-called ontologies, which combine a taxonomy of terms with a set of domain-specific rules for linking objects within the taxonomy. A number of such ontologies coexist in the field (33). Among them, the Gene Ontology (34) is widely used. The main roles of such ontologies are to help unify annotation efforts, permit the integration of data from ontologies and databases in other areas of biological research, and to build of software tools that interpret and use this information.

2.2. The BioMaze Model: An Integrated Solution

An integrated solution to the problem of representing rich information on complex heterogeneous processes is provided by the BioMaze data model, which is a recent extension of the original model of the aMAZE database (35). This solution enables us to archive this information, as well as to readily extract it for use in a range of specialized biological analyses aimed at investigating different properties of these processes.

A major consideration in designing this data model has been that it should reflect the biologist's view of the domain, while enabling efficient and flexible programmatic access to the data. The choice was therefore made to adopt the Extended Entity Relationship modeling paradigm (36). This paradigm generalizes the well-established Entity Relationship model used in the relational databases through the addition of the inheritance concept taken from the object-oriented world, thereby providing the best of both worlds. This made it possible to build a conceptual model of the data, which enables the biologist to readily manipulate the information for query and analysis purposes. Furthermore, using the conceptual model, specialized software tools are able to generate the corresponding relational model and implement it in a relational database management system, as will be illustrated in section 5.

In the following sections, the main features of the BioMaze model are summarized. This model comprises 3 main layers, the biochemical, systemic, and functional layers that map into one another, as illustrated in Figure 1.

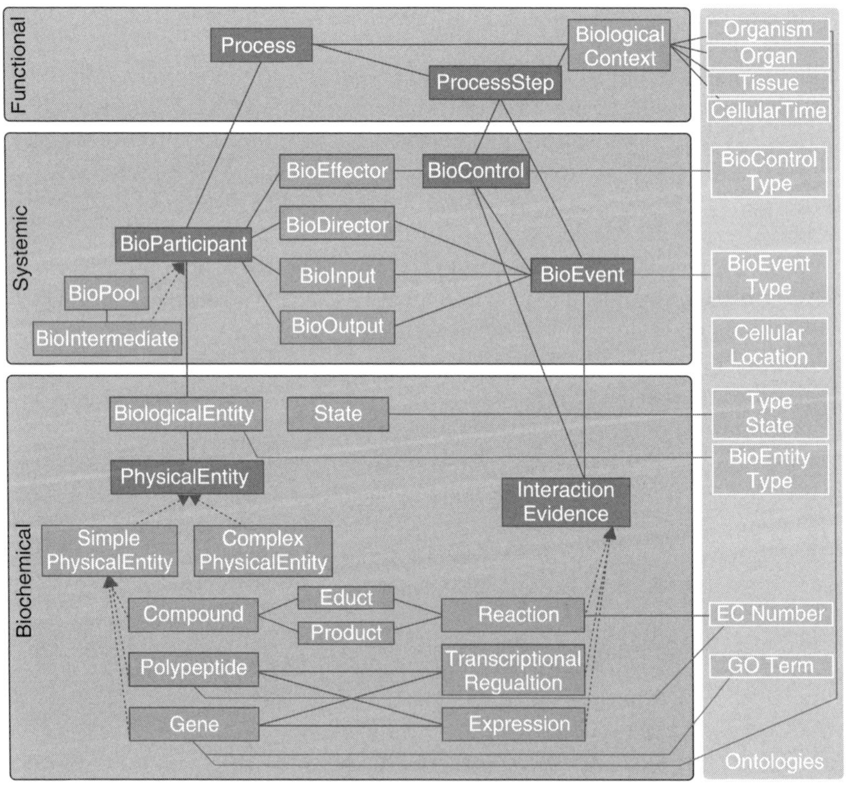

Figure 1. Overview of the BioMaze data model outlining its three main layers: the Biochemical layer (bottom), the Systemic layer (middle), and the Functional layer (top).

2.3. Biochemical Layer

The biological/biochemical knowledge layer of the BioMAZE model (Figure 1, bottom) contains the descriptions of the physical entities that are the basic building blocks of biochemical processes. Physical entities can be "simple" entities such as compounds (small molecules), polypeptides, and genes, or "complex" entities, such as biologically active assemblies (complexes) composed of several simple entities. Both simple and complex physical entities are unified as PhysicalEntity. Because the function of a molecule or a molecular assembly depends not only on its basic chemical composition but also on its state (conformational state, chemical modification, etc.), the PhysicalEntity is combined with the State description to yield the BioEntity, representing the biological entity actually involved in the biochemical process.

In addition to the physical building blocks, the BioMaze model also describes basic chemical and biological processes as building blocks, a feature that was already introduced in several of its precursor data models (35). These comprise Reaction, Expression, and Transcriptional Regulation, which are defined in the model as Interaction-Evidences. Reaction represents chemical reactions (catalyzed or not); Expression represents a shortcut for the process whereby a gene "leads" to the expression of the proteins it codes for; and Transcriptional Regulations is a shortcut for the process whereby the expression of the gene coding for a particular protein is up- or down-regulated (see following section).

2.4. Systemic Layer

At the heart of the BioMAZE data model is the Systemic layer (Figure 1, middle). It is the feature of the BioMaze model that enables it to represent in an integrated fashion very different types of basic biochemical processes, such as metabolic reactions, gene regulation, and signaling.

In this layer, BioParticipant represents a BioEntity in a specific cellular location, with the latter also being described by a cellular location Ontology (Figure 1) (37). BioEvent represents simple (elementary) biological events and involves interaction between several BioParticipants, which can play different roles (BioInput, BioOutput, BioDirector, and Bio-Effector). To understand this representation, it is useful to illustrate how the biochemical layer maps into the systemic layer.

2.4.1. Biochemical Reaction

Figure 2 illustrates this mapping for a biochemical reaction, where Educt and Product map into BioInput and BioOutput, the Polypeptide that catalyzes the reaction (enzyme) maps into BioDirector, and a compound that inhibits the enzyme maps into BioEffector. In addition, a distinction is made between BioParticipants that act as reaction intermediates (Bio-Intermediates) and those acting as pool compounds (BioPool). This distinction is useful for the layout and analysis of metabolic pathways, which represent subgraphs of metabolic networks built by stringing together biochemical reactions.

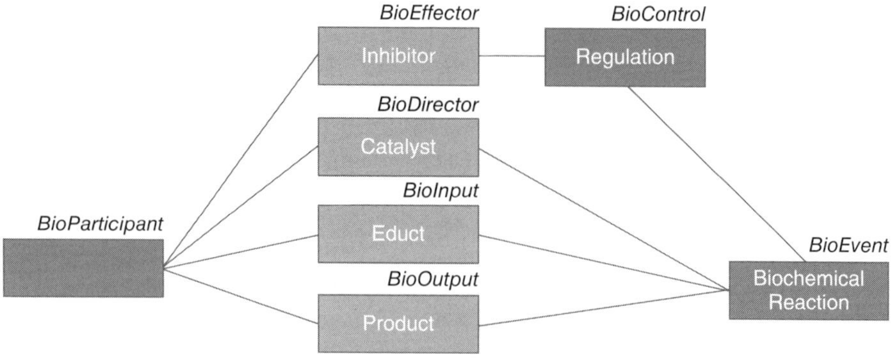

Figure 2. Biochemical reaction data model in BioMaze.

2.4.2. Protein Expression

The process of expressing a protein coded by a given gene can likewise be readily mapped into the systemic layer, as shown in Figure 3. In this process, which is a condensed 1-step representation of a complex multistep process comprising transcription and protein synthesis, the amino acids and the complete polypeptide are the BioEntities that map into BioInput and BioOutput, respectively, the gene is the BioDirector, whereas the BioEffectors can be protein regulators (transcription factors) or small molecules acting as RNA switches. Whenever required, and provided sufficient information is available, this condensed representation can be readily replaced by a detailed description of each individual step, without altering the model.

2.4.3. Transport

Transport across membranes and assembly/disassembly of several physical entities to form a complex can also be readily mapped. The transport "reaction" (Figure 4) takes a BioParticipant (representing a BioEntity in a given cellular location) as BioInput and another BioParticipant

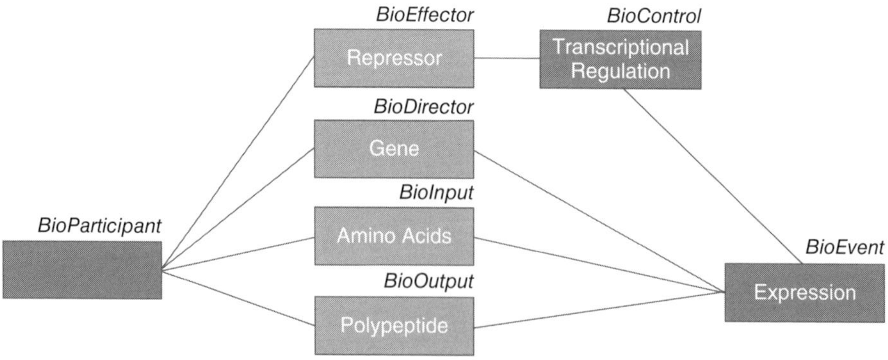

Figure 3. Protein expression data model in BioMaze.

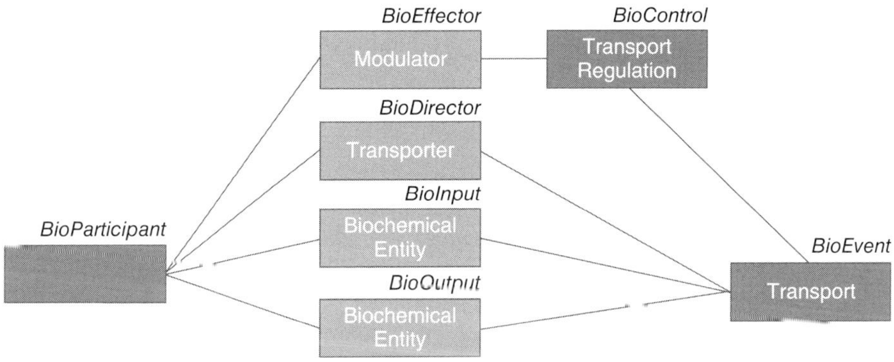

Figure 4. Transport data model in BioMaze.

(representing the same BioEntity as in that of the input, but in another cellular location) as BioOutput; the transporter (protein) is the BioDirector, and any modulator of the transport process is the BioEffector.

2.4.4. Assembly and Disassembly

Finally, mapping of the assembly/disassembly processes (Figure 5) is analogous to that of mapping the biological reaction. Here, the

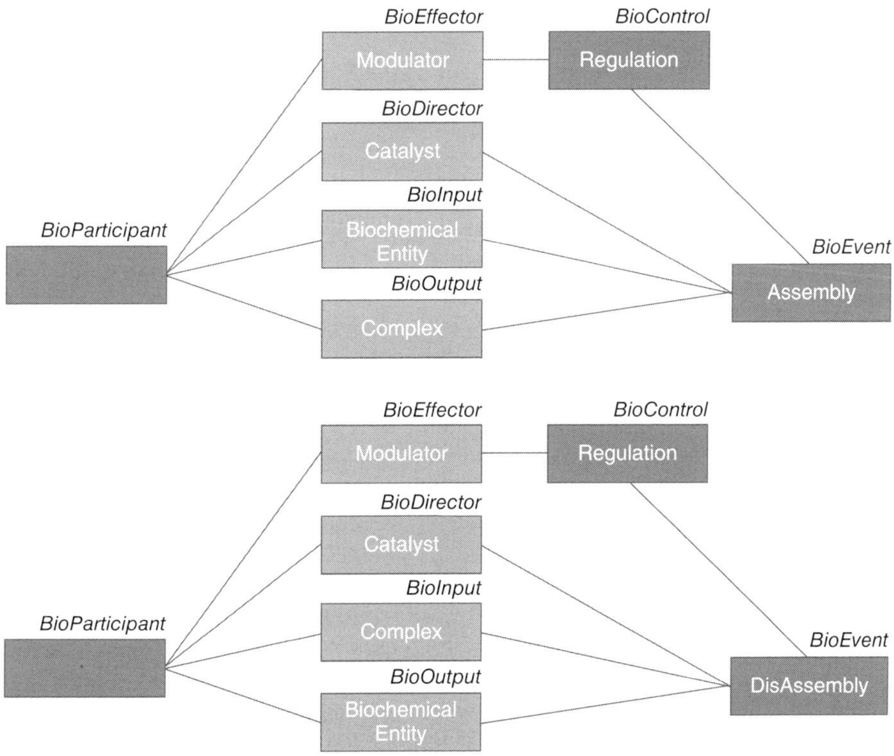

Figure 5. Assembly/disassembly data model in BioMaze.

assembled complex is the BioOutput for the assembly reaction and the BioInput for the inverse reaction (disassembly), with various BioParticipants playing the role of catalysts or effectors.

2.4.5. Signaling

Finally, a signal transduction can likewise be mapped into this layer (Figure 6). To that end, we define the Signaling BioEvent as the primary event (or process) of a signaling cascade. Recalling that a BioParticipant is a BioEntity in a specific cellular location, whereas a BioEntity is a PhysicalEntity in a certain State (the pertinent states here are the various phosphorylation states), we take the signal, as the Director of the conversion of a PhysicalEntity in a certain state to the PhysicalEntity in another state (for example, from inactive to active and vice versa). The BioInput and BioOutput refer to the same PhysicalEntity, but in different states.

Hence, the encoding of the different types of BioEvents discussed above follows the same set of rules that relies on the underlying biochemistry to define the input, output, and director and effector roles. The major difference lies in the interpretation, and, more specifically, in the navigation through the network. Each type of BioEvent has a natural way to be traversed. In BiochemicalReaction, the output of one event is the input of the next. The navigation in a metabolic pathway thus follows the sequence: BioParticipant–BioInput–BioEvent–BioOutput–BioParticipant. In SignalTransduction, the output of one

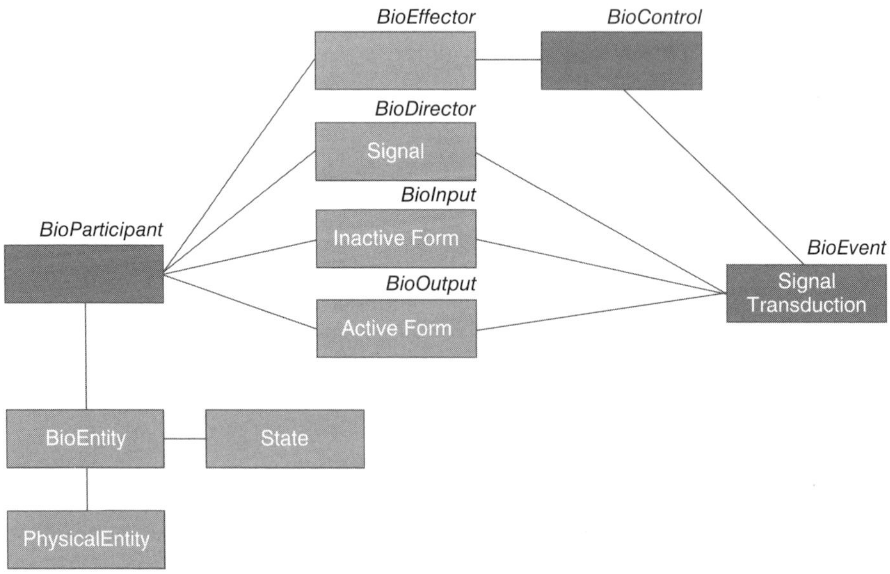

Figure 6. Signal transduction event data model in BioMaze.

event is not the input but the signal of the next event, and the navigation along the pathway follows a different sequence, namely, BioParticipant–BioDirector–BioEvent–BioOutput–BioParticipant.

The combination of the systemic and biochemical layers, thus provides the framework for a unified representation for a wide range of cellular processes. This is achieved through a unified model for the BioEvent, allowing the integration of different types of BioEvents in the same network and supporting process-specific representations by using appropriate navigation rules.

2.5. Functional Layer

The third layer of our model corresponds to the Processes (Figure 1, top). A Process is a subgraph of the BioEvent/BioParticipant graph. We use an intermediate entity, ProcessStep, to represent each BioEvent in the context of the described Process.

Although the data in the BioEntity/BioParticipant graph are tightly integrated, in contrast, Processes are less constrained. The model allows us to annotate different versions of the same biological pathway and stores partially overlapping pathways.

2.6. Ontologies

In addition to the three layers described above, the BioMaze model contains a fourth section (Figure 1, left) that groups all the so-called ontologies. Those are of two types: i) the external ontologies, which include established classifications and controlled vocabularies used by the biologists for describing organism taxonomy, organs, tissues, and cellular location, as well as the Enzyme classification (38) and Gene Ontology (34), and ii) the internal ontologies, such as the BioEntityType, StateType, BioEventType, BioControlType, which are controlled vocabularies that extend the type hierarchy in a flexible manner, allowing us to qualify the generic entities (i.e., a BioEvent representing the catalytic transformation of a compound into another will be qualified as Catalytic-Reaction, a BioEvent representing the biosynthesis of a polypeptide under the "direction" of a gene will be qualified as Expression, etc.).

3. Data Models for Analyzing Biochemical Networks

Most (nontrivial) analyses of biochemical networks cannot be performed through routine database queries. Usually, the network under study is extracted from the database, represented in a suitable data model, and processed by a specialized analysis tool.

Various types of data models can be used for the analysis of biochemical networks. These data models have been reviewed and classified by Deville et al. (39) using a unified framework. A summary of this review is proposed in this section.

It is impossible to determine which model is the best; each model presented in the following subsections has its own advantages and enables specific types of analysis. Models differ either by the chosen view, the coverage (different types of interactions that they can represent), their precision, or their granularity (resolution of the basic information: atomic, molecular, or supramolecular).

The models will be described using a graph framework. A graph G(V,E) is a mathematical object, where V is the set of nodes (vertices) and E is the set of edges connecting pairs of nodes. An edge is an ordered pair of nodes (directed or oriented graph) or an unordered pair of nodes (undirected graph). Object-oriented models can be seen as a natural extension of graphs, where the nodes are typed, and different relations are defined between specific types of nodes. Objects also allow inheritance.

3.1. Compound and Reaction Graphs

The objective of employing compound or reaction graphs is to model a set of chemical reactions. In a compound graph, nodes are the chemical compounds. A directed edge connects compound A to compound B if A occurs as a substrate and B as a product in the same reaction.

A reaction graph is a dual form of the compound graph. Here, the nodes are the reactions. There is an edge between reactions R1 and R2 if a compound is both a product of R1 and a substrate of reaction R2. The graph can be directed or undirected, depending on whether the reactions are considered as reversible or not. It is also possible to extend the definition of an edge by considering edges between two reactions when they share a compound (40).

As an example, let us consider the following simple reaction:

Reaction 1:

glutamate + ATP → gamma-glutamyl phosphate + ADP
EC 2.7.2.11, catalyzed by gamma-glutamyl kinase

The compound and reaction graphs of reaction 1 are shown in Figure 7. The reaction graph is reduced to a single node, as it involves only one reaction.

The use of graph theory, and in particular compound graphs, is a well-established representation technique in biochemistry and chemical engineering (41). Compound and reaction graphs have recently been used in the analysis of topological properties (connectivity, length, statistical properties, etc.) (42,40). The authors stress the small-world character of metabolic networks; their compound graphs are sparse, but much more highly clustered than an equally sparse random graph.

The equivalent of compound graphs can be defined for signal transduction networks, as well as for transcriptional regulation networks. In a transcriptional regulation graph, nodes represent genes, and a directed arc between gene A and gene B means that gene A codes for a transcription factor, which regulates gene B.

In the signal transduction graph (43), nodes are usually signaling molecules, and an edge represents a process relating two signaling molecules. Such a representation is used for path searching.

Although compound and reaction graphs or their equivalent can be used to represent and analyze metabolic, regulatory, or signaling

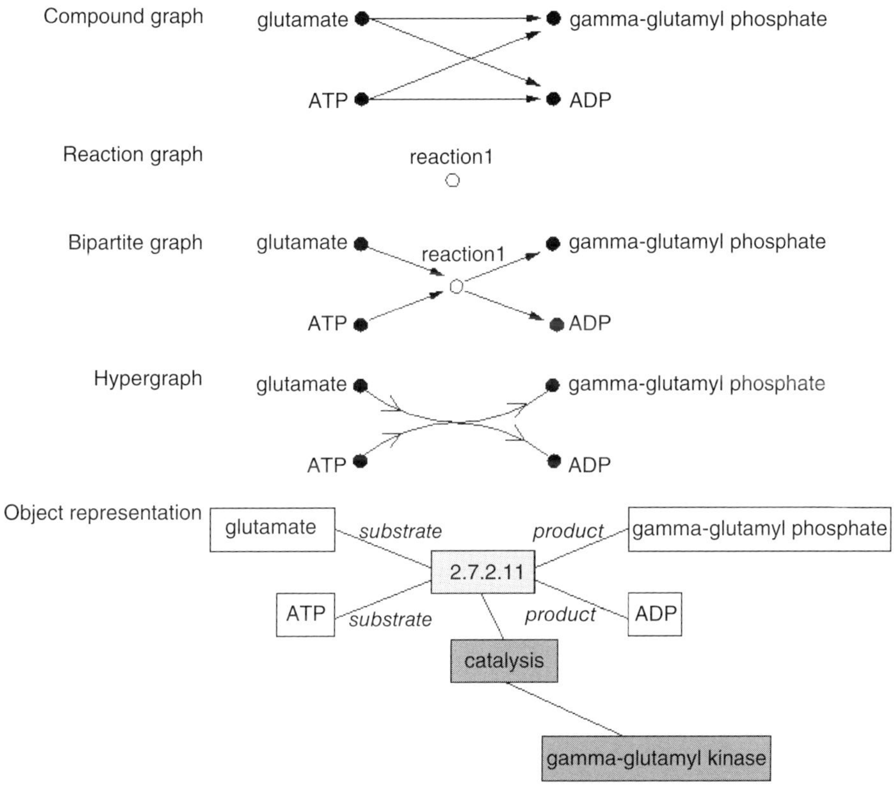

Figure 7. Reaction 1: glutamate + ATP → gamma-glutamyl phosphate + ADP (EC 2.7.2.11, catalyzed by gamma-glutamyl kinase)

pathways, this data model cannot combine these different pathways. A combination of compound graphs requires, for example, a distinction to be made between nodes representing compounds and nodes representing genes, and to distinguish arcs representing a reaction from arcs representing the regulation of some signaling process. Compound and reaction graphs also have obvious limitations in their coverage because they represent only reactions within pathways and contain no information about the enzymes catalyzing these reactions.

The coverage is even more limited for regulatory and signaling pathways because of the large number of different types of interactions that occur in these pathways (assembly, transcriptional regulation, protein–protein interaction, translocation). The descriptive power of compound and reaction graph is also very poor because the structure of the reaction is lost in compound graphs. In a compound graph, one can no longer distinguish if two substrates or two products are involved in the same reaction. In a reaction graph, it is impossible to determine if products produced by two reactions and consumed as substrates by another reaction are identical or not. As a consequence, different sets of reactions can lead to the same compound or reaction graph (39,44).

Nonetheless, although compound and reaction graphs only offer a partial and sometimes ambiguous view of biochemical networks, such representations turn out to be sufficient and useful for some simple analyses such as topological and statistical properties, or the discovery of basic patterns. Such representations can also be helpful in some specific applications, such as the detection of functionally related enzyme clusters (45).

The models described in the next sections extend the above basic graph approaches, and overcome some of their limitations.

3.2. Bipartite Graphs and General Graphs

In a bipartite graph, there are two classes of nodes, and no edges can relate nodes from the same set. In the context of biochemical networks, there are *compound nodes* and *reaction nodes*; an edge, thus, necessarily relates a compound node and a reaction node. Edges can be undirected or directed. A directed edge from a compound node to a reaction node denotes a substrate, whereas an edge from a reaction node to a compound node denotes a product of the reaction. Bipartite graphs can represent reactions without any ambiguity. The bipartite graph representation of our example is provided in Figure 7.

The bipartite graph is a classic data model for the analysis of metabolic pathways. For instance, they have been used by Jeong (46), where metabolic networks of 43 organisms are modeled as bipartite graphs, enabling a systematic comparative analysis that showed that these metabolic networks have the same topological scaling properties.

Bipartite graphs offer an unambiguous representation of the reactions and compounds in biochemical networks. Their coverage is limited however, as possible controls of reactions (catalysis, inhibition) cannot be explicitly represented. This simple data model is appropriate when the analysis is limited to reactions and compounds. This includes applications, such as the analysis of the topological properties of the network, path finding, and building pathways from collections of reactions. Clearly, however, without extensions, bipartite graphs cannot simultaneously model metabolic, regulatory, and signaling pathways.

Bipartite graphs can be generalized to graphs with multiple classes of nodes (and arcs). Instead of considering types for the nodes and the arcs, it is usually easier to attach a set of properties to nodes and arcs, one of these properties being considered as type information if needed. This allows the incorporation of additional information that is needed for analyzing the graph. Graphs with nodes and arc attributes are especially suitable for a graph representation of data extracted from an object-oriented model, such as in BioMAZE. Such a graph representation of Reaction 1 is provided in Figure 7; the type information is visualized here through the color of the nodes, and an attribute (substrate/product) is attached to some arcs. The control of the reaction can also be integrated into the graph.

In the BioMAZE framework, it is possible to extract, from the database, a graph (with attributes) of a specific subset of processes. This is achieved using a query language similar to SQL. Various analysis tools can then process the resulting graphs.

4. Analysis of Biochemical Networks

This section describes different methods that can be used for the analysis of biochemical network. It starts with an overview of standard graph techniques and proceeds to show that some advanced artificial intelligence techniques can provide more elaborated analysis capabilities. The described techniques are illustrated through specific examples. These techniques have been integrated into the BioMAZE workbench Figure 8.

4.1. Standard Graph Techniques

A first type of standard analysis involves deriving classic statistical properties of the graph, such as minimum, maximum, average, and standard deviation on different characteristics of the network: nodes, arcs, degree of nodes, connected components. These properties, together with other global properties, such as the clustering coefficient or eccentricity and closeness of nodes, can be used for the analysis of topological properties (connectivity and length) (40,42,47,48). Other techniques used for network analysis, such as graph decomposition and clustering, which were developed in the Pajek program (49), can also be applied on biochemical networks.

The computation of the shortest paths between pairs of nodes in the context of the global network is another common operation. For instance,

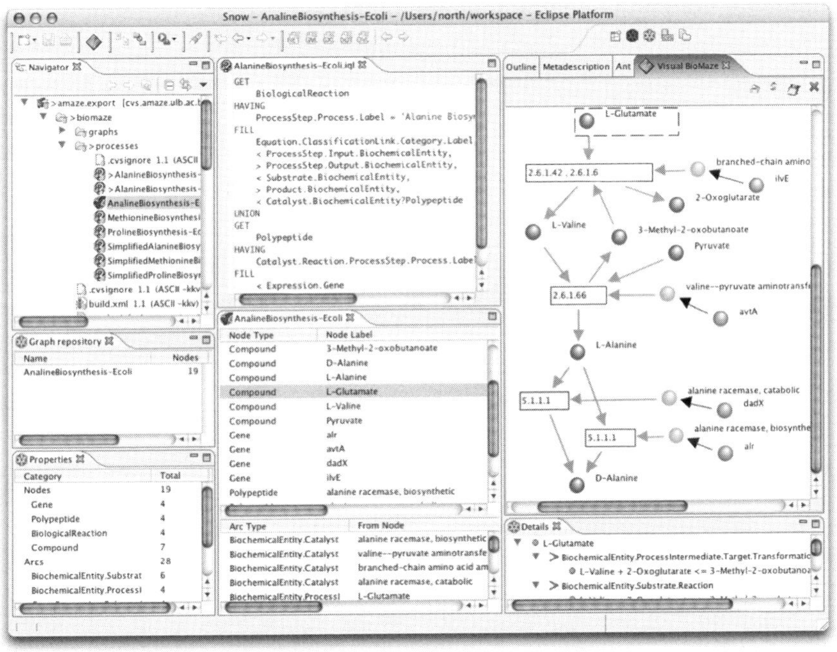

Figure 8. Screen shot of the BioMaze workbench, featuring a metabolic pathway diagram (right).

the functional distance between two genes or proteins can be estimated by computing the length of the shortest path between the corresponding nodes in the network graph. Analysis of the shortest paths in the graph of all biodegradation processes in microbes recently suggested that these processes have evolved through gradual adaptation of enzymes involved in essential metabolism (50). For biochemical networks, such as those representing metabolic reactions, care must be taken that the computed paths are biochemically meaningful. When dealing with metabolic networks, the path-finding problem is usually expressed as the search for a path between a pair of start and end nodes, both usually compounds. In doing so, however, care must be taken to avoid meaningless paths through ubiquitous compounds, such as H_2O or ADP, which play the role of pool metabolites in many metabolic reactions. A recent study has shown (10) that this can be achieved with a simple weighting approach. Each compound is assigned a weight equal to the number of reactions in which it participates in the full network. Path finding is then performed in this graph by searching for one or more paths with lowest weight, effectively disfavoring passage through the highly connected pool metabolites. This search is known as the k-shortest paths problem. It is, however, necessary to restrict the search to paths that satisfy some constraints. For instance, a pathway cannot contain a reaction and its reverse reaction. It can also be useful to apply other constraints on the paths, such as: the path must be simple (path without loop), some nodes are mandatory, and the resulting paths are disjoint, constraints on the minimal/maximal length, etc.

Two approaches can be followed to solve the problem of k-shortest path with constraints. First, classic k-shortest path algorithms, such as backtracking search, can be enhanced to tackle specific constraints. Second, simple algorithms can be applied after an efficient standard k-shortest path algorithm to filter the solutions violating the constraints until the expected number of paths is obtained. This approach allows the combination of several filters, and does not require difficult extensions or sophisticated search algorithms. Using pathfinding algorithms implemented in the BioMaze workbench showed the feasibility of this approach on biochemical networks, such as KEGG. Examples of efficient k-shortest path algorithms can be found in (51,52).

It is often useful to describe or visualize the context of a known pathway or discovered path within the larger biochemical network; e.g., to map/show all nodes in the larger network positioned at up to a specified distance from (a node in) the pathway. Here, too, it is useful to define a weight policy to enhance the meaning of the context. Context extraction is not restricted to simple pathways and can be performed on any biochemical (sub)networks. Context extraction can also be done simultaneously on several pathways to analyze their interactions.

Standard graph techniques can furthermore be used to identify basic building blocks (modules) within the network, which might carry out well-defined functions. At the coarsest level, the network is simply segmented into substructures using clustering procedures. Several procedures for performing such graph segmentation, such as MCL (53), MCODE (54), or RNSC (55), have been developed. Such techniques have been applied on protein–protein interactions to identify larger complexes (56), as well as on metabolic networks (57).

At the next level, the classic graph techniques can be used for motif discovery and analysis. Alon and co-workers (58,59) analyzed transcriptional regulation networks of *Escherichia coli* to uncover its underlying structural design, by means of the discovery of network motifs. They defined a network motif as a pattern of interconnections occurring in networks with a significantly higher frequency than what would be expected in random networks. This analysis relies on sophisticated algorithms for the generation of random networks that have been applied to other networks in neurobiology, ecology, and engineering. This approach is supported by mfinder, a network motifs detection tool (60). A similar approach, supported by a faster algorithm is described in (61). Finally, MAVisto is a tool for the exploration of motifs in network (62). It provides a flexible motif search algorithm and different views for the analysis and visualization of network motifs.

4.2. Advanced Artificial Intelligence Techniques

Constraint programming is a programming paradigm derived from artificial intelligence that uses constraints as basic computational elements. In this approach, the user specifies a set of constraints that a solution must meet, rather than computation steps to obtain such a solution. Constraints thus allow stating complex relationships between objects without having to consider how to enforce them, which reduces development time and maintenance significantly. Constraint programming has been successfully applied in a number of areas, including molecular biology, electrical engineering, operations research, and numerical analysis (63,64,65).

Many analysis problems of biochemical networks can be expressed as combinatorial graph problems. Although such problems can be modeled within constraint programming, such modeling is not always easy, as graphs are not a standard computation domain of constraint programming. Recently, constraint programming has been extended by introducing a new computation domain focused on graphs, including a new type of variables (graph domain variables), and providing constraints over these variables (66). This declarative framework allows sophisticated, constrained subgraph extraction in biochemical networks. Possible examples are as follows: find all pathways traversing a set of specified compounds or reactions; given a set of coregulated enzyme coding genes, find a pathway that can be formed with the reactions catalyzed by these enzymes; find all genes whose expression is directly or indirectly affected by a given compound; show which paths or pathways may be affected when one or more gene/proteins are turned off or missing. In such a constrained path/subgraph extraction, various constraints or properties on the resulting path/subgraph can thus be modeled: properties of nodes, size of the path/subgraph, context of the path (e.g., no enzyme-coding gene regulated by a given set of genes), and relation between different parts of the subgraph.

Comparing different biochemical networks is an important topic in systems biology. Typical problems include the comparison of biochemical pathways from different organisms and tissues, or at different stages of annotation, the highlighting of common features and differences, and the prediction of missing elements. These problems are instances of graph-matching problems. An example of a method for aligning metabolic pathway is described in Pinter et al. (67). Other examples can be found in the PPI networks, where key problems on graphs include aligning multiple graphs, finding frequently occurring subgraphs in a collection of graphs, discovering highly conserved subgraphs in a pair of graphs, and finding good matches for a subgraph in a database of graphs. An example of such techniques can be found in Koyuturk et al. (68), where other relevant references can also be found. Another example is pathBLAST (69), a tool offering a general strategy for aligning two protein interaction networks to elucidate their conserved linear pathways.

Given the incompleteness and the potential lack of reliability of existing biochemical networks, a challenging issue is to handle *approximate* graph matching. Constraint programming, and its extension on graphs, enables approximate graph matching where various constraints can also be stated upon the graph pattern (70). Potential approximations are declaratively stated in the pattern graph as mandatory/optional nodes/edges; forbidden edges, that is, edges that may not be included in the matching, can be declared on the pattern graph. Other constraints between nodes can also be stated.

The extraction of relevant subgraphs can also be achieved through data-mining techniques. A typical problem is the following: given a set of nodes in a biochemical network (e.g., a set of genes), extract a subgraph that best captures the relationships between the given nodes of interest. Simplistic approaches, such as extracting the shortest distance or maximal flow paths between each pair of nodes of interest, do not really capture the relationship between all the given nodes. A better approach, based on electrical network interpretation, has been proposed in (71), but is restricted to two nodes of interest. A more general approach, based on commute time distance and spectral graph analysis, allows a direct solution to the general problem with any number of nodes of interest (72).

5. Implementation Aspects

The BioMAZE workbench is an environment for the representation and analysis of biochemical networks. It integrates the BioMAZE database, which implements the data model described in section 2.2, as well as various analysis and visualization tools. This section briefly sketches the architecture of the BioMaze workbench and describes the available functionalities

5.1. The Architecture

The architecture of the BioMAZE workbench can be described as follows:

- The **Snow** system, a workbench for graph management.
- The **Igloo** database management system dedicated to network data.
- The **VisualBioMAZE** component, which is a biochemical networks visualization tool.
- The **BioEdges** component, which is a collection of tools dedicated to the analysis of biological network.

The flexibility of the system comes from the integration of these tools as plug-ins into the Eclipse environment (73). Figure 8 illustrates a view of the BioMAZE workbench. Its Application Programming Interface (API) is public, enabling programmers to implement their own plug-ins to extend the system and provide new functionalities in a seamless way. The Visualization aspects of BioMaze are outside the scope of this chapter and will not be described here.

Most of the BioMAZE components are independent of the biological data and could therefore be used in different application domains dealing with network data. The BioMAZE environment thus provides an open extensible environment for network analysis and network representation.

5.2. Functionalities

5.2.1. Graph Management

The Snow system is the kernel of the workbench, interconnecting the different tools and providing the basic features. Snow is in charge of data import and export, basic graph edition functionalities, and data browsing. It also provides the user interface for the Igloo DBMS.

5.2.2. Database Management

Igloo is the database management layer in charge of querying and editing (creating and modifying) the entities stored in the database. It also handles the process of loading and annotation of the data. In addition, Igloo provides an Extended Entity Relationship API and the Igloo Query Language for database interrogation. Moreover, Igloo hides the underlying data storage organization, which is a PostgreSQL relational database. The data model is not hard-coded inside the application but is retrieved from an external repository. Igloo can therefore be used in many other application domains, by simply retrieving alternative data models. It is furthermore an independent database management system available as a stand-alone Java library.

Until recently, there were no readily accessible universal sources for pathway information. The current data content of the BioMaze database has therefore been assembled by semiautomatic data loading from diverse external sources and through manual curation.

This situation seems to be evolving, however, with the creation of the BioPAX consortium (21). This consortium is a collaborative effort to develop a file format suitable for exchanging data on biological pathways. The BioPAX format now covers metabolic pathways, PPI, and signal-transduction data. It is expected to evolve into the accepted standard for data-exchange format for biochemical networks. Several pathway resources (KEGG, MetaCyc, and Reactome) already output data in BioPAX format, and most other database providers (TranPath, IHNO, etc.) are currently working on such export software. The BioMAZE database exports data in BioPax format, and will be able to import any data provided in BioPAX format and correctly merge it using the provided unification information.

5.2.3. BioEdge

The BioEdge component implements most of the graph analysis functionalities described in section 4. BioEdge is organized as an extendable set of Eclipse plug-ins, providing Eclipse views in the BioMaze workbench. Each view is thus an independent tool for analyzing a biochemical network. The tools can be combined, enabling the output of one analysis to be processed further by other tools. The constraint satisfaction tools are implemented in the gecode generic constraint development environment (a C++ library) (74) and interfaced with the Eclipse plug-ins.

The tolls currently available in BioEdge are path and graph properties, context explorer, constrained k-shortest path extractor, subgraph extraction, constrained path and subgraph finding, path and graph matching, approximate path and graph constrained matching, motif extraction, and analysis.

6. Concluding Remarks

This chapter discussed two important challenges that need to be met if Systems Biology is to exploit the vast amounts of new data on biochemical networks that are being derived worldwide. One deals with deriving appropriate conceptual frameworks for representing the data on biochemical networks of different types and the other deals with methods of analyzing various global and local properties of these networks. Both are key to developing systems biology approaches that are firmly grounded on state of the art biological data.

The availability of resources from which relevant network graphs and molecular properties can be retrieved and seamlessly pipelined into modeling or simulation software, such as Gepasi (75) or SmartCell (76), or for that matter into any analysis software, should become routine. But clearly more work is needed to make this level of integration possible. In particular, ways need to be found (through the extension of existing models or deriving new dedicated ones) for representing information on rate and equilibrium constants associated with various elementary biochemical processes (catalysis, association, degradation, etc.).

But other, more difficult, problems loom large on the horizon. Currently, the biochemical networks stored in the databases represent a compendium of possible events and interactions: those that can occur under different experimental conditions at different time points, different cellular compartments, and in different cellular populations. The resulting networks and modules derived from them therefore represent, at best, some kind of cumulative time–space ensembles of what might be taking place in the cell. Increasingly aware of these shortcomings, researchers are starting to look for meaningful ways of deconvoluting these ensembles. An obvious way is to incorporate temporal considerations. But this is a difficult task because accurate temporal parameters are not readily available for phenomena, such as PPIs, transcription, or degradation. However, efforts in collecting temporal data are under way, with the fast spreading use of transcription profiling, mRNA decay times, or protein expression profiling. Quite a number of resources dealing with archiving and analyzing genes expression data are currently available and standard exchange formats for these data have also been derived.

In the future, genome-wide experimental procedures will be geared to measuring gene and protein expression levels, PPIs, and metabolite concentrations in a time-dependent fashion, and under strictly controlled experimental conditions. Better ways of synchronizing cell populations might also be available. This should allow us to deconvolute the static networks that we are examining today into more biologically relevant, time-dependent, and condition-dependent networks, which may also vary according to the cell type and the cellular compartment. These networks will include many different types of nodes and edges simultaneously (gene–protein, protein–protein, and compound–enzyme), but specific nodes and edges will vary as a function of time and conditions. This will be the key step that will make systems-level modeling and simulation of cellular processes a realistic and useful undertaking.

Acknowledgments: The authors thank the Région de Bruxelles-Capitale (Belgium) and by the Région Wallonne (Belgium) for support of the research described in this Chapter. The BioMaze workbench has been developed by researchers associated with the aMAZE, BioMaze, and TransMaze projects. We further acknowledge the long-lasting collaboration with Jacques van Helden on various aspects of the BioMaze data model and path finding tools.

References

1. Eisen MB, Brown PO. DNA arrays for analysis of gene expression. *Methods Enzymol* 1999;303.
2. Harbison CT, Gordon DB, Lee TI, et al. Transcriptional regulatory code of a eukaryotic genome. *Nature* 2004;431(7004):99–104.
3. Wang Y, Liu CL, Storey JD, et al. Precision and functional specificity in mRNA decay. *Proc Natl Acad Sci USA* 2002 Apr 30;99(9):5860–5865.
4. Ge H, Walhout AJ, Vidal M. Integrating "omic" information: A bridge between genomics and systems biology. *Trends Genet* 2003;19(10):551–560.
5. Tong AH, Lesage G, Bader GD, et al. Global mapping of the yeast genetic interaction network. *Science* 2004 Feb 6;303(5659):808–813.
6. van Helden J. Regulatory sequence analysis tools. *Nucleic Acids Res* 2003; 31(13):3593–3596.
7. Tompa M, Li N, Bailey TL, Church GM, et al. Assessing computational tools for the discovery of transcription factor binding sites. *Nat Biotechnol* 2005 Jan;23(1):137–144.
8. Romero P, Wagg J, Green ML, et al. Computational prediction of human metabolic pathways from the complete human genome. *Genome Biol* 2005; 6(1):R2.

9. Karp PD, Riley M, Saier M, et al. The EcoCyc Database. *Nucleic Acids Res* 2002;30:56–58.
10. Croes D, Couche F, Wodak SJ, et al. Inferring meaningful pathways in weighted metabolic networks. *J Mol Biol* 2005;356(1):222–236.
11. Huynen MA, Snel B, von Mering C, et al. Function prediction and protein networks. *Curr Opin Cell Biol* 2003;2:191–198.
12. Donaldson I, Martin J, de Bruijn B, et al. PreBIND and Textomy—mining the biomedical literature for protein-protein interactions using a support vector machine. *BMC Bioinformatics* 2003;27:4(1):11.
13. Kaufman M, Thomas R. Emergence of complex behaviour from simple circuit structures. *C R Biol* 2003 Feb;326(2):205–214. Review.
14. Bader GD, Cary MP, Sander C. Pathguide: a pathway resource list. *Nucleic Acids Res* 2006 Jan 1;34(Database issue):D504–D506.
15. van Helden J, Naim A, Mancuso R, et al. Representing and analysing molecular and cellular function using the computer. *Biol Chem* 2000;381(9–10):921–935.
16. Karp PD, Paley S, Romero P. The Pathway Tools software. *Bioinformatics* 2002;18 Suppl 1:S225–S232.
17. Kanehisa M, Goto S, Kawashima S, et al. The KEGG databases at GenomeNet. *Nucleic Acids Res* 2002;30:42–46.
18. Overbeek R, Larsen N, Pusch GD, et al. WIT: integrated system for high-throughput genome sequence analysis and metabolic reconstruction. *Nucleic Acids Res* 2002;28:123–125.
19. Joshi-Tope G, Gillespie M, Vastrik I, et al. Reactome: a knowledgebase of biological pathways. *Nucleic Acids Res* 2005 Jan 1;33(Database issue): D428–D432.
20. Demir E, Babur O, Dogrusoz U, et al. An ontology for collaborative construction and analysis of cellular pathways. *Bioinformatics* 2004 Feb 12;20(3): 349–356.
21. BioPax http://www.biopax.org
22. Xenarios I, Salwínski L, Duan XJ, et al. The database of interacting proteins: a research tool for studying cellular networks of protein interactions. *Nucleic Acids Res* 2002;30:303–305.
23. Bader GD, Betel D, Hogue CWV. BIND: The Biomolecular Interaction Network Database. *Nucleic Acids Res* 2003;31:248–250.
24. Zanzoni A, Montecchi-Palazzi L, Quondam M, et al. MINT: a Molecular INTeraction database. *FEBS Lett* 2002;513:135–140.
25. Hermjakob H, Montecchi-Palazzi L, Lewington C, Mudali S, Kerrien S, Orchard S, Vingron M, Roechert B, Roepstorff P, Valencia A, Margalit H, Armstrong J, Bairoch A, Cesareni G, Sherman D, Apweiler R. IntAct: an open source molecular interaction database. *Nucleic Acids Res* 2004a Jan 1;32(Database issue):D452–D455.
26. Hermjakob H, Montecchi-Palazzi L, Bader G, et al. The HUPO PSI Molecular Interaction Format–a community standard for the representation of protein interaction data. *Nature Biotechnol* 2004b;22:177–183.
27. Matys V, Fricke E, Geffers R, et al. TRANSFAC®: transcriptional regulation, from patterns to profiles. *Nucleic Acids Res* 2003;31:374–378.
28. Salgado H, Santos-Zavaleta A, Gama-Castro S, et al. RegulonDB (version 3.2): transcriptional regulation and operon organization in *Escherichia coli* K-12. *Nucleic Acids Res* 2001;29:72–74.
29. Krull M, Voss N, Choi C, Pistor S, et al. TRANSPATH®: an integrated database on signal transduction and a tool for array analysis. *Nucleic Acids Res* 2003;31:97–100.
30. Takai-Igarashi T, Nadaoka Y, Kaminuma T. A database for cell signalling networks. *J Comput Biol* 1998;5:747–754.

31. INOH. http://www.inoh.org
32. Aladjem MI, Pasa S, Parodi S, et al. Molecular interaction maps–a dliagrammatic graphical language for bioregulatory networks. *Sci STKE* 2004 Feb 24;2004(222):pe8. Review.
33. Rison SC, Hodgman TC, Thornton JM. Comparison of functional annotation schemes for genomes. *Funct Integr Genom* 2000 May;1(1):56–69.
34. Harris MA, Clark J, Ireland A, et al. Gene Ontology Consortium. The Gene Ontology (GO) database and informatics resource. *Nucleic Acids Res* 2004 Jan 1;32(Database issue):D258–D261.
35. van Helden J, Naim A, Lemer C, et al. From molecular activities and processes to biological function. *Brief Bioinform* 2001;2(1):98–93.
36. Teory TJ, Yang D, Fry JP. A logical design methodology for relational databases using the extended entity-relationship model. *Computing Surveys* 1986;18(2):197–222.
37. Huh WK, Falvo JV, Gerke LC, et al. Global analysis of protein localization in budding yeast, *Nature* 2003 Oct 16;425(6959):686–691.
38. http://www.chem.qmul.ac.uk/iubmb/enzyme/
39. Deville Y, Gilbert D, van Helden J, Wodak S. An Overview of Data Models for the Analysis of Biochemical Pathways. *Brief Bioinform* 2003;4(3): 246–259.
40. Wagner A, Fell D. The small world inside large metabolic networks. *Proc R Soc Lond B Biol Sci* 2001;268(1478):1803–1810.
41. Mah RSH. Application of graph theory to process design and analysis. *Comput Chem Eng* 1983;7:239–257.
42. Fell DA, Wagner A. Animating the cellular map. Structural Properties of Metabolic Networks: Implications for Evolution and Modeling of Metabolism. In: Hofmeyr JHS, Rohwer JM, Snaep JL: Model integration An overview of data models for the analysis of biochemical pathways. Stellenbosch University Press, Stellenbosch, 2000:79–85.
43. May GHW. A graph-based pathway searching system over a signal transduction database. Information technologies. University of Glasgow. 2002.
44. Friedler F, Tarjan K, Huang YW, Fan LT. Graph-theoretic approach to process synthesis: Axioms and theorems. *Chem Eng Sci* 1992;47(8):1973–1988.
45. Ogata H, Fujibuchi W, Goto S, Kanehisa M. A heuristic graph comparison algorithm and its application to detect functionally related enzyme clusters. *Nucleic Acids Res* 2000;28(20):4021–4028.
46. Jeong H, Tombor B, Albert R, et al. The large-scale organization of metabolic networks. *Nature* 2000;406:651–654.
47. Gunduz C, Yener B, Gultekin SH. The cell graphs of cancer. *Bioinformatics* 2004;20 Suppl.1:i145–i151.
48. Brandes U, Erlebach T. Network Analysis: Methodological Foundations. Lecture Notes in Computer Science, No. 3418, Springer 2005.
49. Batagelj V, Mrvar A. Pajek. Analysis and Visualization of Large Networks. Jünger, M., Mutzel, P, (Eds.) Graph Drawing Software. Springer, Berlin 2003;77–103.
50. Pazos F, Valencia A, De Lorenzo V. The organization of the microbial biodegradation network from a systems-biology perspective. *EMBO Rep* 2003;(10):994–999.
51. Eppstein D. Finding the k shortest paths. *SIAM J Comp* 1998;28(2): 652–673.
52. Jiménez VM, Marzal A. Computing the K Shortest Paths: A New Algorithm and an Experimental Comparison. Algorithm Engineering: 3rd International

Workshop, WAE'99, London, UK, July 1999; LNCS 1668, 1999, Springer Verlag.
53. van Dongen S. A cluster algorithm for graphs. Technical Report INS–R0010, National Research Institute for Mathematics and Computer Science in the Netherlands, Amsterdam, May 2000. http://micans.org/mcl
54. Bader GD, Hogue CW. An automated method for finding molecular complexes in large protein interaction networks. *BMC Bioinformatics* 2003;4:2. Epub 2003 Jan 13.
55. King AD, Przulj N, Jurisica I. Protein complex prediction via cost-based clustering. *Bioinformatics* 2004 Nov 22;20(17):3013–3020. Epub 2004 Jun 4.
56. Krogan NJ, Cagney G, Yu H, et al. Global landscape of protein complexes in the yeast *Saccharomyces cerevisiae*. *Nature* 2006;440(7084):637–643.
57. Gagneur J, Jackson DB, Casari G. Hierarchical analysis of dependency in metabolic networks. *Bioinformatics* 2003 May 22;19(8):1027–1034.
58. Milo R, Shen-Orr S, Itzkovitz S, et al. Network motifs: Simple building blocks of complex networks. *Science* 2002;298:824–827.
59. Shen-Orr S, Milo R, Mangan S, Alon U. Network motifs in the transcriptional regulation network of Escherichia coli. *Nature Genet* 2002;31:64–68.
60. Kashtan N, Itzkovitz S, Milo R, Alon U. Efficient sampling algorithm for estimating subgraph concentrations and detecting network motifs. *Bioinformatics* 2004;20 no. 11:1746–1758.
61. Wernicke S, Rasche F. FANMOD: a tool for fast network motif detection. *Bioinformatics* 2006 May 1;22(9):1152–1153.
62. Schreiber F, Schwöbbermeyer H. MAVisto: a tool for the exploration of network motifs. *Bioinformatics* 2005;21:3572–3574.
63. Van Hentenryck P. The OPL Optimization Programming Language. Cambridge: The MIT Press; 1999.
64. K. Apt. Principles of Constraint Programming. Cambridge University Press; 2003.
65. Backofen R, Gilbert D. Bioinformatics and constraints. *Constraints* 2001: 6(2/3).
66. Dooms G, Deville Y, Dupont P. CP(Graph): Introducing a Graph Computation Domain in Constraint Programming. International Conference on Principles and Practice on Constraint Programming, Sitges, Barcelona, Spain, October 2005.
67. Pinter RY, Rokhlenko O, Yeger-Lotem E, Ziv-Ukelson M. Alignment of Metabolic Pathways. *Bioinformatics* 2005;21,16:3401–3408.
68. Koyuturk M, Kim Y, Topkara U, et al. Pairwise alignment of protein interaction networks. *J Comp Biol* 2006;13(2):182–199.
69. Kelley BP, Yuan B, Lewitter F, et al. PathBLAST: a tool for aligment of protein interaction networks. *Nuc Acids Res* 2004;32:W83–W88.
70. Zampelli S, Deville Y, Dupont P. Approximate Constrained Subgraph Matching. International Conference on Principles and Practice on Constraint Programming, Sitges, Barcelona, Spain, October 2005.
71. Faloutsos C, McCurley KM, Tomkins A. Fast discovery of connection subgraphs. 10th ACM Conference on Knowledge Discovery and Data Mining (KDD). 2004;2:118–127.
72. Vast S, Dupont P, Deville Y. Automatic extraction of relevant nodes in biochemical networks. Learning and Bioinformatic Workshop, CAp 2005; Conférence d'Apprentissage, Nice: 21–31.
73. Eclipse: http://www.eclipse.org.
74. Gecode: Generic constraint development environment, 2005. http://www.gecode.org.

75. Mendes P. GEPASI: a software package for modelling the dynamics, steady states and control of biochemical and other systems. *Comput Appl Biosci.* 1993;5:563–571.
76. Ander M, Beltrao P, Di Ventura B, et al. SmartCell, a framework to simulate cellular processes that combines stochastic approximation with diffusion and localisation: analysis of simple networks. *Syst Biol* 2004;1:129–138.

Appendices

Appendix I

Software, Databases, and Websites for Systems Biology

Category	Name	Function	Platform	Developer/Provider
Algorithms	Clover	Search for Transcription Factor Binding Sites	Web (http://zlab.bu.edu/clover)	Boston University (free)
	DBRF-MEGN	Deducing Minimum Equivalent Gene Networks from Large-Scale Gene Expression Profiles		Koji Kyoda and Shuichi Onami (free)
	Gossip	Automated functional interpretation of gene groups	Web (gossip.gene-groups.net)	Nils Blüthgen, and MicroDiscovery (free)
	OptGene	Platform for in silico metabolic engineering through evolutionary programming	Web (www.cmb.dtu.dk) UNIX/Windows	Kiran R. Patil and Jens Nielsen
	OptKnock	Platform for in silico metabolic engineering through bilevel linear optimization	Web (http://maranas.che.psu.edu/)	Anthony Burgard and Costas Maranas
	Reporter metabolites	Find metabolites around which most changes in expression occur	Web (www.cmb.dtu.dk) UNIX/Windows	Kiran R. Patil and Jens Nielsen
	TFGossip	Association of Transcription Factor Binding Sites with Factors using GOSSIP	Web (http://tfgossip.gene-groups.net)	Nils Blüthgen, Humboldt University
Data formats	mzData	The aim of mzData is to unite the large number of current mass spec data formats into one	XML Schema	Proteomics Standards Initiative
	mzXML	A standardized output format for mass spectrometry	XML Schema	Institute for Systems Biology
Databases	BIND	Biomolecular Interaction Network Database, containing a curated set of interactions	Web (http://www.bind.ca/Action)	Mount Sinai Hospital
	BioModels	Curation of quantitative biological models	Web (http://www.ebi.ac.uk/biomodels) (http://biomodels.net)	EMBL-EBI (free)

Category	Name	Function	Platform	Developer/Provider
	BLAST	Similarity Search	Web (http://www.ncbi.nlm. nih.gov/BLAST)	NCBI (free)
	DBTSS	Database of transcription start sites	Web (http://dbtss.hgc.jp/)	University of Tokyo
	DIP	Database of Interacting Proteins, containing a curated set of protein-protein interactions	Web (http://dip.doe-mbi.ucla.edu/)	UCLA
	EnsEMBL	Annotated genomes and Perl interface	Web (http://www.ensembl.org)	European Bioinformatics Institute
	EPD	Eukaryotic promoters	Web (http://www.epd.isb-sib.ch)	Swiss Institute of Bioinformatics
	FANTOM	Annotated mouse transcriptome	Web (http://fantom3.gsc.riken.jp)	Riken
	GPMDB	A repository for data from tandem mass spectrometry	Web accessible	The Global Proteome Machine Organization
	GRID	Comprehensive database of genetic and physical interactions for yeast, fly, and worm	Web (http://biodata.mshri.on.ca/grid/)	B.J. Breitkreutz, C. Stark, M. Tyers
	HomGL	Storing, Mapping, and Comparison of gene groups	Web (www.gene-groups.net)	Nils Blüthgen (free)
	HPRD	Visually depict and integrate information pertaining to domain architecture, posttranslational modifications, interaction networks, and disease association for each protein in the human proteome	Web (http://www.hprd.org)	Commercial (free for research)
	iHOP	Text-mining. Information Hyperlinked Over Proteins	Web (http:// www.ihop-net.org/UniPub/iHOP/)	Robert Hoffmann and Alfonso Valencia
	Jaspar	Profiles of transcription factor binding sites	Web (http://jaspar.cgb.ki.se)	Karolinska Institute (free)
	Karma	Mapping of diverse identifiers from multiple array platforms and organisms	Web (http://biryani.med.yale.edu/karma/cgi-bin/mysql/karma.pl)	Yale
	Open Proteomics Database	OPD is a public database for storing and disseminating mass spectrometry-based proteomics data	Web accessible	University of Texas at Austin
	PEDRoDB	Storing, searching, and disseminating experimental proteomics data	Web accessible	University of Manchester
	PRIDE	PRIDE is a centralized, standards compliant, public data repository for proteomics data	Web accessible	European Bioinformatics Institute

Category	Name	Function	Platform	Developer/Provider
	ProbematchDB	Mapping between two array platforms	Web (http://brainarray.mhri. med.umich.edu/ brainarray)	University of Michigan
	Resourcerer	Mapping of human, mouse and rat microarray gene identifier	Web (http://www.tigr.org/ tigr-scripts/magic/r1.pl)	TIGR
	SGD	*Saccharomyces* Genome Database. Very complete resource of genomic information for *S. cerevisiae*	Web (http://www. yeastgenome.org/)	SGD Project
	STRING	Database of known and predicted protein-protein interactions	Web (http://string.embl.de)	EMBL. Peer Bork Group
	SWISS2D-PAGE	Two-dimensional polyacrylamide gel electrophoresis database	Web accessible	Swiss Institute of Bioinformatics
	Transfac	Profiles of transcription factor binding sites	Web (http://www.gene-regulation.com)	Partially free, commercial, BioBase
	YEAST protein complex database	Data set of systematic analysis of yeast protein complexes with TAP tag	Web (http://yeast.cellzome.com.)	Cellzome AG
Gene centered approach	MiCoViTo	Visualization of groups of genes having similar expression in two sets of microarray experiments	Web (http://transcriptome. ens.fr/micovito)	Gaëlle Lelandais
	yMGV	Data mining interface for microarray data with easily interpretable and mostly graphical outputs	Web (http://www. transcriptome. ens.fr/ymgv)	Philippe Marc
Gene ontology	GO	The Gene Ontology project provides a controlled vocabulary to describe genes and gene product attributes in any organism	Web (http://www. geneontology. org)	Gene Ontology Consortium
	GoMiner	GoMiner is a tool for biological interpretation of gene expression microarrays using GO annotations	Web (http://discover.nci. nih.gov/gominer) Windows/MacOS/ Linux	
	Protégé	Ontology editor and knowledge-base framework	Web (http://protege. stanford.edu)	Free, open-source
Modeling	CellDesigner	Modeling/simulation tool of biochemical networks with graphical user interface	Web (http://systems-biology.org/002/001. html) (http://www. celldesigner.org) Windows/MacOSX/ Linux	Systems-biology.org; Funahashi A, Kitano H (free)

Category	Name	Function	Platform	Developer/Provider
	Cellerator	Computer-algebra based conversion of biochemical arrows to differential equations	Web (http://xcellerator.info)	Caltech/Univ. of California, Irvine
	CellML 1.1	Biological model specification and reuse	UNIX/Windows	Catherine M. Lloyd (free)
	CellML2SBML	Converts CELLML files to SBML files	Web (http://sbml.org/software/cellml2sbml)	Univ. Hertfordshire/ EMBL-EBI
	Kegg2SBML	Converts KEGG files to SBML with CellDesigner tags	Web (http://sbml.org/kegg2sbml.html)	Systems-biology.org
	libSBML	Library providing an API for SBML file manipulation	Web (http://sbml.org/libsbml.html)	Caltech
	MathSBML	Package for Manipulating SBML-based Biological Models in Mathematica	Web (http://sbml.org/mathsbml.html) UNIX/Windows/Macintosh/VMS	Caltech, Bruce E. Shapiro (free)
	SBML (System Biology Markup Language)	Computer-readable format for representing models of biochemical reaction networks	Web (http://www.sbml.org)	Free, standard driven by community needs
	SBML Editor	SBML Model Editor	Web (http://www.ebi.ac.uk/compneur-srv/SBMLeditor.html)	EMBL-EBI
	SBML Toolbox	MatLab package for using SBML models	Web (http://sbml.org/software/sbmltoolbox)	Univ. Hertfordshire
	Sigmoid	Cellerator-based pathway database management	Java/Web (http://sigmoid.sf.net)	Univ. of California, Irvine
	Virtual Cell	Modeling and Simulation Framework	Web (http://www.nrcam.uchc.edu)	Data are freely available if shared by submitter
	XPPAUT	ODE solver and phase plane analysis	Web (http://www.math.pitt.edu/~bard/xpp/xpp.html)	Univ. Pittsburgh
Networks	Osprey	Platform for visualization of complex interaction networks	Web (http://biodata.mshri.on.ca/osprey) UNIX/Windows/ Macintosh	The GRID team (free for academic use)
	PathBLAST	Alignment of Protein Interaction Networks		Whitehead Institute
	ProViz	Tool for visualization of protein-protein interaction graphs	Web (http://cbi.labri.fr/eng/proviz.htm)	LaBRI. David Sherman
	SBGN	Graphical notation for SBML	Web (http://sbgn.org)	Systems-biology.org
Orthology detection method	INPARANOID	Program that automatically detects orthologs (or groups of orthologs) from two species	Web (http://inparanoid.sbc.su.se)/Linux	Erik Sonnhammer

Category	Name	Function	Platform	Developer/Provider
	OrthoMCL	OrthoMCL provides a scalable method for constructing orthologous groups across multiple eukaryotic taxa	Web (http://www.cbil.upenn.edu/gene-family)	Li Li
Pathways	BioPax	Data exchange format for biological pathway data	Web (http://www.biopax.org)	Free
	SigPath	Information system designed to support quantitative studies on the signaling pathways and networks of the cell	Web (http://www.sigpath.org)	Data and code are freely available
	Ingenuity Pathway database	Model, analyze, and understand complex biological systems	Web (http://www.ingenuity.com)	Commercial
Proteomics tools	ProteinProspector	Proteomics tools for MS data mining, including database search and programs simulating MS pattern	Web (http://prospector.ucsf.edu/)	The University of California San Francisco
	Software tools	Data mining tools for MS analysis, including validation identification and quantification of peptide and protein	Web (http://www.proteomecenter.org/software.php)	The Institute for Systems Biology
Search engine	MASCOT	Comparing the recorded MS or MS/MS spectrum with theoretical masses from protein database	Web (http://www.matrixscience.com/home.html)	Matrix Science Ltd.
	Sequest	Comparing the recorded MS/MS spectrum with theoretical masses from protein database	Web (http://fields.scripps.edu/sequest/index.html)	The Scripps Research Institute

Appendix II

Glossary

Apoptosis	**Apoptosis** (programmed cell death) is the process by which a cell commits suicide.
Application Programming Interface	**Application Programming Interface (API)** is the interface that a computer system, library, or application provides in order to allow requests for services to be made of it by other computer programs, and/or to allow data to be exchanged between them.
Biochemical reaction	**Biochemical reaction** is a process by which one or more components are transformed by a biochemical system. A biochemical system often consists of one enzyme, with or without cofactors.
Bistabilility	**Bistabilility** means that a dynamical system has two stable steady-states for a certain set of parameters.
Bistable	A system that exhibits two unique, stable steady-states is said to be **bistable**.
Boolean search terms	**Boolean search terms** are the logic terms AND, OR, and NOT, which are used to make database searches precise.
Cellular automata	**Cellular automata**, composed of massively simple, autonomous, and interacting computational components (cells), are an important parallel computational paradigm. They were first introduced by J. von Neumann and S. Ulam and are widely used to model dynamics of large, parallel systems. Standard cellular automata are rule-based and discrete in space, time and value. Various nonstandard extensions have been developed to meet particular requirements of parallel computation in specific fields. Language-based cellular automata use a programming language, instead of rules, to describe computation in the cell.

Chemical reaction — **Chemical reaction** is a process by which one or more components are chemically transformed. Mass and charge are conserved through a given chemical reaction. A chemical reaction may proceed spontaneously.

Collision-Induced Dissociation — In **Collision-Induced Dissociation (CID)**, accelerating voltage provides ions with energy of motion, and the ions collide with the inert gas molecules in the collision cell of MS instrument. Collision energy is converted to internal energy to induce dissociation of the ion. CID is divided into two classes, lower-energy CID and high-energy CID, depending on MS equipment, and the former is frequently used for proteomics analysis.

Compendium — A **compendium** of expression profiles is an expression matrix composed of a large number of DNA microarray experiment results.

Component — **Component** is a molecule, an ion, or an arrangement of molecules and ions that participate in interactions in a signaling pathway. Components are objects.

Data standard — **Data standard** is a documented agreement on the format of data.

Diffusion-limited rate — **Diffusion-limited rate** is the first encounter rate.

Diffusivity — **Diffusivity** is the proportionality constant used to describe diffusive flux as linearly proportional to the negative of the concentration gradient (Fick's law).

DNA microarrays — **DNA microarrays** are tools providing direct access to transcriptome analysis.

Efficiency — **Efficiency** is the quick and complete response in adaptation to different needs.

Emergent properties — **Emergent properties** are shown only by collective systems. They are created by the emergent interaction among entities in systems.

Endocytosis — **Endocytosis** is the process in which areas of the plasma membrane invaginate and pinch off to form intracellular vesicles.

Exciton — **Exciton** is an excited state of an insulator or semiconductor that allows energy to be transported without transport of electric charge; may be thought of as an electron and a hole in a bound state.

Extensible Markup Language — **Extensible Markup Language (XML)** is a W3C-recommended general-purpose markup language for creating special-purpose markup languages, capable of describing many different kinds of data. XML is a way of describing data, and an XML file can contain the data too, as in a database. Its primary purpose is to facilitate the sharing of data across different systems, particularly systems connected via the Internet.

First principle modeling — **First principle modeling** is modeling based on known physical, chemical, and biological information. In subcellular processes, it often starts from the stochastic kinetic equations.

Flexibility — **Flexibility** is the ability a cell has to adapt to a wide range of environmental conditions.

Fractal kinetics — **Fractal kinetics** is a kinetic law for dimension-restricted reactions that do not follow traditional mass-action kinetics. The kinetic coefficient of fractal kinetics is not a constant, but a time-dependent function. When a reaction occurs under steady-state conditions, kinetic order reflects the dimensional restriction of reaction.

Gel electrophoresis — **Gel electrophoresis** is a separation method in which a protein mixture is loaded onto a gel and subjected to an electric current, causing individual proteins to migrate a particular distance depending on a property, such as their molecular weight.

Gene Ontology — **Gene Ontology (GO)** is a structural network consisting of defined terms and relationships between them that describe three attributes of gene products, which are Molecular Function, Biological Process, and Cellular Component.

Gene regulatory network — **Gene regulatory network** is an arrangement of genetic interaction, in space and time, to produce a given function.

Genotype — **Genotype** is the specific genetic makeup of an individual, i.e., the specific genome encoding the total potential inventory of cellular resources of the organism.

GenPept — **GenPept** is a comprehensive protein database that contains all of the translated coding regions of GenBank sequences.

GNU Lesser General Public License — **GNU Lesser General Public License (LGPL)** is a free software license published by the Free Software Foundation. It was designed as a compromise between the strong-copyleft GNU General Public License (GPL) and simple permissive licenses, such as the BSD licenses and the MIT License. The LGPL is intended primarily for software libraries, although it is also used by some stand-alone application.

Graphical User Interface	**Graphical User Interface (GUI)** is a method of interacting with a computer through a metaphor of direct manipulation of graphical images and widgets, in addition to text. GUIs display visual elements such as icons, windows, and other gadgets.
GTPase–activating protein	**GTPase–activating protein (GAP)** is a protein that facilitates the GTP hydrolysis by a GTP-binding protein.
Guanine nucleotide exchange factor	**Guanine nucleotide exchange factor (GEF)** is a protein that catalyzes the exchange of GDP for GTP for a GTP-binding protein.
Hill curve	**Hill curve** is a sigmoidal curve often used to describe the reaction rates for uncooperative enzymes . It is also used to fit sigmoidal stimulus-response curves.
Homologs	**Homologs** are genes coming from a common evolutionary ancestor.
Hoppers	**Hoppers** are species that hop from the first "trapping" point to the next in discontinuous manner, called "hopping conduction," when the species are electrons.
Hysteresis	**Hysteresis** is a property of a system where the state of the system is not independent of its history. Different steady-states are reached, depending on whether a bifurcation parameter increases or decreases (a kind of "memory"). As a parameter is increased, the system jumps to the alternative state at a particular value of the parameter. However, if the parameter decreases, the system jumps back to the original state at a lower parameter value. Bistable systems exhibit hysteresis.
***In vivo* and *in vitro* differences**	***In vivo* and *in vitro* differences** are the differences in molecular parameters between the values measured in "test tube" (*in vitro*) and in native biological environment (*in vivo*).
Interaction	**Interaction** is a chemical and biochemical reaction (e.g., phosphorylation, protein cleavage) or biochemical process (e.g., transport across a membrane, transcription, translation).
Ion-trap analyzer	In the **ion-trap analyzer**, the ions are first captured in the trapping space, and then ejected by increasing the voltage to obtain the MS spectrum. For MS/MS analysis, the selected ion is isolated and dissociated into fragment ions with low-energy CID in the trapping space. MS/MS spectrum is obtained in the same manner described above.

Java	**Java** is an object-oriented programming language developed by James Gosling and colleagues at Sun Microsystems in the early 1990s. Unlike conventional languages, which are generally designed to be compiled to native code, Java is compiled to a bytecode which is then run by a Java virtual machine.
Java Native Interface	**Java Native Interface (JNI)** is a programming framework that allows Java code running in the Java virtual machine to call and be called by native applications and libraries written in other languages, such as C, C++ and assembly.
Java Runtime Environment	**Java Runtime Environment (JRE)** is the software required to run any application deployed on the Java Platform.
Java Web Start	**Java Web Start (JWS)** is a framework developed by Sun Microsystems that enables starting Java applications directly from the Web using a browser.
Law of Mass Action	**Law of Mass Action** is that the speed of a chemical reaction is proportional to the quantity of the reaching substances.
Liquid chromatography	**Liquid chromatography** is one of the chromatography methods to separate the molecules using liquid as mobile phase. Cation-exchange and reverse-phase methods are widely used for peptide analysis.
Lysis	**Lysis** is the developmental phase in which the bacterium is eaten by the phage.
Lysogeny	**Lysogeny** is the developmental phase in which the phage lives together with the bacterium.
Mass Spectrometry	**Mass Spectrometry (MS)** is a technique for determining the mass of a substance. Mass spectrometry is frequently used in proteomics to identify proteins.
Metabolic engineering	**Metabolic engineering** is the application of directed genetic modifications to improve the properties of a given cell, e.g., to improve yields or productivities, to expand substrate range utilization or to insert heterologous pathways for the production of novel products.
Metabolic flux	**Metabolic flux** is the rate of conversion of one metabolite into another by an enzyme catalyzing the corresponding metabolic reaction.
Metabolic network	**Metabolic network** is a set of connected metabolic reactions. The term can be used when referring to part or to the whole set of reactions occurring in a cell.

Michaelis-Menten kinetics	**Michaelis-Menten kinetics** describes the rate of enzyme mediated reactions for many enzymes. This is valid only in the particular case of steady-state, where the concentration of the complex enzyme-substrate is constant. The Michaelis-Menten constant KM determines the substrate level at which the reaction rate reaches half the maximal value.
Minimum Information about a Microarray Experiment	**Minimum Information about a Microarray Experiment (MIAME)** is a microarray experiment international standard that encloses all the description of the experiment needed to understand how data have been processed.
Minimum quantitative modeling	**Minimum quantitative modeling** is the minimum detailed physical and chemical modeling, which captures the essential biological feature and allows a quantitative comparison between theoretical calculations and biological data. Its physical and chemical parameters are fixed by other independent experiments. In the present case, in the minimum modeling, only one operon site was considered with a reduced configuration. All others are treated effectively as possible contributions to *in vivo* and *in vitro* differences.
Monte Carlo methods	**Monte Carlo methods** are a class of computational algorithms for simulation using random numbers.
Neofunctionalization	A **neofunctionalization** event creates after a duplication event a new copy of the gene duplicated with a new function not found in the ancestor gene.
Network evolution	**Network evolution** can be phylogenetic, meaning signaling network changes during species evolution from simple to complex organisms, or ontogenetic, meaning signaling network changes during embryonic development from an egg to a multicellular adult. Here, it means the evolution of wiring among active genes and proteins during development.
Newtonian fluid	**Newtonian fluid** is a fluid in which shear stress is linearly proportional to the velocity gradient in the direction perpendicular to the plane of shear. The constant of proportionality is known as the viscosity.
Object-oriented programming	**Object-oriented programming** is a computer programming paradigm. The idea is that a computer program is composed of a collection of individual units, or objects, as opposed to a list of instructions. It means to package, or

encapsulate, data and functionality together into units, and each object exposes an interface to allow other objects to interact with it. Usually, objects interact and communicate with each other using message passing, and their action depends on both the current state and the received messages.

Ontology — **Ontology** is a digital representation of knowledge, usually comprising a collection of controlled terms with agreed definitions.

OpenMP — **OpenMP** is a set of programming protocol to run a program on a computer with shared memory and multiple CPU to enhance performance.

Ordinary Differential Equation — **Ordinary Differential Equation (ODE)** is a relation that contains functions of only one independent variable, and one or more of its derivatives with respect to that variable. Many scientific theories can be expressed clearly and concisely in terms of ordinary differential equations.

Orthologs — **Orthologs** are two genes that predate a speciation event and that code functionally equivalent proteins that arise from evolution.

Orthology — **Orthology** defines the relationship between genes in different species that originate from a single gene in the last common ancestor of these species.

Paralogs — **Paralogs** are two genes that have arisen by duplication events and whose function generally have diverted from the original ancestor.

Parameter — **Parameter** is a fixed quantity in a mathematical model, as opposed to a variable.

Percolation — **Percolation** concerns the movement and filtering of fluids through porous materials.

Phage λ — **Phage λ** is the bacterium-eating virus living on the bacterium *E. coli*.

Phenotype — **Phenotype** is the observable physical or biochemical characteristics of an organism, as determined by both genetic makeup and environmental influences. The observable sometimes results from the activation of a cellular process, as determined by genetic makeup of the cell under study or by environmental influence (cell medium, temperature, and other experimental conditions). Apoptosis is an example of a phenotype that can be observed at the cell level (cells die at the completion of the apoptosis program) and at the biochemical level (DNA fragmentation is a marker for apoptosis).

Potential landscape	**Potential landscape** is the visualization of the potential function in the stochastic dynamical structure analysis. It provides a graphic picture of robustness and stability.
Principal component analysis	**Principal component analysis** is a decomposition method that linearly transforms a high-dimensional dataset into a low-dimensional space. The original space is transformed into a new coordinate system in such a way that the data with the greatest variance defines the first axis (also called first principal component), and the second principal component is the vector orthogonal to the first principal component that captures most of the remaining variance (and so on for the other axis).
Proteomics	**Proteomics** is the simultaneous investigation of all the proteins in a cell or organism.
Quadrupole-TOF	In the hybrid type of **quadrupole-TOF**, the ion is selected with quadrupole mass-filter and dissociated in the collision cell located behind the quadrupole. The masses of fragment ions are recorded with TOF (Time-of-flight) analyzer to obtain the MS/MS spectrum.
Reaction-diffusion equation	**Reaction-diffusion equation** is a partial differential equation that contains partial derivatives in respect to two or more independent variables and describes both the temporal behavior and diffusion in space.
Reporter metabolites	**Reporter metabolites** are metabolites in the metabolic network of an organism around which most changes in expression occur.
Reverse two-hybrid	**Reverse two-hybrid** system is a modification of Y2H bearing a suicide reporter gene to select against protein-protein interactions, in contrast with conventional Y2H. It is useful to isolate interaction-defective alleles.
Robustness	**Robustness** is the insensitivity of biological functions to various disturbances.
Sodium Dodecyl Sulfate-Polyacrylamide Gel Electrophoresis	**Sodium Dodecyl Sulfate-Polyacrylamide Gel Electrophoresis (SDS-PAGE)** is a method widely used for separation of proteins according to their molecular weights, in which proteins denatured with SDS are electrophoresed in a polyacrylamide gel.
Soft ionization methods	**Soft ionization methods** have been developed to enable the ionization of large biomolecules, including proteins, without any destruction. They include MALDI (Matrix-Assisted Laser Desorption Ionization) and ESI (Electro-Spray Ionization).

Spatial concentration gradient — **Spatial concentration gradient** is a gradual change in the concentration over a specific distance.

Species — **Species** is a synonym for component.

Steady state — **Steady state** is a dynamic system state that does not change over time. If a system is described by differential equations, a steady state is determined by equating the time derivatives of all variables to zero.

Stochastic dynamical structure analysis — **Stochastic dynamical structure analysis** is the tentative name for the novel mathematical structure that emerges from the λ switch study. It has four dynamical elements: the potential function, the degradation matrix, the transverse matrix, and the stochastic drive. Among those four dynamical elements, the degradation matrix is constrained by the stochastic drive.

Stochastic kinetic equations — **Stochastic kinetic equations** are the chemical rate equations that describe the molecular processes inside a single cell.

Stochasticity — **Stochasticity** is the random aspect of the sub-cellular dynamical processes. It may originate from the randomness embedded in dynamics (intrinsic noise) or from fluctuating environmental condition (extrinsic noise).

Sub-functionalization — A **sub-functionalization** event separates after a duplication event two gene functions in the separated copies that were originally present in the ancestor gene.

Supertrapps — **Supertrapps** are species that cannot transfer freely and are localized by a trapping point of lattices, such as impurities or a broken lattice of a crystal.

Systems Biology Graphical Notation — **Systems Biology Graphical Notation (SBGN)** is a visual notation for network diagrams, such as biochemical reaction and gene-regulatory networks, which is commonly used in the field of computational systems biology. The goal of the SBGN effort is to help standardize a graphical notation for computational models in systems biology. For example, it will add rigor and consistency to the usually ad hoc diagrams that often accompany research articles in publications. It will also help bring consistency to the user interfaces of different software tools and databases.

Systems Biology Markup Language — **Systems Biology Markup Language (SBML)** is a machine-readable language, derived from XML, for representing models of biochemical

	reaction networks. SBML can represent metabolic networks, cell-signaling pathways, regulatory networks, and other kinds of systems studied in systems biology.
Systems Biology Workbench	**Systems Biology Workbench (SBW)** is a software framework that allows heterogeneous application components, written in diverse programming languages and running on different platforms, to communicate and use each other's capabilities via a fast binary-encoded message system. SBW enables applications (potentially running on separate, distributed computers) to communicate via a simple network protocol.
Temporal dynamics	**Temporal dynamics** is a quantitative description of how a system changes over time.
Transcriptome	The **transcriptome** is the expression level of all the genes expressed in a cell at any given time.
Ultrasensitivity	**Ultrasensitivity** describes a response that is more sensitive than a Michaelis-Menten curve. Often used synonymously with sigmoidality.
XML Schema	**XML Schema** is a language for defining the structure and content of XML documents.
Yeast two-hybrid	**Yeast two-hybrid (Y2H)** system is a molecular genetic method to detect protein-protein interactions. In Y2H, proteins X and Y are expressed as hybrid proteins with DNA-binding domain and transcription activation domain, respectively. The former and latter hybrids are often called bait and prey, respectively. An interaction between bait and prey, or X and Y, reconstitutes a transcription factor activity, which can be readily detected by use of reporter gene.
λ switch	**λ switch** is the gene regulatory network in phage λ deciding the switching from lysogeny to lysis.

Index

T